Driving and Engine Cycles

Evangelos G. Giakoumis

Driving and Engine Cycles

Springer

Evangelos G. Giakoumis
Department of Thermal Engineering,
School of Mechanical Engineering
National Technical University of Athens
Athens
Greece

ISBN 978-3-319-49033-5 ISBN 978-3-319-49034-2 (eBook)
DOI 10.1007/978-3-319-49034-2

Library of Congress Control Number: 2016956823

Printed on acid-free paper

This Springer imprint is published by Springer Nature
The registered company is Springer International Publishing AG
The registered company address is: Gewerbestrasse 11, 6330 Cham, Switzerland

Preface

Environmental pollution has been one of the most significant threats faced by mankind in the last decades. It affects severely the whole planet, resulting in millions of premature deaths every year, degradation of the human life level, and considerable financial burden on individuals and the whole society. Passenger cars, light- and heavy-duty trucks and buses, and non-road equipment, i.e., vehicles powered by fossil-fueled internal combustion engines, are among the major contributors to the anthropogenic-related environmental issues, particularly in large cities. The authorities, having acknowledged this fact from the early 50s, proceeded to various measures, such as legislation of gradually stricter emission limits, cleaner fuels, inspection and maintenance tests.

Many books have been published in the last decades dealing with vehicular-related environmental issues. A lot of these books have focused on emission limits and emission legislation in general. Others deal with pollutant emissions analyzing their production mechanisms; additionally, a significant portion has been concerned with after-treatment control. None of these subjects are dealt with in this book. In contrast, it seems that there has been no book dealing exclusively with certification cycles, i.e., the test schedules employed at type approval level to test the vehicle, being representative of the concerned vehicle's or engine's duty cycle. Instead, with few exceptions, driving cycles have been presented in the past mainly as 'accompanying text' to manuscripts dealing with pollutant regulations and emission limits. Since, to the best of the author's knowledge, no comprehensive analysis of drive cycles has ever been attempted, this is the gap in the open literature that the present book aims to fill.

In the following chapters, an effort has been made to cover all possible subjects related to drive cycles, namely describe what a test cycle is, how it is constructed, and provide detailed historical information on the development of the most influential certification test schedules employed in the past. Moreover, the shortcomings of the various cycles with respect to their representativeness (in terms of driving activity and emission results) will be highlighted, as well as their most important technical specifications. The book also aims to combine this information with typical performance and emission results from real vehicles operation, in order for the reader to draw a more complete picture on how the vehicle/engine behaves during a test cycle regarding development of engine speed, torque, power, as well as pollutant emissions. The results that will be presented stem from both (experimentally validated) simulations and experiments. Moreover, an extensive part of the book is dedicated to engine-dynamometer cycles, for heavy-duty and non-road engines/vehicles, which have been dealt with in the past even more scarcely.

It should be pointed out that it has never been the intention to provide an exhaustive list of all test cycles or their technical specifications. This holds true as regards the numerous non-legislated test schedules. For legislated cycles, on the other hand, a detailed presentation of all of them will be provided. Owing to the availability of large amount of data regarding U.S., European, and (recently developed) worldwide regulations, these cycles will be presented in more detail, particularly vis-à-vis their historical background. On the other hand, owing to language limitations and reluctance of the relevant authorities in Japan to provide background information, the analysis of Japanese test cycles will be considerably shorter. The discussion of drive cycles will be primarily from a mechanical engineer's point of view and only secondarily from a traffic engineering perspective.

The book is organized as follows: Chap. 1 serves as an introduction to driving cycles and test procedures in general. Test cycles are defined and classified with the focus on why certification cycles for all kinds of vehicles should be transient. A description of the main attributes follows for all cycle types (chassis and engine, modal and transient). Emphasis is given here on representativeness issues, also providing and analyzing various cycle metrics. The procedure followed when constructing a driving cycle is also discussed at the end of the chapter. Chapters 2 and 3 provide details on driving cycles for passenger cars and light-duty trucks, which form the biggest and oldest category (Chap. 2), and motorcycles (Chap. 3). These have been exclusively of the chassis-dynamometer type. Test cycles for heavy-duty engines/vehicles are discussed in Chap. 4. An interesting characteristic for these test schedules is that both chassis and engine-dynamometer cycles have been employed in the past, the latter on a stationary or transient form. All of them are discussed in detail. Chapter 5 deals with non-road engines/vehicles, such as those utilized in agricultural and construction equipment, marine and locomotive applications. The relevant cycles here are exclusively of the engine-dynamometer type. Lastly, Chap. 6 serves as an introduction to the experimental procedure during vehicle or engine certification in the laboratory, discussing the driving cycle test. Various topics are dealt with such as type approval issues, dynamometers, the coast-down test, and, primarily, the experimental setup and the equations used to calculate emissions and fuel consumption during the cycle execution in the laboratory.

One inherent feature of test cycles is their numerous technical specifications. These are provided collectively in the Appendix, for the most important schedules, and on a relatively detailed basis. It was intentional to not incorporate many 'numbers' inside the text. In any case, some technical specifications are important to supply context, particularly when comparing cycles; such comparisons form an important component of the book. In order to understand the basis under which these technical characteristics have been derived, the interested reader is advised to check initially the introductory text in the Appendix, where the exact equations/logic is provided.

Although the biggest part of the book deals with driving cycles, i.e., those executed on a chassis-dynamometer where the whole vehicle is under test, it was decided to title the book 'Driving and Engine Cycles' so as to cover the second important category as well, namely engine-dynamometer cycles, which are also discussed in detail.

The present book is intended to serve as a reference for engineers and researchers, but it should also be useful to students as a supplementary text on exhaust pollution courses. Although for much of its length, it does not require specific technical knowledge and can be rather easily conceived by most people involved in the research and study of engine emissions and fuel economy, it is expected that the reader is already familiar with emission regulation matters and with some basic aspects of internal combustion engines operation.

Lastly, and bearing in mind that the broad subject of emission legislation is constantly evolving, the reader should always have in mind that a regulation valid when preparing this book (2016) might have been superseded at the time of reading it.

Athens, Greece
November 2016

Evangelos G. Giakoumis

Acknowledgments

The author wishes to acknowledge several individuals who helped in various ways the writing of this book, namely Prof. Emeritus C.D. Rakopoulos, Prof. D.T. Hountalas, Dr. A. Dimaratos, Prof. D.C. Kyritsis, Ms. S. Schmidt, Ms. Anna Eleni Giakoumi, Messrs R. Brezny, Z. Chen, K. Chun, M. Cox, T. Drongitis, K. Engeljehringer, A. Kerkhof, C. Mi, N.-O. Nylund, S. Park, S. Zhang, T. Zheng, as well as various current and former employees of the U.S. Environmental Protection Agency (C. Courtois, C. France, J. France, J. Spieth, and T. Wysor).

Work conducted over the past few years by some of my students, under my supervision, is also presented in the next chapters. The contribution of A. Alafouzos, A. Alysandratou, S. Lioutas, N. Kakouras, G. Triantafyllou, E. Velidakis, C. Vitsas, and A. Zahiotis is greatly acknowledged.

I would also like to express my gratitude to the supporting staff at Springer for their continuous help.

Lastly, the various publishers and companies that have granted permission to reproduce figures and photographs from their publications are gratefully acknowledged; they are also credited in the text.

Contents

Notation

a	Acceleration (m/s^2)
A	Area (m^2)
C	Concentration (ppm or %)
E	Emission (g or g/km or g/kWh)
f	Sampling frequency (Hz)
F	Force (N)
g	Gravitational acceleration (m/s^2)
H	Humidity (g water/kg dry air)
K_h	NO_x humidity correction factor
K_r	Multiplicative regeneration factor
m, M	Mass (g or kg)
N	Rotational speed (rpm) or Number
p	Pressure (Pa or bar)
P	Power (kW)
S	Distance (m or km)
t	Time (s)
T	Absolute temperature (K) or Torque (Nm)
V	Volume (m^3) or vehicle speed (km/h or mph)
W	Work (J) or weight (kg)

Greek symbols

θ	Road grade (%)
Θ	Moment of inertia (kgm^2)
ρ	Density (kg/m^3)
σ	Standard deviation

Subscripts

a	Aerodynamic
act	Actual
d	Dry or drag
da	Dilution air
deg	Diluted exhaust gas
eg	Exhaust gas
f	Frontal or fuel
fr	Friction
g	Gas
gr	Gravitational or grade
i	Each pollutant
in	Inertia
j	Each mode
norm	Normalized
PM	Particulate matter
r	Rolling or regeneration
ref	Reference
tr	Traction
V	Vehicle
w	Wet or weighted

Abbreviations

ACEA	Association des Constructeurs Européens d' Automobiles (European Automobile Manufacturers Association), www.acea.be
AMA	Approved mileage accumulation (cycle)
CAA	Clean Air Act
CAAA	Clean Air Act Amendments
CARB	California Air Resources Board, http://www.arb.ca.gov
CBD	Central business district
CFR	Code of federal regulations (USA), www.ecfr.gov
CFV	Constant flow venturi
CH_4	Methane
CI	Compression ignition
CNG	Compressed natural gas
CO	Carbon monoxide
CO_2	Carbon dioxide
CUEDC	Composite urban emissions drive cycle
CVS	Constant volume sampling (a universally applied technique for measuring pollutants mass emissions during driving cycles—Chap. 6)
CWF	Carbon weight fraction
DF	Dilution factor

DI	Direct injection (fuel injection takes place inside the cylinder after the intake valves have closed)
DOC	Diesel oxidation catalyst (a catalytic converter employed in diesel engines for oxidation of CO and HC to CO_2 and H_2O)
DPF	Diesel particulate filter (the DPF traps particulates; these are periodically burned through an ECU-initiated brief regeneration event involving high-load operation that generates exhaust gas temperatures in the range of 500–550 °C capable of burning the particulates and restoring the trap)
EC	European Commission, http://ec.europa.eu
ECU	Engine control unit (an electronic unit that gathers engine and vehicle data from various sensors, processes them based on stored maps, and determines, through various actuators, the exact timing of fuel injection, spark plug ignition, etc.)
EEA	European Environment Agency, http://www.eea.europa.eu
EEC	European Economic Community (the predecessor of European Union, 1958–1993)
EEV	Enhanced environment-friendly vehicle
EGR	Exhaust gas recirculation (part of the exhaust gas leaving the cylinder, after being cooled, is inducted into the engine in order to limit in-cylinder combustion temperatures and reduce the production of NO_x)
ELR	European Load Response
EPA	Environmental Protection Agency (USA), https://www3.epa.gov
ESC	European steady-state cycle
ETC	European transient cycle
EU	European Union, http://europa.eu
EUDC	Extra urban driving cycle
EV	Electric vehicle
FC	Fuel consumption
FIGE	Forschungsinstitut für Geräusche und Erschütterungen
FR	Federal register
FTP	Federal test procedure
GDI	Gasoline direct injection (type of spark ignition engine that applies a stratified charge (rich in the vicinity of the spark plug and leaner toward the cylinder walls), primarily aiming to reduce fuel consumption)
GDP	Gross domestic product
GHG	Greenhouse gas
GPF	Gasoline particulate filter
GRPA	Group de Rapporteurs sur la Pollution de l'Air
GRPE	Group de Rapporteurs sur la Pollution et l' Énergie
GTR	Global technical regulation
GVW	Gross vehicle weight (maximum operating weight of a vehicle including the vehicle's chassis, body, engine, engine fluids, fuel, accessories, driver, passengers, and cargo, according to the legislation)

HC	Hydrocarbons
HD	Heavy duty
HDE	Heavy-duty engine
HDV	Heavy-duty vehicle
HEV	Hybrid-electric vehicle
HFET	Highway fuel economy test
HHDDT	Heavy heavy-duty diesel truck (cycle)
I.C.	Internal combustion
I/M, IM	Inspection and maintenance
ICCT	International Council on Clean Transportation
IDC	Indian driving cycle
ISO	International Organization for Standardization, http://www.iso.org
LD	Light duty
LDT	Light-duty truck
LDV	Light-duty vehicle
LEV	Low-emission vehicle
LNT	Lean NO_x trap, also known as NO_x adsorber (deNOx after-treatment system employing zeolite as adsorbent, where NO_x is trapped; the filter requires some kind of periodic regeneration to be restored/emptied)
LoS	Level of service (measure used to describe the quality of traffic conditions)
LPG	Liquefied petroleum gas
MDPV	Medium-duty passenger vehicle
MOVES	Motor vehicle emission simulator
mph	Miles per hour
MVEG	Motor vehicle emissions group
MY	Model year
NEDC	New European driving cycle
NG	Natural gas
NHTSA	National Highway Traffic Safety Administration (USA), http://www.nhtsa.gov
NMHC	Non-methane hydrocarbons
NMOG	Non-methane organic gases (the sum of non-oxygenated and oxygenated hydrocarbons contained in a gas sample, including, at a minimum, all oxygenated organic gases containing 5 or fewer carbon atoms (i.e., aldehydes, ketones, alcohols, and ethers), and all known alkanes, alkenes, alkynes, and aromatics containing 12 or fewer carbon atoms, excluding methane)
NO_x	Nitrogen oxides
NRSC	Non-road steady-state cycle
NRTC	Non-road transient (composite) cycle
NTE	Not to exceed
NYCC	New York city cycle
OBD	On-board diagnostics

OCE	Off-cycle emissions
OCTA	Orange County Transportation Authority
OICA	Organisation Internationale des Constructeurs d'Automobiles (International Organization of Motor Vehicle Manufacturers), www.oica.net
OPEC	Organization of the Petroleum Exporting Countries
PDP	Positive displacement pump
PEMS	Portable emissions measurement system
PFI	Port fuel injection (conventional spark ignition engine, where the fuel is injected onto the inlet valve and is mixed with the incoming air)
PFSS	Partial flow sampling system
PI	Positive ignition
PKE	Positive kinetic energy (m/s^2)—Eq. (1.3)
PM	Particulate matter
PMR	Power-to-mass ratio (kW/t)
PN	Particle number
RDE	Real driving emissions
RMC	Ramped modal cycle
RPA	Relative positive acceleration (m/s^2)—Eq. (1.2)
SAE	Society of automotive engineers
SAFD	Speed/acceleration frequency distribution
SBC	Standard bench cycle
SCR	Selective catalytic reduction (a, usually, vanadium-based catalyst, where NO_x is reduced to N_2 employing ammonia NH_3; in most cases, a liquid urea solution is injected in the exhaust gas stream prior to the SCR that is then converted to ammonia)
SET	Supplemental emission test, or sulfate emission test
SFTP	Supplemental federal test procedure
SI	Spark ignition
SRC	Standard road cycle
THC	Total hydrocarbons
TRIAS	Test Requirements and Instructions for Automobile Standards (Japan)
UDC	Urban driving cycle
UDDS	Urban dynamometer driving schedule
UN	United Nations, http://www.un.org
UNECE	United Nations Economic Commission for Europe, http://www.unece.org
VPR	Volatile particle remover
VSP	Vehicle specific power (kW/t)—Eq. (1.4)
wf	Weighting factor (%)
WHDC	Worldwide harmonized Heavy-Duty Certification Procedure
WHSC	Worldwide harmonized Steady-State Cycle
WHTC	Worldwide harmonized Transient Cycle

WLTC	Worldwide harmonized Light-Duty Driving Test Cycle
WLTP	Worldwide harmonized Light-Duty Vehicles Test Procedure
WMTC	Worldwide harmonized Motorcycle Emissions Certification/Test Procedure
WOT	Wide open throttle
WTVC	Worldwide Transient Vehicle Cycle

Chapter 1
Introduction

Abstract This chapter provides an introduction to driving cycles. Driving cycles are defined and classified with respect to each application and each application's special attributes. A detailed discussion on the characteristics of each category (chassis dynamometer and engine dynamometer, modal and transient) is performed, with emphasis placed on cycle representativeness issues. Furthermore, the various metrics used to assess the technical properties and shortcomings of test cycles are detailed. The last part of the chapter describes the procedure followed when developing a test cycle; both the collection of driving data as well as the cycle construction techniques are discussed.

1.1 Motor Vehicles: Environmental Pollution, Emission Legislation and Recent Trends

Over the last decades, one of the major risks faced by mankind has been the anthropogenic contribution to environmental pollution, and degradation in general. This has posed substantial burden on both human health and economy. Particulate matter (PM), nitrogen dioxide (NO_2) and ground-level ozone (smog) are now generally recognized as the three pollutants that affect human health most significantly [1]. Long-term and peak exposures to these pollutants range in severity of impact, from increases in the risk of non-allergic respiratory symptoms and disease to premature death particularly from cardiopulmonary causes. For example in 2010, air pollution in Europe was responsible for over 600,000 premature deaths, the economic cost of which is estimated at 1.4 trillion €, approximately one tenth of EU's 2013 GDP [2]. Overall, the World Health Organization estimates that 3.7 million deaths worldwide could be attributed to outdoor air pollution in 2012. The majority of these deaths is registered in developing countries, where the relevant epidemiological studies indicate that air pollution and the associated degradation in quality of life impose a significant burden on people in all sectors of society, especially the poor [2].

E.G. Giakoumis, *Driving and Engine Cycles*,
DOI 10.1007/978-3-319-49034-2_1

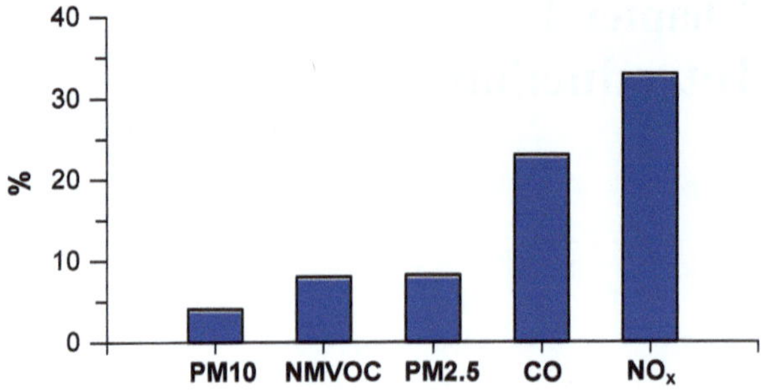

Fig. 1.1 Contribution of the road transport sector to total (includes other non-road and non-transport sectors) emissions in Europe (*Source* European Environment Agency 2016)

Further to the above, one relatively recent issue is global warming, being responsible for various negative environmental effects such as sea level rise and extreme weather conditions. The main culprit here is excessive greenhouse gas (GHG) emissions, with carbon dioxide (CO_2) assuming the largest share.

There is a growing awareness that on-road, as well as non-road, vehicles powered by fossil-fueled internal combustion (I.C.) engines are amongst the major contributors to the environmental degradation in the form of both toxic pollutants and CO_2 emissions. This has been particularly true in the last decades, when the use of all kinds of automobiles and I.C. engines has increased exponentially.

For example in Europe, the road transport sector alone is responsible for emitting one third of the total nitrogen oxide (NO_x) emissions in 2015 (this percentage has dropped from 42 % in 2009), one quarter of total carbon monoxide (CO), and 8 % of PM2.5 and NMVOC (non-methane volatile organic compounds), as Fig. 1.1 demonstrates. Furthermore, road transport contributes about one fifth of the EU's total emissions of CO_2 [1], as is also the case in Japan [3]. 75 % of the GHG emissions in the EU stem from light-duty and 25 % from heavy-duty vehicles. Non-road mobile machinery, on the other hand, is responsible for 15 % of the NO_x and 5 % of the PM emissions in the EU [1].

Similarly in the United States, the transportation sector (includes also railway, marine and air traffic) accounted for one quarter of total GHG emissions in 2014 (increased by almost 17 % since 1990), being second only to the electricity sector, Fig. 1.2; 84 % of these emissions derive from on-road transport vehicles.

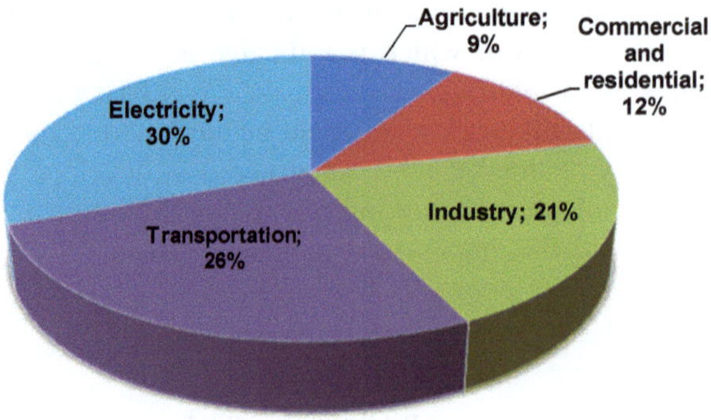

Fig. 1.2 Total U.S. greenhouse gas emissions in 2014 by economic sector (*Source* Inventory of U.S. Greenhouse Gas Emissions and Sinks: 1990–2014)

The term road transport covers vehicles that are designed for the transportation of passengers (passenger cars, motorcycles, buses, coaches) and freight (vans, trucks etc.). On-road or highway vehicles are typically fueled by gasoline, diesel or even alternative fuels or biofuels (biodiesel or alcohols). Non-road vehicles, on the other hand, are those principally employed for agricultural and construction purposes, primarily powered by diesel engines. Road transport poses a dilemma in the sense that it contributes significantly to environmental pollution but is essential for economic and social growth and development.

The earliest indication of *vehicle-related* effects on urban pollution appeared in the 40s in Los Angeles in the form of photochemical smog (a linguistic blend of the words smoke and fog[1] [4, 5]). In order to address the severe smog problem, the first *emissions standards* from motor vehicles were established in California in 1959 by the Department of Public Health regarding CO and HC emissions. The concerned vehicles were gasoline-engined, the number of which had, in the meantime, increased substantially [5, 6]. Four sources of motor-vehicle created pollution were identified, namely exhaust, crankcase, carburetor and fuel tank. Based on the conditions projected to prevail in the area in 1970, it was estimated that a reduction of the order of 80 % in HC and 60 % in CO emissions would be required to achieve the air quality that existed in Los Angeles in 1940. The emission limits set to achieve that target were 275 ppm for HC and 1.5 % for CO. These referred to a composite test sequence encompassing eleven driving modes [7]. The setting of these first limits was a historic milestone, as it practically opened a new and huge chapter in the development of all kinds of engines and vehicles by introducing the terms 'emission level' and, more importantly, 'emission control'. Moreover, it rendered the United States one of the pioneers in dealing with environmental issues,[2] and particularly California, the region in the world with the strictest emission limits. Other areas, such as Europe and Japan, followed in the mid to late 60s. Since one obvious measure from manufacturers to cope with the first emission regulations of CO and HC was to shift the (gasoline) engine operation to leaner mixtures, the amount of emitted nitrogen oxides increased substantially; NO_x soon (1971) became the third controlled pollutant in California [5] and the United States, and later in Europe and Japan [6].

A second critical issue related to environmental pollution from vehicles appeared in the late 60s, and concerned diesel-engined trucks, which at the time were becoming turbocharged at increasing rates. Diesel trucks emitted a large amount of black smoke during (cold) starting as well as at each acceleration or load increase.

[1]The first occurrence of ground-level ozone, or smog, in Los Angeles was in 1943 and concerned, apart from vehicular internal combustion engines, domestic and industrial combustion processes. From the transportation point of view, the major smog culprits are hydrocarbons (HC), NO_x, CO, sulfur dioxide (SO_2), and VOCs. These react with sunlight, heat and various other compounds to form the noxious vapors, ground level ozone, and particles that are constituents of smog.

[2]A subsequent key event in the air-pollution history of the United States was the deadly smog in Donora Pennsylvania in 1948. The town's heavy industrial production and a temperature inversion resulted in high air-pollutant concentrations for a period of one week [4].

This led to a widening of the emission regulations to also cover diesel-engined vehicles, and of the regulated pollutants to also include particulate matter [8].

In the following years, the number of countries adopting emission regulations grew at a highly dynamic rate, and so did the tightening of the prescribed levels, as Figs. 1.3 for passenger cars and 1.4 for heavy-duty engines characteristically demonstrate. In order to meet these targets, the use of various after-treatment systems, such as 'catalysts' or traps, became mandatory, together with cleaner fuels.

Unfortunately, air quality has not always improved in line with the general decrease of anthropogenic emissions of air pollutants or with the increasing stringency of the emission limits. One notable example is the increase in NO_2 emissions from motor vehicles in the 2000s, most probably owing to the extensive use of diesel particulate filters to control PM. As pointed out by the Health Effects Institute, 'the rapid growth of the world's motor-vehicle fleet due to population growth and economic improvement, the expansion of metropolitan areas, and the increasing dependence on motor vehicles because of changes in land use, resulted in an increase in the fraction of the population living and working in close proximity to busy highways and roads. This, to some extent, counteracted the expected benefits of pollution-control regulations and technologies' [9]. Other factors that have contributed to the inconsistency between decreasing emission limits and not adequately improving air quality are: persistence of old and outdated technology in the vehicle fleet, poor maintenance of the vehicles in use, unavailability or improper use of appropriate fuels, as well as atmospheric, topographic and climatological aspects of metropolitan areas where pollution is concentrated [10].

To implement the emission targets set by the legislation, one key issue is to establish the exact procedures under which vehicles and engines are to be tested, measured and if the emitted pollutants are found within the prescribed limits, ultimately certified. Since it is dissuasive from a financial point of view, and also extremely time consuming, to conduct in-use tests on a statistically appropriate number of vehicles, inevitably regulation focuses on tests representative of the vehicle's/engine's duty cycle at the certification level (test cycles). This is actually the subject of the present book, namely certification test cycles, and will be detailed in Chaps. 2–5 for all kinds of vehicles and engines. Information regarding the measurement procedure and the corresponding emission calculations will be provided as well, in Chap. 6. Generally speaking, the relevant regulatory documents concerning emission legislation usually comprise four main subject matters, namely

- emission limits, such as those illustrated in Figs. 1.3 and 1.4;
- test cycles during which these limits are measured;
- measurement procedure and specifications; and
- emissions calculation and report.

One of the key challenges for the legislation worldwide is to ensure that emissions from vehicles measured during the certification procedure—also known as homologation or type approval—are in line with real-world driving emissions. This is a very timely matter, and will be dealt with during the analysis of the various test

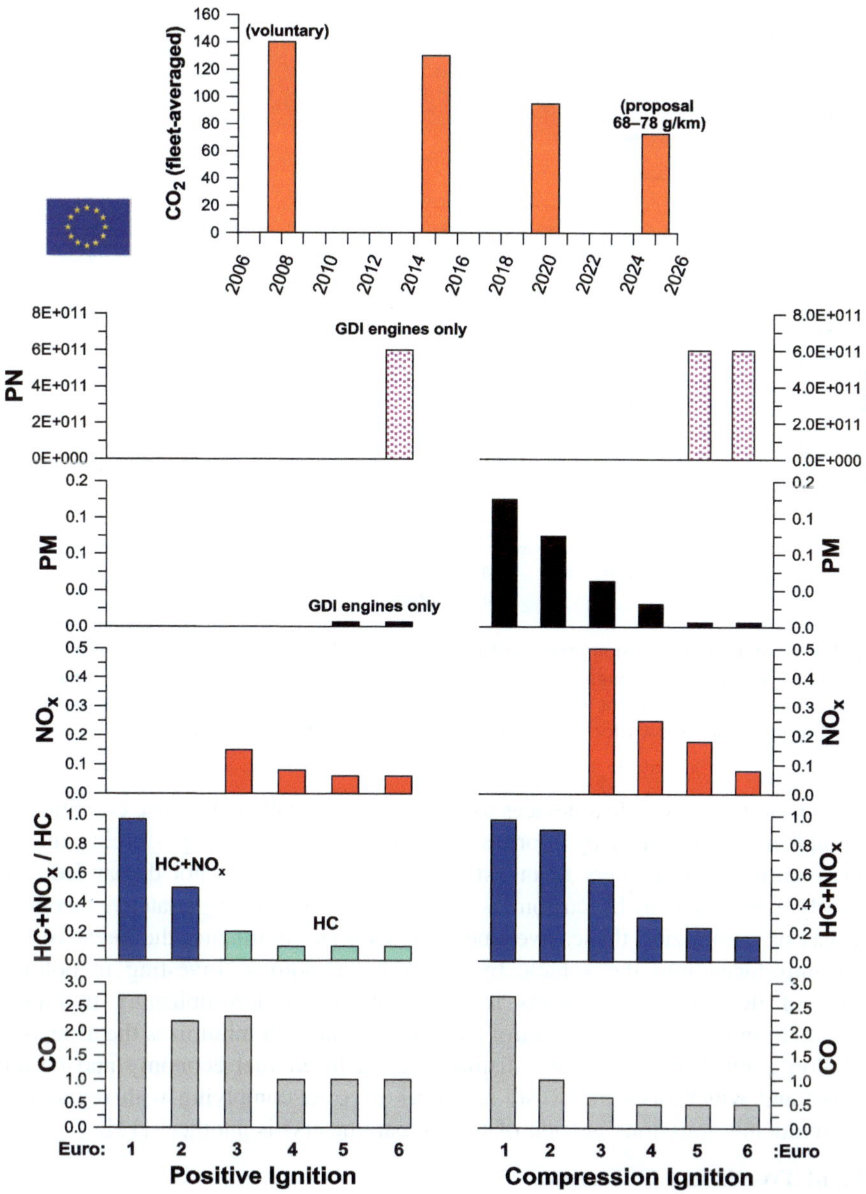

Fig. 1.3 Evolution of 'Euro' emission levels in the European Union (g/km) for category M_1 vehicles comprising passenger cars (for Euro 5 and 6 of PI-engined vehicles, a level of 0.068 g/km of NMHC is valid; CO_2 limits are implemented gradually)

cycles in the next chapters, as experience has shown that type approval results and in-use emissions do not necessarily go hand in hand. Historically, the regulators' approach has been to set emission standards on a specific test procedure, and then

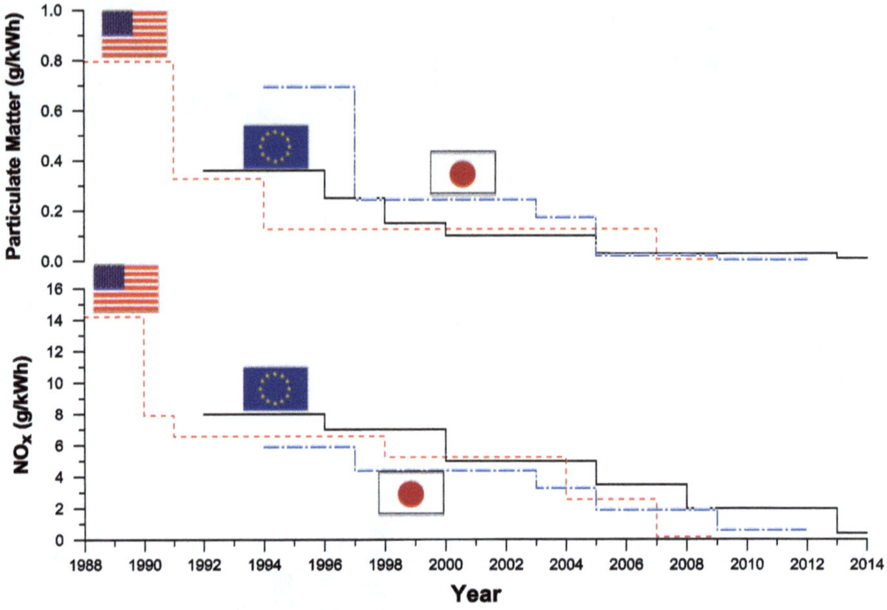

Fig. 1.4 Evolution of emission levels in Europe, USA and Japan regarding PM and NO_x from heavy-duty on-road diesel engines (in Europe, the emission levels correspond to steady-state testing up to 2000 and transient testing from 2000 onwards; earlier than 1994 limits in Japan were defined in ppm; data shown for the United States correspond to federal limits)

rely on prohibition of defeat devices to ensure that control on all in-use conditions is accomplished. As argued by Morriss et al., however, 'technology-forcing regulations require manufacturers to invest in developing features not demanded (and possibly even rejected) by customers. Their profit-maximizing strategy, therefore, consists of minimizing those investments as well as minimizing the negative features introduced into the vehicle by the new technology. Investing in political means of delaying requirements is one strategy for accomplishing this goal. Investment in new emissions control technology that both minimizes the aspects of pollution control that customers dislike (e.g., reduced fuel economy and performance) and which does not waste resources by over-complying with regulations (e.g., reducing emissions outside of testing parameters) is another' [11].

Recent Trends in I.C. Engines

Today, vehicles with spark ignition (SI[3]) engines, mostly used in passenger cars and light-duty vans, are equipped with multipoint injection systems, employing port fuel injection (PFI). A three-way catalyst with precise air-fuel equivalence ratio (lambda) control at stoichiometric conditions to simultaneously limit CO, HC and

---

[3]European regulations use the term positive ignition (PI); U.S. regulations refer to Otto-cycle (heavy-duty) engines.

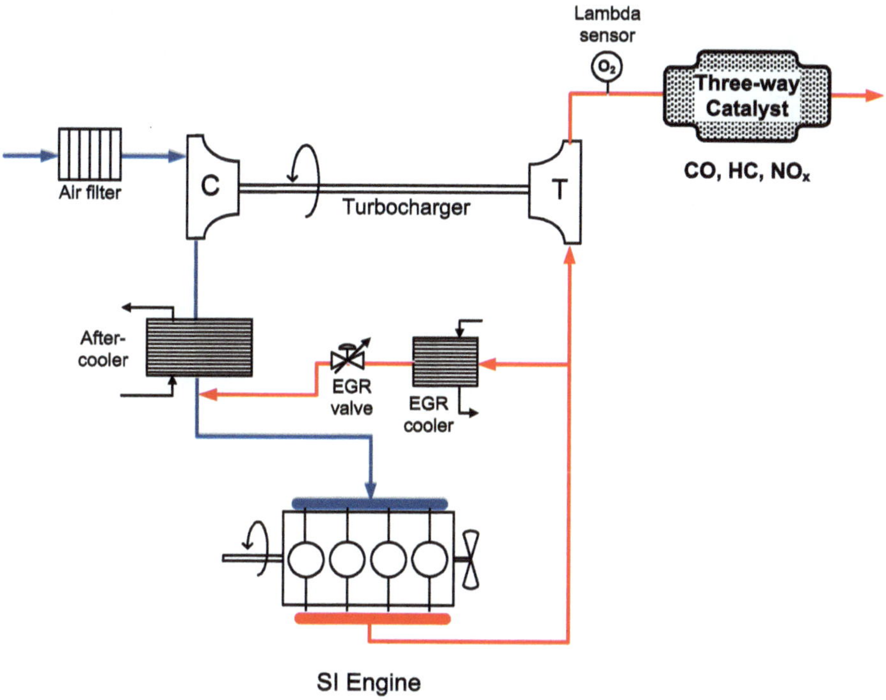

Fig. 1.5 Modern, turbocharged, automotive SI engine with EGR for NO_x control and three-way catalyst with lambda sensor

NO_x is utilized; see for example Fig. 1.5. An alternative, gasoline direct injection (GDI), technology is based on lean-burn stratified charge, primarily aiming at reducing fuel consumption. Exhaust gas recirculation (EGR) for NO_x control is usually applied in both SI variants for many years now. Alternative fuels, mainly in the form of ethanol, are also utilized in many countries (Brazil, USA, Sweden) either as principal fuel (e.g., E85, meaning 85 % ethanol mixed with 15 % gasoline), or in smaller, 5–10 %, proportions. For heavy-duty SI (gas) engines, natural gas (NG) and liquefied petroleum gas (LPG) are the two usually applied fuels.

On the other hand, turbocharged diesel/compression ignition (CI) engines have been traditionally employed as the principal prime mover in the medium and medium-large transport sector (trucks, buses, coaches; non-road applications as well). Major contributing factors have been the CI engine's superior fuel economy over its spark ignition counterpart, coupled with high torque and increased reliability. In contrast, discrepancies in the form of exhaust smokiness, cold starting difficulty and combustion noise radiation delayed the infiltration and broad acceptance of diesel engines in the highly competitive passenger car market. The extensive use, however, of electronically-controlled common-rail direct injection systems and sophisticated after-treatment control, most notably in the form of diesel particulate filters (DPF), has changed this trend in the recent years, with the

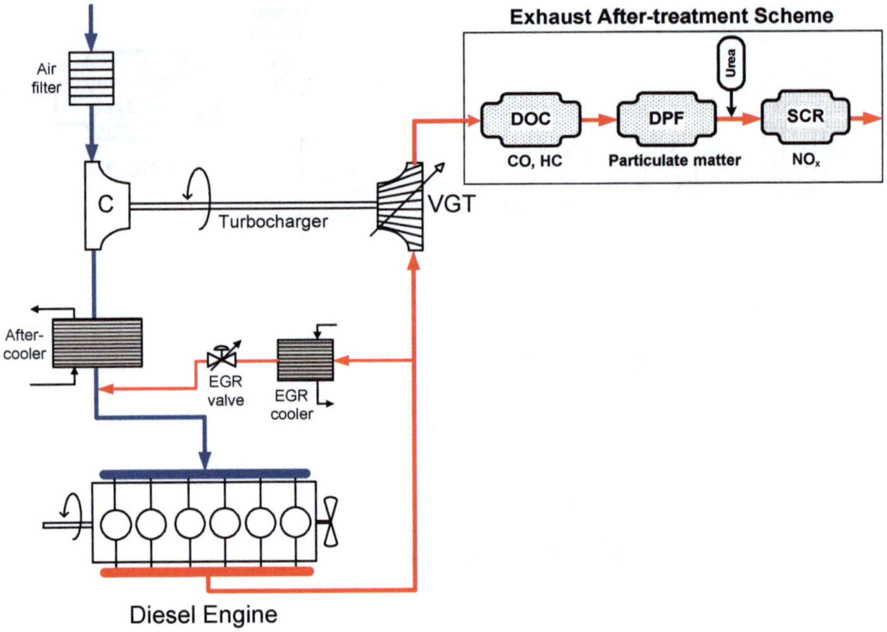

Fig. 1.6 Modern, variable geometry turbocharged (VGT) truck diesel engine with diesel oxidation catalyst (DOC) for CO and HC control, diesel particulate filter (DPF) for PM, and selective catalytic reduction (SCR) and exhaust gas recirculation (EGR) for NO_x control

majority of passenger cars sold, for example in Europe, being now compression-ignition engined [12]. Figure 1.6 illustrates a state of the art diesel engine configuration typical of Euro 6/VI or U.S. 2010 emission certification, and Fig. 1.7 focuses on the exhaust after-treatment scheme of a modern bus.

The development needs for diesel engines concentrate on the reduction of NO_x and particulates, which unfortunately develop in an inverse manner, with different combinations of in-cylinder measures and after-treatment control applied, depending also on the specific emission standard. For example, one option is to reduce engine-out NO_x using EGR, and apply a combination of DOC + DPF for CO/HC and PM reduction. Another is to reduce engine-out PM through in-cylinder measures (most notably, employing higher gas temperatures, which are also expected to favor fuel economy), and comply with the NO_x targets based solely on after-treatment control, e.g., in the form of SCR or LNT (lean NO_x trap). Certain diesel after-treatment control, such as DPFs and NO_x adsorbers, do not function in the traditional way of continuously reducing emissions; instead, pollutants are trapped inside the filter. Therefore, periodic regeneration is required to restore the device. This is accomplished by brief engine operation at rich mixtures in order to burn the trapped emissions. Such after-treatment systems, which rely on filtering to control pollutants, tend to increase the back pressure in the exhaust manifold, hence adversely affect engine fuel consumption by increasing the cylinder pumping losses.

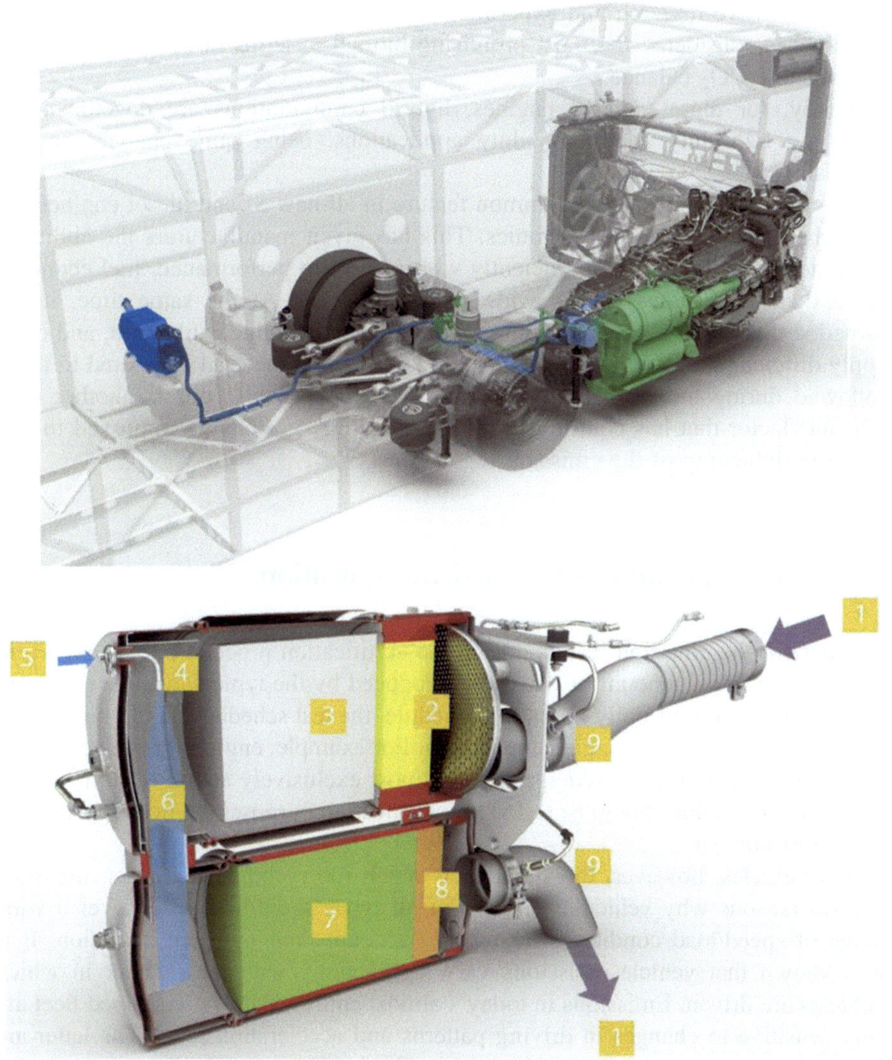

Fig. 1.7 Euro VI technology for city buses (SCRT system): (*1*) Stream of exhaust gas; (*2*) Diesel-oxidation catalytic converter; (*3*) Closed particulate-filter system CRT; (*4*) Injection of AdBlue®; (*5*) AdBlue® supply line; (*6*) AdBlue® conversion tract; (*7*) SCR catalytic converter module; (*8*) Ammonia-blocking catalytic converter; (*9*) NO_x sensors (*Source* MAN Truck & Bus AG)

For the future, alternative diesel combustion processes (e.g., homogeneous charge compression ignition—HCCI, or premixed control compression ignition—PCCI) are under vigorous research in order to simultaneously reduce NO_x and PM. However, complex, process-controlled engine management issues prevent, for the moment, their practical implementation. The use of alternative fuels is another field where extensive research and experimentation are carried out. For diesel engines, biodiesel is considered the most promising biofuel in terms of reduced emissions and positive CO_2 balance.

Lastly, for diesel and SI engines, hybrid configurations are developed too regarding light-duty and heavy-duty applications, being quite challenging and promising.

For many years now, one common feature in all new SI and diesel engines has been the extended use of electronics. This has given manufacturers the ability to calibrate their engines more efficiently with respect to performance, fuel economy and exhaust emissions over a wide operating range. At the same time, it has provided them with a powerful tool to manipulate the engine calibration, and often apply different strategies when the vehicle is driven on the road compared to those followed during the certification procedure. Most probably, this is another (significant) factor that has contributed to the air quality not having improved in line with the tightening of the emission regulations.

1.2 The Importance of Transient Operation

In order for the emission results during the certification procedure to be representative of the actual amount of emissions produced by the typical/average engine or vehicle throughout its daily operating schedule, the test schedule should be adapted according to the specific engine application. For example, engines driving electrical generators function at a fixed speed and almost exclusively at full load. Thus, the operating points that should be tested for certification purposes are both specific and limited in number.

For vehicles, however, the situation is much more complex. In fact, there are various reasons why vehicular engines of all types should be tested over a wide range of speed/load conditions taking into account their transient operation. It is well-known that vehicle emissions vary significantly with the manner in which vehicles are driven. Emissions in today's almost entirely catalyst-equipped fleet are very sensitive to changes in driving patterns and acceleration rates. The latter are inherently related to the types of roads, traffic conditions, and drivers or vehicles being considered.

Many measurement campaigns of pollutant emissions from older technology gasoline-engined vehicles have suggested that CO and HC are generally decreasing functions of speed (thus, low speeds should be contained in the test procedure). For NO_x, the trend is reversed, with the emissions being increasing function of speed (thus, higher engine speeds should be tested as well). In general, over typical speed ranges, unit emissions can vary by a factor of 1–5, which underlines the necessity for as wide as possible a speed (and load) range coverage of the engine's operation during the certification procedure [13]. By doing so, likely attempts by manufacturers to deactivate exhaust after-treatment at operating points not covered by the legislated procedure is avoided, or at least limited.

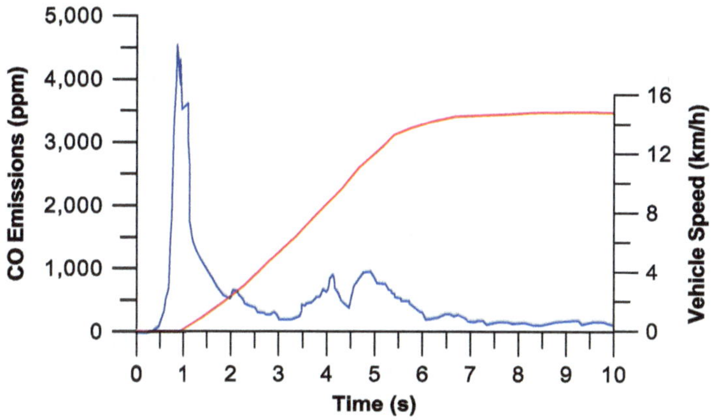

Fig. 1.8 Development of CO emission trace after a steep low-load transient of a PFI gasoline engine highlighting the emission overshoot

Furthermore, owing to the continuously changing traffic, road and even weather conditions, the operating cycle of vehicular engines involves constantly varying speed and load (transient) operation. Unfortunately, during transients, emissions and fueling, hence CO_2, which is inter-related to fuel consumption, typically peak well above their steady-state values. The following paragraphs discuss this issue more extensively for both SI and diesel engine operation.

For *gasoline* passenger car engines, accurate fuel control during transients has traditionally been difficult because of inaccuracies in predicting the air and fuel flow under rapidly changing throttle conditions. Because of air and fuel dynamics, e.g., fuel evaporation in the intake manifold and air-flow behavior, and the time delay between the air-flow measurement and the injection of the calculated fuel mass, temporary excursions from the stoichiometric ratios occur during transients that can cause engine hesitation, poor drivability and increase in emissions [14]. This is typically highlighted in Fig. 1.8, which illustrates the CO emission peak during an abrupt acceleration of a vehicle powered by a port fuel injection SI engine. Enrichment of the (otherwise homogeneous) mixture at or near full throttle during hard accelerations, causes substantial increases in the emitted CO (and HC) due to lower than stoichiometric mixtures. Moreover, during these very frequently occurring on the road transient phases, the exhaust gas recirculation is shut off to aid the acceleration, however at the expense of NO_x emissions.

Another condition that makes it difficult to operate an engine at stoichiometric mixture is high engine loads (a third one will be discussed later and concerns cold engine operation). When operating an engine at high engine loads and at stoichiometry, in-cylinder temperatures increase considerably. This is then reflected into elevated NO_x emissions, higher knock tendency as well as danger of catalyst damage. To avoid these, PFI SI engines are usually operated at slightly rich mixtures when at high load (e.g., lambda around 0.90). This, however, results in increased CO and HC emissions owing to the rich mixtures involved.

Things are even more complicated when turbocharged engines are involved, which is the vast majority of diesel and a steadily increasing portion of SI vehicular engines. In a diesel engine, owing to the heterogeneous combustion, it is inevitable that some amount of soot/PM will be produced. Specifically, the lower the air-fuel equivalence ratio, the higher the amount of emitted PM. Emission of particulates was not a problem with port-injected SI engines but is for GDI ones; in fact, the European legislation, starting from the emission standard Euro 5, has set the same PM limit for both GDI and DI diesel engines (Fig. 1.3).

Contrary to steady-state *diesel engine* operation, where performance is primarily determined by the amount of fuel burned (hence a properly calibrated engine should not be polluting, as regards the carbonaceous emissions CO, HC and PM), during transients, it is the incoming air-supply, the EGR management and the, gradually adjusting to the new fueling, cylinder wall temperature that play the most influential role. Further, whereas breathing in a naturally-aspirated engine is mainly determined by the engine speed, in a turbocharged engine it is strongly influenced by the operation of the turbocharger. The latter is responsible for an unfavorable engine behavior pattern during transients through a well-established and complex phenomenon known as turbocharger lag (although there are other delays in the system apart from the turbocharger). Turbo lag leads on the one hand to problematic torque build-up and poor vehicle drivability, and, on the other, to a considerable increase in engine-out particulate, gaseous and even noise emissions from diesel engines [8, 15]. Probably the most notable example has been the PM peak evidenced as a cloud of black smoke coming out of the exhausts of accelerating, older-technology diesel-engined vehicles. Figure 1.9 highlights this emission overshoot during acceleration of a Euro II turbocharged diesel engine illustrating both engine-out smoke (indicative of PM which are difficult to measure instantaneously) and NO emission traces; similar development is noticed during load increases. In general, the emission overshoot is amplified the lower the initial engine load/selected gear. This means that it is during urban driving that it is mostly pronounced; in fact, the acceleration in Fig. 1.9 commenced from no load conditions. It is also more prominent the harder/steeper the acceleration, i.e., it directly relates to the individual driving style [15–17].

During the early engine cycles after a higher speed (or load) is demanded (tenths of a second), the amount of injected fuel rises almost instantly to its new higher level. Nonetheless, the turbocharger compressor is not capable of instantly providing the appropriately higher air-mass flow rate, primarily owing to the turbocharger lag. Consequently, a mismatch between fueling and air-supply is established, resulting in the instantaneous lambda assuming very low values (even lower than unity). This is then reflected in the soot spike demonstrated in Fig. 1.9. There are many reports arguing that during these transient phases, the number of the emitted particles increases as well, particularly those in the nuclei mode (diameter 5–50 nm), which are considered the most dangerous [15].

For NO_x emissions, it is firstly the high gas temperatures due to close to stoichiometric operation that elevate the emissions during transients, as well as the engine's ECU standard strategy to shut off the EGR system in order to aid the

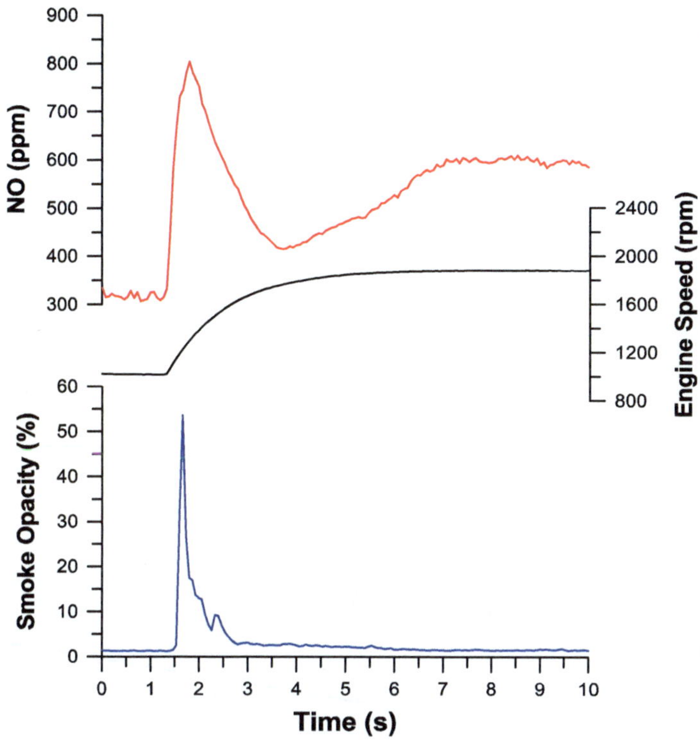

Fig. 1.9 Overshoot in engine-out smoke and NO emissions during acceleration from no load of a turbocharged diesel engine (reprinted from [17], copyright 2010, with permission from Elsevier)

acceleration and limit PM. Nitrogen oxide spikes are of notably lower magnitude than their PM counterparts, as Fig. 1.9 indicates, hence they are, so to speak, sacrificed for the sake of lower PM. Interestingly, however, the NO_x spike extends over a wider time period. In this regard, the use of biofuels (e.g., biodiesel, but also alcohols) enabling considerable decrease in the emitted PM, facilitates stricter EGR control during transients [18].

A particularly hard case, mainly in terms of PM/soot emissions, is experienced during a vehicle's up-gear shift schedule, e.g., when entering a highway. Primarily at the beginning of the acceleration, where the turbocharger speed is very low and the available exhaust gas energy almost zero, the highest amount of PM is produced. The latter behavior in terms of soot and CO_2 is demonstrated in Fig. 1.10 for a turbocharged, diesel-engined medium-duty vehicle. After the first gear has been engaged, the gradual increase in the engine and turbocharger speed aids the vehicle's acceleration; however, temporary overshoots in (engine-out) soot emissions/PM are noticed practically during every gear change. Incorporation of some kind of fueling control can limit the phenomenon but adversely affects vehicle response. Similarly, transient operation leads to sharp fuel consumption (and CO_2)

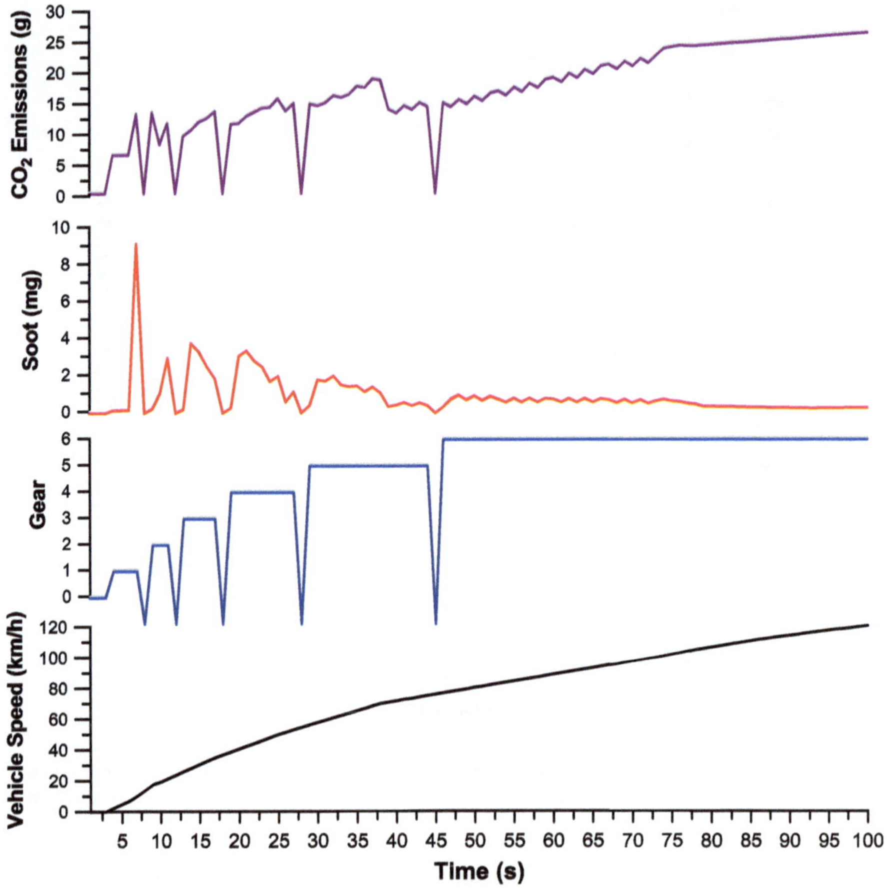

Fig. 1.10 Development of instantaneous CO_2 (indicative of fuel consumption) and engine-out soot during continuous up-gear change of a medium-duty turbocharged diesel-engined vehicle

increases, owing to the need to overcome the engine and vehicle inertia, as well as the various losses in the transmission system. The heavier the vehicle, therefore the higher the number of engaged gears during accelerations, such as the one depicted in Fig. 1.10, the harsher the acceleration pattern and the emission overshoot.

Similar arguments hold for HC and CO emissions, although the concentration of both in the exhaust of modern diesel-engined vehicles is very small (contrary to SI engines, in diesel engines the mixture is always lean). In any case, instantaneous transient CO and HC have been measured several times the respective quasi-steady values, highlighting once again the different evolution pattern and magnitude of exhaust emissions during transients compared to steady-state operation [15]. On the contrary, when the vehicle is cruising, HC and CO emissions decrease substantially; PM is not a problem when cruising at high speeds either.

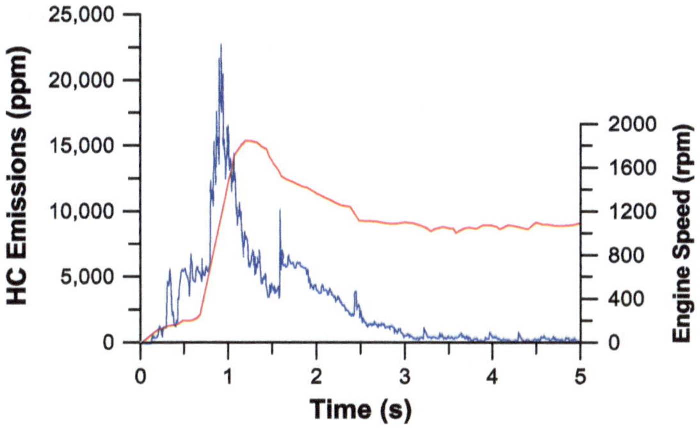

Fig. 1.11 Engine-out HC emissions during cold starting of a 1.6 L GDI engine (courtesy of Cambustion)

Another factor that significantly contributes to elevated gaseous and particulate (both in mass and number) emissions is the thermal state of the engine. Gasoline and diesel engines emit considerably more CO and HC right after cold starting compared to their fully warmed-up conditions, as Figs. 1.11 for SI and 1.12 for diesel engine operation typically demonstrate. Fuel enrichment in SI engines and misfire in CI ones owing to the still cold manifold and cylinder walls are the probable causes here; the phenomena are amplified the lower the ambient temperature. Fuel consumption and CO_2 follow a similarly increasing pattern. Mismatch between fueling and air-supply in turbocharged diesel engines is again the reason for elevated cold-started PM. For example, it has been found that the few

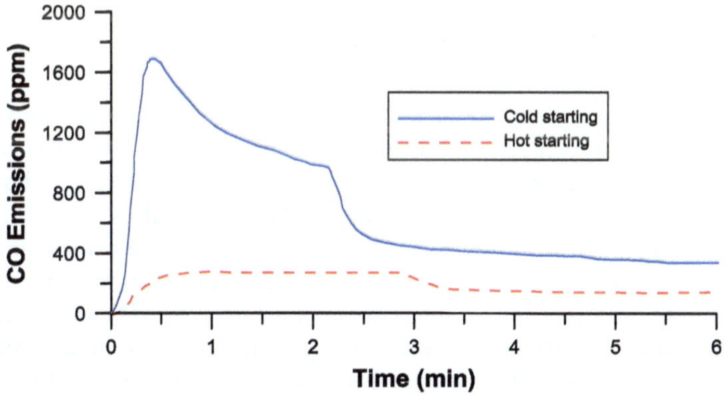

Fig. 1.12 CO emissions during cold and warm starting and idle run of a 2 L, four-cylinder, high-speed, turbocharged diesel engine (adapted from Bielaczyc et al. [20]); CO emissions exceed the 1000 ppm threshold for almost 2 min after cold start, and, at the same time, peak cold-start CO is almost six times its hot-start counterpart

minutes after cold start are responsible for more than 70 % of the total amount of emitted particles for both CI and SI engines during a typical daily driving schedule [19].

In addition to the engine-out cold-starting overshoots observed in Figs. 1.11 and 1.12, the fact that the catalyst (three-way or oxidation/SCR) has not yet reached its operating temperature, further elevates the overall tailpipe or vehicle-out emissions, as most of the exhaust gas is rejected practically without any after-treatment control [14, 15].

Apart from the arguments raised above regarding the need to (a) test the engine over a wide speed/load operating range, and (b) include transients in the form of accelerations and (cold) starting to capture the overshoot in exhaust emissions, there is a third reason why the test procedure for vehicular engines should include as wide as possible an operating pattern. Old technology vehicles with no electronic management had emission-related components that operated on a 'continuum'. For example in a gasoline engine, if the air-fuel mixture at idle was too rich, then the air-fuel mixture was likely to be too rich across much of the operating range of the vehicle (i.e., cruise, acceleration, deceleration). For this reason, a test performed only at idle or only at a certain engine or vehicle speed was likely to identify such malfunctioning vehicles to a sufficient degree. In modern, electronically-controlled vehicles, on the other hand, the engine ECU can include discrete instructions for the air-fuel mixture at idle that have little bearing on the mixture at other speeds or loads, or during an acceleration. For this reason, a vehicle with low emissions at idle or 2500 rpm or 30 km/h could in principal have unacceptable emissions during other modes, and this holds true for both SI and diesel-engined vehicles. Transient testing, on the other hand, is responsive to changing emission levels during different modes of vehicle operation and can thus overcome the limitations of steady-state tests on electronically-controlled vehicles [21].

1.3 Fundamentals of Test Cycles

Beginning in the early 60s, vehicles have been tested for compliance with the emission standards (type approval) using standardized tests; these have been known as test cycles or driving/drive cycles (for the common case where the whole vehicle is being tested), or transient cycles.[4] A test cycle is defined as a sequence of test points, each with a defined vehicle speed to be followed by the vehicle under study, upper sub-diagram of Fig. 1.13, or with a defined rotational speed/torque to be followed by the engine, lower sub-diagram of Fig. 1.13. These test points are divided into time steps, usually seconds, during which acceleration and deceleration are assumed constant. Sometimes, apart from the vehicle speed versus time schedule,

[4]These three terms, test, transient and driving/drive cycle, are used interchangeably in the literature; nonetheless, not all the cycles are transient and not all of them are exactly driving.

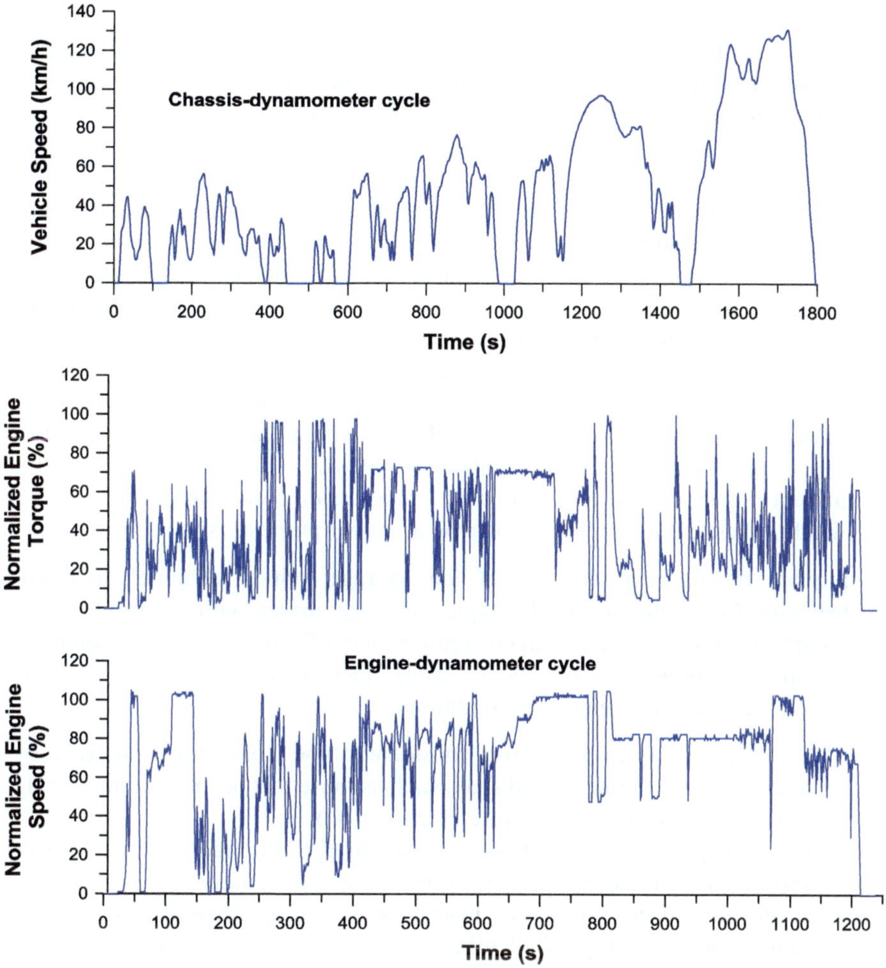

Fig. 1.13 Typical vehicle speed versus time cycle for light-duty vehicles (upper sub-diagram), and engine speed/torque versus time cycle for engine-dynamometer testing (lower sub-diagram)

the road grade versus time profile is provided as well. Driving cycles typically last from a few up to 30 min, although even longer schedules have been developed.

The fundamental advantage when applying a transient cycle for the testing of new vehicles/engines is that a relatively wide engine operating range (in terms of speed and load) is tested and not just specific points such as the maximum power or torque. Moreover, the serious discrepancies that are experienced during abrupt speed and load changes, as discussed in the previous section, are taken into account. It should be pointed out, however, that the primary objective of a transient cycle procedure is to establish the total amount of exhaust emissions and fuel consumption and not indicate the specific parts or conditions under which these emissions are produced.

Test cycles are important for various reasons. First and foremost, they serve in many areas of the world as standardized measurement methods for pollutant/CO_2 emissions and fuel consumption/economy during the certification procedure. Actually, emission standards strongly depend on the cycle and test procedure employed, with manufacturers designing and calibrating their vehicles so that they meet the standards set by the relevant test procedures; often, even designing their engines strictly to the limits of the test procedure. The standardization makes it also possible to compare different vehicles/engines that fulfill the same operation. Further to the above, test cycles are utilized in many countries in the world for inspection and maintenance purposes, i.e., to verify that the after-treatment devices function properly, so vehicles retain their low emission profiles in actual use. Test cycles, serving as proxy for 'average' or typical driving or test conditions, are utilized in many research endeavors or projects, e.g., for modeling and inventory purposes, and for implementing emission control strategies. It goes without saying that the 'quality' of a drive cycle, in terms of representativeness of real-world driving activity, is of paramount importance for any air quality and emission reduction targets.

Ideally, a test cycle employed for type approval purposes should:

- provide results representative of the average vehicle's/engine's real-world behavior (as is driven by the typical/average driver with regard to driving 'aggressiveness' and gear-shift changing pattern). To this aim, and depending on the specific application, test cycles should incorporate some or all of the following daily driving/operating conditions:
  - cold starting (an engine start is considered to be 'cold' if it is preceded by a relatively long uninterrupted soak, such as those that occur after an overnight soak, typically at a temperature of 20–30 °C),
  - hot starting (an engine start is considered 'hot' if it is preceded by a relatively short soak of the order of a few minutes up to one hour),
  - accelerating decelerating, and load increase and decrease phases,
  - idling conditions typical of urban driving, e.g., when waiting at the stop light,
  - sub-urban or rural driving schedule, and
  - extra-urban driving representative of driving on the highway or expressway.
- be practical, i.e., not too long or too complicated for its execution; for example, 20 or 30 min duration seems long enough to include a reasonable amount of data representative of all driving activity and, at the same time, short enough to be reproduced cost-effectively.
- be repeatable and have reproducible results.
- be fully controlled.

Furthermore, all vehicles intended to be tested on a specific test cycle should be able to follow closely the cycle pattern.

As pointed out earlier, test cycles serve as standardized measurement procedures for pollutant and CO_2 emissions, with manufacturers calibrating their engines so as they meet the standards on the specific test. In view of that, there is a very strong relation between legislated test and anti-pollution measures or engine/vehicle design in general. For example, incorporating cold starting in the test procedure has proven influential in driving the engine design to stricter antipollution control and the research to lower light-off catalyst temperatures. A driving cycle with high speeds results in higher aerodynamic forces, hence elevated CO_2 emissions, initiating the research for better aerodynamic measures. Test schedules with frequent stop-and-go, transient, conditions are expected to be more polluting, and are expected to compel engine design to stricter emission control or to hybrid vehicles. On the other hand, cycles with increased idling phases are more likely to promote fuel-consumption reducing technologies such as start and stop.

Vehicles (and engines) have always been tested for compliance with the emission standards in a laboratory. There is a good reason for this. In a vehicle laboratory, technicians can control important influencing factors, such as ambient temperature and vehicle speed trace, while eliminating the effects of wind speed (and weather conditions in general), thereby ensuring reproducibility and comparability of the obtained results, which is of utmost importance for the certification procedure. Execution of a driving cycle in a laboratory requires sophisticated experimental facilities such as fully automated test-bed with electronically controlled dynamometer, (fast response) exhaust gas analyzers, dilution tunnels, etc. (discussed in Chap. 6).

There are many classifications used to categorize test cycles. The primary one is whether the whole vehicle or just its engine is being tested, hence reference is made of

- Chassis dynamometer, and
- Engine-dynamometer cycles.

Chassis-dynamometer cycles are either of simplified 'modal' type or 'true' transient (as the one depicted in Fig. 1.13). Similarly, engine-dynamometer cycles are either steady-state, or, nowadays almost exclusively, transient.

Another classification is whether the cycle is legislated, hence employed in the certification procedure, or non-legislated, in which case the main purpose is in various research projects and for inventory purposes (often in comparison to the legislated ones).

Test cycles may be developed on a local/national basis, suited to the specific traffic conditions/habits and vehicle usage in a certain city or country, e.g., USA, Europe or Japan; the latter have been the pioneering regions in the world in setting emission targets and legislating the appropriate test procedures. On the other hand, test cycles may be *worldwide*, that is cycles designed to be adopted by many countries in the world. Harmonizing the test cycle (and the test methodology in general) reduces certification costs to the benefit of both manufacturers and consumers. A harmonized approach also makes it easier to compare emission limits and fuel consumption across regions, eventually improving the effectiveness of air quality targets and CO_2 reduction strategies. In the case of worldwide cycles, it is the World Forum for Harmonization of Vehicle Regulations of the United Nations

Economic Commission for Europe (UNECE) that is responsible for the development, under the working party on pollution and energy (GRPE),[5] as shown in Fig. 1.14, aided by experts from the participating countries and the industry. Owing to vast dispersion in traffic conditions, road categories, driving habits and vehicle technical specifications, worldwide cycles are further divided into sub-categories or 'classes' (for example, with respect to the vehicle maximum speed, or power-to-mass ratio).

Passenger cars, motorcycles and light-duty trucks undergo a vehicle speed versus time test cycle, such as the one depicted in Fig. 1.13, on a *chassis dynamometer*, and the results are expressed in g/km (or g/mile). The speeds of the cycle are fixed, hence the same cycle is used for each vehicle. This means that high-powered vehicles will

[5]The World Forum for Harmonization of Vehicle Regulations is a working party (WP.29) of the Inland Transport Division of the United Nations Economic Commission for Europe (UNECE—established in 1947—see also Fig. 1.14). WP.29 was created on 6 June 1952 as 'Working party of experts on technical requirement of vehicles'; the current name was adopted in 2000. The forum works on regulations aiming at improving vehicle safety, protecting the environment, promoting energy efficiency and increasing anti-theft performance. It is tasked with creating a uniform system of regulations for vehicle design to facilitate international trade [22].

The core of the Forum's work is based around the '1958 Agreement', formally titled *'Agreement concerning the adoption of uniform technical prescriptions for wheeled vehicles, equipment and parts which can be fitted and/or be used on wheeled vehicles and the conditions for reciprocal recognition of approvals granted on the basis of these prescriptions'* (E/ECE/TRANS/505/Rev.2, amended on 16 October 1995). This forms a legal framework wherein participating countries (contracting parties) agree a common set of technical prescriptions and protocols for type approval of vehicles and components. These were formerly called 'UNECE Regulations' or, less formally, 'ECE Regulations' in reference to the Economic Commission for Europe. As of 2015, there are 135 UN Regulations appended to the 1958 Agreement; most regulations cover a single vehicle component or technology.

The 'Agreement concerning the Establishing of Global Technical Regulations for Wheeled Vehicles, Equipment and Parts which can be fitted and/or be used on Wheeled Vehicles', or 1998 Agreement, is a subsequent to the 1958 Agreement. Following its mission to harmonize vehicle regulations, the UNECE solved the main issues preventing non-signatory countries to the 1958 Agreement to fully participate to its activities. The 1998 Agreement was born to produce meta-regulations called Global Technical Regulations (GTRs) without administrative procedures for type approval and so, without the principle of mutual recognition of type approvals. The 1998 Agreement stipulates that contracting parties will establish, by consensus vote, United Nations Global Technical Regulations (UN GTRs) in a UN Global Registry. The UN GTRs contain globally harmonized performance requirements and test procedures. Each UN GTR contains extensive notes on its development. The text includes a record of the technical rationale, the research sources used, cost and benefit considerations, and references to data consulted. The contracting parties use their nationally established rulemaking processes when transposing UN GTRs into their national legislation [22].

The Working Party on Pollution and Energy (GRPE—Group de Rapporteurs sur la Pollution et l' Énergie) is the subsidiary body of the WP.29 of the UNECE that prepares regulatory proposals on pollution and energy efficiency to WP.29. This group of experts, whose meetings are attended by representatives from governments, manufacturers etc., conducts research and analysis to develop emission and energy requirements for vehicles [22].

UNECE WP.29 regulations can be accessed online through the following link http://www.unece.org/trans/main/welcwp29.html.

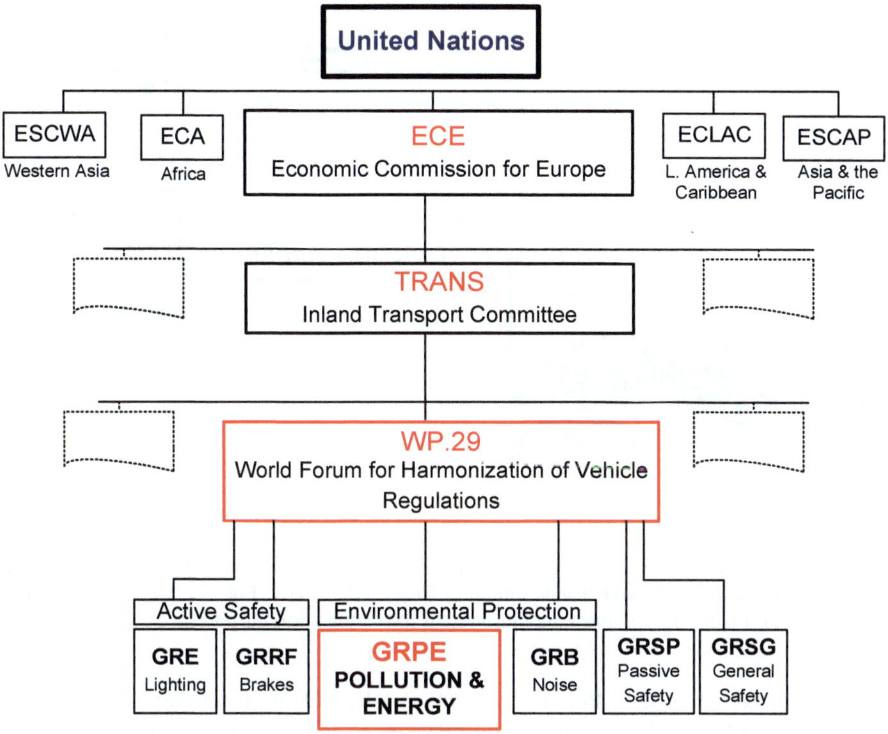

Fig. 1.14 Organizational chart of the UNECE WP.29 forum [22]

be less challenged by the cycle compared to under-powered ones. The vehicle remains stationary during the test, but at least two of its wheels are spinning on the dynamometer to simulate driving at different speeds, thus the whole drive-train is under test, Fig. 1.15. The tested car is driven by a trained driver who follows closely the defined cycle schedule within certain legislated tolerances.

The dynamometer rolls, on which the car rests, have the task to simulate the resistances the vehicle experiences when being driven. These forces are the aerodynamic, the rolling resistance, the grade-dependent, as well as the inertia force. Total traction force F_{tr} is given by (detailed analysis is provided in Sect. 6.3)

$$F_{tr} = \overbrace{c_{fr} m_V g}^{rolling} + \overbrace{m_V g \sin\theta}^{gravitational} + \overbrace{\frac{1}{2}\rho_a c_d A_f V^2}^{aerodynamic} + \overbrace{m_V \frac{dV}{dt}}^{acceleration} \tag{1.1}$$

where c_{fr} tire friction coefficient, m_V vehicle loaded mass, g gravitational acceleration, θ road grade, ρ_a air density, c_d aerodynamic resistance coefficient, A_f vehicle frontal area, and V vehicle speed [23].

It is the above described forces that the dynamometer has to simulate, however, with the exception of the grade component. Legislative test cycles assume straight

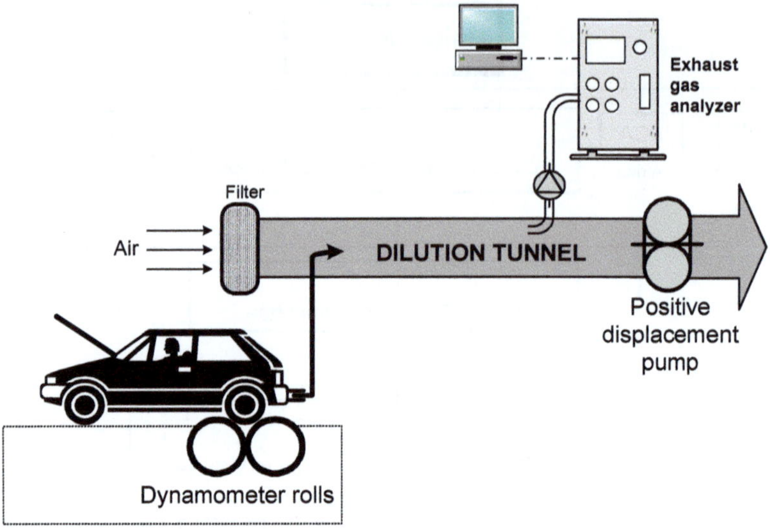

Fig. 1.15 Vehicle on a chassis-dynamometer for emissions measurement; emissions are measured either continuously or through batch sampling (bags, filters) (see Fig. 6.7 for a more detailed illustration of a constant volume sampling system for certification purposes)

roads with zero gradient, thus no account is taken of the road-dependent gravitational term in Eq. (1.1). These driving resistances are determined during a separate test (the coast-down test) that takes place prior to the certification on the dynamometer (Sect. 6.4). They are then simulated in the laboratory by adjusting the resistance of the chassis-dynamometer rollers. Obviously, the simulation of vehicle motion on the road is not dynamically identical with the process of the vehicle tied down and its power absorbed through contact with a rotating drum. During the test, the exhaust of the vehicle is collected and analyzed to calculate emission levels and fuel consumption, with a test protocol containing the details of the cycle and the test procedure. The term test *procedure* describes the exact methodology of the vehicle testing with respect to preconditioning (e.g., whether the test cycle is run with the engine hot or cold started), determination of vehicle mass and road load, ambient conditions, analyzers calibration etc. The main aspects of chassis-dynamometer cycles will be introduced in Sect. 1.3.1, and the cycles will be detailed in Chap. 2 (passenger cars and light-duty vans), Chap. 3 (motorcycles) and Chap. 4 (heavy vehicles).

Vehicle testing is not always suitable for heavy-duty or non-road vehicles, thus the exhaust emission certification procedure for the latter usually makes use of an engine instead of a vehicle cycle, such as the one depicted in the lower-sub-diagram of Fig. 1.13. This is realized on an *engine test bed*, where the engine under study follows a prescribed speed/torque versus time pattern; emission results in this case are usually expressed in g/kWh. Since the cycle is expressed in terms of normalized and not fixed points, it is practically adapted to each tested engine. Section 1.3.2

serves as an introduction to these cycles that will be detailed in Chap. 4 for on-road and Chap. 5 for off-road vehicles.

1.3.1 Chassis-Dynamometer Cycles

As mentioned earlier, chassis-dynamometer cycles are either simple-modal or true transient. The former were the initially developed cycles, with the latter adopted at a later stage.

In order to develop a driving cycle, driving surveys are first conducted. Such surveys in the 50s, at least in part owing to measurement capability constraints, were only able to identify limited driving modes, namely: idle, constant rate acceleration from and up to specific vehicle speeds, driving at certain steady velocities and deceleration at similarly constant rates. The most frequent of these modes were then selected and combined together. Table 1.1 details such a sequence of driving modes based on a driving survey in Tokyo in the early 70s, lasting 135s for a traveled distance of 664 m [24].

From Table 1.1, Fig. 1.16 is derived, being a graphical illustration of the driving modes in a sequence (i.e., a cycle), and is a good example to highlight the characteristics of a typical simply-structured, 'stylized' or 'synthesized' or 'polygonal' test cycle, in this case the Japanese J10 (Sect. 2.3). Similar to all chassis-dynamometer test procedures, the cycle depicted in Fig. 1.16 is plotted in

Table 1.1 Description of the sequences/modes of a typical stylized test cycle [24] (see Fig. 1.16 for a graphical illustration)

Driving mode	Description	Acceleration (m/s^2)	Speed (km/h)	Duration (s)	Cumulative time (s)	Distance covered (m)
1	Idle	–	0	20	20	0
2	Acceleration	0.79	0–20	7	27	19.4
3	Constant speed	0	20	15	42	102.8
4	Deceleration	−0.79	20–0	7	49	122.2
5	Idle	0	0	16	65	122.2
6	Acceleration	0.79	0–40	14	79	200.0
7	Constant speed	–	40	15	94	366.7
8	Deceleration	−0.55	40–20	10	104	450.0
9	Constant speed	–	20	2	106	461.1
	Acceleration	0.46	20–40	12	118	561.1
10	Deceleration	−0.55	40–20	10	128	644.4
		−0.79	20–0	7	135	663.9

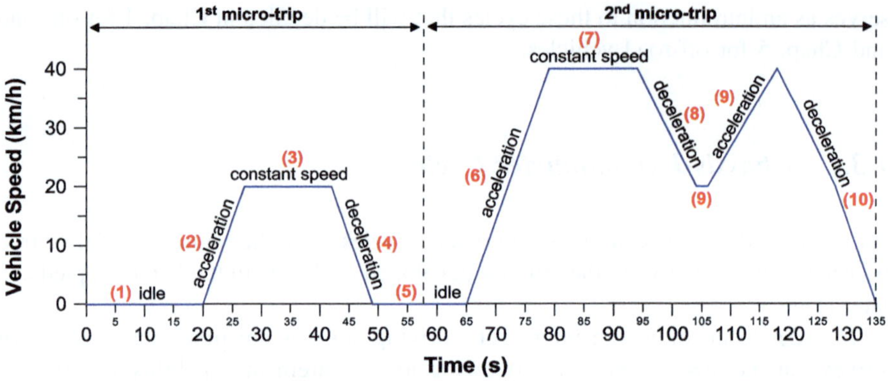

Fig. 1.16 Vehicle speed versus time trace of a typical modal cycle simulating urban driving conditions, with constant accelerations and decelerations and long steady-speed segments; numbers in parentheses denote the modes of the cycle detailed in Table 1.1; notice that mode 10 combines two deceleration operations

terms of fixed vehicle speed (in km/h or mph) vs. time (in seconds). An alternative graphical illustration in terms of vehicle speed vs. distance is provided in Fig. 1.17. Since this test cycle is intended to simulate urban driving, vehicle speeds are intentionally kept rather low; with modal cycles, the same was usually the case for the accelerations, although in real-world conditions, accelerations are usually both more frequent and steeper during urban driving.

The initial idling period of 20 s in Table 1.1 may represent the engine start-up. Afterwards, the driver accelerates (linearly) for 7 s up to 20 km/h speed, having engaged the first gear. This speed is maintained constant for 15 s and then, possibly because the vehicle approaches a crossroad, the driver decelerates in an equally linear manner. At t = 49 s, the vehicle comes to a stop, where it remains idling for 16 s, e.g., while waiting at a crossroad or stop light. The sequence described so far, that began with an idle and ended at the next idling phase, is called micro-trip or short-trip. A similar, somewhat more complicated, sequence of driving events is performed during the second micro-trip depicted in Fig. 1.16.

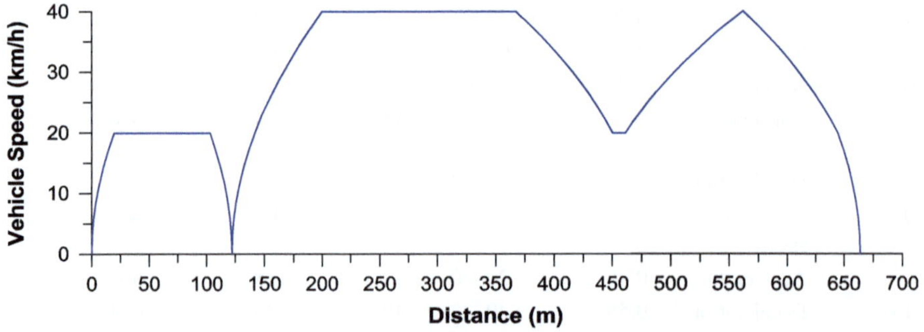

Fig. 1.17 Vehicle speed versus traveled distance for the synthesized cycle of Fig. 1.16

This type of simplistic cycle of 'polygonal' form may contain more similarly-structured micro-trips, perhaps with even higher vehicle speeds, or just one (in this case, maybe more elaborate than the ones depicted in Fig. 1.16). Such synthetically structured cycles do not last long, typically less than 2 or 3 min, and are usually run more than once during the certification procedure. Thus, they are referred to as 'repetitive' cycles. They are also called modal[6] because they are formed from specific driving modes of constant speed or acceleration, as is evident in Table 1.1. Notice in Fig. 1.16 that the cycle exhibits acceleration discontinuities, meaning that there is an instantaneous change from acceleration to cruise, a fact that might actually present difficulty during cycle reproduction on the dynamometer. Undeniably, this is not a feature compatible with the actual driving experience, which is characterized by decreasing acceleration as the demanded speed is approached.

The first driving cycles for the certification of passenger cars and light-duty trucks that appeared in the early 60s werc all of the 'repetitive' or modal type, such as the one depicted in Fig. 1.16 (or better still, such as the one described in Table 1.1). The California 7 mode was historically the first, followed by the European ECE-15 and the Japanese 4-mode, J10, J11 and J15 (Chap. 2). Similarly structured cycles have been used for motorcycle testing in many areas of the world (Chap. 3).

Repetitive cycles are structured in a way that guarantees easy reproduction on the test bench. They are also much simpler to develop since they do not require enormous real-world driving data from operating vehicles. Nonetheless, they are clearly not compatible with daily driving practice, thus most probably lead to erroneous emission results. Drivers do not achieve constant accelerations, and certainly do not maintain constant velocities for a long time. Owing to the continuously changing nature of both traffic and road conditions, the accelerator pedal is rarely retained at a fixed position. Even the slightest change in the gas pedal position, however, results in changes in the air-fuel ratio and/or the amount of mixture entering the cylinders; the latter, however subtle, are responsible for slight but non-negligible changes in the engine torque (and emissions). Consequently, vehicle speed and acceleration are affected, virtually changing continuously during (at least urban) driving. Thus, in order to represent the true driving pattern more accurately, chassis-dynamometer cycles evolved from the simply structured to a sophisticated transient type; the latter are developed based on data collected from instrumented or chase vehicles, i.e., real driving data, that are analyzed appropriately during the cycle development process (Sect. 1.4).

Application of *transient* cycles for testing the compliance of vehicles with the emission regulations began in the United States in the early 70s with the FTP-72 cycle (Fig. 2.16) followed by other areas of the world in the next decades. The lower sub-diagram of Fig. 1.18 demonstrates a typical transient cycle (in this case,

[6]This term is also used for the steady-state cycles of heavy-duty vehicles (Sect. 1.3.2) that are defined by a series of steady-state points or modes (in terms of engine speed and torque).

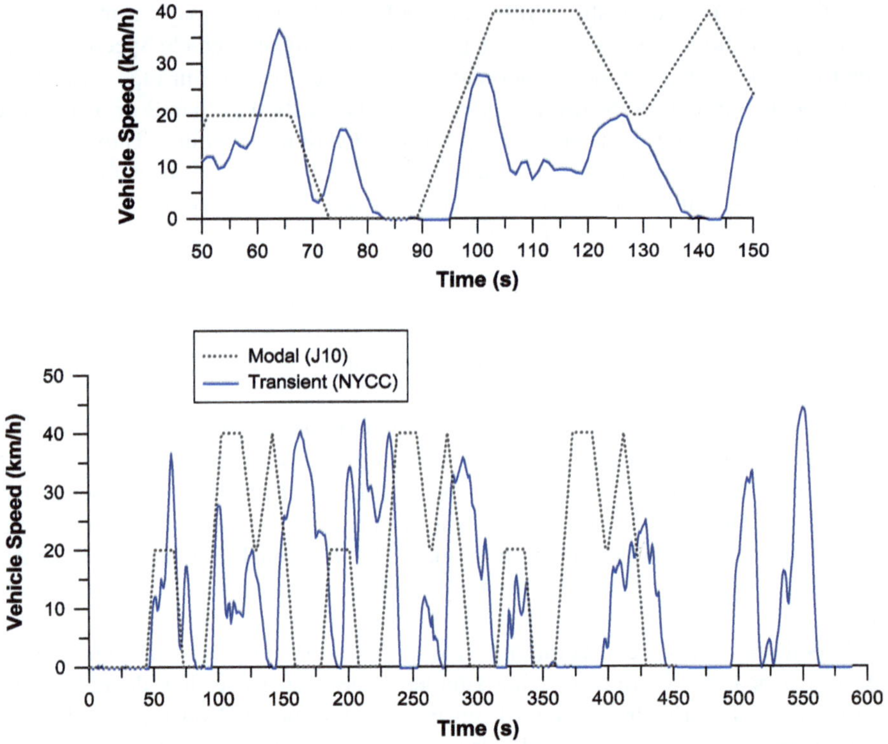

Fig. 1.18 Typical transient cycle simulating urban driving conditions (*solid line*); *dotted line* represents three repetitions of the 'stylized' cycle of Fig. 1.16. The *upper* sub-diagram focuses on the 50–150 s segment of both cycles (for an even shorter increment, see Fig. 1.20)

the New York City cycle—NYCC, discussed in Sect. 2.2). In order for the differences with the previous 'stylized' cycle to be highlighted, the speed traces of both are demonstrated in Fig. 1.18[7]; the upper sub-diagram of this figure further focuses on a 100-s segment of both cycles, pinpointing the different speed traces more clearly. The previously mentioned driving modes can be identified in the transient cycle, but the vehicle speed trace is neither linear nor simplistic, the accelerations and decelerations are not constant either, and the constant-velocity segments have been minimized (in this case, 2 % instead of 24 %). Moreover, and consistent with the daily driving practice, acceleration decreases as the demanded vehicle speed is approached. Notice also in Fig. 1.18 that the transient cycle lasts longer than its 'repetitive' counterpart; therefore, for the purposes of emission sampling, such cycles are not run repeatedly.

[7]A fundamental difference between the modal and the transient cycle is that the former is defined in a tabular form containing specific driving modes, while the latter as a sequence of vehicle speed at each time point (second).

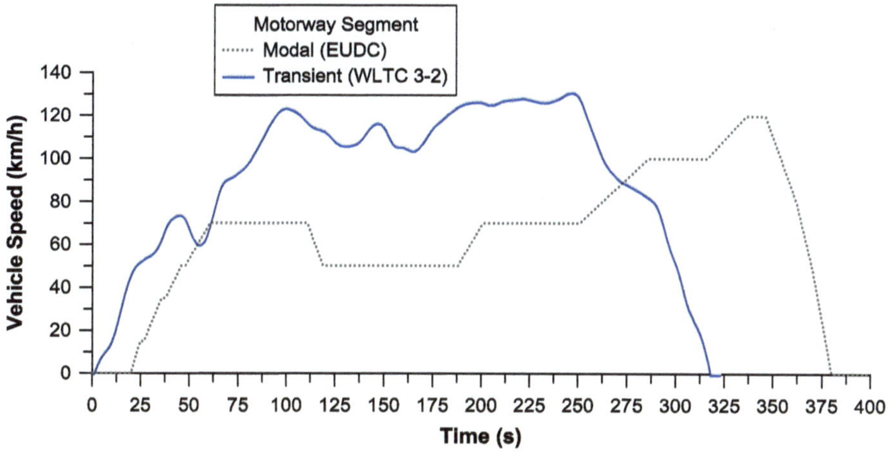

Fig. 1.19 Typical drive cycle motorway segment of the transient and the modal type

To ensure that the testing procedure, therefore the emissions and fuel consumption, are compatible with the whole range of typical driving activity, drive cycles often include a rural and/or a motorway or highway segment; Fig. 1.19 represents such a typical motorway driving schedule. Again, two speed traces are demonstrated, one for a typical modal (in this case the European extra-urban cycle), and one for a transient (in this case, the extra high-speed segment of the newly developed worldwide WLTC). As expected, the general structural profile of each highway cycle type retains the attributes discussed previously for the urban segments in terms of vehicle speed trace, acceleration form etc.

A continuous up-shift gear schedule characterizes the beginning of the motorway parts. Predictably, high vehicle speeds are encountered and maintained throughout the cycle (much higher than during the urban part); modern cycles typically manifest maximum vehicle speeds of the order of at least 100 km/h during their highway segments. Another common feature of motorway drive cycles is the absence of intermediate idle phases, as well as relatively fewer and, most importantly, milder accelerations. High-speed motorway segments are actually the only driving phases that approximate steady-state engine operation. Nowadays, it is standard for the (primary) driving cycle for the certification of vehicles to include both an urban and a motorway part, either combined together in a composite cycle or run separately.

In order to shed more light into the differing microscopic speed development during driving cycles, Fig. 1.20 demonstrates speed traces from a few typical modal and transient, urban and highway cycles during a 10-s acceleration increment.

The most prominent legislated transient cycles today are the U.S. FTP-75 and HFET, the Japanese JC08, and the worldwide WMTC for motorcycles and WLTC for light-duty vehicles, all detailed in Chaps. 2 and 3; on the other hand, the ARTEMIS-project cycles are amongst the most well-established, and frequently

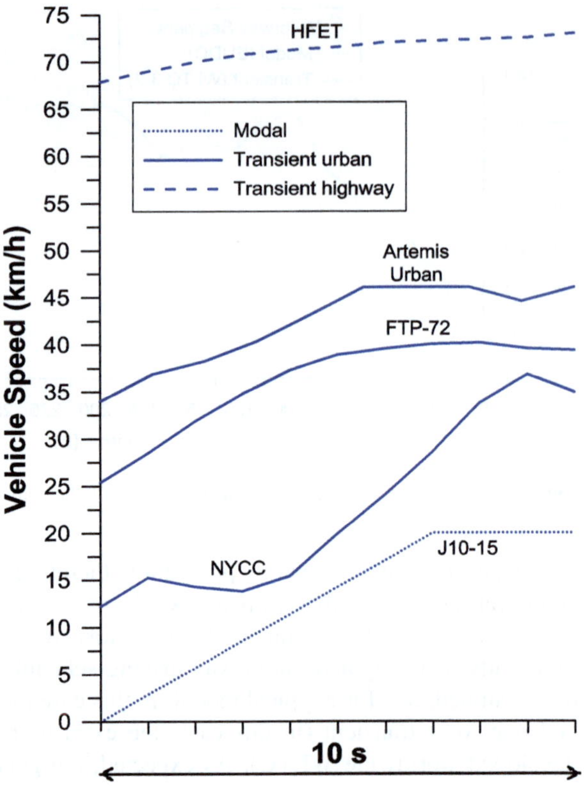

Fig. 1.20 Vehicle speed trace of various modal and transient cycles during a 10-s acceleration increment

employed in research studies, non-legislated transient cycles. Apart from the latter, special-purpose chassis-dynamometer cycles have been developed as well, e.g., for durability, for inspection and maintenance reasons, or specifically intended for electric vehicles.

Although most of the chassis-dynamometer drive cycles have been legislated with the intent to be used for the testing of light-duty vehicles (passenger cars, light-duty trucks and motorcycles), chassis-dynamometer cycles for heavy vehicles have been developed and employed, as will be discussed in Chap. 4; the most prominent ones being the U.S. heavy-duty UDDS and the Japanese JE05. The Appendix provides detailed data for the most important chassis-dynamometer cycles developed over the years, with the emphasis placed on legislated ones.

During the early years of vehicle certification on the chassis dynamometer in the 60s, a preconditioning phase was usually applied, which secured that the engine—at that point, exclusively spark ignition—was already fully warmed up when the sampling began. The controlled pollutants were initially only CO and HC, and it was their concentration (ppm) that was originally measured. Sampling took place throughout all or certain runs of the repetitive/modal cycle, with specific weighting factors assigned to each run or certain parts of the cycle.

In the next decades, the testing was extended to diesel-engined vehicles and later to motorcycles, whereas the authorities in many countries gradually changed the testing procedure with respect to two important aspects. Firstly, the constant volume sampling (CVS) method was adopted (Sect. 6.5), with the results expressed now in g/test, or g/km (or g/mile) for the pollutants concerned. The latter are usually HC/NMHC, CO, NO_x, PM (in Europe particle number (PN) too), with fuel consumption and CO_2 also measured. Second, sampling begins at the moment the vehicle is cold started. Through cold-starting test procedures, it is acknowledged that a significant amount of pollutants is emitted during the early minutes of the engine's operation before the after-treatment devices reach their operating temperature and begin to function efficiently (cf. Figs. 1.11 and 1.12). Regardless, since it is fairly unnatural to enter a motorway with the engine cold, highway drive cycles (if separately run from the urban part) are always executed with the engine fully warmed up.

Some regions, such as Europe and lately Japan, rely on a single certification cycle aiming to cover all aspects of daily driving activity. In other countries, such as the United States, the certification procedure includes two or more driving cycles, each one targeting specific vehicle activity, e.g., urban, motorway, or driving following a cold-start. When a test procedure incorporates different driving cycles, it is common for different weighting factors to be assigned to each schedule, as will be discussed in Chaps. 2 and 3. Consequently, the total amount of emissions, and fuel consumption and CO_2 for that matter, is calculated using the weighted averages from each cycle.

Cycle Metrics

One of the inherent 'internal' characteristics of a test cycle is the numerous technical specifications that can be derived from the speed trace signal. These technical attributes are used to describe a chassis-dynamometer driving cycle and assess its capabilities or shortcomings. The most common are [25–27]:

- duration of the cycle.
- distance covered by the vehicle throughout the test.
- average speed (including or excluding idling phases, in which case reference is made of the overall average speed or the average *driving* speed respectively); average speed is one of the most influential cycle properties, with low values usually indicating more transient cycles, and higher ones highway, thus more steady-state oriented cycles.
- maximum speed; following Eq. (1.1), both average and maximum speeds are the cycle attributes that have the greatest influence on aerodynamic technologies. As will be discussed in Sect. 6.3, vehicle speeds are also influential on the rolling resistance term.
- maximum and average positive acceleration; by its nature, the distribution of acceleration is centered at zero and tends to be symmetric and unimodal. Thus, the average is not useful for comparing two or more distributions. Positive

accelerations, on the other hand, are indicative of the transient schedules throughout the cycle, hence are interrelated to emissions.

- standard deviation σ of vehicle speed and acceleration; is a measure used to quantify the amount of dispersion of speeds and accelerations throughout the cycle, with low values (modal cycles) indicating data points close to the average, whereas high values indicate that the data points are spread out over a wider range.
- idling; this can be expressed in various ways, such as average stop duration, total number of stops throughout the cycle, or, most importantly, stops per km. The latter metric is often used to assess the dynamic behavior of a cycle, since the amount of stops during driving influences decisively the accelerating profile, hence fuel consumption. The greater the number of stops, the greater the number of no-load accelerations, which are the most demanding for the engine and, at the same time, the most polluting as the figures in Sect. 1.2 indicate.
- percentage of time spent accelerating, decelerating, cruising and idling; cycles with increased percentage of cruising time are either highway ones or, in general, modal cycles, i.e., less transient oriented schedules.
- percentage of time spent driving with speeds lower (or higher) than a specific threshold, e.g., 50 km/h, or 100 km/h, or, for that matter, percentage of time spent accelerating faster or slower than a specific threshold.
- speed/acceleration distribution and/or speed acceleration frequency distribution (SAFD), being very useful metrics in assessing the degree to which a cycle covers the typical daily driving activity.
- relative positive acceleration, RPA (see Eq. (1.2) later in the text), positive kinetic energy PKE (Eq. (1.3)), vehicle specific power (Eq. (1.4)) etc.

From the parameters mentioned above, some denote physical (or 'level') properties such as time length, distance and maximum speed or acceleration; some are more complex properties such as the relative positive acceleration or PKE; some are 'extensive' or distance-related properties (distance traveled, stops per km), while others are 'intensive' (do not depend on the cycle length) such as acceleration; lastly, some refer to two-dimensional or three-dimensional distributions (SAFD). Figure 1.21 illustrates a typical speed and acceleration frequency distribution (three-dimensional speed/acceleration frequency distributions will be provided in the next chapters).

Further, Fig. 1.22 compares the two motorway segments from Fig. 1.19 in terms of their speed/acceleration distribution. From such figures, it is easy to establish whether a cycle is biased towards specific engine speeds or speed regimes, and whether it encompasses an adequate mixture of driving conditions. In both figures it is evident that the transient cycle incorporates wider range of accelerations, being thus considerably more compatible with the everyday driving behavior. Modal cycles, on the other hand, being characterized by an increased amount of cruising, predictably exhibit poorer representation of the vehicle's possible operating conditions. Obviously, the technical specifications mentioned above differ substantially from cycle to cycle as well as between the cycle's intermediate segments.

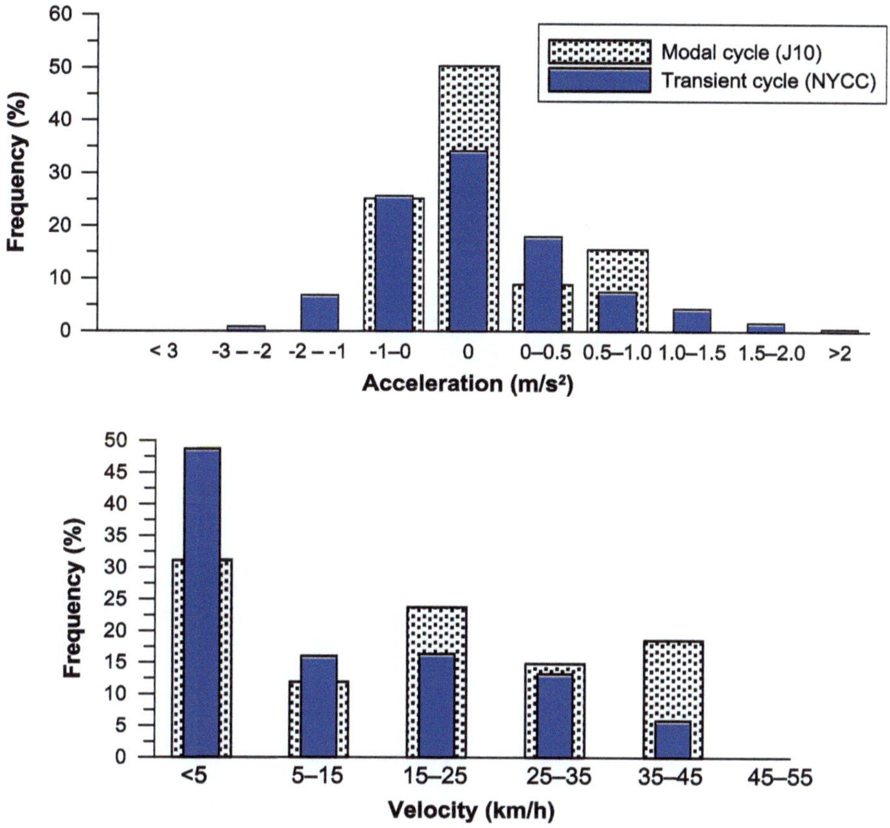

Fig. 1.21 Comparison of the speed and acceleration frequency distributions between the modal and transient urban cycles of Fig. 1.18; notice that for half of the time, the modal cycle exhibits zero acceleration, primarily owing to increased cruise section

A critical feature of a cycle is its dynamics or harshness, being strongly inter-related to the emitted pollutants and fuel consumption/CO_2. Consequently, it assumes a key role in assessing the extent to which a cycle may be considered realistic. In order to estimate the harshness of a cycle, a typical parameter employed is the frequency of (positive) accelerations as well as their absolute values (indicative, but not adequate, measure is the maximum acceleration noticed throughout the cycle). For the two urban cycles discussed earlier, Fig. 1.23 provides their respective acceleration profiles.

Obviously, the transient cycle exhibits steeper and more frequent accelerations compared to the constant ones of the modal cycle, being thus much more realistic.

A further indication of a cycle's dynamics, commonly employed in the relevant studies, is the relative positive acceleration or RPA, being related to the vehicle power [28]. The power needed from an engine to accelerate a vehicle is proportional to both the vehicle speed and the acceleration. Thus, neither variable alone is

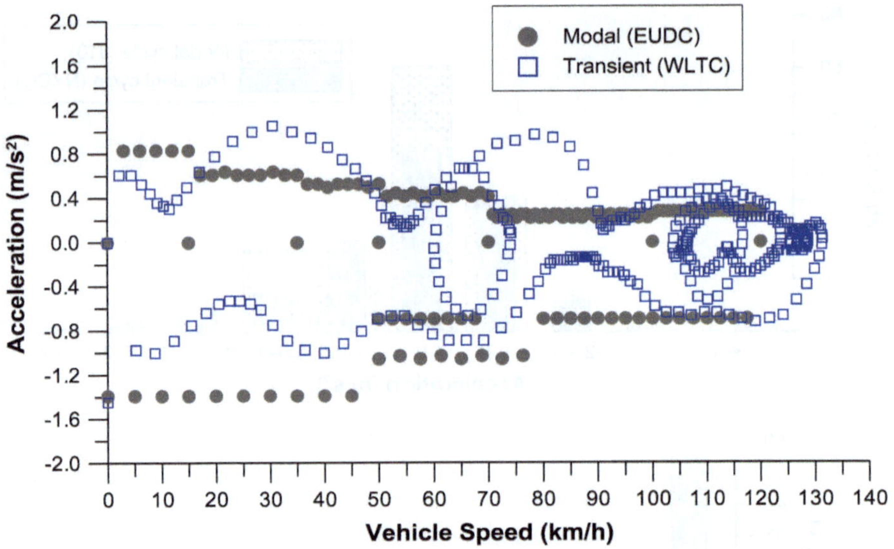

Fig. 1.22 Comparison of the speed/acceleration distribution between the modal and transient motorway segments of Fig. 1.19 (dense parts indicate high frequency in the specific speed and/or acceleration occurrences)

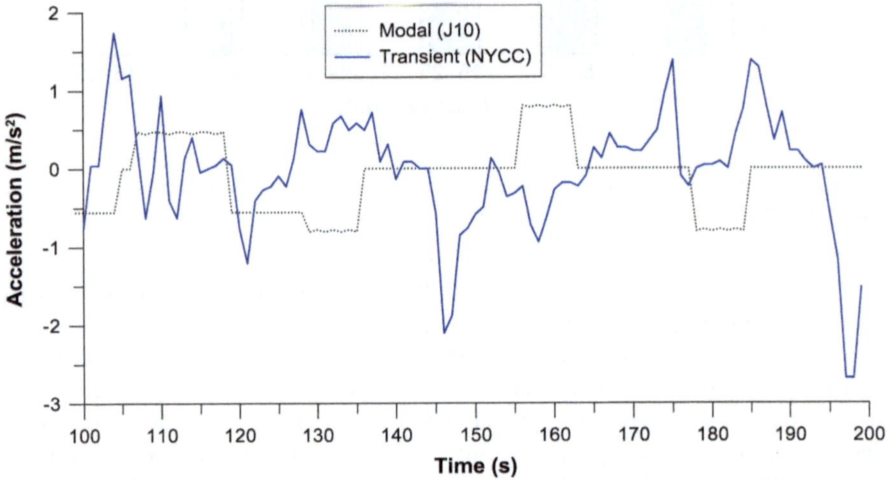

Fig. 1.23 Comparison of the urban acceleration profile between the modal cycle of Fig. 1.16 and the transient of Fig. 1.18; it is evident that a considerable amount of high-frequency speed fluctuations occurring during real-world driving are not captured by the modal cycle

adequate to describe the load placed on the engine during acceleration. In contrast, their combination provides perhaps the best measure. RPA is defined as the integral of the product of instantaneous speed and instantaneous *positive* acceleration over a defined section of the test cycle or route

$$RPA = \frac{1}{S}\int_{0}^{t} V(t) \cdot a(t) \cdot dt \tag{1.2}$$

with V the vehicle speed, a the acceleration, t the duration of the cycle (or the cycle segment) and S the corresponding traveled distance; only positive accelerations are taken into account in Eq. (1.2).

Routes with RPA values lower than 0.10 are considered 'soft' (this usually applies to modal cycles, as well as to highway ones, as in both cases there are few and rather moderate accelerations). On the other hand, RPA values higher than 0.20 indicate more dynamic/harsh cycles with frequent and steep accelerations, representative of more aggressive driving. These cycles are also expected to be the most polluting, at least as regards engine-out emissions. Transient cycles simulating urban driving, or, in general, transient segments with frequent accelerations and low speeds, are the ones usually exhibiting the highest values of RPA. For the modal urban cycle of Fig. 1.16, RPA is 0.20 (quite high for a modal cycle), while for its transient counterpart in Fig. 1.18, 0.34. If the intermediate micro-trips are accounted for separately, RPA of the transient cycle reaches up to 0.53 during the third micro-trip. For the extra-urban segments in Fig. 1.19, RPA is 0.094 for the repetitive cycle and 0.13 for the transient one. In general, RPA decreases with increasing speed, owing to longer traveled distance (denominator in Eq. (1.2)) and generally fewer and milder accelerations; Fig. 1.24 is indicative of this trend. In order to assess and compare RPA values between different cycles, ideally the average driving speed should also be taken into account in the comparison.

An alternative to RPA property of a cycle is the positive kinetic energy (m/s^2) defined in Eq. (1.3) for those points in the cycle where $V_i > V_{i-1}$, i.e., when the vehicle accelerates (i denotes each time-point, e.g., second, in the cycle). PKE is a measure of the acceleration work required throughout a driving pattern [29]

$$PKE = \frac{1}{S}\sum_{i}^{t}\left(V_i^2 - V_{i-1}^2\right) \tag{1.3}$$

For the two urban cycles discussed earlier, the respective PKE values are 4.82 for the modal one and 8 for the transient. PKE and RPA practically describe the same thing, as Fig. 1.24 confirms. In the same figure, various cycle metrics are compared to each other for many of the LD and HD chassis-dynamometer cycles that will be discussed in the next chapters.

From research conducted in [26] on a large number of chassis-dynamometer cycles, it was concluded that for the diesel-engined heavy vehicles studied, average

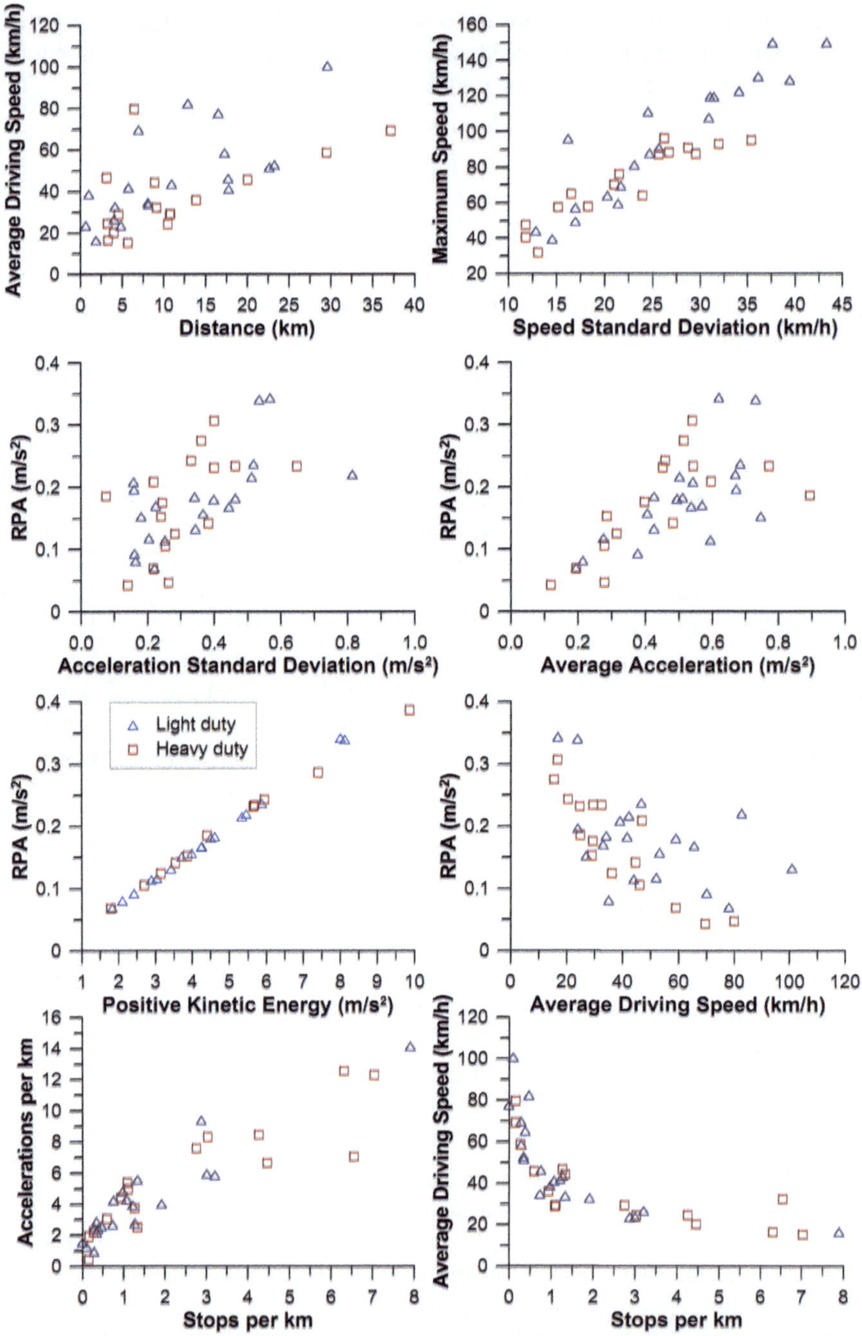

Fig. 1.24 Correlations between cycle metrics for various light-duty and heavy-duty chassis-dynamometer test schedules that will be discussed in the next chapters

speed and stops/km were the most influential cycle parameters correlated to engine emissions and fuel consumption. Specifically, CO_2 emissions (and fuel consumption) were found to exhibit a strong negative correlation with average speed, meaning that higher vehicle speeds (hence, fewer transients) result in lower CO_2 emissions; the number of stops/km were also found to correlate well with carbon dioxide emissions. The number of stops/km, on the other hand, being indicative of the transient nature of a cycle, were strongly related to CO and PM emissions. NO_x emissions were affected by percentage idling time, average speed and stops/km. Lastly, average speed and stops/km were identified as the most important correlation properties for HC emissions. On the other hand, average acceleration during the cycle, was found to have a rather weak correlation with emissions and fuel consumption [26]. Further to the above, the RPA (or PKE), being indicative of the harshness of a cycle is, in general, strongly related to engine-out emissions from SI and turbocharged diesel engines, as will be discussed in the next chapters (see for example, Fig. 4.35).

From Eq. (1.1), the vehicle specific power (VSP) can be also calculated, which is often used in fuel consumption and emission estimates studies. At a given time t, the instantaneous VSP (kW/kg) represents the vehicle's tractive power normalized to its weight, i.e., the instantaneous power required to overcome the various resistances plus acceleration

$$VSP = \frac{\overbrace{F_{tr} \cdot V}^{Power}}{m_V} = \frac{\left[c_{fr} m_V g + m_V g \sin\theta + \frac{1}{2}\rho_a c_d A_f V^2 + m_V \frac{dV}{dt}\right] V}{m_V} \tag{1.4}$$

Positive values of VSP indicate a tractive effort from the vehicle, while negative ones (owing to deceleration or negative grade) indicate the need for power absorption by the power-train, e.g., through mechanical brakes or regenerative braking. VSP is not a 'classic' cycle metric, as is RPA or PKE, since it requires vehicle/road input data, but is often used in the relevant analyses and, in particular, in predictive emission models. It has been employed to assess the degree to which a candidate drive cycle resembles the data set from which it was derived.

Cycle Representativeness

The most important quality of a driving cycle is the degree to which it reflects true driving behavior, i.e., its representativeness of real driving. Undeniably, no test cycle can ever encompass all possible driving conditions, thus be 100 % representative of all driving activity. As mentioned earlier, test cycles often serve as proxy for typical or 'average' driving or test conditions in many research endeavors or projects [30, 31]. Models such as EMFAC (in California) [5] and MOVES (in the rest of the United States) [32], or COPERT in the EU [33] estimate macroscopic on-road vehicle emissions by coupling emission rates with corresponding activities through the use of emission factors; the mobile emission rates are generally developed from dynamometer tests using driving cycles. Consequently, the representativeness of driving cycles is crucial for the correct emission assessment and subsequent policy planning/decision making, as it is, of course, for certification purposes.

It is without doubt that the simply-structured modal cycles cannot be considered representative, particularly so for today's vehicles and driving habits; transient cycles, on the other hand, can be far more realistic, as Figs. 1.20 and 1.23 demonstrated. It was the authorities in the United States that moved sooner towards legislating a truly transient cycle, taking also into account both cold and hot-started emissions already from the early 70s. The Japanese followed in the mid 2000s, while the Europeans will embrace a 'real' transient cycle practically with the adoption of the WLTC from September 2017. At least in part, this delay in the European legislation should be attributed to the much more complex decision-making scheme, as there are many countries involved in the policy making, a fact that generally slows things down.

Nonetheless, as many recent analyses demonstrate, official laboratory test results reflect less and less the actual experience of average drivers on the road [34–38], and this is something that affects negatively consumers, manufacturers and governments. For example in [36], certain tested vehicles easily satisfied the Euro 6 NO_x limit of 80 mg/km during all chassis dynamometer measurements in the laboratory (including test cycles other than the official test protocol) but in real-world driving tests on the road, they measured an average NO_x emission of more than 600 mg/km (detailed info will be provided in Chap. 2). Such discrepancies have been confirmed by numerous tests in various countries, primarily concerning 'Euro'-certified cars, and seem to be manufacturer-independent.

There are various reasons that explain these discrepancies or gaps between real-world and certification results. The first has to do with the test cycle itself. Particularly modal cycles are not representative of real driving behavior in terms of aggressiveness, vehicle speeds and overall driving profile, therefore they underestimate both pollutant emissions and fuel consumption. The discrepancy between real-world and certification results is, generally, enhanced with the increase in the vehicle power-to-mass ratio, hence it has become more pronounced during the last years (see, for example, Fig. 2.8 regarding CO_2 emissions). Vehicles equipped with high-performance engines are those that are capable of achieving steeper accelerations during driving, hence the difference compared to the 'soft' driving pattern of many certification cycles is amplified. Even recently developed cycles, such as the worldwide WLTC (Sect. 2.5), have been criticized for not being aggressive enough. We should not forget, however, that a legislated driving cycle is a compromise between various driving profiles and habits. Whereas many drivers (mostly younger ones) usually drive aggressively, there are many drivers who follow a more conservative driving approach. If the legislation acknowledges only the former, e.g., by defining steep and frequent accelerations throughout the cycle, a considerable part of the driving community will not be represented in the legislation. This means that the results from the laboratory test procedure will not form a statistically representative average.

The second reason is related to the certification procedure itself. Dynamometer testing is an artificial way of measuring emissions, and its results may differ from the actual on-road pollutants because it eliminates several factors that influence emissions on the road, such as road gradient, road roughness, use of air conditioning—and other appliances such as headlights or audio system—and varying traffic or weather (temperature, humidity, wind speed) conditions. Furthermore, the road-load test used to determine the resistances when the vehicle is on the road, often limits the 'accuracy' of chassis dynamometer testing owing to test track configuration, ambient conditions and 'representativeness' of the test car (Sect. 6.4). Undeniably, manufacturers make use of the flexibilities and loopholes in the legislation for their fullest advantage.

The third reason that contributes to the serious discrepancies between real-world and certification results is ECU-embedded 'cycle beating' or cheating techniques. The extensive use of electronics in modern cars allows manufacturers practically limitless calibration of the engine ECU, hence manipulation of the engine operation. More specifically, it is the operating points outside the test cycle, i.e., at conditions that will not be tested during the certification procedure that are usually influenced through cycle bypass techniques. This is actually cycle-related as well, in the sense that a more representative cycle leaves narrower range for cycle bypass to be implemented. The Volkswagen emission scandal of 2015 is perhaps the most prominent manifestation of such cycle-beating behavior. It is perhaps the most extreme too, in the sense that the diesel engine ECU in certain vehicles had been programmed in such a way to detect when the car is being certified on the dynamometer (hence not moving), and to only in this case activate certain antipollution measures with regards to NO_x emissions, otherwise not employed during normal driving on the road. According to the German automaker, as many as 11 million vehicles had been affected by the illegal software ('defeat device') [39]. The pollutant that is usually manipulated through cycle beating techniques is NO_x owing to its inverse relation with fuel consumption, and primarily when internal measures, e.g., in the form of EGR, are employed. Consequently, for the engine speed/load region not covered by the legislated test cycle, manufacturers opt for better fuel consumption (a feature greatly appreciated by customers), or better performance (e.g., faster accelerations) in lieu of emission abatement.

In order for the regulators to diminish the risk of cycle beating, one option is to 'extend' the cycle so that it includes different driving styles and loads. A second, more promising and perhaps less time consuming, procedure is to switch to a 'non-exceed' (NTE) approach to testing. This would provide for emissions to be tested at any feasible engine operating point (any combination of engine speed and power not known to the manufacturer prior to the test), against a threshold for emissions that must never be exceeded. Consequently, the engine manufacturer will be obliged to apply the emission optimization strategy on a much broader engine operating range, as it is not known a priori which extra points will be tested during the certification procedure. For example, for SI engines, the mixture enrichment at high loadings discussed in Sect. 1.2 may not be applicable anymore if such NTE approach is adopted. Testing the vehicle on the road under real driving conditions, as part of the certification procedure, is another very promising option, and this is programmed to be implemented in Europe from 2017. The use of on-board diagnostic systems (OBD) is another tool. OBD is a diagnostic system for emission control which must have the capability of identifying the likely area of malfunction by means of fault codes stored in computer memory. The OBD system must be so designed, constructed and installed in a vehicle, that it is able to identify types of deterioration or malfunction over the entire life of the vehicle. The OBD system must indicate the failure of an emission-related component or system when that failure results in emissions exceeding the threshold limits [34].

Specific representativeness issues, for many of the legislated cycles, will be discussed in more detail in the next chapters.

1.3.2 Engine-Dynamometer Cycles

As was discussed in the previous section, for the majority of vehicular applications (passenger cars, light-duty vans and trucks, and motorcycles), a test procedure is employed where the whole vehicle is being tested on the chassis dynamometer. Vehicle testing is, however, much more difficult for heavy-duty vehicles (HDVs), such as buses and trucks, or non-road ones (tractors, excavators, wheel loaders). This is due to great variety in engine, drive-train and vehicular combinations, as well as to large size and weight. Despite the fact that heavy-vehicle chassis-dynamometer tests do exist, the exhaust emission certification for HDVs is usually realized on an *engine*, rather than a vehicle test cycle, performed on an engine dynamometer. The main point being that engines and trucks are, for the most part, not designed together (lack of vertical integration). Since an engine may be used in different truck applications, and a truck model may be available with multiple engines, whole HDV testing would require considerable more time and cost compared to engine testing [11]. Table 1.2 lists some advantages and disadvantages of whole vehicle chassis vs. engine testing for heavy vehicles.

During engine-dynamometer testing, the engine under study follows a prescribed sequence of operating points in the form of speed and torque with respect to time. Such sequences of engine speed/torque vs. time operating points derive from a (heavy) vehicle speed versus time pattern, as the ones described in the previous section in Figs. 1.18, 1.19 and 1.20, applying a suitable drive-train model, Fig. 1.25. These models take into account various drive-train and vehicle properties e.g., frontal area, coefficient of aerodynamic resistance, vehicle weight, number of gears and gear ratios etc. The emission results during engine testing are usually expressed in g/kWh or g/HPh, i.e., the amount of emitted pollutants is reduced to the total work produced by the engine during the test (instead of the distance covered by the vehicle, which is the case with chassis-dynamometer tests).

From a regulatory point of view, a work-specific emissions unit is acceptable to compare engines and to eliminate those that over-emit. Results of engine testing prove, however, rather impractical for the calculation of emission inventories [41] as well as for fuel consumption and CO_2. Imagine, for example, the same engine used in a 6000 and a 10,000 kg HD vehicle. The engine will most probably operate

Table 1.2 Advantages and disadvantages of testing a heavy vehicle on a chassis or engine dynamometer (some data derive from [40])

Test Method	Advantages	Disadvantages
Chassis dynamometer	• Ability to test any vehicle configuration, including hybrids and vehicles with advanced transmissions • Ability to test all of the vehicle components as a system	• Limited availability of chassis dynamometers due to high capital cost[a] • Coast-down testing is prerequisite for developing road-load inputs—limited availability of adequate test facilities, and variability based on ambient conditions • Not consistent with existing emission targets, which are based on engine dynamometer testing • Each HDV category would practically require separate (g/km) emission limits
Engine dynamometer	• Reduced cost as only one engine is tested that can then be used in a variety of chassis/applications • One set of emission limits for a variety of applications	• Setup of the engine on a dynamometer is more time consuming • Cannot test drive-train sub-systems such as the transmission • Test does not guarantee that the engine coupled to a specific drive-train and fitted on a specific vehicle will behave in the same way as during the certification

[a]According to the U.S. EPA, the cost of a HDV chassis facility ranges between \$1.3 to \$4.0 million (Federal Register Vol. 80, July 13, 2015)

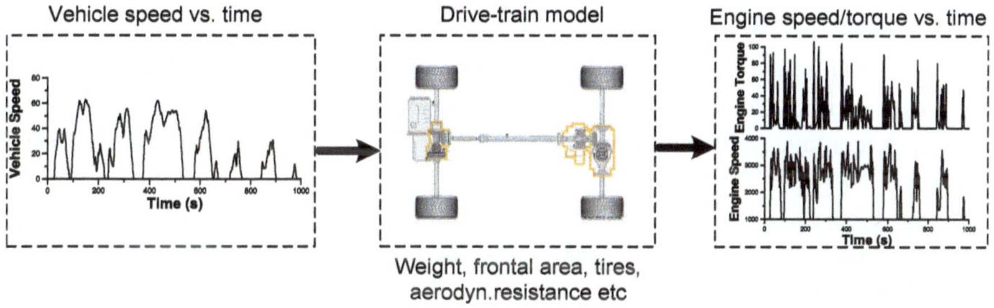

Fig. 1.25 Transformation of a vehicle speed versus time trace into an engine speed/torque versus time one applying a drive-train model

at both higher engine speeds and loads when installed on the heavier vehicle, a fact that is expected to influence accordingly both pollutant and CO_2 emissions.

In order for different engines to be tested on the same cycle, the engine speed and torque are provided in normalized form (%), with the denormalization procedure depending on the mapping of the tested engine. Perhaps the most significant difference between chassis and engine testing is located here. The engine cycle is actually scaled to the capabilities of each engine, something that would not be true if a chassis-dynamometer testing procedure was followed. If a cycle with an inflexible sequence of speeds were applied, it would be rather easy for high-powered vehicles and probably undrivable by under-powered ones. In contrast, every engine should be equally challenged by the engine-dynamometer cycle's power demands. Another key difference between engine and chassis testing is the extent of the driver input. An engine dynamometer facility controls all inputs, including engine accelerator pedal; a chassis-dynamometer facility, on the other hand, usually relies on an experienced driver to operate the vehicle throughout the test [41].

Cycles employed for testing the compliance of heavy-vehicle engines with the emission regulations are further classified into:

- Steady-state or modal cycles, where the engine is tested on *specific* (steady-state) operating points, known as modes, and
- Transient cycles consisting of a *continuous* sequence of engine speed/torque operating points on a second-by-second basis.

A typical example of a steady-state cycle for heavy vehicles is illustrated in Fig. 1.26. Testing of the engine is accomplished at three intermediate speeds, A, B and C, calculated based on the specific engine mapping, and for certain engine loads; in this case, four for each tested speed. For other cycles, such as non-road ones, two speeds, the maximum-power and the maximum-torque one, are under test. According to the specific application, one test point may be the no-load idle condition. The no-load condition is important for on-road heavy-vehicle engines, since idling constitutes a non-negligible portion of their operation, but would be irrelevant for engines driving electrical generators. During steady-state testing, the engine is allowed to stabilize at each tested point before sampling commences.

The load (torque) in Fig. 1.26 is provided in percentage form. This means that at each tested speed, the loads examined depend on the torque mapping of the engine. Hence, percentage values represent, for a given test mode, the ratio of the required

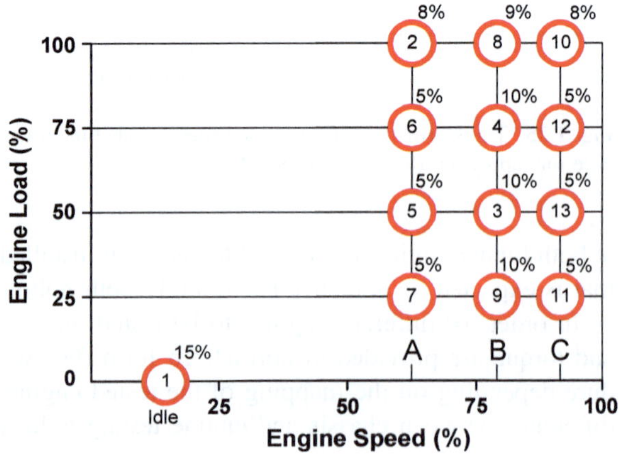

Fig. 1.26 Steady-state cycle employed for the certification of heavy-duty engines in the EU during the 2000s comprising 13 modes; percentages denote weighting factors for each mode

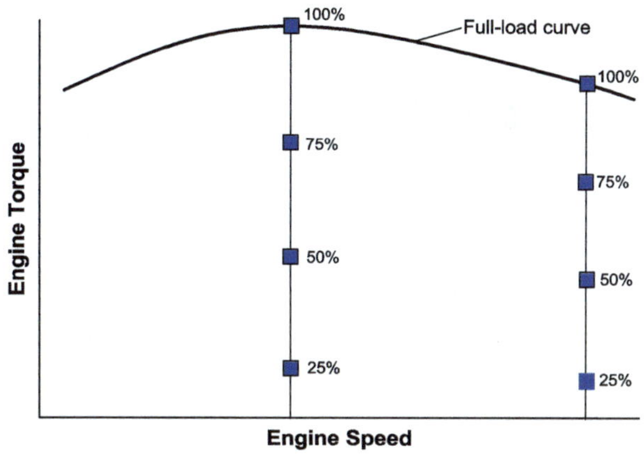

Fig. 1.27 Torque scales: percentage of full-load torque at two engine speeds (in this case, the maximum-torque and maximum-power ones)

torque to the maximum possible torque at this specific speed, as Fig. 1.27 illustrates.

For the cumulative emission results to be representative of the vehicle's driving pattern (or in general, of the engine's operating cycle), weighting factors are assigned to the results from each test point. For the case depicted in Fig. 1.26, and since this cycle is designed for the certification of heavy-duty on-road trucks, a rather high percentage (15 %) has been assigned to idle operation, whereas the weighting factors for the full load operating points are smaller (8–9 %). Similarly structured, to the one depicted in Fig. 1.26, steady-state cycles have been employed for the certification of non-road engines (such as agricultural tractors, crawlers, skid-steer loaders, excavators, electrical generators, railway locomotives or marine), and will be presented in Sect. 5.1.

Obviously, the transient overshoots that are experienced during the continuously changing operating conditions of the engine cannot be captured efficiently by stationary cycles.[8] Another peculiarity of steady-state cycles is the fact that the engine is always tested hot (since stabilization to each test point is prerequisite before sampling), therefore no cold-start effects are taken into account. Regeneration effects from modern engines equipped with diesel particulate filters or NO_x adsorbers, which result in brief overshoot of emissions, cannot be accounted for either. Furthermore, specific engine speeds (two or three) are only tested, despite the fact that during real-world driving, the engine covers a very broad range of speeds (and loads). From a purely experimental point of view, the procedure for conducting a steady-state test has two disadvantages. Firstly, there are many sample start and stops, which means that the previous mode's emissions may be incorrectly included in the next mode due to an unavoidable dead volume in the sampling system. Secondly, the short sampling period of each mode results in very small

[8]Particularly for soot, which notoriously peaks during transients more than NO_x, the legislation often included a special soot or smoke test even when steady-state cycles were the only applicable option for the other exhaust pollutants.

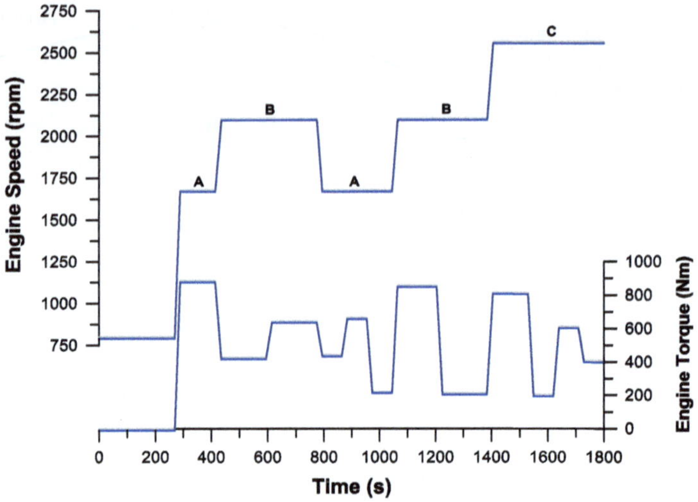

Fig. 1.28 Ramped modal version of the steady-state cycle of Fig. 1.26 for a specific engine and for 30 min total duration; notice the similarities with the overall structure of the modal chassis-dynamometer cycle in Fig. 1.16

amount of PM deposited onto the filter. This might prove problematic because the nowadays applicable PM standards approach the minimum detection limits for many PM microbalances. In light of the above, it is not surprising that steady-state cycles were gradually replaced by more sophisticated tests, primarily owing to their inability to correlate satisfactorily with real transient emissions.

Firstly, steady-state cycles may be run in a continuous 'ramped' mode, meaning that the exhaust emissions are not measured discretely at each test point only. Instead, measurement is continuous, also including the transitions between the modes; the transitions are to be accomplished in a linear manner within a specified (e.g., 20s) time period. For the steady-state cycle of Fig. 1.26, a ramped version is illustrated in Fig. 1.28. There are no weighting factors now but the time length of each mode defines its weighting in the whole emission sample. Such ramped engine-dynamometer cycles resemble the chassis-dynamometer modal cycles discussed in the previous section, exhibiting an overall simple form with constant speed/load changes and extended periods of operation at steady speed and torque.

Further to the ramped modal cycles, engine-dynamometer cycles have evolved to sophisticated true transient form as is typically illustrated in Fig. 1.29. Similar to many chassis-dynamometer cycles, the cycle demonstrated in Fig. 1.29 consists of various phases simulating different parts of daily driving, namely urban, rural and highway. Obviously, the cycle is highly transient with continuous speed and load changes, as one would expect the engine of a heavy truck or bus to operate during daily driving. Taking into account that the vast majority of such heavy vehicles are diesel-engined and turbocharged, it is beyond doubt that the overshoots in exhaust

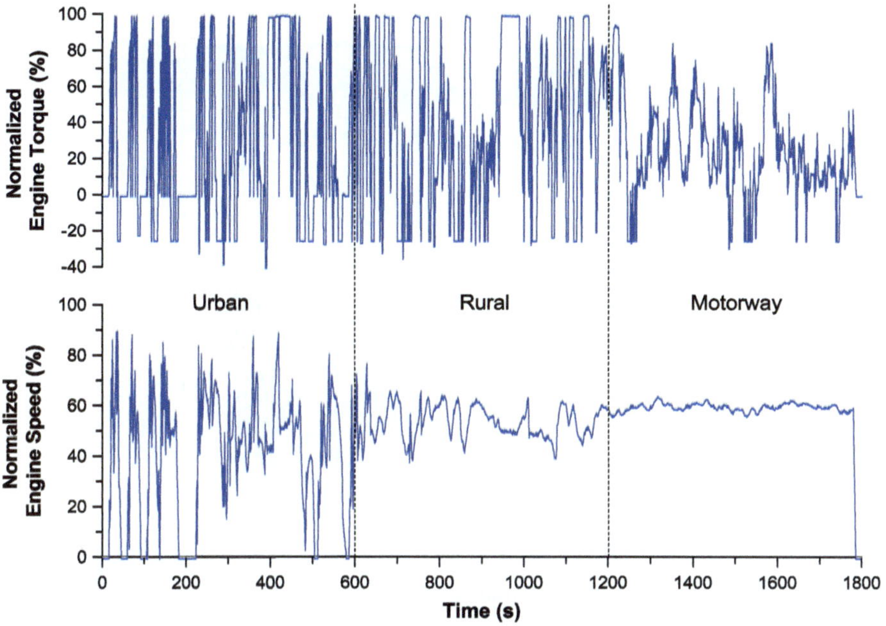

Fig. 1.29 Transient cycle employed during the 2000s for certification of heavy-duty engines in the European Union composed of three segments (600s duration each)

emissions (in particular, PM/soot) during transients, as discussed in Sect. 1.2, can be captured by the cycle.

An important aspect of transient engine-dynamometer cycles is their motoring segments, i.e., phases where the engine is driven by the dynamometer (identified as negative torque points in Fig. 1.29). These motoring segments are used to test the engine in conditions where, on the road, fuel delivery would not occur although the engine would still be rotating (e.g., traveling downhill with the accelerator pedal closed). In such cases, it is important to test for fuel dribble from the injector nozzles. However, modern-technology nozzle, having low sac volume or valve covering the orifice devices, suffer from this to a lesser extent, and the need for motoring may become open to question in the future [42].

Figure 1.30 highlights the differences between a steady-state and a transient cycle for the testing of heavy-duty engines. It is clear how through the transient cycle not only the dynamic operation of the engine (through the transitions between the various test points), but also a much wider operating range is under test.

Apart from the European ETC cycle depicted in Fig. 1.29, two other prominent cycles for heavy trucks are the American FTP and the worldwide WHTC. Furthermore, a transient cycle for the certification of non-road vehicles has been developed, the NRTC. These cycles will be detailed in Chap. 4 (on-road vehicles) and 5 (non-road ones).

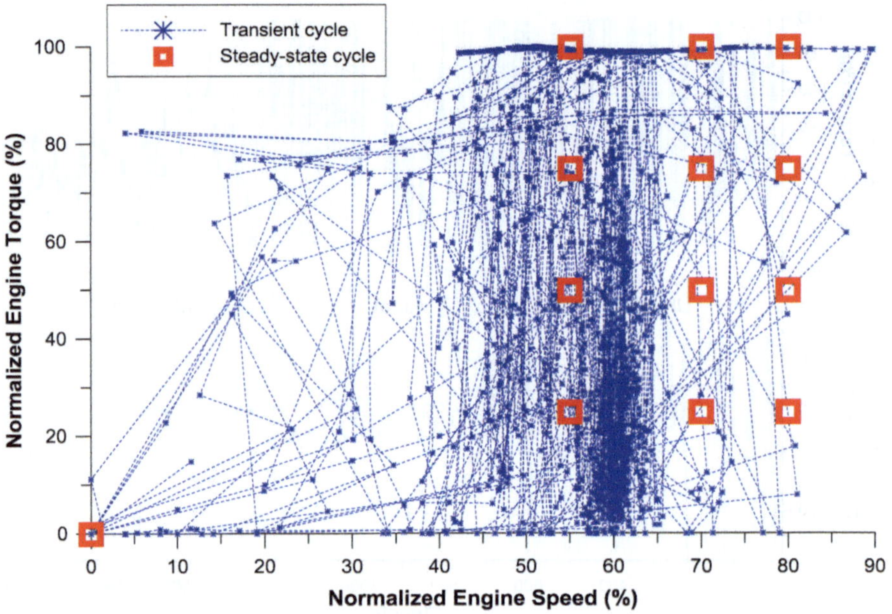

Fig. 1.30 Comparison between the steady-state cycle of Fig. 1.26 and the transient of Fig. 1.29

Typical attributes used to study and assess an engine dynamometer cycle are:

- Duration of the test;
- Average (normalized) engine speed and torque;
- Maximum (normalized) engine speed and torque;
- Maximum speed and load increase;
- Average load increase;
- Time spent accelerating or decelerating or with steady speed;
- Time spent with increasing or decreasing or with constant load;
- Idle time expressed as total number of 'stops' throughout the cycle, average 'stop' duration, and percentage of idling time with respect to total;
- Motoring phase expressed as percentage of total time, and frequency (torque reversals per minute);
- Time spent with the engine operating in specific speed and/or load regimes (e.g., lower than 30 %, higher than 70 % etc.);
- Frequency of load increases (per minute);
- Engine speed/torque distribution and engine/speed torque frequency distribution.

Particularly the frequency of load and speed increases, as well as the average load/speed increase are the most influential metrics with respect to engine-out emissions (g) from turbocharged diesel engines. On the other hand, the average level of speed and torque and the percentage idle time define the total amount of

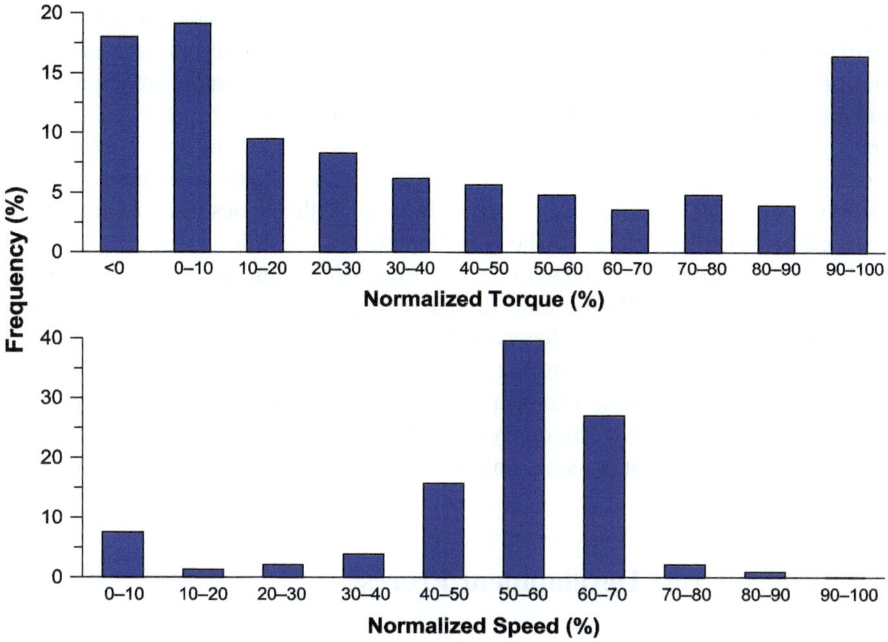

Fig. 1.31 Normalized speed/torque frequency distribution for the transient cycle of Fig. 1.29

work (kWh) produced by the engine during the test; the latter is the denominator in the 'reduced' amount of emitted pollutants (g/kWh).

Notice that since the respective vehicle's traveled distance cannot be calculated, the various specifications of the engine cycle can only be reduced to its time duration. Actual speeds, hence accelerations, cannot be calculated from the normalized data alone; they depend on the specific tested engine's speed range, as well as the exact denormalization expression Two-dimensional and three-dimensional frequency distributions with regard to (normalized) engine speed and torque can be plotted for transient engine cycles, and provide valuable information as to the exact profile of the cycle. For the cycle depicted in Fig. 1.29, Fig. 1.31 demonstrates such a frequency distribution for various speed and torque bins. It seems that the cycle is biased to speeds around 40–70 % (normalized values), with the majority of the torque occurrences at low load (<10 %) and high-load (>90 %) conditions. From Fig. 1.31, an interesting conclusion drawn is that the cycle seems to underestimate low engine speeds, i.e., urban driving.

Since heavy-duty and non-road engine calibration has traditionally focused on low fuel consumption, it is without doubt a very challenging task to maintain this objective while at the same time dealing with increasingly stringent exhaust emission targets. This holds particularly true when NO_x emissions are concerned since these exhibit an inverse relation with fueling. For heavy-duty vehicles, a gap between real-world and certification emissions, mentioned earlier for the chassis-dynamometer tests, unsurprisingly exists as well; the main gap with respect to cycle bypass measures, and since only the engine is being tested, is the ability to design the power-train of a vehicle in such a way that the most frequently used engine speed range lies outside the range currently tested for type approval [34]. For the cycle depicted in Figs. 1.26 and 1.29, this would be the low-speed engine operation. The consent decree of 1998 in the United States is perhaps the most famous case of HD engines alleged cycle-beating, with a series of court settlements reached between the U.S. EPA, the U.S. Department of Justice, the California ARB and various engine manufacturers[9]; the total settlement package cost for the industry was over $850 million. Engine manufacturers were accused to have manipulated emissions from heavy-duty diesel engines during certain off-cycle driving conditions such as constant speed highway cruising [4]. Again, the concerned pollutant was NO_x. The arguments mentioned above highlight once again the necessity for the certification cycle to be as much representative as possible, incorporating the widest possible engine speed and load range.

1.4 Drive Cycle Development Process

The process of developing a driving cycle is complex and comprises various phases; the following can, in general, be identified, with reference to Fig. 1.32:

(a) A detailed collection of driving data from the concerned vehicle category is first conducted. It is important to gather data from a large number of drivers and driving conditions, so that the developed database is representative, and all driving patterns (Fig. 1.33) are contained. For example, data from older and younger drivers should be incorporated, from various technology vehicles, from various times during the day, from various streets, cold-started and hot-started trips, under different weather conditions, as well as (according to the application) from both urban and motorway driving. Data collection is nowadays realized using either instrumented vehicles or 'chase' cars; GPS systems are also employed. Especially for the case a worldwide cycle is to be developed, it follows that the driving data should originate from different areas of the world, a fact that poses an extra burden on the second and third stage of the cycle development process.

(b) The second step is to analyze the data in order to describe or characterize the driving conditions. This stage often involves defining an analysis level (i.e., a scale on which driving patterns are analyzed) and variables describing these driving patterns [43]. Depending on the specific vehicle category for which the

[9]Caterpillar Inc., Cummins Engine Company, Detroit Diesel Corp., Mack Trucks Inc., Navistar International Transportation Corp., Renault Vehicles Ind. S.A. and Volvo Truck Corp; these seven companies together manufactured more than 95 % of U.S. heavy-duty diesel engines at the time [11].

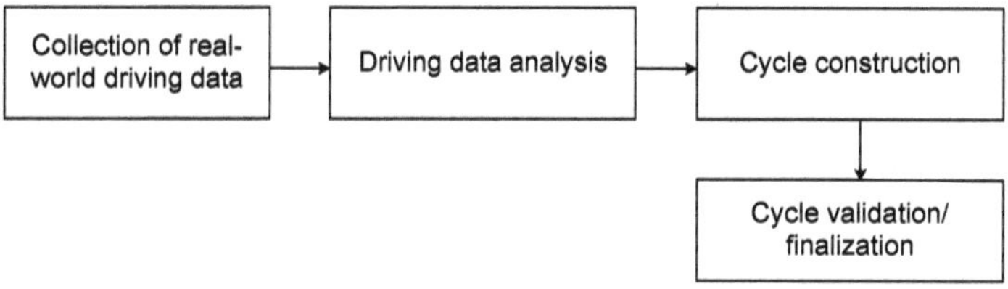

Fig. 1.32 Stages during the development of a driving cycle

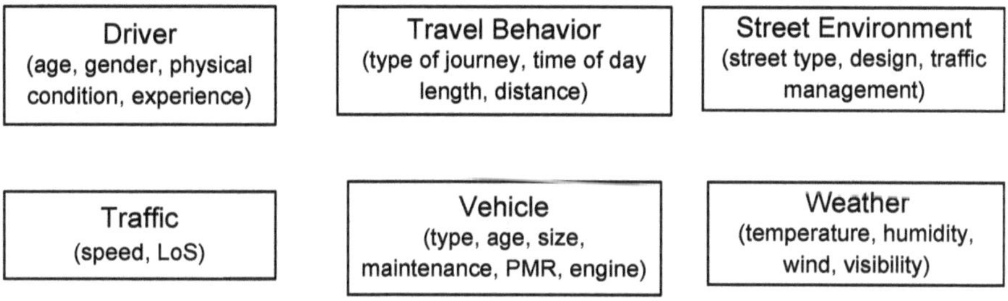

Fig. 1.33 Factors influencing driving patterns (adapted from [44])

cycle is built, various cycle segments (sub-cycles) may be identified, e.g., with reference to the road category (urban, rural and motorway) or speed (low, medium and high).

(c) Transformation of data into a cycle in such a way that it matches the overall summary characteristics of the dataset is the next phase. Historically, this process evolved from a simple selection of the 'best' route to complex statistical/stochastic techniques. An important aspect here is the time length of the cycle; this should be long enough to incorporate all possible driving patterns, and, at the same time, short enough to be cost-effectively reproduced in the laboratory.

(d) The developed driving cycle is then validated, among others, in terms of emission results on various vehicles in the laboratory and in comparison to other existing cycles. Finalization may be also performed to address certain problems or irregularities, such as vehicle drivability, ease of reproduction in the laboratory (e.g., avoidance of tire slip on the dynamometer rolls) etc.

For the case a (transient) engine-dynamometer cycle is to be constructed, the previously described procedure is further extended so as to obtain the engine speed/torque profile (from the vehicle speed vs. time one). This is realized applying a suitable drive-train model (Fig. 1.25). The engine speed/torque versus time pattern for the reference engine is then normalized (percentage speed and torque at each time point) in order for the engine cycle to be applicable to every engine of the specific category.

Usually, it is discrepancies in points (a) and (c) from which most of the problems associated with the existing driving cycles may arise, that is, either lack of representative driving data with respect to the driving behavior encapsulated within a cycle, and/or problematic cycle construction methodology [45]. Both the data collection process and the cycle construction procedure will be discussed in the next two sections.

The general profile of the chassis-dynamometer cycle has experienced substantial variation over the years. With the gradual increase in the power of the vehicles (better still, vehicle power-to-weight ratio), higher maximum speeds have been enforced in the legislation combined with steeper accelerations. Geographic peculiarities play a vital role here, in the sense that driving habits differ considerably around the world based on local tradition as well as status of vehicle fleet, road and traffic conditions. For example, drive cycles in Japan traditionally feature relatively low maximum speeds and also prolonged idle periods. In India, the majority of vehicles are low powered and the traffic conditions typically congested; both result in even lower vehicle speeds. In contrast, in Europe and the United States, the average power of vehicles is much higher, and so are the driving speeds. Watson and Milkins [29], on the other hand, found that neither European nor U.S. nor Japanese cycles adequately simulate the observed driving activity in Australia as regards speeds and accelerations, and, by extension, fuel consumption and emissions. In any case, whereas the early cycles limited maximum vehicle speeds to 50 or 70 km/h, many recent drive cycles demonstrate speeds of more than 100 or even 130 km/h. As regards acceleration, earlier cycles were limited to values lower than 1.5 m/s^2 (largely owing to limitations in the testing infrastructure), with recently legislated cycles exhibiting values of the order of 2 m/s^2 or even more. Accelerating at 1 m/s^2 (=3.6 km/h/s) is equivalent to reaching 100 km/h from standstill in almost 28s, which is rather slow for today's vehicles and (most) driver habits in many areas in the world.

1.4.1 Collection of Driving Data

The first stage for the construction of a test cycle is the gathering of the driving data from the concerned vehicle category. Obviously, it is of utmost importance that the database used to create the cycle be as much representative as possible. Thus, the gathering of the driving activity is crucial for the success of the cycle development. This data collection process has progressed substantially over the years.

During the 50s and 60s, when attempts to construct a driving cycle for certification purposes were initiated, researchers characterized real-world driving (of gasoline-engined cars) on the proportion of time spent in specific engine speed/manifold pressure bins; the latter indicative of SI engine load. These proportions were then used to define *modes* of driving. For example in 1950, an initial survey was carried out in Los Angeles on a single vehicle along a single route with stopwatches used to time the various driving modes. The results obtained indicated

Table 1.3 Comparison between Los Angeles and Paris driving during the 50s and 60s [46]

Mode	Proportion of time in driving mode (%)	
	Los Angeles—1950	Paris (UTAC)—60s
Acceleration	18	15.6
Cruise	46	52
Deceleration	18	13.4
Idle	18	19

that a vehicle accelerated, decelerated and idled for 18 % of the time, and cruised for 46 % (Table 1.3). Another survey, this time in Paris during the 60s, identified quite similar percentages of time spent during each driving mode, also provided in Table 1.3 [46].

A more elaborate survey was conducted in Los Angeles in 1956 under the auspices of the (then called) Automobile Manufacturers Association, employing this time seven vehicles for 60 runs on arterial, residential, freeway and business district routes. The experimental equipment consisted again of a manifold vacuum sensor for classifying engine operation into rapid acceleration, moderate acceleration, cruising, idle, mild deceleration and rapid deceleration. Counters for totalizing time in ten increments of car speed, fuel consumed in each speed increment, time in each manifold vacuum condition, and time of positive and negative acceleration were also employed. Table 1.4 summarizes the results from this survey, highlighting the surprisingly small differences observed between the involved drivers [46, 47].

Table 1.4 Results from the Los Angeles driving survey of 1957 [47] (notice that totals do not exactly equal 100 due to rounding of intermediate data)

Mode/speed	Proportions of average time in each mode					Driver Diff. Mean deviation from average (%)
	Route IA	Arterial road	Residential road	Freeway	Business district	
Total acceleration	19.0	12.3	15.6	12.5	16.7	±3.8
Road load	37.3	51.7	35.5	56.6	33.4	±5.7
Idle	19.4	11.3	20.6	12.7	29.8	±1.3
Mild deceleration	9.0	8.3	14.0	9.6	11.7	±0.9
Rapid deceleration	15.4	16.7	13.8	8.6	8.2	±2.8
Total	100	100	100	100	100	
Average speed (mph)	22.3	24.6	16.4	29.6	13.2	±0.5
Test runs	25	21	3	6	5	

In the following years, the process of gathering driving data changed considerably. A vehicle equipped with a fifth wheel and a strip-chart recorder to record the actual speed/time history of the car was initially developed by federal officials in the United States. During work on the development of a new drive cycle in the late 60s, it was discovered that it was actually much easier to work with the vehicle speed/time strip chart rather than the engine speed/manifold vacuum bins. More importantly, it was observed that the vehicle's speed/time history contained a considerable amount of low-amplitude, high-frequency speed fluctuations, which could not be captured by the engine speed/manifold vacuum instrumentation tools. For the first time, it was apparent that the effect these speed deviations have on emissions might be significant. Immediately, an attempt was made to follow the actual speed vs. time history on the chassis dynamometer, where it was established that the drivers could effectively follow these minor speed changes. Not surprisingly, the previous efforts to record engine speed/manifold vacuum data were abandoned, and the federal air-pollution control personnel (soon to become part of the new U.S. Environmental Protection Agency) set out to develop a new driving cycle using actual speed/time traces generated along a road route [48]. This cycle was the FTP-72, employed for the testing of new vehicles in the United States since model year 1972, discussed in detail in Sect. 2.2.2.

This first attempt to gather (and process) actual speed/time driving data had two interesting characteristics. First, a single road route in Los Angeles was selected, which, based on previous modal-type research efforts, produced the same average mode distribution with a variety of drivers using the same test vehicle. Second, a single vehicle was used for the data gathering process (a third unique characteristic is related to the cycle construction method, or rather in this case, cycle selection, and will be discussed in the next section). The approach of 'best route' selection utilized for the development of the FTP-72 cycle has been followed during the construction process of other driving cycles in the following decades, mainly local ones, i.e., cycles intended to simulate (mostly urban) driving in a specific city, with the aim to estimate emissions and fuel consumption within the respective area. Owing to different driving behavior, vehicle fleet composition and road network, driving characteristics in each city are unique, eventually leading to different local drive cycles developed. The most representative route is usually selected based on vehicle mileage traveled, traffic volume, traffic density or even experience/knowledge of the local traffic conditions.

In the next years, the process of vehicle driving data collection was expanded accounting now for a variety of road-types (including the parameter of road gradient), drivers and vehicles. Vehicles are selected in such a way to represent as much as possible the actual fleet in each region based on sales, age, technology and usage. By doing so, the volume of data increased exponentially but so did the representativeness of the data set [49]; the latter being of paramount importance when the goal is to develop a nationwide or worldwide driving cycle. There are two approaches that have been utilized for the collection of the driving data. The first involves instrumented vehicles and the second 'chase' cars (a combination of the

Table 1.5 Comparison between instrumented vehicle and chase car approach [49]

	Instrumented car	Chase car
Sources of bias		
Driver does not know he/she is being monitored		✓
Representative of vehicle population		✓
Representative of drivers		✓
Non-intrusive installation of data loggers		✓
Representative of all types of trip segments	✓	
Technical		
Monitors each car continuously	✓	
Contains high frequency information	✓	✓
Measures instantaneous speed	✓	✓
Measures average speed	✓	✓
Measures accelerations	✓	✓
Measures number of cold and warm starts	✓	
Measures during cold and warm starts	✓	
Measures hot-soak time	✓	
Measures distance traveled	✓	
Contains load information	✓	
May include road gradient, or traffic info		✓
Cost		
Low installation cost		✓

two methods is sometimes used); a comparison between the two approaches is summarized in Table 1.5.

For the instrumented car case, various vehicle types of the concerned category are instrumented with on-board data acquisition systems or data loggers recording vehicle speed, engine speed, engine load (e.g., manifold vacuum for SI engines or injection pressure for CI ones) etc., usually in steps of 1–10 Hz. For example, vehicle speed can be measured from the engine ECU, the speedometer or the speed control system in the ABS brake system. Road grade information can also be calculated from longitudinally mounted accelerometer and vehicle speed data [50]. The instrumented vehicles are driven by their owners and for their normal daily purposes, and data is collected usually for a period of a few days up to a few weeks. The apparent advantage here is the fact that data from a variety of drivers, cars and, more importantly, driving conditions is gathered, including different hours of the day, as well as both hot and cold-started trips. The instrumented vehicle method also obtains information on the start and finish time of trips, driving behavior following the start of a trip, and the amount of time vehicles are shut off between trips (soak time). One possible disadvantage, apart from the cost to instrument a large number of vehicles, is that the driver, knowing that his/her driving profile is being recorded, might not behave entirely normally.[10] A typical case where instrumented vehicles have been utilized for the collection of driving data is the worldwide harmonized cycles, i.e., the worldwide cycle for light-duty vehicles (WLTC—Sect. 2.5) and motorcycles (WMTC—Sect. 3.5).

Recently, the use of GPS (global positioning system) has provided an attractive alternative to the instrumented vehicle approach. Calculations of road grade on a second-by-second basis is feasible applying GPS-based altitude information [50]. GPS systems can also capture facility types on which the sampled vehicles are driven [51]. Drivers are also more familiar with GPS units and may behave more 'normally'.

In contrast, in the 'chase' car approach, an appropriately instrumented 'patrol' vehicle follows closely other randomly selected cars and records their speed/time profile. To this aim, a laser beam is installed in the dashboard that monitors the distance between chase and target vehicle, associating it with the chase car speed to calculate the instantaneous target vehicle velocity. While it is not feasible to gather large amount of data from each target vehicle, the main advantage is that data from many drivers is collected, a fact that aids the build-up of a comprehensive database. Another major advantage is the relatively low cost, as only one vehicle is practically instrumented (see also Table 1.5). In the absence of target vehicles, the chase car, flowing through traffic, can be used to record its own operating data, which means that the ability of the chase car drivers to approximate speed and acceleration of other vehicles is very influential. This is particularly true in light traffic conditions when there may be few vehicles to emulate. When chase car data is used in place of missing target data, however, this does not increase the overall sample size of the data set. Instead, only one driver/vehicle profile—that of the chase driver—is added to the sample, thereby increasing the sample size by one. Since this single profile may contain more speed/time data points than those of the target vehicles, there is great potential for the chase vehicle data biasing cycles developed from the data [52]. An interesting feature of the chase car approach is that information regarding the road (gradient) and traffic conditions may be manually submitted by the chase car personnel. Inherent to the chase car approach is the inability to measure vehicle activity at the very beginning and very end of a trip. Difficulties are accentuated when a vehicle having totally different driving characteristics from the chase-vehicle is being monitored. Also, it is not always possible to identify the target vehicle type and load from a chase situation. Predictably, the technology used for data collection influences significantly the success of the chase car protocol. The U.S. HFET cycle (Sect. 2.2.3) [53] and the California LA-92 (Sect. 2.2.6) [48] are typical examples where the 'chase' car approach was used. Part of the U.S. SFTP data collection in Baltimore and Spokane (that eventually led to the

[10]Interestingly, for the U.S. EPA SFTP data collection in Baltimore in 1992 (one of the biggest in history), the sub-contractor did not explicitly discuss the monitoring of the vehicle speed with the car owners involved in the survey [49].

development of the US06 cycle discussed in Sect. 2.2.4) involved chase cars [49], as well as a small part of the driving data for the worldwide WLTC cycle [54].

Other than the above discussed types of data collection, called 'floating car' studies (primarily conducted by transportation planning agencies), collect speed measurements on specific roadways using a vehicle where the driver is directed to 'float' along at what is judged to be the average speed of surrounding vehicles. This floating car-based approach has limited usefulness for development of emission-oriented cycles because it does not capture the actual speed/time profiles that affect resulting vehicle emissions [50].

Regarding heavy-duty vehicles, the process of data collection is, more or less, the same as the one described for light-duty cars, with instrumented vehicles of various heavy-duty applications monitored for a time period; GPS systems are often used. Again, the data collection covers a variety of drivers, vehicle types, road-types and, in this case, loadings as well. Typical examples here is the European FIGE and the U.S. heavy-duty UDDS (Chap. 4). More details about this procedure will be discussed in Sect. 4.5 regarding the development of the world-wide HDV transient cycle (WTVC), where also the additional difficulties due to the diversity in the driving data from different regions of the world will be highlighted, a rather common issue when a worldwide cycle is to be developed.

It should be noted that, in general, the measured speed data obtained from the instrumented or chase cars cannot always reproduce the real speed profiles perfectly since it exhibits unevenness due to measuring errors. This problem becomes more intense when trying to compute the acceleration profile by differentiating the speed curve [44]. For example, Fig. 1.34 illustrates such an acceleration curve obtained from 'raw' speed data. Certain techniques can prove helpful in this regard by smoothing the speed trace.

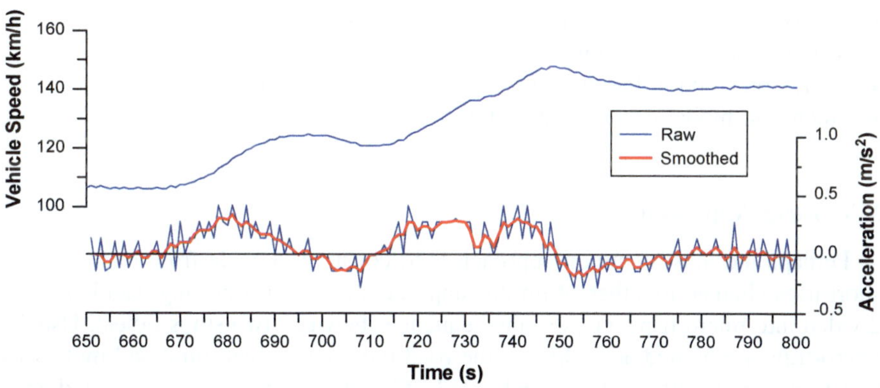

Fig. 1.34 Speed and acceleration (raw and smoothed) profile during part of the ARTEMIS cycle motorway segment

1.4.2 Data Analysis and Cycle Construction

The gathered data from the early driving collection surveys during the 50s and 60s (consisting of engine speed/inlet manifold vacuum bins), were subsequently used to define driving modes representative of average vehicle use. For example, from the data presented in Table 1.4, the derived modes were: idle, cruising at 20, 30, 40 and 50 mph, acceleration from 0–25, 0–60 and 15–30 mph, and deceleration from 50–20, 30–15 and 30–0 mph. Weighting factors were also assigned to these modes based on two factors: frequency of mode of operation in urban traffic over 24 h, and exhaust volume produced in that mode. From these driving modes, the California 7 mode, the first certification driving cycle in the world was built (Sect. 2.2.1) [7]. In a similar manner, the European urban ECE and motorway EUDC (Sect. 2.1.1) and the Japanese J10, J11 and J15 (Sect. 2.3) were constructed; all the above, being synthetic, 'polygonal' cycles.

On the other hand, the first *transient* driving cycle in the world based on transient speed/time traces was the U.S. FTP-72 or LA4 (Sect. 2.2.2), already discussed in the previous section regarding its data collection process. Its construction was based on six speed/time traces of the same vehicle on the same route. From these six traces, the one closest to the average was initially selected as the most representative speed/time trace. This trace, after shortened, and 'softened' in terms of maximum accelerations/decelerations, became the principal cycle of the U.S. exhaust emissions certification program [55].

During the next decades, the process of cycle construction progressed considerably, with statistical and stochastic methods playing for many years now a key role in the development. Depending on the type of driving activity that is being used to build the cycle, driving cycle construction methodologies (for light-duty vehicles) can be classified into four types, namely [56]

- micro-trip cycle construction;
- segment-based cycle construction;
- cycle construction based on pattern classification; and
- stochastic modal cycle construction.

Micro-trip Approach

Perhaps the most common approach is to construct a cycle from micro-trips or short-trips chained together. With this approach, real-world driving data is initially divided into micro-trips, i.e., driving sequences between two stops (idles). Usually, micro-trips are binned according to the road type, i.e., urban, rural and motorway; however, for the light-duty vehicles WLTC suite of cycles, it was found more rational to bin micro-trips with respect to vehicle speed. The driving cycle is then constructed by splicing 'representative' micro-trips, with the goal that the developed cycle closely matches the originating data.

A crucial stage in this technique is the selection of the representative micro-trips. First of all, a series of elimination criteria should be applied to the short-trip and idle databases for determining those not representative segments to be excluded from the analysis. For example, for the case of the WLTC cycle these were: idle periods with duration longer than ten minutes, short-trips with duration shorter than ten seconds, micro-trips with maximum speed lower than 3.6 km/h, micro-trips with accelerations higher than 4 m/s^2 (14.4 km/h/s) and lower than –4.5 m/s^2. The reasons for these eliminations were linked to considerations of statistical representativeness, and also to the feasibility of the test in a laboratory [54]. From the remaining micro-trips, the representative ones can be selected using random selection, 'best incremental' (incrementally searching for and adding a micro-trip with specific modal characteristics), or a hybrid of both approaches [49, 56]. Candidate cycles are then assessed using parameters such as average and maximum speed, average acceleration, stop duration, proportion of time spent idling, accelerating, decelerating and cruising, and, most importantly, speed/acceleration frequency distribution. Vehicle emissions are greatly influenced by specific second-by-second speed variations as well as the frequency with which these occur. Consequently, compiling driving data into joint frequency speed/acceleration distributions provides a useful means to represent second-by-second driving patterns [50].

As an example, for the 1992 California Unified Cycle/LA-92 (Sect. 2.2.6), a 'quasi-random' approach for micro-trip selection was employed. Firstly, 'seed' micro-trips were selected, completely at random, to complete a seeding period forming the first 120-s start phase. Subsequent micro-trips were randomly selected in such a way that they improved the match to the sample's SAFD. Each time a micro-trip was selected, it was removed from the micro-trip set. The remaining micro-trips were scanned, and subsequent micro-trips were again selected such that they improved the match of the cycle's SAFD to the sample's SAFD. This procedure repeated until a pre-determined driving cycle time was achieved. A driving cycle constructed this way is not entirely a random combination of micro-trips (i.e., 'quasi-random') because the micro-trips were, instead of at random, selected to incrementally improve the cycle SAFD, and therefore this reduces the total number of all possible combinations of micro-trips if otherwise selected at random. The final driving cycle best matched the SAFD of the entire data set within 22 % of the sum of differences [45, 48].

The major limitation of the micro-trip based method is that micro-trips do not differentiate with respect to the various types of driving conditions such as roadway type or Level of Service. Studying the sequences between stops does not always makes it possible to analyze 'local' driving conditions (sequences of several tenths or even hundreds of kilometers without any stop can be recorded), thus the danger exists of biasing the analysis of urban conditions (very high percentage of short sequences). As a result, replicating driving activities under a particular driving condition is difficult. An instructive example has been given in [45] highlighting the inability of the micro-trip approach to always successfully replicate a modal activity's actual frequency, duration and intensity. 'Suppose there are two micro-trips: one follows an average driving pattern and the other contains some irregular driving. We further assume that both micro-trips equally incrementally improve the cycle's SAFD during the course of cycle construction, meaning that adding either of the two micro-trips to the driving cycle equally reduces the sum of differences between the cycle's and the sample's SAFDs, regardless of where the reduction in difference occurs. Theoretically speaking, there is an equal probability that either micro-trip can be chosen using, for example the construction method described above for the LA-92 cycle. Intuitively, however, irregular driving rarely occurs in reality, and therefore is less likely to be presented in a driving cycle. The micro-trip construction method discussed previously is, however, unable to differentiate the two micro-trips based only on the SAFD, because it does not differentiate between modal events (e.g., cruise, idle, acceleration and deceleration) within a micro-trip nor does it consider the temporal interdependence of modal activities' [45].

Use of micro-trip based methods are typically employed when developing cycles designed to represent a single type of trip, e.g., strictly urban or highway, or cycles designed to replicate region-wide driving conditions; the worldwide WLTC, WMTC and WTVC for passenger cars, motorcycles and heavy vehicles respectively are typical examples of (recently) developed cycles based on the micro-trip approach.

Segment-based Approach

Segment-based cycle construction is another approach for constructing a drive cycle. A trip 'segment' is obtained by partitioning vehicle speed/time profiles using changes in roadway (facility) type or LoS, in addition to stops. Therefore, overall vehicle driving activity can be stratified by roadway type or LoS, and cycles can be constructed to represent driving activity for specific roadway types and traffic conditions [56]. One major application of the segment-based cycle construction method is the U.S. EPA facility-specific speed correction cycles employed in the MOBILE and MOVES emission simulators along with the legislated FTP cycles [50, 57]; another, the Australian CUEDC (Sect. 2.4). Similar to the micro-trip based method, trip segments are selected and concatenated using a hybrid of random and best-incremental logic. However, unlike a micro-trip, trip segments can start and end at any speed. Therefore, connecting such segments must be done more carefully in order to achieve realistic driving behavior. The chaining of segments requires certain constraints on speed and acceleration between two connecting seconds respectively of previous and succeeding micro-trips. For instance, in constructing EPA's facility-specific cycles for the MOVES emission inventory model, the differences in speeds between two connecting seconds was required to be within 0.5 mph, and the difference in acceleration within 0.5 mph/sec [50]. The 'best' cycle is then selected using two primary parameters: (a) the sum of difference in SAFD between the test cycle and the target population, and (b) the amount of operation occurring in high power mode. For example, for the MOVES facility-specific cycles, a hybrid random/best incremental cycle construction approach was followed. As discussed earlier, a 'seed' sample of trip segments for some subset of the desired cycle time was selected completely at random. After the seed sample time was reached, subsequent trip segments were selected and successively built onto the seed sample by finding the segment that produced the best match of the cycle SAFD to the population SAFD for the driving group when added to the existing segments. This latter trip segment sampling phase is referred to as the 'best incremental' phase because successive trip segments built on to the cycle after the random phase are selected by producing the best marginal improvement in matching the cycle's SAFD to that of the driving population [50]. The limitation of segment-based cycle construction is that data is stratified strictly from a transportation engineering perspective and might be less related to emissions [56].

Pattern Classification

The cycle construction method *with pattern classification* has been widely applied in constructing various European (non-legislated) driving cycles [58, 59] that will be discussed in Sect. 2.1.2. Under this approach, 'kinematic sequences' (similar to micro-trips) are classified into heterogeneous classes using statistical methods. Also, the approach uses succession probabilities to estimate and consider the likelihood that one class of activity precedes or follows a different activity class. Driving cycles are constructed by re-connecting kinematic sequences randomly selected from each of the activity classes in accordance with the probability and chronology of kinematic sequences [56]. In constructing a series of European urban cycles, kinematic sequences were described by 20 variables including duration, idle time and distance throughout the sequence; the means, maximums, standard deviations, 20 and 80 % percentiles of driving modes, and other measures such as instantaneous speed and acceleration, and the distance between two accelerations. Principal Component Analysis (PCA, i.e., linear analysis method based on correlated variables) was applied to these 20 variables, followed by cluster analysis on the kinematic sequences [58]. Four distinctive classes were identified, respectively, representing congested and free-flow urban traffic, road and motorway driving conditions. Next, a trip was viewed as a series of kinematic sequences. All observed trips were classified based on the frequency of sequences in each kinematic sequence class and the number of transitions between two classes of kinematic sequences. Three major types of trips were identified, namely urban, road (rural) and motorway trips.

The method was further improved in constructing the driving cycles under the European ARTEMIS project [43]. To avoid bias due to the varying duration of sequences between two stops, speed/time traces were segmented into sequences of homogeneous size rather than partitioned by stops (distance varies from a few hundred meters at low speeds to a few kilometers at higher velocities). These kinematic segments were described by their idle duration and cross-distribution of the instantaneous speeds and accelerations instead of synthesized parameters such as average speed. Due to the distance of integration, these segments were considered as local (but not punctual) measurements of the driving conditions [60].

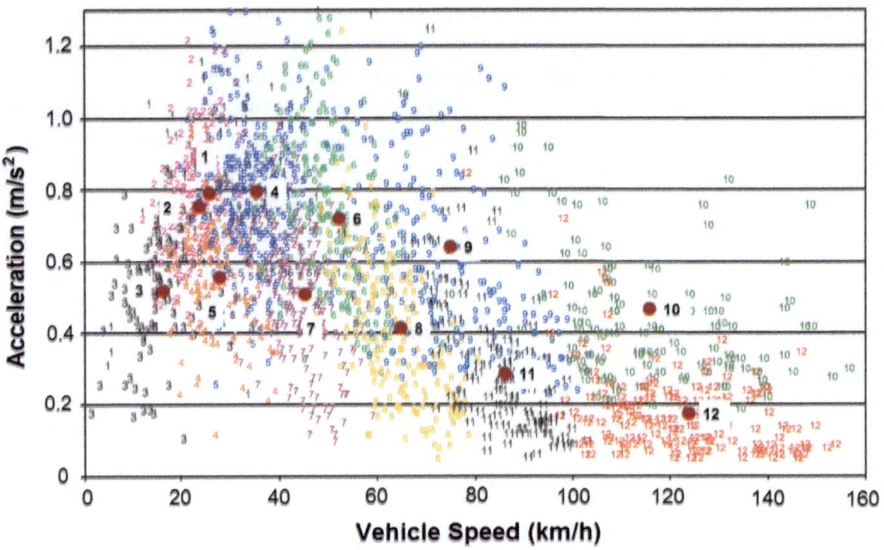

Fig. 1.35 Variability of the driving conditions and positioning of the twelve classes centers amongst a sample of observations (*1* congested urban; *2* urban dense; *3* urban, low speed; *4* urban, free flowing; *5* urban, unsteady; *6*: secondary road, unsteady; *7*: secondary rural roads; *8*: rural road, steady speed; *9*: main road, unsteady; *10*: main road steady; *11*: motorway, unsteady; *12*: motorway, steady speed); adapted from [43]

Correspondence analysis (based on chi-squared distance) and clustering tools were used to classify these segments according to their speed/acceleration distribution. The principle was to group segments that are similar or close to each other and to differentiate groups that are contrasted or far from each other. Eventually, segments were classified based on their speed/acceleration distribution into twelve typical driving conditions, as illustrated in Fig. 1.35.

This figure highlights the high variability of the observations (i.e., the driving conditions) and how clustering made possible the definition of a scatter diagram considering a limited number of pertinent classes. It also shows the difficulty of describing such a scattered reality through a limited number of classes. It should be noted that the classification does not correspond to an exact geographic location (i.e., one cannot say that the so-called motorway driving was recorded on a motorway, etc.), as the recorded data does not enable the location to be deduced. The resulting driving characteristics allow the type of driving conditions to be assumed (i.e., the heading of the class); but what is particularly important is that a pertinent partition of the data was achieved, based on the speed profiles. This kind of analysis made it possible to develop a typology of local driving conditions into classes, with the characterization based on the speed profiles themselves, rather than on arbitrary considerations [43].

The limitations of the pattern classification methods are: (a) the basic units of cycle construction, including micro-trip and uniform sequences, are not directly related to emissions, and potentially not the best units to be used in defining emission-related driving activity, and (b) the classification of sequences is based on the chi-square distance of speed/acceleration joint distribution. Although such classification differentiates the kinematic driving activity, it does not necessarily differentiate emissions associated with these activities [43, 56].

Stochastic Modal Approach

Lastly, the stochastic modal cycle construction technique can be employed. Studies by Lin and Niemeier [45, 61] applied a mode-based cycle construction method. This method is actually a stochastic approach that divides a speed trace into a sequence of modal activities (cruise, idle, acceleration, or deceleration), and describes the sequencing of those events using Markov process theory. That is, occurrence of modal event k is dependent upon the previous modal event k–1. The Markov process facilitated replication of the average (global) driving characteristics while preserving micro-transient events (i.e., small timescale speed fluctuations) that contain the information related to driving variability. Modal cycle construction comprises four basic steps: (a) the real-world driving data is partitioned into snippets[11] of various durations based on acceleration using a maximum likelihood estimation (MLE) clustering method; (b) snippets are classified into different modal bins, again using the MLE clustering method. This time, the clustering variables include average, minimum, and maximum speeds and acceleration rates; (c) the third step creates a transition matrix that contains the succession probabilities between different modes; (d) finally, snippets are chained into one cycle applying Markov chain; to add one additional snippet, the next modal bin is predicted based on the modal nature of the current snippet and the transition matrix. One snippet is selected from the predicted bin without replacement. The snippet selection requires that the selected snippet best improves the match to the observed SAFD, and that the start speed of the snippets matches the end speed of the previous modal snippets with an acceptable difference (e.g., 0.2 km/h). Snippet selection is repeated until the desired cycle length is achieved. The final driving cycles are selected using a composite assessment measure, which integrates parameters such as differences in average speed, differences in minimum and maximum speeds, and percentage of idling operation.

The shortcomings of the modal approach include: (a) the criterion used to connect the snippets is arbitrary, and (b) cycles are constructed for a specific facility and LoS, and the number of cycles needed to represent the emission related driving activity are not well studied [45, 56].

The U.S. EPA heavy-duty cycle, FTP transient, was developed in the late 70s based on Monte Carlo simulation [62]. For the latter, a model was established for describing how actual driving occurs, including both deterministic and probability, or 'stochastic' elements to account for known physical properties of driving as well

[11] A snippet is a change in traffic density for a given freeway or a change in the physical link to or from an area transportation network [51].

as unexplained random phenomena. Using random number generation to simulate the probability component, a time sequence of speeds was produced with characteristics resembling those of the underlying model. The typical model was estimated by fitting a set of assumptions to in-use data. For example, in developing the FTP cycles, 50s-to-second speed transitions were assumed to obey a Markov process, in which the probability of a given speed at the end of the second depends only on the vehicle's speed at the end of the previous second. These conditional probabilities were estimated from in-use driving data frequency distributions. However, evidence exists that the model used in that work was not sufficiently complex to capture all important emission-related elements of real driving, such as the duration of acceleration events [61]. The Monte Carlo simulation approach offers considerable flexibility in modeling driving behavior and generating cycles. It is less popular than the micro-trip or the segment-based method because the generated cycles are 'artificial' (the second-to-second sequences have not actually been driven in-use). This means that a simulated cycle may not be drivable or, in emission testing, may yield emissions that do not or cannot occur in real-world driving conditions [49].

Further to the above, engineering cycles also exist. An engineering cycle is a speed-time trace that satisfies certain criteria not generally based on the frequency distribution of in-use driving behavior. For example, it may be the designer's judgment that a cycle should include certain extreme levels of speed, acceleration, or specific power. These objectives can be achieved without reference to in-use data, possibly using simple 'straight-line' construction methods. This approach to cycle development is useful as a way of forcing certain conditions that may occur infrequently, or to implement feasibility testing [49].

References

1. European Environment Agency website. http://www.eea.europa.eu. Accessed 12 July 2016.
2. Economic cost of the health impact of air pollution in Europe: Clean air, health and wealth. Copenhagen: WHO Regional Office for Europe, 2015.
3. Japan Automobile Manufacturers Association. 2011 Report on environmental protection efforts—Promoting sustainability of road transport in Japan, 2011.
4. Reitze AW Jr. Air pollution control law: compliance and enforcement. Washington DC: Environmental Law Institute; 2001.
5. California Air Resources Board website. http://www.arb.ca.gov. Accessed 8 July 2016.
6. Berg W. Legislation for the reduction of exhaust gas emissions. In: Gruden D, editor. The handbook of environmental chemistry, Vol. 3, Traffic and environment. Berlin: Springer Verlag; 2003. p. 175–253.
7. Clarkson D, Middleton JT. The California control program for motor vehicle created air pollution. J Air Pollution Control Assoc. 1962;12:22–8.
8. Watson N, Janota MS. Turbocharging the internal combustion engine. London: MacMillan; 1982.
9. Health Effects Institute. Special Report No 17 - Traffic-related air pollution: a critical review of the literature on emissions, exposure, and health effects, 2010. http://pubs.healtheffects.org/getfile.php?u=553. Accessed 14 July 2016.

10. Gorham R. Air pollution from ground transportation—an assessment of causes, strategies and tactics, and proposed actions for the international community. United Nations, 2002.
11. Morriss AP, Yandle B, Dorchak A. Regulating by litigation: the EPA's regulation of heavy-duty diesel engines. Admin L Rev. 2004;56:403–518.
12. ACEA official website. http://www.acea.be/statistics/tag/category/share-of-diesel-in-new-passenger-cars 1 27 Apr, 2016.
13. Eggleston HS, Gaudioso D, Gorissen N, Joumard R, Rijkeboer RC, Samaras Z et al. CORINAIR working group on emission factors for calculating 1990 emissions from road traffic - Vol. 1: methodology and emission factors. CEC Report, Contract B4-3045 (91)10 PH. Commission of the European Communities, Brussels, 1991.
14. Heywood JB. Internal combustion engine fundamentals. New York: McGraw-Hill; 1988.
15. Rakopoulos CD, Giakoumis EG. Diesel engine transient operation. London: Springer; 2009.
16. Hagena JR, Filipi ZS, Assanis DN. Transient diesel emissions: analysis of engine operation during a tip-in. SAE Paper No. 2006-01-1151, 2006.
17. Rakopoulos CD, Dimaratos AM, Giakoumis EG, Rakopoulos DC. Investigating the emissions during acceleration of a turbocharged diesel engine operating with bio-diesel or n-butanol diesel fuel blends. Energy. 2010;35:5173–84.
18. Giakoumis EG, Rakopoulos CD, Dimaratos AM, Rakopoulos DC. Exhaust emissions of diesel engines operating under transient conditions with biodiesel fuel blends. Progr Energy Combust Sci. 2012;38:691–715.
19. Mathis U, Mohr M, Forss AM. Comprehensive particle characterization of modern gasoline and diesel passenger cars at low ambient temperatures. Atmos Environ. 2005;39:107–17.
20. Bielaczyc P, Merkisz L, Pielecha J. Investigation of exhaust emissions from DI diesel engine during cold and warm start. SAE Paper No. 2001-01-1260, 2001.
21. Pidgeon WM, Dobie N. The IM240 transient I/M dynamometer driving schedule and the composite I/M test procedure. Report EPA-AA-TSS-91-1, January 1991.
22. United Nations website. http://www.unece.org/wp29.html. Accessed 9 Sept. 2016.
23. Gillespie TD. Fundamentals of vehicle dynamics. Warrendale PA: SAE International; 1992.
24. Kimura T. General resources energy investigation committee, Energy savings standards section, Automobile standards judging subcommittee, LPG car fuel efficiency standards evaluation group, Final Report, January 2006. https://www.eccj.or.jp/top_runner/pdf/tr_lpg_passenger.pdf. Accessed 14 July 2016.
25. Clark N, Thompson G, Delgado O. Modeling heavy-duty vehicle fuel economy based on cycle properties. Final Report: West Virginia University; 2009.
26. Tu J, Wayne WS, Perchinschi MG. Correlation analysis of duty cycle effects on exhaust emissions and fuel economy. J Transp Res Forum. 2013;52:97–115.
27. Bata R, Yacoub Y, Wang W, Lyons D, Gambino M, Rideout G. Heavy duty testing cycles: survey and comparison. SAE Paper No. 942263, 1994.
28. Van de Weijer C. Heavy duty emission factors. Development of representative driving cycles and prediction of emissions in real life. PhD Thesis, Graz University of Technology, 1997.
29. Watson H, Milkins E, Preston M, Chittleborough C, Alimoradian B. Predicting fuel consumption and emissions-transferring chassis dynamometer results to real driving conditions. SAE Paper No. 830435, 1983.
30. U.S. Environmental Protection Agency. A comprehensive analysis of biodiesel impacts on exhaust emissions. Draft Technical Report, EPA 420-P-02-001, Washington DC, 2002.
31. Giakoumis EG. A statistical investigation of biodiesel effects on regulated exhaust emissions during transient cycles. Appl Energy. 2012;98:273–91.
32. U.S. EPA MOVES website. https://www3.epa.gov/otaq/models/moves/. Accessed 13 August 2016.
33. Ntziachristos L, Samaras Z. COPERT III computer program to calculate emissions from road transport—methodology and emission factors. Technical report No. 49, November 2000.
34. Reducing NO_x emissions on the road. European conference of ministers of transport. Report prepared by the ECMT group on transport and the environment in co-operation with the

OECD environment policy committee's working group on transport, 2006. http://www.itf-oecd.org/sites/default/files/docs/06nox.pdf. Accessed 27 July 2016.
35. Franco V, Sánchez FP, German J, Mock P. Real-world exhaust emissions from modern diesel cars—Part 1: aggregated results. ICCT White Paper, October 2014.
36. Kadijk G, Ligterink N, Spreen J. On-road NO_x and CO_2 investigations of Euro 5 light commercial vehicles. TNO Report. 2015;2015 R10192.
37. Tietge U, Zacharof N, Mock P, Franco V, German J, Bandivadekar A, et al. From laboratory to road—A 2015 update of official and 'real-world' fuel consumption and CO_2 values for passenger cars in Europe. ICCT White Paper, September 2015.
38. Ntziachristos L, Mellios G, Tsokolis D, Keller M, Hausberger S, Ligterink NE, et al. In-use vs. type-approval fuel consumption of current passenger cars in Europe. Energy Policy. 2014;67:403–11.
39. BBC.com website. http://www.bbc.com/news/business-34324772. Accessed 27 July 2016.
40. Sharpe B. Testing methods for heavy-duty vehicle fuel efficiency: Trends from regulatory programs around the world and implications for India. ICCT Working Paper 2015-3, 2015.
41. Brown JE, Harris DB, King FG Jr. Heavy-duty truck test cycles: combining driveability with realistic engine exercise. Heavy Vehicle Syst. 2000;7:299–316.
42. Aarnink TJ, Flöysand SÅ, Gadd P, Le Breton D, Mann N, Marotta A et al. Motor vehicle emission regulations and fuel specifications—Part 2: Detailed information and historic review (1970-1996). Prepared for the CONCAWE automotive emissions management group, 1997.
43. André M. The ARTEMIS European driving cycles for measuring car pollutant emissions. Sci Total Environ. 2004;334–335:73–84.
44. Ericsson E. Variability in urban driving patterns. Transp Res Pt. D. 2000;5:337–54.
45. Lin J, Niemeier DA. An exploratory analysis comparing a stochastic driving cycle to California's regulatory cycle. Atmos Environ. 2002;36:5759–70.
46. Watson HC. Vehicle driving patterns and measurement methods for energy and emissions assessment. Canberra: Australian government publishing service; 1978.
47. Stonex KA. Survey of Los Angeles traffic characteristics. 36th Annual Meeting, US Highway Research Board, pp. 509–538, 1957.
48. Austin TC, diGenova FJ, Carlson TR, Joy RW, Gianolini KA, Lee JM. Characterization of driving patterns and emissions from light-duty vehicles in California. California Air Resources Board, Final report, Contract No A392-185, November 1993.
49. FTP Review Project—Preliminary technical report. U.S. EPA 420-R-93-007, May 1993.
50. Carlson TR, Austin TC, McClement D, Yoon S-H. Development of generic link-level driving cycles. Report SR2009-05-02, prepared for U.S. EPA, May 2009.
51. Yu L, Wang Z, Shi Q. PEMS-based approach to developing and evaluating driving cycles for air quality assessment. Southwest Region University Transportation Center report SWUTC/10/169300-1, April 2010.
52. Morey JE, Limanond T, Niemeier DA. Validity of chase car data used in developing emissions cycles. J Transp Statist. 2000;3:15–28.
53. Austin T, Hellman K, Paulsell CD. Passenger car fuel economy during non-urban driving. SAE Paper No. 740592, 1974.
54. Ciuffo B, Marotta A, Tutuianu M, Anagnostopoulos K, Fontaras G. Pavlovic J et al. The development of the world-wide harmonized test procedure for light duty vehicles (WLTP) and the pathway for its implementation into the EU legislation. Transportation Research Board, 94th Annual Meeting, Washington DC, United States, 2015.
55. Kruse R, Huls T. Development of the federal urban driving schedule. SAE Paper No. 730553, 1973.
56. Dai Z, Niemeier D, Eisenger D. Driving cycles: a new cycle-building method that better represents real-world emissions. U.C. Davis-Caltrans air quality project, Task order No. 66, September 2008.
57. Brzezinski D, Hart C, Enns P. Final facility specific speed correction factors. U.S. EPA Final report, EPA420-R-01-060, November 2001.

58. André M. Driving cycles development: Characterization of the methods. SAE Paper No. 961112, 1996.
59. André M, Hickman J, Hassel D, Joumard R. Driving cycles for emission measurements under European conditions. SAE Paper No. 950926, 1995.
60. Joumard R, André M, Vidon R, Tassel P, Pruvost C. Influence of driving cycles on unit emissions from passenger cars. Atmos Environ. 2000;34:4621–8.
61. Lin J, Niemeier DA. Regional driving characteristics, regional driving cycles. Transp Res Pt. D. 2003;8:361–81.
62. Smith M. Heavy-duty vehicle cycle development. EPA-460/3-78-008, U.S. EPA, Washington, DC, 1978.

Chapter 2
Light-Duty Vehicles

Abstract This chapter presents the most important driving cycles used for testing passenger cars and light-duty trucks, which are all of the chassis-dynamometer type. European, U.S., Japanese, Australian and worldwide modal and transient cycles are presented, including those intended for battery and electric vehicles, with graphic illustration of the speed profiles together with a detailed historical background. Main technical specifications are provided, as well as identification of the shortcomings, and representative results from real vehicles operation. An extensive comparison of the most important legislated cycles is also presented and discussed at the end of the chapter.

Passenger cars and light-duty trucks/vans (collectively referred to as light-duty vehicles—LDVs), were the first vehicle types for which emission standards and test cycles were legislated in the late 60s, limited to gasoline engines. Initially, the employed cycles were modal but later evolved to more sophisticated transient form, and covered diesel-engined vehicles too. Both urban and suburban/motorway segments have been usually included it the test cycle, with varying duration, aggressiveness and maximum speed depending on the specific region. United States, Europe (through UNECE regulations) and Japan have been the pioneering regions in the world in legislating certification test cycles for LDVs. The most important of them, all of the chassis-dynamometer type, will be detailed in the following sections. The respective drive cycles for another light-vehicle category, namely motorcycles, are discussed in Chap. 3.

2.1 European Union

With a yearly production of more than 18.4 million vehicles (20 % of global motor vehicle production) in 2015, from which 98 % light duty, European Union (EU) is the second biggest car manufacturer in the world today after China. One out of four passenger cars sold worldwide is produced or imported in the EU [1]. It is not

E.G. Giakoumis, *Driving and Engine Cycles*,
DOI 10.1007/978-3-319-49034-2_2

surprising then that the European regulations on automobile emissions affect the biggest manufacturers and many other non-European countries worldwide.

The legislative function in the European Union regarding emission regulations and test cycles/procedures is exercised in the form of directives or, more recently, regulations, by three regulating bodies:

(a) the European Parliament (elected by the peoples of the Member States),
(b) the Council of the EU (representing the governments of the EU Member States), and
(c) the European Commission (the executive body of the European Union responsible for proposing legislation and implementing decisions).

Through the years, the European Economic Community (EEC) and later the EU have produced a series of directives and regulations,[1] usually based upon the technical recommendations of the UNECE. The first emission limits were set in 1970 with Directive 70/220/EEC, concerning HC and CO emissions from gasoline vehicles; the limits (g/test) were defined with respect to the vehicle's reference weight. The directive applied to 'any vehicle with a positive-ignition engine, intended for use on the road, with or without bodywork, having at least four wheels, a permissible maximum weight of at least 400 kg and a maximum design speed at least 50 km/h, with the exception of agricultural tractors and machinery, and public works vehicles'. Directive 70/220/EEC harmonized draft national exhaust emission legislations from Germany ('Straßenverkehrs-Zulassungs-Ordnung' from 1968) and France ('Composition des gaz d'échappement émis par les véhicules automobiles équipés de moteur à essence' from 1969). These had been passed on to the Commission under the Standstill Agreement of 1969 General Program for the elimination of technical barriers to trade. With Directive 70/220/EEC, the first European test cycle, the ECE-15, was also legislated.

Since both Directive 70/220/EEC and its first amendment 74/290/EEC focused exclusively on HC and CO, the obvious reduction measure from the manufacturers was to adjust the SI engine operation towards lean mixtures. This, however, resulted in an increase in the emitted NO_x from gasoline passenger cars. As of October 1977 (Directive 77/102/EEC), NO_x emission limits were defined too, expressed as NO_2 equivalent g/test, again with respect to the vehicle's reference weight. Diesel engine emissions were covered from 1988 (Directive 88/76/EEC), with PM taken also into account (Directive 88/436/EEC), whereas fuel consumption measurement was introduced with Directive 80/1268/EEC [2].

The 'Euro' standards began in 1992, a few months before the European Union was established through the Treaty of Maastricht, with Euro 1 for passenger cars (Directive 91/441/EEC). This triggered the use of catalysts in cars and unleaded gasoline with a delay of more than a decade compared to the United States; evaporative emission standards were covered too. Light-duty trucks followed two years later, in 1993 (Directive 93/59/EEC). In September 2014, the last stage, Euro 6,

[1]EU directives and regulations can be accessed online through http://eur-lex.europa.eu.

came into force (Regulations 715/2007/EC and 692/2008/EC). Nowadays, the controlled pollutants from LDVs are mass emissions of CO, NO_x, HC/NMHC and PM, as well as particle number (see also Fig. 1.3). Passenger cars fall into categories M_1 and M_2, with light-duty vans into categories N_1 and N_2; reference mass for all these vehicles is lower than 2610 kg. Light-duty vans of category N_1 are further divided into three classes, I, II and III depending on the reference weight [3]. Detailed analysis of the complicated EU environmental policy-making and the often conflicting objectives between Member States and between car makers that led to the formulation of the EU automobile emission legislation is available in [4].

2.1.1 European Driving Cycle ECE+EUDC/NEDC

The driving cycles that have been employed for many decades in the European Union for the certification of passenger cars and LD vans were the ECE (initially) and the ECE+EUDC beginning with the Euro 1 emission standard in 1992, from 2000 known as NEDC. Although originally intended for gasoline-engined vehicles, the cycles have been also employed for the testing of diesel-engined vehicles, as well as to estimate the electric power consumption and driving range of hybrid and battery-electric cars. It is the intention of the EU authorities to adopt the WLTC (Sect. 2.5) from September 2017 together with the Euro 6c standard.

Work about emission test procedures for automobiles started in Europe in the mid 50s. For example in Germany, the VDA sub-committee 'Abgase von Otto-Motoren' (Exhaust gases from gasoline engines) was assigned to establish emission standards, evaluate possibilities for pollutant reduction, and develop necessary measurement techniques. Until October 1958, the German Ministry of Traffic had distributed various research assignments on automobile emissions, for example air quality measurements in German cities, and pollutant reduction in the exhaust gas from gasoline engines. However, it soon became clear that these activities had to be coordinated with similar ones conducted at the time in Sweden and France, in order for the results to be more effective [2].

In France, research on the early development of a cycle to simulate Paris driving is reported in [5]. Two routes, a north–south (8.4 km) and an east–west (11.25 km) were selected, and continuous traces of engine speed, inlet manifold vacuum, brake usage, and gear selection were recorded in two vehicles driven over the routes. Analysis of the resulting traces yielded data similar to that obtained in a Los Angeles research of 1957 (Sect. 2.2.1). An 11-mode cycle was then constructed by UTAC (Union Techniques de l' Automobile, du Motorcycle et du Cycle), which contained mode times as indicated in Table 2.1, and further detailed in Table 2.2 also showing weighting factors for continuous emissions analysis. Under the auspices of the UNECE, GRPA (later GRPE) carried out studies of driving patterns in ten European cities and recommended modifications to the UTAC cycle [2, 6].

The driving cycle discussion had been monitored by the WP.29 working group, which in its 20th session on December 20, 1965, assigned the BPICA ('Bureau

Table 2.1 Comparison between Paris driving, UTAC cycle and ECE-15 cycle [6]

Mode	Proportion of time in driving mode (%)		
	Paris	UTAC	ECE-15
Acceleration	32	15.6	18.5
Cruise	13	52	32.3
Deceleration	22	13.4	18.5
Idle	33	19	30.7

Early results from Germany in 1965 showed a markedly higher percentage of idling in German cities of the order of 45 %, but this figure was subsequently revised to 35 % [2]

Table 2.2 UTAC cycle modes and weighting factors [6]

Mode	Speed (km/h)	Weighting factor (%)
1	Idle	7.3
2	0–20	33.1
3	20	6.4
4	20–40	36.2
5	40	3.3
6	40–25	7.1
7	25	0
8	25–60	5.2
9	60	1.1
10	60–25	0.3
11	25–0	0

Permanent International des Constructeurs d'Automobile') to propose a unified European driving cycle. The first draft of this cycle was presented by BPICA during the 1st session of the GRPA in Paris on July 6–8, 1966. After some modifications,—e.g., a reduction of the average speed from 21.2 to 18.9 km/h which was requested by Great Britain—and after an evaluation in the London laboratories of the BPICA, the cycle was eventually accepted during GRPA's 2nd session on January 9–11, 1967 [2]. The resulting cycle was the ECE-15, which was the first drive cycle to be legislated in the EU (EEC at the time), and is illustrated in Fig. 2.1. The name ECE-15 corresponds to UNECE Regulation No. 15 published in April 11, 1969.[2]

The cycle was adopted by the European Economic Community initially on 20 March 1970 (Directive 70/220/EEC) concerning CO and HC emissions from

[2]Regulation No. 15 was replaced by No. 83, which introduced the extra urban segment of the cycle, No. 84 as regards fuel consumption measurement, and No. 101 as regards CO_2 emission and fuel consumption measurement. The above UN regulations, as well as the respective EU documents, provide detailed information on the driving cycle, i.e., the exact gear-shift strategy, guidelines for the measuring procedure, calibration of the test equipment, reference fuels as well as detailed description of all the applicable type-approval documents.

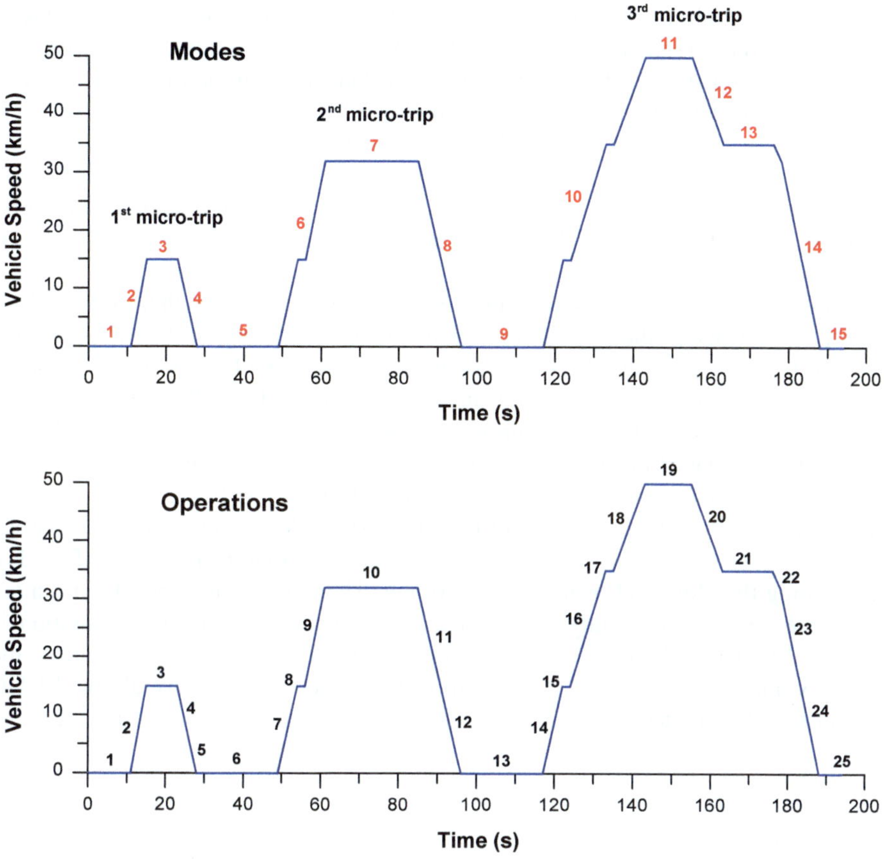

Fig. 2.1 Speed profile of the ECE-15; the *upper sub-diagram* identifies the 15 modes of the cycle and the *lower sub-diagram* the operations to be followed by the driver (for example, mode *10* consists of accelerations (operations) No. *14*, *16* and *18* and gear changes No. *15* and *17*). Notice that the points the gears are to be changed are explicitly defined in the legislation. During the certification procedure, the ECE-15 is run four times consecutively

gasoline cars only using the single bag measuring procedure; the concept of modal weighting in the UTAC cycle from Table 2.2 was abandoned in favor of collection of all the exhaust gases from the cycle. The sampling method changed in 1983 to constant volume sampling (Directive 83/351/EEC), and from the late 80s covered diesel-engined vehicles too.

The urban cycle ECE-15 or ECE (also known as urban driving cycle UDC) is illustrated in Fig. 2.1 and is a typical modal/'synthetic' cycle. Each micro-trip comprises an initial idling phase, acceleration—depending on the specific micro-trip, there are one or two intermediate gear changes—steady speed, and deceleration. Overall, the elementary urban cycle encompasses 15 modes and 25 operations to be followed by the driver. The cycle is repeated four times for a total duration of 13 min (4 × 195 = 780 s) and a total distance of 4045 m. The UDC is characterized by frequent gear changes, relatively low vehicle speeds (and loads) up to 50 km/h, and several stops, with a rather prolonged idling period of the order of 31 %; further, the cruise section is very high at 32 %. The average driving speed is quite low, at 18.7 km/h. It should be noted that, as is the case with all modal chassis-dynamometer cycles, the ECE-15 is defined in terms of specific modes and operations to be followed by the driver. From these modes, provided in tabular form in Directives 70/220/EEC and 91/441/EEC, the graphical illustration of Fig. 2.1 is derived.

After pressure from the Netherlands, who presented evidence that over 70 % of European mileage in the 80s was driven at vehicle speeds higher than 70 km/h, the extra urban driving cycle, EUDC, was introduced in 1989 by UNECE Regulation No. 83 and adopted by the European Community on June 26, 1991 (Directive 91/441/EEC). The cycle was a compromise between the West German and British proposals and that of the consultative Committee of Manufacturers of the Common Market [7, 8]. The modal EUDC, graphically illustrated in Fig. 2.2 from the tabulated driving operations detailed in Directive 91/441/EEC, represents extra urban driving, with much higher vehicle velocities up to 120 km/h maintained for 11 s, accounting thus for rural or motorway driving. The aim was to replicate in a more complete way the real 'duty cycle' of a typical passenger car. Total duration is 400 s, of which 332 s (83 %) is spent with the fourth or fifth gear engaged in the gearbox; interestingly, for 54 % of the time, the vehicle cruises. Overall, the EUDC comprises 13 modes, namely idle, accelerations, steady speed driving, and

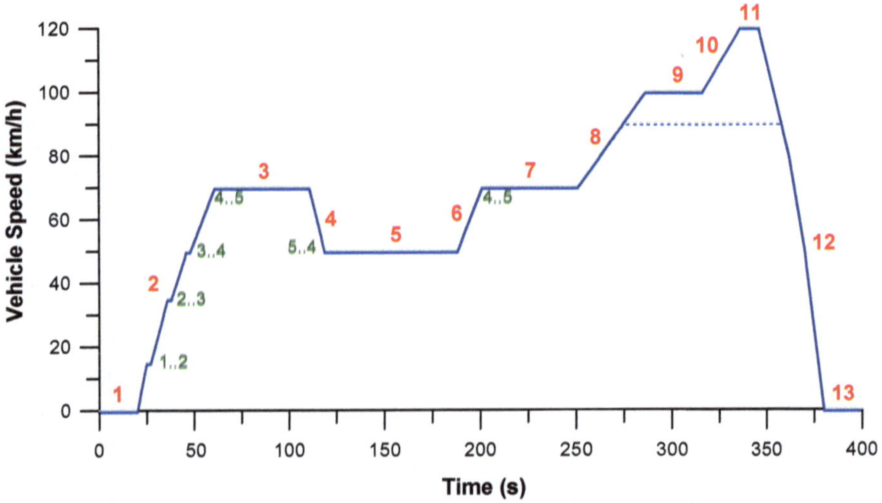

Fig. 2.2 Speed profile of the motorway EUDC segment with the dotted line corresponding to the low-powered version of the cycle (numbers from *1* to *13* denote the cycle modes; the gear changes throughout the cycle are also indicated for the case of a five-speed gearbox)

Table 2.3 ECE and EUDC breakdown by use of gears (in s) (Directive 91/441/EEC)

Cycle segment	Total	Idling	Idling, vehicle moving, clutch engaged on one combination	Gear shift	1st gear	2nd gear	3rd gear	4th gear	5th gear
UDC	195	60	9	8	24	53	41	–	–
EUDC	400	20	20	6	5	9	8	99	233

decelerations. Predictably, there are no intermediate idle periods in the cycle's speed trace but there is an idle phase of 20 s both at the beginning and the end. Table 2.3 provides some data for the urban ECE and motorway EUDC with reference to the time spent with each engaged gear.

Upon introduction of the EUDC in 1991, the full version of the European driving cycle was formulated, demonstrated in Fig. 2.3; it comprised two parts: the urban ECE segment formed the first part and the motorway EUDC the second. An alternative version of the EUDC was also defined at that time, where the maximum vehicle speed during the cycle was limited to 90 km/h. This was employed for low-powered vehicles having a maximum engine power less than 30 kW, or 30 kW/t for LD vans, and a maximum vehicle speed lower than 130 km/h. According to Directive 93/59/EEC, the low-powered version of the cycle was to be employed until 1 July 1994 for M-category vehicles, 1 January 1996 for N_1-category Class I, and 1 January 1997 for N_1-category Classes II and III. After that dates, vehicles which do not attain the acceleration and maximum speed values required in the cycle must be operated with the accelerator control fully depressed until they once again reach the required operating curve.

Beginning with emission standard Euro 1 in 1992, passenger cars in Europe were tested on the combined ECE+EUDC, Fig. 2.3, using the CVS system. This combined version of the urban ECE and the motorway EUDC is known as MVEG-A.[3] The cycle has a total duration of 1180 s (=4 × 195 + 400) with 11 km traveled distance. It is the Type I test in the type approval, as originally defined in Directive 70/220/EEC.

For compliance with the Euro 1 and 2 emission standards, the vehicle (run-in and driven for at least 3000 km) was kept for at least 6 h before the test in a room with a constant temperature between 20 and 30 °C. Especially for compression-ignition engined vehicles, and with regard to their PM measurement, Directive 91/441/EEC established a further preconditioning requirement. Specifically, the motorway EUDC part of the cycle was to be run three times between 6 and 36 h prior to the test. After the preconditioning, the vehicle was started and kept idle for 40 s before the cycle was run and the emissions sampled. During the test, the cell temperature was

[3]The Motor Vehicle Emissions Group—MVEG, has been an expert working group that played a central role in the development of the European automobile emission regulations.

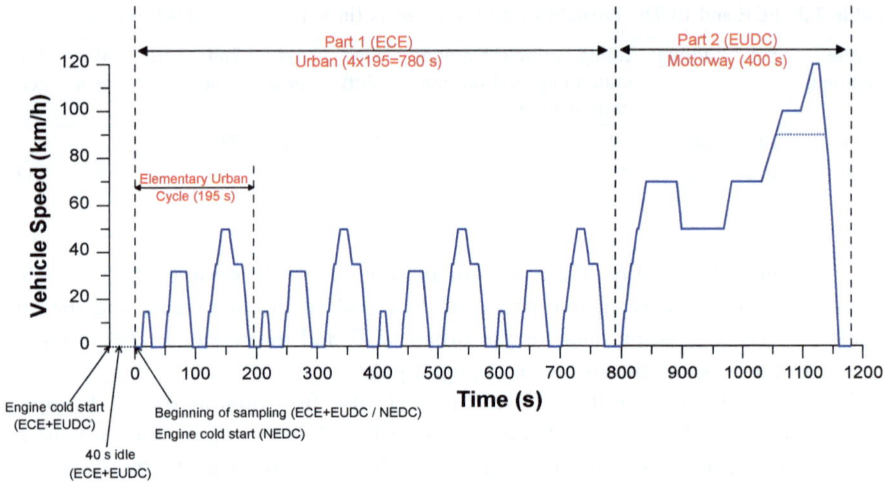

Fig. 2.3 Speed profile of the ECE+EUDC/NEDC driving cycle valid in the EEC/EU from 1970 to 2017 (*dotted line* designates the low-powered version of the cycle). Initially, there was a 40-s idling period before the cycle commenced (and sampling began), hence total duration was often cited as 1220 s. From 2000 onwards, the cycle runs and the sampling begins with the engine cold started (NEDC). A tolerance of ±2 km/h between indicated and theoretical speed is allowed as well a time tolerance of ±1 s at certain operations during the cycle

again between 20 and 30 °C, and the absolute humidity between 5.5 and 12.2 g of water per kg dry air. Directive 98/69/EC of October 13, 1998, implemented a slight but important change in the procedure, valid from the year 2000 with the transition to emission standard Euro 3. Emission sampling commences now immediately, i.e., without the 40-s warm-up period. This slightly modified cold-started procedure is known as the New European Driving cycle (NEDC) or MVEG-B, Fig. 2.3. Obviously, the first UDC run is responsible for higher amount of pollutants compared to the other three, as during the first minutes after cold start the after-treatment devices have not reached their operating temperature. Figure 2.4 eloquently illustrates this for CO and HC emissions of a Euro 4 passenger car.

The same testing procedure is employed for CO_2 emissions. The EU does not directly set fuel consumption standards but regulates CO_2 emissions, from the late 90s following a voluntary agreement with car manufacturers, and from 2009 on a mandatory basis, also including penalty payments in case of exceedingly high fleet-averaged CO_2. Fuel consumption is also measured during the NEDC; urban (Part 1) and extra-urban (Part 2) values are calculated and reported too, without applying any weighting factors. Emissions are sampled during the whole 1180 s duration of the cycle according to the constant volume sampling technique detailed in Sect. 6.5. For the low-temperature (Type VI) test of spark-ignition engined

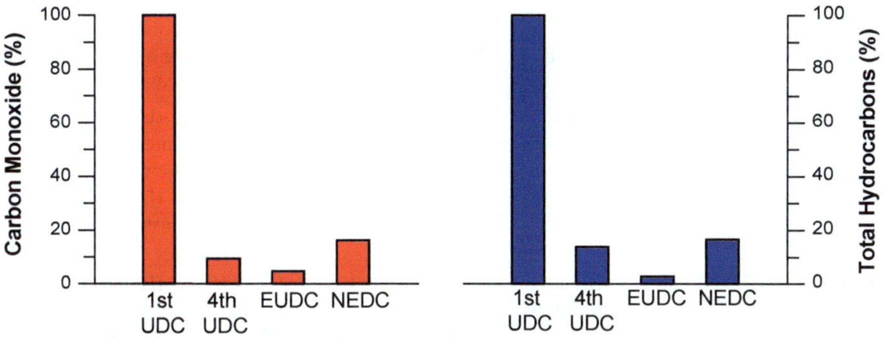

Fig. 2.4 Comparison of the relative CO and HC engine-out emissions during the NEDC for a Euro 4 diesel passenger car; the first UDC segment is the base (=100). During the fourth UDC, CO emissions are less than 10 % of those during the cold-started first UDC run (data from [9])

vehicles at −7 °C, however, initially legislated in Directive 98/69/EC to be valid from the Euro 3 standard, only the urban ECE part applies.[4]

Table 2.4 summarizes the driving cycles valid in the EU over the years, and Table 2.5 provides some of their important technical specifications. More detailed data is provided in the Appendix. Furthermore, Fig. 2.5 illustrates the frequency distribution for the speeds and accelerations encountered during the NEDC, where an increased density at zero acceleration is evident owing to the high percentage of time spent idling and cruising throughout the cycle. Figure 2.6 expands on the previous figure by highlighting the different speed/acceleration ranges between the urban and motorway segments of the NEDC.

Representative results during the NEDC are illustrated in Fig. 2.7 for a large passenger car. A strong influence between vehicle speed and traction force can be established from this figure. Aerodynamic resistance force follows closely the vehicle speed pattern; during the urban part of the cycle (0–780 s), where the vehicle speeds are maintained overall low, the rolling resistance term (not shown) generally prevails over the aerodynamic one. During motorway driving (781–1180 s) on the other hand, the aerodynamic force assumes much higher values. The points in the cycle where accelerations occur, lead to instantaneous sharp increases in CO_2 emissions, which are also indicative of the fueling rate.

Apart from the NEDC, Directive 91/441/EEC and Regulation 692/2008/EC defined special purpose cycles, namely the AMA and the SRC/SBC respectively, to be used for vehicle full useful life (durability) testing. These will be discussed in Sect. 2.2.7 as they also form part of the U.S. regulation.

---

[4] *For the type approval in the EU, the tests conducted are: Type I (tailpipe emissions after a cold start), Type II (CO emission at idling speed—gasoline, LPG and natural gas PI engines), Type III (emission of crankcase gases—PI engines only), Type IV (evaporative emissions—gasoline PI engines only), Type V (durability of anti-pollution control devices), Type VI (low temperature CO and HC tailpipe emissions after a cold start—gasoline PI engines only).*

Table 2.4 Driving cycles legislated in Europe (1970–2016)

Cycle	Cycle type	Traffic type	Procedure/Duration	EU Directive (UNECE Reg.)
ECE (4 identical elementary sub-cycles)	Modal	Urban	Cold starting + 40 s (idling) + 780 s (sampling)	70/220/EEC (UNECE R15/00) ('Single 'big' bag sampling/PI-engined vehicles) 83/351/EEC (UNECE R15/04) (constant volume sampling)
ECE + EUDC	Modal	Urban + Motorway	Cold starting + 40 s (idling) + 1180 s (sampling)	91/441/EEC (R83/01) (passenger cars) 93/59/EEC (R83/02) (light-duty trucks)
NEDC	Modal	Urban + Motorway	1180 s (cold started)	98/69/EC (R83/05)
WLTC (Sect. 2.5)	*Transient*	*Urban + Motorway*	*Cold started*	*To be finalized (GTR No. 15)*

Table 2.5 Summary of technical specifications of the European driving cycle (1970–2016)

Cycle/Segment	Duration (s)	Distance (m)	Maximum speed (km/h)	Average speed (km/h)	Maximum accel. (m/s^2)	Idling time (%)	RPA (m/s^2)
ECE-15	780	4045	50	18.7	1.04	30.8	0.154
EUDC	400	6955	120	62.6	0.83	10.0	0.094
NEDC	1180	11,000	120	33.6	1.04	23.7	0.116
NEDC (low power)	1180	10,656	90	32.5	1.04	23.7	0.097

Discussion—Criticism

As is made obvious from the previous figures, the European regulatory test cycle is quite simplistic, with long constant-speed phases. This is entirely unrealistic of real driving, where changes in the throttle position are practically continuous even when cruising, a fact that affects both the air-fuel ratio and the emissions from the vehicle. Moreover, constant accelerations are established throughout the cycle. The maximum speed (120 km/h) might be considered low by the standards of current European cars and drivers, although it is higher compared to other legislated test cycles that will be discussed in the next sections. What is undeniably low is the maximum acceleration, being only 1.04 m/s^2 or 3.74 km/h/s, i.e., much lower than would be expected during daily driving. In other words, and based on the cycle's specifications, almost 27 s are needed to reach 100 km/h from standstill. This acceleration lasts for 4 s during the first brief peak at the beginning of each UDC. Similarly unrealistic are the values of the accelerations throughout the beginning of the EUDC. Consequently, the NEDC is run with the vehicle actually operating at

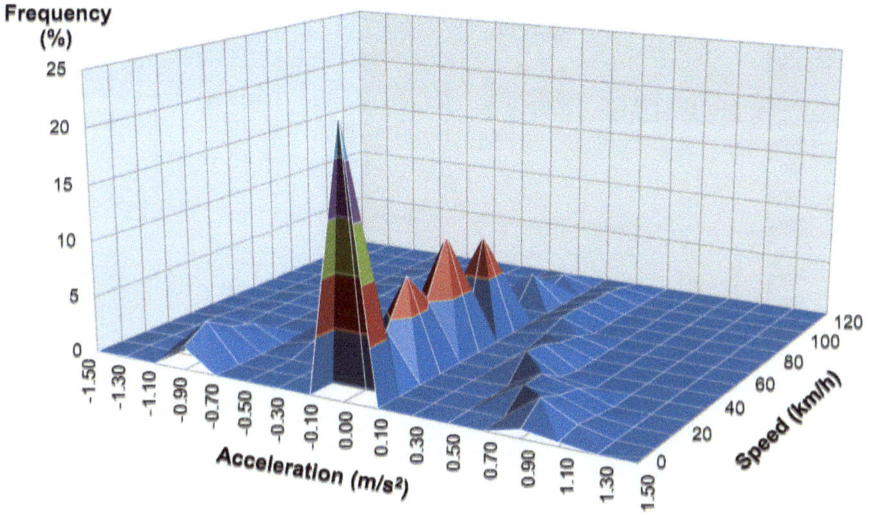

Fig. 2.5 3D speed/acceleration frequency distribution of the NEDC; notice the high density at zero acceleration, and the absence of penetration in many speed/acceleration combinations

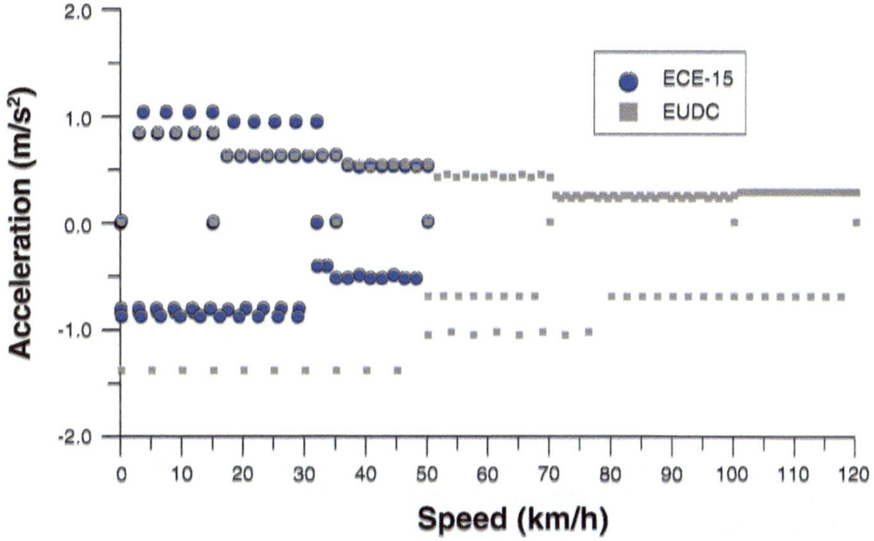

Fig. 2.6 Speed/acceleration distribution of the ECE-15 and EUDC segments of the European NEDC; the modal profile of both sub-cycles is evident

relatively low engine loads (cf. Figs. 2.69 and 2.70 at the end of the chapter). It follows that it is not difficult for the manufacturers to calibrate the engine ECU so that operation outside the tested cycle points, that is medium to high speeds/loads, is

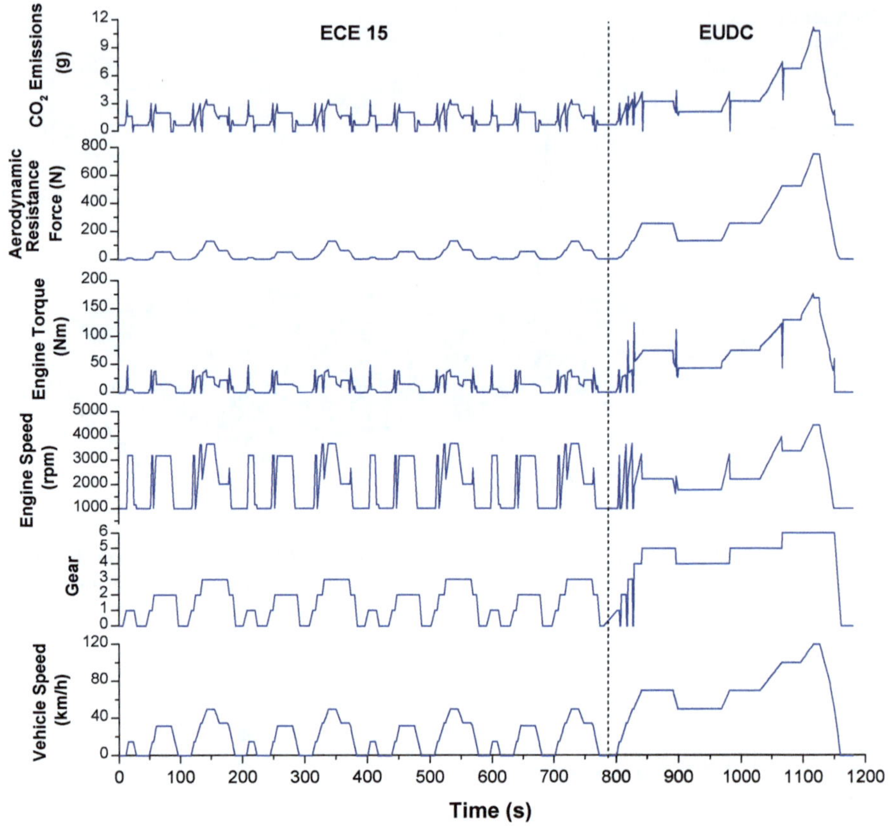

Fig. 2.7 Development of various engine and vehicle parameters during the European NEDC for a diesel-engined vehicle (reprinted from [10], copyright 2010, with permission from Elsevier)

strictly performance/fuel consumption oriented, with no concern for the emitted pollutants.

Further constraints/loopholes originate in the legislation itself (some of these loopholes exist in the U.S. and Japanese legislation too):

- The exactly defined profile of the gear-shift schedule makes it easy for the manufacturers to implement 'cycle beating' techniques.
- No account is taken for the use of air-conditioning, which is nowadays fitted to almost every car (or any other accessory for that matter).
- The wide range, as well as the rather high ambient temperatures during the test (20–30 °C), is another cause for inconsistencies and unrepresentative results.
- There is a 2 % allowable margin between the achieved and target velocity profiles, which can be exploited to achieve better fuel economy.

- Particularly as regards CO_2 emissions, manufacturers can declare up to 4 % lower values compared to the measured results.
- Instead of using its actual weight, the vehicle is categorized for the test into a discrete inertia class; this made sense when mechanical dynamometers were used but nowadays seems unreasonable with electronic dynamometers being capable of simulating practically any weight. As a result, manufacturers design their vehicles so as they belong to the lower class.
- Certain flexibilities also exist with regard to the coast-down test that determines the resistances of the dynamometer, discussed in Sect. 6.4 [11].

On the other hand, the test procedure only accounts for cold starting and disregards the fact that a typical daily driving schedule includes hot-started trips too. This fact, combined with the relatively short distance of the cycle, results in overestimation of cold-started emissions.

As a result of the above, much has been reported about the extent to which the NEDC fails to represent the real-world driving behavior of cars as regards both pollutant emissions and fuel consumption/CO_2 [12–23]; a few examples follow. Based upon analyses of more than 600,000 vehicles from eleven data sources and six countries, a study published by ICCT (International Council on Clean Transportation) revealed that the divergence, or 'gap', between real-world and certification CO_2 emissions increased from approximately 8 % in 2001 to 40 % in 2014, as the lower sub-diagram of Fig. 2.8 demonstrates [18]. Almost half of this gap is attributed to the non-realistic test cycle. Similar gaps were reported as regards fuel consumption measurements from 924 cars by Ntziachristos et al. [19].

For another research program, NO_x emissions from fifteen Euro 5 and 6 certified diesel passenger cars equipped with three different deNO_x technologies were measured using portable equipment on the road. It was found that, on average, about seven times higher NO_x was emitted than indicated by the official laboratory test results, as the middle sub-diagram of Fig. 2.8 indicates. Some individual vehicles performed significantly worse, although few exhibited an 'acceptable' and one a very good behavior [20].

A detailed study on ten Euro 5 certified light commercial N_1 Class III vehicles was conducted by TNO in 2015 regarding NO_x and CO_2 emissions. Tests were conducted on the road with the vehicles loaded at 28 and 100 %. For CO_2, a difference (increase) of the order of 7–52 % was measured between real-world and certification results. For NO_x, on the other hand, the results were much more impressive, as the upper sub-diagram of Fig. 2.8 illustrates, with on-road results found five to six times higher than the certification limit of 280 mg/km [21]. Interestingly, the same study concluded that driving behavior, vehicle payload and external circumstances caused less than 15 % variation in the obtained results.

Lastly, based on data collected from various passenger cars and vans in [22], it was found that for speeds up to 100 km/h, the NEDC covers only half the range of the accelerations of real-world driving. It follows that this leaves a wide area of operating conditions of daily driving conditions practically uncontrolled.

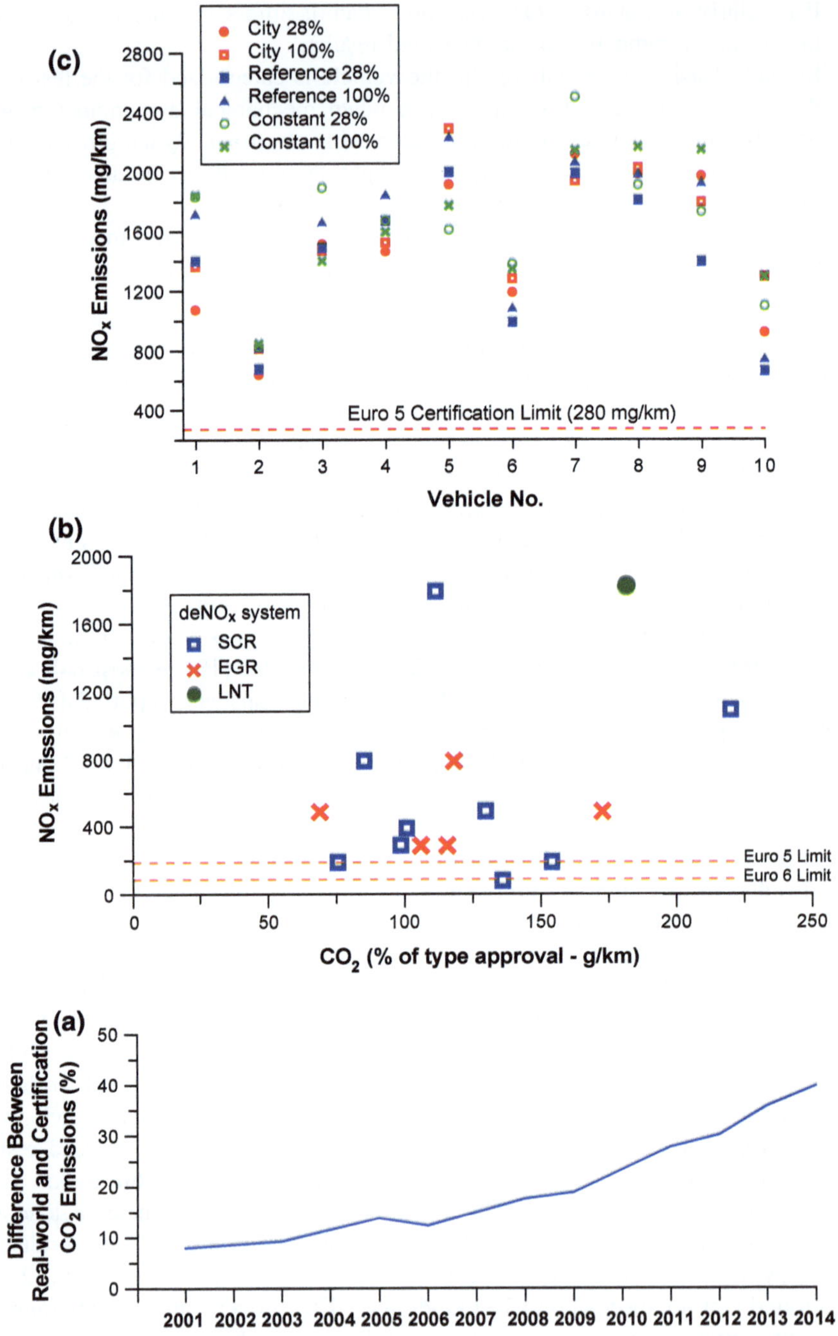
(c)
City 28%
City 100%
Reference 28%
Reference 100%
Constant 28%
Constant 100%
NOx Emissions (mg/km)
2800
2400
2000
1600
1200
800
400
Euro 5 Certification Limit (280 mg/km)
1 2 3 4 5 6 7 8 9 10
Vehicle No.
(b)
deNOx system
SCR
EGR
LNT
NOx Emissions (mg/km)
2000
1600
1200
800
400
0
Euro 5 Limit
Euro 6 Limit
0 50 100 150 200 250
CO2 (% of type approval - g/km)
(a)
Difference Between Real-world and Certification CO2 Emissions (%)
50
40
30
20
10
0
2001 2002 2003 2004 2005 2006 2007 2008 2009 2010 2011 2012 2013 2014

◀ **Fig. 2.8** Discrepancies between certification and real-world emissions in Europe regarding: **a** CO_2 emissions from passenger cars (data from [18]); **b** NO_x emissions from fifteen Euro 5 and 6 certified passenger cars (data from [20]); and **c** NO_x emissions from ten Euro 5 certified N_1-class III commercial vehicles ['city' refers to exclusively urban route, 'reference' to an urban/rural/highway mix, and 'constant' to exclusively highway driving (data from [21])]

Prospect

Taking into serious consideration the above-mentioned emission discrepancies, and acknowledging (even with considerable delay[5]) the shortcomings of the NEDC, the EU authorities established in January 2011 a working group to address the severe inconsistency between real-world and certification emissions from passenger cars, and propose best new strategy. One component of the new approach is the adoption of a much more representative cycle, which will be the worldwide WLTC (Sect. 2.5), effective September 2017. Following the WLTC suite of cycles presentation in Sect. 2.5, a detailed comparison between the NEDC and the WLTC will be performed in Sect. 2.7.

In parallel, the introduction of the Real Driving Emission (RDE) test in the European legislation will take place. This means that, for the first time, driving of the car on the road under real and varying traffic conditions will form part of the certification process; measurements will be conducted using PEMS. It was decided that the RDE test should be introduced in two stages to allow manufacturers to gradually adapt. During a first transitional period, the procedure will only be applied for monitoring purposes. Afterwards, binding quantitative RDE requirements will be set. These will be in the form of conformity/multiplicative factors CF with respect to certification Euro 6 limits (Regulation 2016/646/EU). The not-to-exceed NTE limit during the RDE is

$$\text{NTE} = \text{CF} \times \text{Euro}\,6 \tag{2.1}$$

More specifically, for NO_x, which is the usually manipulated pollutant from diesel engines owing to its inverse relation with fuel consumption, the conformity factor will be temporarily set to 2.1, effective 1 September 2017 ('2nd RDE package'/standard Euro 6d-'Temp'). On January 1, 2020, the factor will drop to 1.0 plus a margin parameter to take into account measurement uncertainties (Euro 6d); the latter margin is initially set at 0.5, and will be reviewed annually, with the aim to be eliminated in the future, at the latest by 2023 (all the above-mentioned dates correspond to new car type approvals). The particle number conformity factor has not been determined yet; CO emissions will be recorded too.

Furthermore, Regulation 2016/427/EU describes the exact procedure for the RDE test. Specifically, the RDE trip, between 90 and 120 min duration, will consist

[5]Reportedly, the European Commission preferred in the mid 2000s to put forward the Euro 5/6 legislation and develop a new procedure at a later stage, over the alternative scenario of adopting at that time a more realistic test cycle and put the Euro 5/6 stricter levels on hold (European Parliament's Committee of Inquiry into Emission Measurements in the Automotive Sector-EMIS).

of approximately 34 % urban (speed up to 60 km/h), 33 % rural (between 60 and 90 km/h) and 33 % motorway (higher than 90 km/h) operation, following random acceleration and deceleration patterns. At least 16 km will be driven during each of the three phases. The average driving speed of the urban phase of the trip including stops, should be between 15 and 30 km/h. Stop periods, defined as vehicle speed of less than 1 km/h, will account for at least 10 % of the time duration of urban operation. Cold-start emissions, although monitored, will not be accounted for in the calculations. The speed range of the motorway segment will cover a range between 90 and at least 110 km/h; the vehicle's velocity should be above 100 km/h for at least 5 min. The regulation also defines that the air-conditioning and other auxiliary systems will be operated in a representative way during the test. It should be noted that the not-to-exceed limits in Eq. (2.1) should be met not only on the whole RDE trip but also its urban part.

The 3rd RDE package will define a procedure for the measurement of the particulates, and includes the effect of vehicle cold starts into the RDE testing. In addition, manufacturers will be obliged to publish the conformity factor of an individual vehicle in its certificate of conformity. In the 4th RDE package, the rules for independent real driving emissions testing of vehicles being in-service will be defined, including the regulatory consequences in case of non-conformity. In order to take into account the statistical and technical uncertainties of the measurement procedures on the road, it may be considered in the future to reflect in the RDE emission limits applicable to individual PEMS trips the characteristics of those trips, described by certain measurable parameters, e.g., related to the driving dynamics or workload.

Following the requirements of Regulation 2016/427/EU, the European Automobile Manufacturers Association (ACEA) has launched a web page providing access to the real driving emission results of new type-approved vehicles.[6]

It is expected that owing to the RDE test, manufacturers will be forced to apply emission optimization strategies over a much broader engine operating range, affecting the exhaust after-treatment strategy. For example, greater volume of exhaust after-treatment systems will be needed, deactivation of the EGR system at high altitudes will not be feasible anymore, most probably LNT will not be adequate for efficient NO_x control, whereas higher urea (AdBlue®) consumption will be required in the SCR-equipped vehicles. Several test cycles have been used by vehicle and component manufacturers to approximate the demands of the RDE test in the laboratory, such as the Artemis-project cycles discussed in the next section, the RTS-95, illustrated in Fig. 2.9, the TNO random cycle generator etc. [23].

Lastly, Fig. 2.10 provides a synopsis of the applicable test cycle, emission regulation, controlled pollutants and typical exhaust after-treatment over the years in Europe for both compression ignition and positive ignition engined passenger cars and light-duty vans.

[6]Accessible via http://www.acea.be/publications/article/access-to-euro-6-rde-monitoring-data.

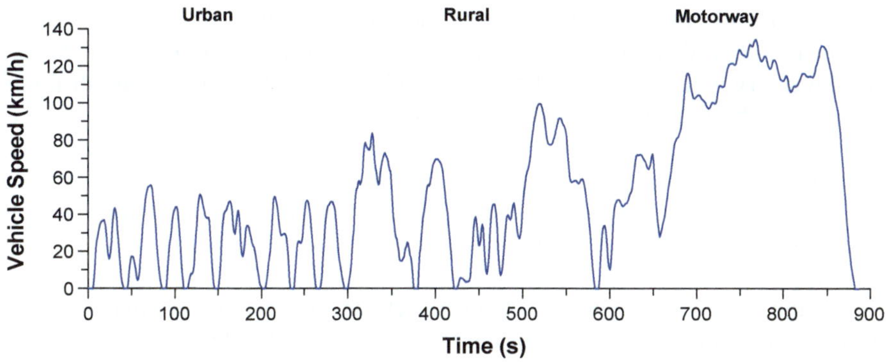

Fig. 2.9 Speed profile of the RTS-95 driving cycle employed for RDE simulation in the laboratory by many component and vehicle manufacturers in Europe (compared to the NEDC, this 886-s and almost 13-km long much harsher test schedule exhibits maximum acceleration of the order of 2.88 m/s^2 with an RPA value of 0.30); the cycle was developed based on driving activity incorporated in the WLTP database discussed in Sect. 2.5, and is actually a non-standardized cycle, therefore different versions might exist or be developed [24, 25]

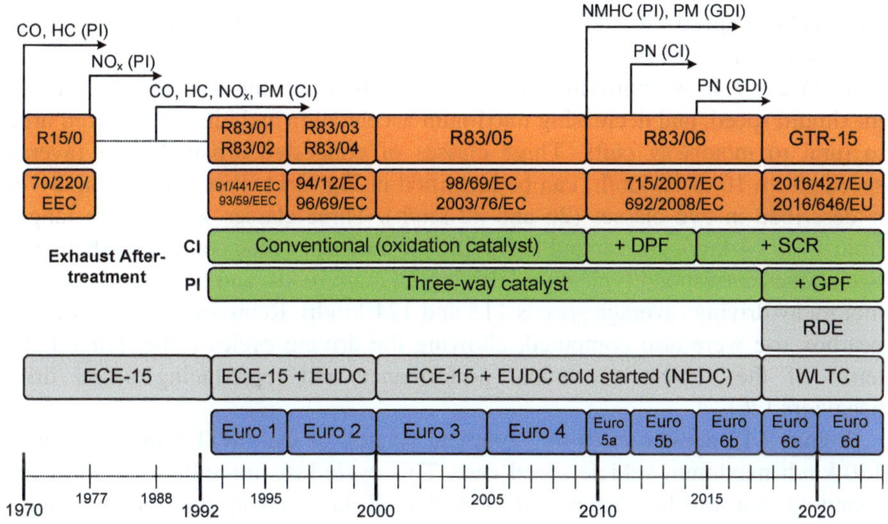

Fig. 2.10 Illustration of European emission standards, regulated pollutants, test cycles and corresponding directives/regulations over the years (adapted from [23]); dates correspond to new vehicle type approvals

2.1.2 *Non-legislated Cycles*

A much more realistic, although never legislated but frequently employed in research studies, approach to simulate real driving conditions in Europe has been the ARTEMIS cycles. These were developed within the frame of the European ARTEMIS (Assessment and Reliability of Transport Emission Models and Inventory Systems) research project funded by the EC within the 5th framework research program DG TREN [26]. Throughout a period of 5.5 years, data on actual driving conditions was collected in 1-s time intervals in four European countries in the frame of the DRIVE-MODEM[7] [27]) and HYZEM[8] [12] research projects. Specifically, 77 instrumented vehicles were monitored in France, the UK, Germany and Greece for 2000 days, 10,300 trips, 88,000 km and 2200 h of driving.

For the purpose of cycle development, elementary periods or kinematic segments with homogeneous sizes (distance varying from a few hundred meters at low speeds to a few kilometers at higher speeds) were defined within the trips. These kinematic segments were described by their idling duration and cross-distribution of the instantaneous speeds and accelerations. Correspondence analysis (based on chi-squared distance) and clustering tools were then used to classify the segments according to their speed/acceleration distribution. Eventually, three fundamental cycles were developed considering 12 different typical driving conditions, in effect forming sub-cycles within the cycle: an urban, a rural road and a motorway one; the latter was expressed in two variants, one with 130 and one with 150 km/h maximum vehicle speed [26]. The speed profiles of these three fundamental cycles are illustrated in Fig. 2.11.

Each cycle lasts approximately 1000 s, with increasing average, driving and maximum speed, and decreasing maximum acceleration and idling time from urban to rural to motorway parts. Three classes of congested urban driving (average speeds from 10 to 16 km/h) can be identified in Fig. 2.11. Free-flow urban driving is described in two classes (26 and 32 km/h). Three classes with speeds ranging from 44 to 64 km/h correspond to driving on secondary roads or in suburban areas, two classes correspond typically to driving on main roads, and two classes describe motorway driving (average speeds 115 and 124 km/h). Representative strategies of gearbox use were also computed, allowing the driving cycles to be monitored in terms of the vehicles' technical performance and reproducing actual driver behaviors [26].

Figure 2.12 demonstrates the complete version of the ARTEMIS cycle for the 150 km/h maximum vehicle speed case. This cycle lasts more than 50 min corresponding to a traveled distance of more than 50 km, making it one of the longest both in duration and traveled distance of the chassis-dynamometer cycles. Unlike the NEDC, this is a true transient cycle, having been composed of real driving data and not just specific driving modes.

[7]MODeling of EMissions and fuel consumption in urban areas research project within the DRIVE initiative (Dedicated Road Infrastructure for Vehicle Safety in Europe), funded by the EC, DG XIII.

[8]European development of HYbrid vehicle technology approaching efficient Zero Emission Mobility research project of BRITE/EURAM 2, EC-DG XII.

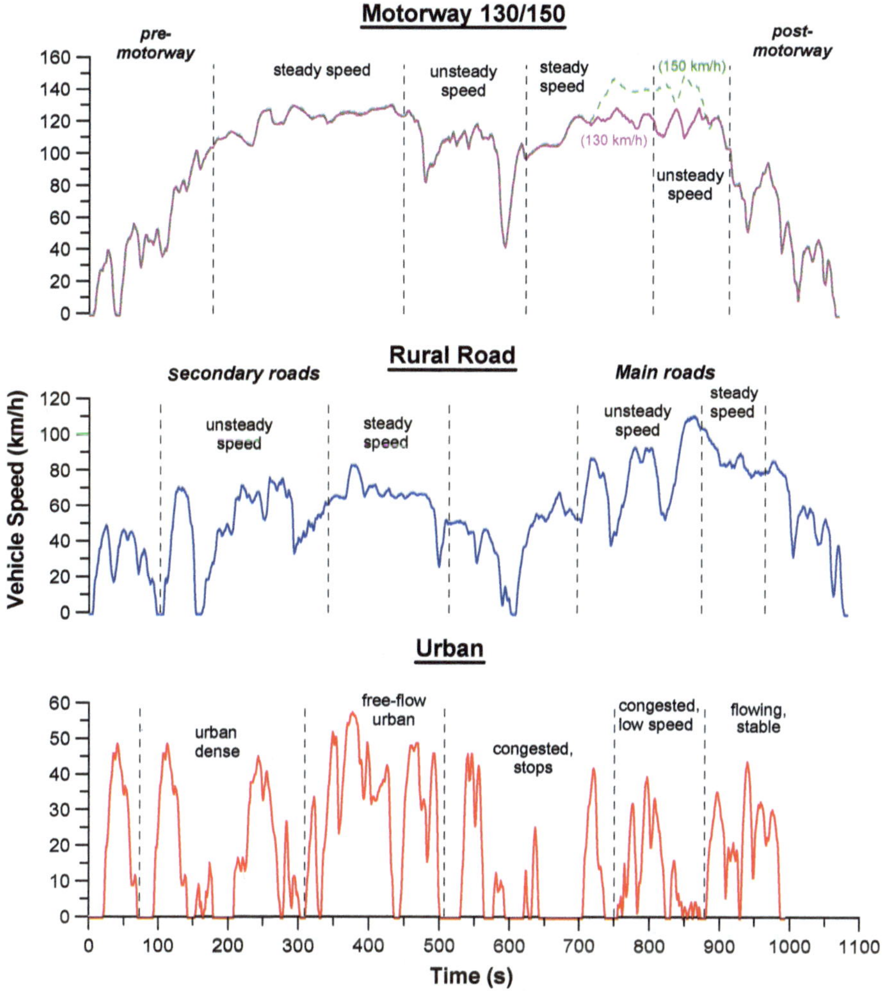

Fig. 2.11 Speed profiles of the ARTEMIS-project urban, rural road and motorway driving cycles with reference to their structure encompassing typical driving conditions (see Fig. 1.35 regarding the different driving conditions incorporated)

Some of the most important technical specifications of the ARTEMIS cycles are summarized in Table 2.6. Comparing this data with that of the NEDC from Table 2.5, it is clear that the ARTEMIS cycles are much harsher, with more frequent and steeper accelerations and higher vehicle speeds. A much more realistic speed profile with almost complete absence of steady-speed phases is also established from Fig. 2.11.

Figure 2.13 further supports these arguments by directly comparing the speed/acceleration distribution between the NEDC and the Artemis cycles; a much broader profile in both vehicle velocity and acceleration can be established for the

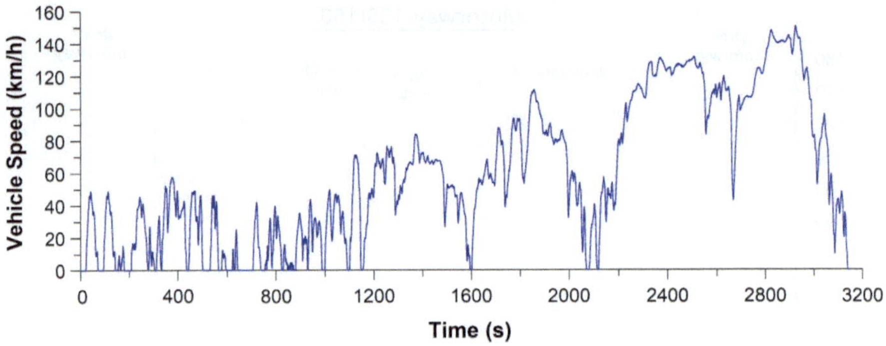

Fig. 2.12 Speed profile of the full version of the Artemis-project driving cycle

Table 2.6 Summary of technical specifications of the ARTEMIS-project cycles

(Sub)cycle	Duration (s)	Distance (m)	Maximum speed (km/h)	Average speed (km/h)	Maximum accel. (m/s^2)	Idling time (%)	RPA (m/s^2)
Urban	993	4870	57.7	17.7	2.86	26.2	0.342
Rural	1082	17,272	111.5	57.5	2.36	2.7	0.182
Motorw. 130	1068	28,736	131.8	96.9	1.92	1.3	0.137
Motorw. 150	1068	29,545	150.4	99.6	1.92	1.3	0.134
URM 150	3143	51,687	150.4	59.2	2.86	9.6	0.170

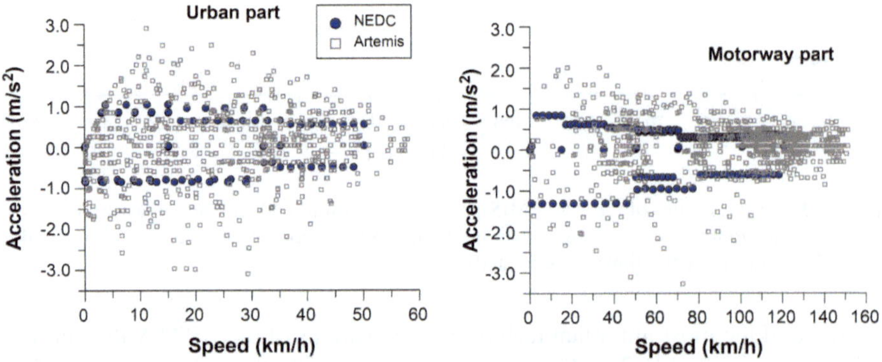

Fig. 2.13 Comparison of the vehicle speed/acceleration distribution between the NEDC and the ARTEMIS cycles (denser areas correspond to higher frequency of vehicle/acceleration points)

latter from this figure. This results in the Artemis cycles being more compatible with the daily driving behavior, and covering wider range of daily driving. Unsurprisingly, as many reports have shown, emission/fuel consumption results compared to the NEDC are much higher [28].

Many non-legislated cycles have been developed based on European driving conditions, with driving behavior data collected in various countries. These have been primarily used in research studies, and are summarized in Table 2.7. Speed traces together with detailed technical specifications for these cycles are provided in Reference No. 29.

Table 2.7 Various European non-legislated driving cycles [12, 29]

Project	Number of cycles	Description	Duration (s)	Average speed (km/h)
INRETS	10	5 urban, 3 rural and 2 motorway cycles derived from actual driving conditions around Lyon, France (Institut National de REcherche sur les Transports et leur Sécurité)	680–1067	3.8–94.5
INRETS	4	Urban and rural short cycles derived from the previous ones to measure cold-start effects	126–208	7.3–41.1
MODEM	14	Urban cycles derived from monitoring of 60 European cars in 6 cities, based on speed, acceleration and trip length parameters over significantly varying time periods	91–1027	5.1–42.5
MODEM-IM	4	Based on the same database as the MODEM ones; they include congested urban, free urban, rural and motorway driving; intended to be used for inspection and maintenance	355–712	14.4–101
MODEM-HYZEM	8	Driving cycles for evaluating hybrid cars	560–1868	18–92.2
EMPA	22	Developed by the Swiss EMPA Laboratory (ETH)	399–2290	3.2–130
Handbook	9	Nine driving cycles	1208–4820	17.2–108.2
Warren Spring Lab	12	Developed by the TRL (Transport Research Laboratory, UK) used for road tests	256–1207	6.7–112.1
OSCAR	10	Defined within the European 5th framework project OSCAR-cars (Optimized expert System for Conducting environmental Assessment of urban Road traffic)	350–455	7.8–35.8

A European cycle of special interest and use is the German ADAC or BAB130, illustrated in Fig. 2.14, being part of the ADAC EcoTest conducted since 2003. The test provides consumers with information on the eco-friendliness of cars (covering CO_2 and pollutant emissions incl. particle number). Since 2012, the EcoTest is performed over the NEDC cold started, the worldwide WLTC Class 3-2 hot-started with the air-condition on (at the moment, for CO_2 measurements only), and the BAB130, again with the air-condition on. Vehicles are tested at their actual weight, with daytime running lights or low beams on. Based on the results from all three tests, each car is assigned an EcoTest rating ranging from one to five stars. The EcoTest rating is calculated from the combined scores for pollutant emissions and CO_2. The pollutant score is indicated as an absolute value, regardless of vehicle size and fuel type. The consumption-dependent CO_2 emissions, on the other hand, are assessed individually for each vehicle class [30].

The BAB130 employed in the ADAC test is a highway cycle, as the EUDC, but with steeper and more frequent accelerations; it lasts 13 min. The cycle was developed in order to also test the vehicle in operating points outside the NEDC. It comprises three parts, a preconditioning segment, and phases 1 and 2. The two phases are identical, and testing the vehicle on both of them is made in order to identify possible actuation of the after-treatment regeneration system (in which case, the test is repeated). As the name suggests, maximum speed is 130 km/h maintained for a long period throughout the cycle; specifically, for approximately three quarters of the time the vehicle cruises. RPA is 0.122 for the 25 km traveled distance throughout the cycle. Notice in Fig. 2.14 the repeated hard accelerations during phases 1 and 2.

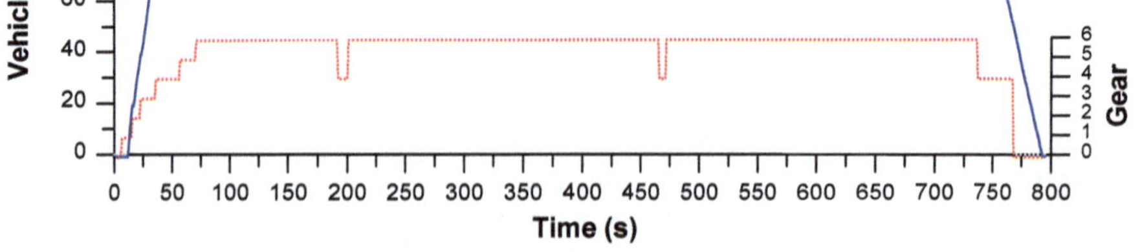

Fig. 2.14 Speed trace and gear schedule of the ADAC/BAB130 driving cycle. The cycle is run with the vehicle hot started, and measurements for the EcoTest rating are taken during phases 1 and 2, each of approximately 10 km length. A modified version of the cycle is employed for testing electric vehicles [30]

2.2 United States of America

In the United States, the research in the field of emissions from motor vehicles was initiated in the early 40s sparked by the Los Angeles photochemical smog. On July 14, 1955, the first federal legislation was signed dealing specifically with air pollution (the ‘Federal Air Pollution Control Act’), with the Clean Air Act (CAA) of 1963 being the first federal legislation regarding air pollution control [2, 31]. The research was expanded considerably with the enactment of the Clean Air Act Amendments on December 31, 1970 (1970 CAAA). Following the 1970 CAAA, test cycles as well as the respective federal emission standards in the United States is the responsibility of the Environmental Protection Agency (EPA), created in December 1970. EPA issues regulations through a process in which regulated entities, and other interest groups, have the opportunity to comment on proposals, seek judicial review of the agency’s procedures and compliance with the statutory framework created by the legislature, and seek action by the political branches to alter the agency’s actions. In effect, the CAA requires EPA to act but also constrains how it may act [32]. The CAA was later amended in 1977 and 1990, introducing, among other things, evaporative emissions testing and on-board diagnostics [2]. Interestingly, California is the only State in the U.S. which has the right to adopt its own standards (CAA, Section 209); these are more stringent than the federal. The pertinent authority here is the California Air Resources Board (CARB), created in 1967. A comprehensive presentation of U.S. air pollution control laws is available in [31].

The first nationwide U.S. light-duty vehicle emission standards were implemented in 1966 for model year 1968 vehicles. Initially, new standards were referred to by the effective model year of the regulation. In order to meet the 1975 and subsequent limits, oxidation catalysts were required combined with unleaded gasoline. Federal legislation in 1981 established new emission standards, retroactively known as ‘Tier 0’ beginning in 1987. The CAAA of 1990 subsequently defined two new levels of standards for LDVs, namely Tier 1 (phased-in progressively between 1994 and 1997), and Tier 2 (phased-in between 2004 and 2009). These applied to vehicles up to 8500 lbs GVWR (3850 kg) and ‘medium-duty’ passenger cars (MDPV), i.e., larger SUVs and passenger vans, between 8500 and 10,000 lbs GVWR (3850–4530 kg) [3]. From 2017–18, Tier 3 standard comes into force to be phased-in from 2017 to 2025. The new standard also extends coverage to HDVs of Class 3 (10–14,000 lbs/4530–6350 kg GVWR). Gross Vehicle Weight Rating (GVWR) is defined by federal regulation in 40 CFR 86.082-2 as ‘the value specified by the manufacturer as the maximum design loaded weight of a single vehicle’. It is the weight of the vehicle completely loaded with the maximum load that the manufacturer states the vehicle is capable of carrying. The controlled pollutants are mass emissions of non-methane organic gases (NMOG), CO, NO_x, PM, and formaldehyde (HCHO).

One marked difference between U.S. and European regulations is that U.S. emission limits are the same irrespective of fuel used. Moreover, each manufacturer in the United States can choose, among several emission 'bins', the one which best fits to his vehicles, provided that a certain NO_x (or $NMOG+NO_x$ from Tier 3 standards) emission target is achieved for the whole fleet each model year. As will be discussed in the next paragraphs, U.S. test cycles are far more representative, encompassing substantial portion of real driving activity compared to the simplistic NEDC that has been in use in Europe from 1970 up to 2017.

Unlike other countries, where the 'best cycle' approach is employed, in the United States a multiple-certification cycle approach is followed. At the moment, four different cycles are used during each light-duty vehicle's certification procedure (regarding tailpipe emissions), or five if we also take evaporative emissions testing under consideration; each one of these cycles covers a specific segment of a vehicle's operation.

2.2.1 *California 7-Mode*

California was not only the first region in the world to establish motor vehicle emission standards but also legislated the first chassis-dynamometer driving cycle for emission certification, together with the relevant test procedure [2]. The first attempts to construct a drive cycle were made in the 50s by the Los Angeles (LA) County Air Pollution Control District Laboratory for emissions measurement of typical LA driving. As was also the case with Europe and Japan at that time, researchers characterized real-world driving (of gasoline-engined cars) on the proportion of time spent in specific engine speed/manifold pressure bins, which were then used to define databases of driving modes. In 1950, an initial survey was carried out in Los Angeles using a single vehicle along a single route, with stopwatches to time the various modes; the results obtained were: idle for 18 % of the time, acceleration 18 %, deceleration 18 % and cruise 46 % of the total time [6].

A more elaborate survey was conducted in 1956 by the (then called) Automobile Manufacturers Association, employing seven vehicles. From this survey (see also Table 1.4), a cycle was recommended, comprising eleven driving modes chosen to represent average urban vehicle usage. These were: idle, cruising at 20, 30, 40 and 50 mph, acceleration from 0 to 25, 0 to 60 and 15 to 30 mph, and deceleration from 50 to 20, 30 to 15 and 30 to 0 mph. Weighting factors were also assigned to these modes, with the 0–25 mph mode's factor being 18.5 %, and that of the 15–30 mph mode 45.5 % [33]. Examination of these eleven modes reveals that it is impossible to combine them in a continuous cycle in the proper time proportions without inserting additional transition modes. Thus, in total 19 sequences were derived for the 11-mode cycle to become drivable. Based on these, the 7-mode cycle, the first certification drive cycle in the world, was subsequently formulated [34, 35]. To obtain the seven modes from the initial eleven, the cruise modes were reduced to two at 30 and 15 mph, the 0–25 acceleration was extended to 0–30 mph, and the 0–60 and 15–30 mph accelerations were replaced by a 15–50 mph acceleration; the 50–20 and 30–0 mph decelerations were combined. The modes were further weighted to represent typical driving based on frequency of mode of operation in urban traffic during a day, and exhaust volume produced in that mode, as detailed in Table 2.8 [6]. The cycle is graphically demonstrated in Fig. 2.15, identifying also the respective modes. The detailed requirements for emissions testing of passenger cars were approved by the recently (1960) created California Motor Vehicle Pollution Control Board on May 19, 1961, and the cycle was later modified in 1964. It was used to test emissions in California from 1966 to 1971 [2, 6, 34–37].

The 7-mode cycle lasts 137 s with a maximum speed of 80 km/h (50 mph) and an average driving speed of 41.8 km/h (25.9 mph). Unlike its European ECE-15 counterpart, which was more urban oriented, the idling phase is much smaller (14.6 % instead of 31 %) and so is cruising (21.9 % instead of 32 %). This results

Table 2.8 Description of modes and weighting factors of the California 7-mode cycle [34, 35]

Mode	Speed (mph)	Acceleration (mph/s)	Duration (s)	Weighting factor (%)
1	Idle	–	20	4.2
2	0–25	2.2	11.5	24.4
	25–30		2.5	Data not read
3	30	–	15	11.8
4	30–15	–1.4	11	6.2
5	15	–	15	5.0
6	15–30	1.2	12.5	45.5
	30–50		16.5	Data not read
7	50–20	–1.2	25	2.9
	20–0	–2.5	8	Data not read

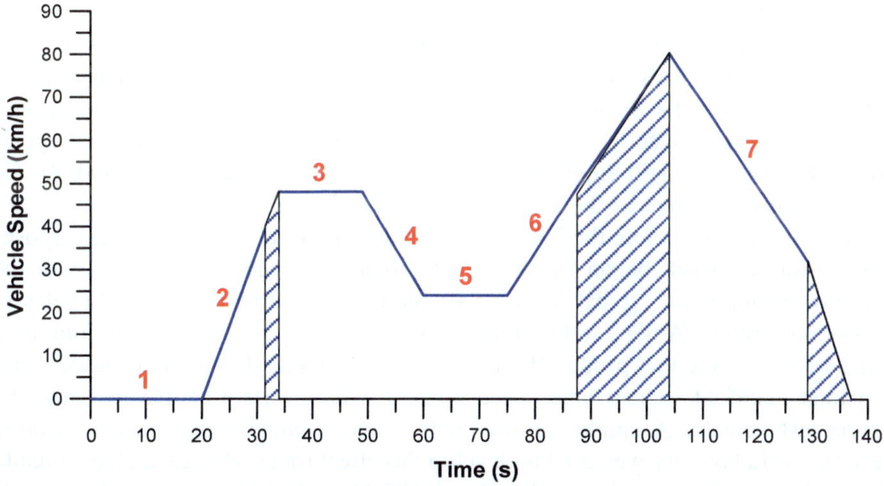

Fig. 2.15 Speed profile of the California 7-mode driving cycle according to Table 2.8 (*shaded areas* correspond to data not read during the test procedure)

in the cycle being much more transient and dynamic compared to the ECE-15; RPA is 0.24 for the California 7 mode in contrast to 'only' 0.15 for the ECE-15.

The California 7-mode cycle, with slight modifications, was adopted by the federal regulation as the 1968 U.S. FTP procedure for passenger cars and light-duty trucks (FR Vol. 31, No. 61, March 30, 1966). In application, the cycle was run 7 times, starting with a cold engine. The 'cold' runs (1–4) were weighted by 0.35 and the 'hot' runs (6 and 7) by 0.65, and the results summed; run 5 of the cycle was discarded. Effective model year 1970/71, the methodology changed to mass emission values; these were converted from the respective concentration measurements applying an empirically determined formula, which assumed the exhaust gas volume as a function of vehicle mass and transmission type for an average vehicle of the 4000 lbs inertia weight class [2]. A truly CVS system was introduced as of model year 1972 together with the FTP-72 cycle (next section).

Even before the cycle was used to certify MY 1966 vehicles in California, some of its serious shortcomings were recognized. For example, since the operating modes included were rather few, the cycle practically encouraged manufacturers to design control systems that would function on the 7 mode but not under other common driving conditions. It was further recognized that the cycle did not represent driving under the morning rush-hour traffic conditions, which were believed to be the most polluting, as the cycle had been designed to represent 24-h average conditions. Another significant drawback was the simplistic form of the schedule being not compatible with the real-life driving habits [36].

2.2.2 FTP-72 and FTP-75

Similar to the European (and Japanese) approaches in the 60s, the initial cycle employed in the United States for testing the compliance of motor vehicles with the emission standards was modal/'stylized', with constant accelerations and a repetitive form. Unlike the European and the Japanese regulations however, the federal legislation soon moved from simple/modal to much more realistic transient cycles (modal cycles were still employed in the United States during the 70s and 80s for special purposes, such as the FSC discussed later in the text).

In 1965 (that is, before the 7-mode cycle was employed for certification purposes in California), work began on the development of a new drive cycle, with the purpose to represent typical morning (home-to-work) driving in Los Angeles in rush-hour traffic. Among other things, the methodology focused on finding a specific road route that produced the same average mode distribution (based again on inlet manifold vacuum and speed ranges) with a variety of drivers using the same test vehicle. A single car driven by various employees of the California Emissions Laboratory was used to develop this street route. The research eventually led to the short-lived, and never legislated, XC-15 cycle. This was a synthetic cycle comprising 18 operating modes, with 22.7 mph average speed and 4 min duration [36]. During this research, a 12-mile commute route, called the LA4, beginning and ending at the California Emissions Laboratory was established. This route met the pre-defined criteria of being representative of typical driving in central LA during morning peak-hour traffic, and was used in a subsequent research effort, this time with the participation of the EPA [36, 38].

With the aim to develop an improved federal test procedure (based on speed/time distributions instead of manifold pressure/rpm ranges, and applying advanced vehicle instrumentation techniques), six different drivers from EPA's West Coast Laboratory drove a specific vehicle over the LA4 route. The six traces were analyzed for idle time, average and maximum speed, and number of stops; total time required ranged from 35 to 40 min One of the six traces demonstrated much harder acceleration rates than the other five and was discarded. The remaining five traces were surprisingly similar. Of those five, the trace with the actual time closest to the average was selected as the most representative speed/time trace. This trace contained 28 micro-trips and had an average speed of 31 km/h.

Subsequently, EPA—based on a 1969 report on driving patterns in Los Angeles [39]—shortened the LA4 from 12 to 7.5 miles (12 km) to represent the average commute length in Los Angeles at that time; this was denoted the LA4-S3. An attempt was made to preserve the proportionate time in each operating mode, so that average speed, proportion of idle time etc. remained unaltered. This paring process removed much of the low-speed driving iterations to compensate for the freeway driving reductions. Further, EPA adapted the cycle for use in the laboratory on a chassis dynamometer by cutting back the accelerations and decelerations to 3.3 mph/s (equivalent to 1.48 m/s^2), which was the maximum design rate of a belt-driven dynamometer at that time to avoid tire slip. Comparison of mass emission tests between the shortened cycle and the full cycle showed very high correlation. This, shortened and modified, final version of the drive cycle is known as the Urban Dynamometer Driving Schedule—UDDS, and is shown in Fig. 2.16. It is also referred to as the LA4-S4, or simply LA4; it was originally published in the U.S. FR Vol. 35, No. 219, November 10, 1970.

The UDDS/LA4 became the standard driving cycle for the certification of passenger vehicles and light-duty trucks in the United States, starting with the 1972 model year, for this reason known as FTP-72. The FTP-72 was the first cycle to employ the CVS sampling method and, as of model year 1973, the test was expanded to also include NO_x emission measurement via the chemiluminescence method [2]. At the time of its implementation, light-duty vehicles were defined as 'any motor vehicle either designed primarily for transportation of property and rated at 6000 lbs GVW or less or designed primarily for transportation of persons and having a capacity of 12 persons or less'.

The much more realistic compared to the California 7 mode, in terms of daily driving conditions, FTP-72 cycle simulates urban/suburban routes. It is a true transient cycle with 74 % of the time spent with the vehicle accelerating or decelerating. The cycle lasts 1372 s (approx. 23 min), and is divided into two segments: a cold-started 'transient' phase after overnight parking, and a second 'stabilization' (in terms of the engine having reached its fully warmed-up condition) phase with frequent accelerations and stops. Table 2.9 details some technical

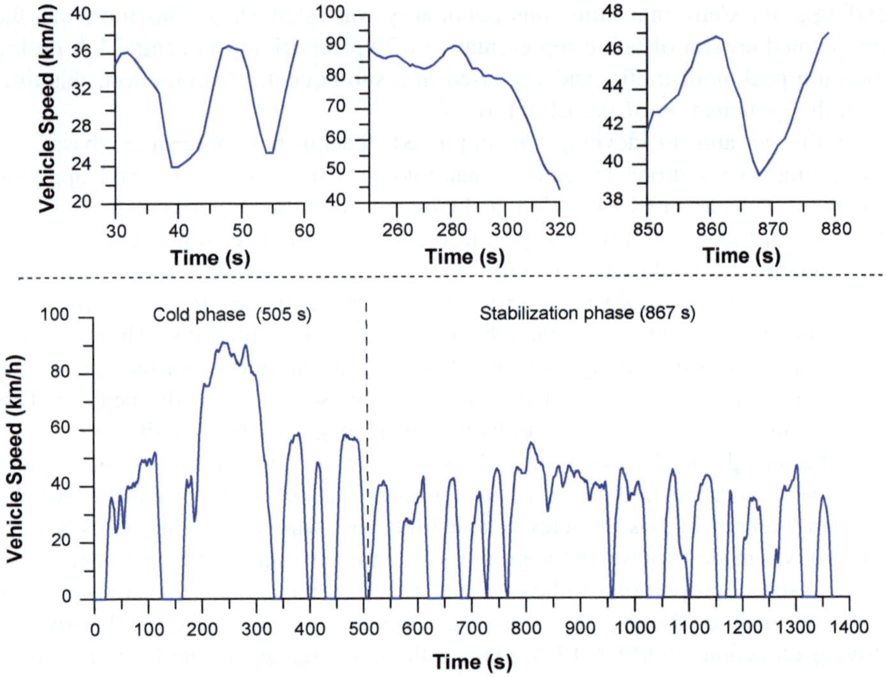

Fig. 2.16 Speed profile of the U.S. FTP-72 (the UNECE Reg. No. 53 refers to the cycle as the 'Test equivalent to Type I test (verifying emissions after a cold start)'; the upper three sub-diagrams illustrate in more detail specific segments of the cycle (40 CFR 86, App. I)

Table 2.9 Specifications of the FTP-72 cycle (max. acceleration for all phases is 1.48 m/s^2)

FTP-72	Duration (s)	Distance (m)	Maximum speed (km/h)	Average driving speed (kmh/h)	Idling phase (%)	RPA (m/s^2)
Phase 1	505	5779	91.2	50.6	18.6	0.181
Phase 2	867	6211	55.2	31.2	17.3	0.190
Whole cycle	1372	11,990	91.2	38.3	17.8	0.185

specifications of the two segments and of the whole cycle. Ambient temperature range is 68–86 °F (20–30 °C) and absolute humidity 75 grains of water per pound of dry air; the vehicle is allowed to soak for at least 12 h before the test.

Overall, maximum speed during the cycle reaches 91.2 km/h, total traveled distance is 12 km with 17.8 % of the time spent idling. Interestingly, the highest speeds in the cycle are experienced during the first minutes after cold start, a fact that influences decisively the warming-up of the catalyst, and is certainly not representative of other areas/cities or routes. Figure 2.17 highlights the speed/acceleration distribution of the cycle.

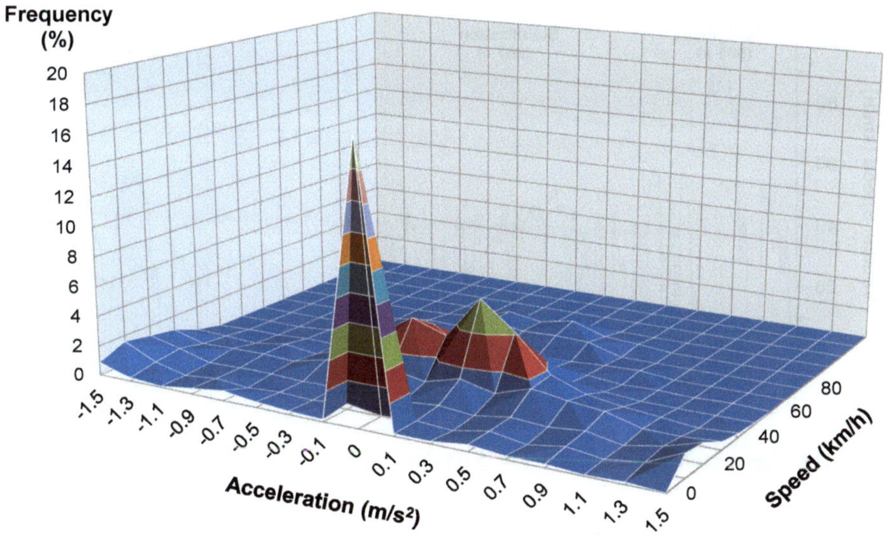

Fig. 2.17 3D speed/acceleration frequency distribution of the FTP-72 driving cycle

The FTP-72 is the applicable cycle for battery electric vehicles testing according to the SAE J1634 standard. Following the standard, the battery is fully charged, the vehicle is parked overnight, and then the following day it is driven over successive UDDS cycles until the battery becomes discharged (and the vehicle can no longer follow the cycle). After that, the battery is recharged from a normal AC source and the 'city' energy consumption of the vehicle is determined (in kW-h/mile or kW-h/100 miles) by dividing the kW-h of energy to recharge the battery by the miles traveled by the vehicle [40].

A subsequent 'variation' of the FTP-72 cycle was the FTP-75 (also known as EPA75), which is based on the FTP-72 adding a third 'transient' phase of 505 s exactly as the first (cold-started) one, but this time hot started; this extended version is demonstrated in Fig. 2.18.

The FTP-75 has been the 'primary' cycle used in the U.S. for testing the compliance of passenger cars and light-duty trucks with the emission standards, beginning with the 1975 model year. Diesel-engined vehicles were included as of model year 1975 for passenger cars and 1976 for LD trucks, with particulates measured from 1982 [2, 36]. It is important to note that during the FTP-75, the third phase starts after the engine has stopped for 10 min. This intends to simulate parking a car and then returning to it after a short period. Thus, emissions are collected during the cycle not only after a cold start (as was the case with the FTP-72 and the European NEDC) but also after a hot one. By so doing, the test provides a more accurate reflection of typical driving experience than running just

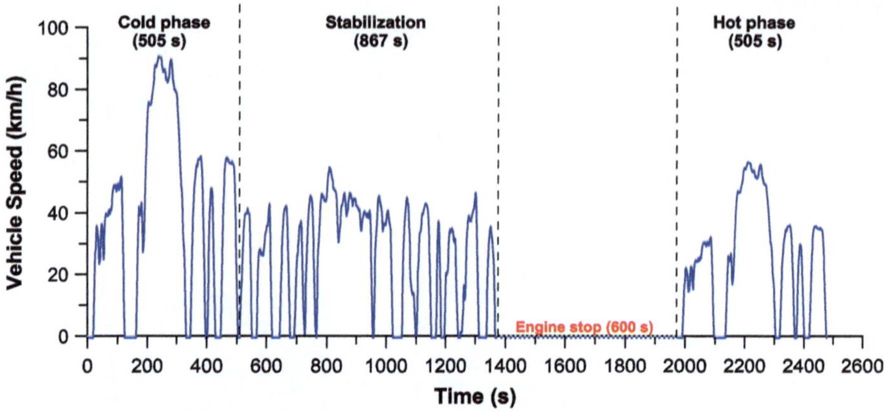

Fig. 2.18 Speed profile of the U.S. FTP-75 driving cycle

one 12-km cycle from a cold start [41]. The emissions from each phase are collected in a separate Teflon bag, analyzed and expressed in g/mile, with weighting factors 0.43 for the cold-started phase, 1.0 for the 'stabilization' and 0.57 for the hot-started one.[9] According to 40 CFR 86.544-90, the final reported 'weighted' distance-related emission result E_{wi} of pollutant i (gaseous or PM, and CO_2), is calculated as follows

$$E_{wi} = 0.43\left(\frac{E_{i-ct} + E_{i-s}}{S_{ct} + S_s}\right) + 0.57\left(\frac{E_{i-s} + E_{i-ht}}{S_s + S_{ht}}\right) \quad (2.2)$$

where E_{i-ct}, E_{i-s} and E_{i-ht} the emission of pollutant i during the cold transient phase, stabilization phase and hot transient phase respectively (Sect. 6.5) in grams per test phase, and S_{ct}, S_s and S_{ht} the respective distances ($S_{ht} = S_{ct}$).

For hybrid vehicles, the stabilization phase is run again after the hot phase, for a total duration of 2744 s. For the cold CO testing, adopted in 1992 for gasoline vehicles, the FTP-75 is also run at 20 °F (approximately −7 °C).

Table 2.10 summarizes some characteristics of the cycle phases.

Among the problems noted in the FTP cycle (besides the relatively high speeds at the beginning of the test, which accelerate the catalyst warming), the underestimation of acceleration and cruise activities between 64 and 80 km/h and above 96 km/h, the underestimation of the time spent in cold transient mode and thus the

[9]During work performed in the late 60s within the APRAC (Air Pollution Research Advisory Committee) project CAPE-10 of the Coordinating Research Council (CRC), it was found that vehicles in Los Angeles were used on average for 4.7 trips per day. From these, two were cold-started (one in the morning) and the rest hot-started. Hence, the cold weighting factor resulted as 2:4.7 = 0.43, and the hot one 2.7:4.7 = 0.53. These factors became part of the test procedure as of model year 1975 [2].

Table 2.10 Specifications of the FTP-75 phases

FTP-75	Duration (s)	Description	Distance (m)	Maximum speed (km/h)	Weighting factor
Phase 1	505	Cold start	5779	91.2	0.43
Phase 2	867	Stabilized	6211	55.2	1.00
Engine stop (600 s)					
Phase 3	505	Hot start	5779	91.2	0.57

emissions, and the overestimation of the time at stop and at cruise between 40 and 56 km/h have been identified [42]. Since the cycle was based on vehicle tests from the late 60s, and following the limited capabilities of the dynamometers at the time, maximum acceleration (and the level of accelerations in general) cannot be considered representative of modern real-world driving.

2.2.3 Highway Fuel Economy Test—HFET

In the 70s, EPA began to publish city fuel economy values based on the FTP cycle (the official fuel economy labeling program began in 1975). This was largely initiated by the 1973 disruption of U.S. oil imports caused by an oil embargo of the OPEC members. The need for a highway cycle became evident at some point, as some manufacturers, complaining about the lack of non-urban fuel economy reports, began advertising 'highway' fuel economy figures based on their own tests.

This new highway cycle was designed with the aim to [43, 44]:

- reflect driving on a variety of non-urban roads;
- be self-weighting (i.e., have the correct proportion of travel on each road type);
- be of a length equal to the average trip in a non-urban area;
- be appropriately transient; and
- have an average speed and number of stops per mile equal to that experienced in non-urban driving.

Unlike with the development of the LA4, which was a minimally shortened version of one particular trip, the developed highway cycle was a composite one; it comprised four different road types, namely locals, collectors, minor and principal arterials. These were selected from different trips made by three drivers of a single vehicle following other cars but also 'flowing along' with traffic over 1700 km of non-urban roads. The data was collected in southern Michigan (Ann Arbor), northern Ohio, and Indiana. In particular, the principal arterial data was collected in the Ohio area only, which had a strictly enforced 55 mph speed limit (drafted in response to oil supply disruptions and price spikes during the 1973 oil crisis). The routes traveled during the development of the cycle did not contain the target mix of operation. Thus, when the shortened version of the cycle was constructed, the length of the speed/time trace from each type of road was selected in such a way to achieve the target mix. Additionally, each segment of the speed/time trace was selected to match the average characteristics of all operation on the same road type in terms of average speed, major and minor speed deviation and stops per km [36, 43, 44]. The resulting drive cycle is the Highway Fuel Economy Test (HFET, also known as HWFET or FET), developed in 1974 and demonstrated in Fig. 2.19; in this figure, the segments corresponding to each of the four road types are also

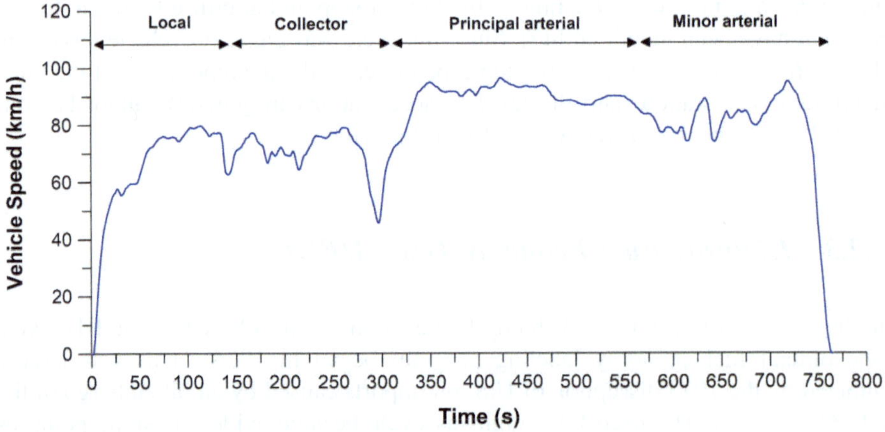

Fig. 2.19 Speed profile of the U.S. HFET cycle (40 CFR 600, App. I)

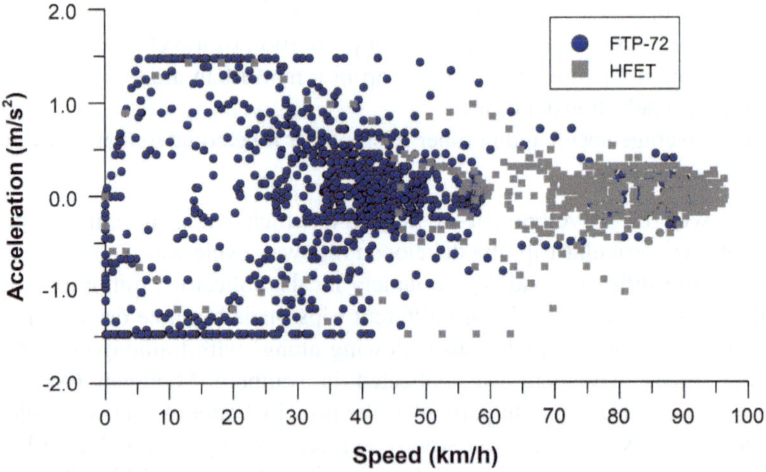

Fig. 2.20 Speed/acceleration distribution of the urban FTP-72 and highway HFET driving cycles (notice the strictly defined maximum acceleration and deceleration values at ±1.48 m/s^2 (=3.3 mph/s) imposed by the capabilities of the Clayton dynamometers at the time)

identified. Further, Fig. 2.20 compares the speed/acceleration distribution between the FTP-72 and the HFET.

Overall, HFET simulates interstate highway and rural driving conditions, with the purpose of a highway fuel economy test (urban fuel consumption is calculated during the FTP-75). The cycle lasts 765 s (approx. 13 min) for a total traveled distance of 16.5 km, with maximum speed 96 km/h (60 mph). There is a steep acceleration at the beginning (indicative of the vehicle entering a highway), and after that speed is almost constantly maintained at values higher than 70 km/h, hence the 78 km/h average driving speed. The initial version of the cycle had accelerations up to 2.2 m/s^2 (4.9 mph/s), which caused tire slip on dynamometers with belt-driven inertia weights. Consequently, the first 10 and the last 20 s of the cycle were modified so as to limit the acceleration and deceleration to lower than 1.48 m/s^2 (3.3 mph/s), as was also the case with the LA4. Although hard acceleration events are well known to significantly affect emissions and fuel consumption, EPA contended that owing to their very small duration, these segments did not affect the total fuel economy results during the cycle. Furthermore, two seconds of idle were added at the beginning and the end to account for the portion of idle operation the analysis indicated would be experienced in this length of non-urban driving (idle time less than 1 %) [44].

The cycle is run twice, with the first run serving to warm-up the engine, so as sampling takes place during the second run with hot engine; a break of 15 s is allowed between the two runs. Alternatively, it is run once immediately after the FTP-75 (40 CFR 1066.840). Although the cycle is named 'highway', nowadays, with the introduction of a true high-speed cycle (the SFTP US06—Sect. 2.2.4), HFET practically represents in the U.S. EPA fuel economy program, driving on lower-speed highways as well as rural and suburban driving.

Interestingly, although EPA developed the HFET, it was the California ARB that initially implemented it in an attempt to investigate whether manufacturers employed control systems that would fail to limit NO_x emissions during extended high-speed operation, outside the range of the FTP-72 [36].

As already mentioned, the primary objective of the HFET is for fuel economy purposes (miles per gallon—mpg, in effect, the inverse of fuel consumption). Up to MY 2007, fuel economy was estimated from FTP-75 and HFET values weighted 55 and 45 % respectively. Equation (2.3) provides the initial (up to 1984) formula

$$\text{EPA combined mpg}_{\text{until 1984}} = \frac{1}{\frac{0.55}{\text{FTP mpg}} + \frac{0.45}{\text{HFET mpg}}} \tag{2.3}$$

Since the late 70s, this fuel economy value has been indicated on a window sticker on the vehicle. Further, it was used for the manufacturers' corporate average fuel economy (CAFE), a program managed by NHTSA (National Highway Traffic Safety Administration). The CAFE achieved by a given fleet of vehicles in a given model year is the production-weighted harmonic mean fuel economy. CAFE was established following the Energy Policy and Conservation Act of 1975, and became effective as of model year 1978 (FR Vol. 42, No. 176, September 12, 1977).

From 1985, adjustment factors of the order of 90 % to the FTP and 78 % to the HFET values were adopted (only for the window sticker—not for CAFE purposes) to account for real-world driving effects such as road, vehicle condition, tire pressure, weather etc. [45]

$$\text{EPA combined mpg}_{1985-2007} = \frac{1}{\frac{0.55}{0.90 \times \text{FTP mpg}} + \frac{0.45}{0.78 \times \text{HFET mpg}}} \tag{2.4}$$

Beginning with model year 2008, a new formula for fuel economy has been adopted, based now on a 5-cycle procedure, that will be discussed in the next section. Formulas for calculation of the FTP and HFET mpg values will be provided in Sect. 6.5.

Apart from fuel economy purposes, and beginning with the Tier 3 standard, the HFET is also used for emission testing. Specifically, the new standard requires that $NMOG+NO_x$ emission limits should be met on the HFET as well apart from the FTP-75 [46].

The HFET cycle, together with the UDDS, are the applicable cycles for battery electric vehicles testing according to the SAE J1634 standard. The procedure described earlier for the UDDS test is applied with the vehicle running on the HFET to calculate highway energy consumption [40].

Figure 2.21 illustrates the development of some characteristic engine and vehicle parameters during the HFET for a turbocharged, diesel-engined large vehicle. Owing to the exclusively motorway pattern, the engine in Fig. 2.21 is kept at relatively high loads, as indicated by driver demand or 'fuel rack' varying between roughly 0.6 and 1.0 during most of the cycle. Consequently, engine speed is also relatively high; it is close to 2000 rpm during the first 300 s of the cycle, while during the high-speed part of the cycle it fluctuates between roughly 2300 and 2400 rpm (2400 being the rated-power engine speed). The upper sub-diagram of Fig. 2.21 shows that turbocharger speed follows these trends, hence during high speed driving, the rotor speed approaches 100,000 rpm. The most critical transient, besides the one at launch, occurs at t = 295 s, when the high speed of the cycle is initiated, and this is expected to produce the highest amount of engine-out soot/PM [47].

2.2.4 *Supplemental FTP US06 and SC03*

Section 206(h) of the Clean Air Act, as amended in 1990, requires that EPA consider actual driving conditions under which motor vehicles are used, including conditions relating to four areas: fuel, temperature, altitude, and acceleration. To this end, on October 22, 1996, the final rule on 'Motor Vehicle Emissions Federal

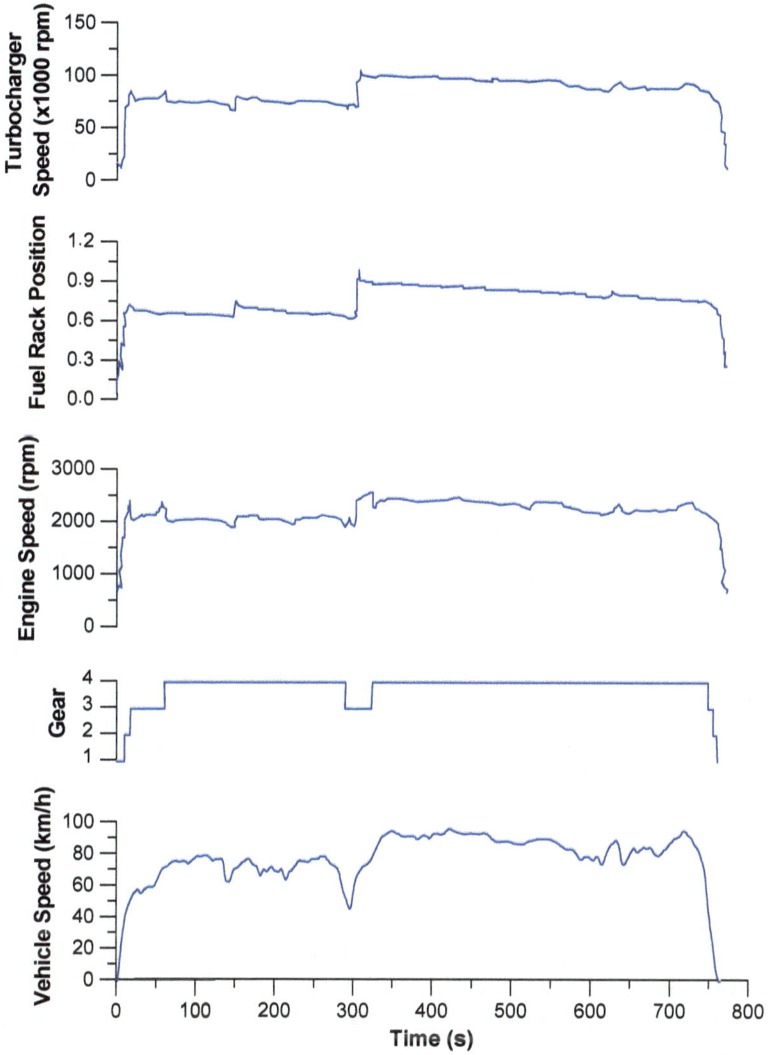

Fig. 2.21 Development of engine and vehicle parameters during the U.S. HFET driving cycle (adapted from [47])

Test Procedure; Final Regulations' was published in the Federal Register [48]. The primary new element of the rulemaking was a Supplemental Federal Test Procedure (SFTP) designed to address shortcomings with the FTP-75 cycle, some of which were mentioned in Sect. 2.2.2. The basis for the SFTP rulemaking was two studies of real-world driving conducted during the early 90s. From the studies, two new driving cycles were legislated, the US06 and the SC03. These cycles were not meant to replace the FTP-75 procedure but rather expand it to also include engine speed/load conditions not covered during the testing of vehicles but experienced by drivers on the road. It is important to note that by the time the SFTP cycles were legislated, dynamometer technology had improved, and higher acceleration rates were possible. Thus, these new drive cycles were not limited to acceleration rates of 1.48 m/s^2 as was the case with the LA4 in the early 70s.

The first study was conducted (initially) in two U.S. cities, Baltimore, MD and Spokane, WA, using a combination of instrumented vehicles and chase cars [41]. Baltimore was chosen to represent a northeast, medium-sized, ozone non-attainment area. Spokane was selected being typical of a CO non-attainment area (a suitable mid-west city could not be found). Overall, 215 vehicles in Baltimore and Spokane were instrumented with three-parameter data logger packages that recorded second-by-second engine speed, vehicle speed and inlet manifold vacuum; 79 more vehicles were equipped with six-parameter data loggers, also recording coolant temperature, throttle position and air-fuel ratio downstream of the catalyst. These 294 instrumented vehicles were observed for a period of one week. On the whole, approximately 1700 h of data was collected. The final count of vehicles for which the data was good was eventually reduced to 217 (168 three-parameter and 59 six-parameter vehicles). A separate chase-car study collected similar speed data in the two cities using a laser device mounted on a patrol car that tracked in-use target vehicles. This produced relatively short sequences of data but on a much larger sample of vehicles (the number of targets from both cities was 1641, and, in total, approximately 40 h of target time was collected). In addition to Baltimore and Spokane, two other cities had already been selected for related studies, namely Atlanta, GA (101 instrumented vehicles) and the South Coast Air Basin around Los Angeles, CA (chase-car work performed on 102 routes) [41]. It should be noted that both studies were conducted prior to the 1995 repeal of the federal 55 mph freeway speed limit.

From the collected data in Baltimore, it was found that driving was characterized by (much) higher speeds and steeper accelerations than the FTP and the HFET. More specifically [41]:

- About 8.5 % of all speeds exceeded the FTP maximum (only in part attributed to newer, higher-powered cars).
- 2.5 % of all driving in Baltimore exceeded the FTP maximum acceleration of 1.48 m/s^2.
- Whereas the proportion of time spent in the four fundamental operating modes (idle, acceleration, constant speed and deceleration) only slightly differed from the FTP-75, the average trip length was found shorter than the 12 km of the LA4, on average 7.9 km. This means that much higher proportion of the overall driving was actually performed with the engine during the warming phase, i.e., before the after-treatment devices reach their normal operating temperature.
- The soak periods (time between the end of a trip and the beginning of the next), were much different too. The LA4 soak periods are 12–36 h before the test and 10 min before the beginning of the third phase. In contrast, almost 40 % of all soak periods in Baltimore were between 10 min and 2 h, which is the most critical period concerning catalyst cool-down. This poses another potential

emission concern as catalysts cool off much faster than engines, and most are completely cold in approximately 1 h. Hence, the implicit assumption of the FTP that 57 % of all starts occur with the catalyst hot was not confirmed (30 % was the new figure found). The other implicit assumption of the FTP that 43 % of the starts are cold was not confirmed too but this time the new figure was found lower (around 25 %) (in any case, it is possible that these two counter-effecting properties might cancel themselves out).

All in all, a significant (18 %) part of driving in Baltimore was found to occur outside the speed/acceleration distribution of the FTP drive schedule, highlighting the fact that the LA4 doesn't (any more) realistically account for the emission effects of real-world aggressive driving behavior and high acceleration rates, all of which contribute significantly to vehicle emissions. Subsequent analysis on the larger 3-city (Baltimore, Spokane, Atlanta) instrumented vehicle database showed the results to be consistent with the Baltimore ones; the 3-city analysis showed that nearly 13 % of vehicle operation time occurred at combinations of speed and acceleration that fell outside the matrix of those found on the LA4 [49].[10]

Following the data collection from these driving surveys, three cycles were initially developed by the EPA, based on selection of actual segments of in-use driving which best matched the joint distribution of speed and acceleration (primarily from the Baltimore database): the start driving (ST01) cycle, the aggressive driving REP05 and the Remnant cycle REM01. The ST01 cycle represented driving immediately following engine start-up; REP05 represented high speed and aggressive driving; the REM01 cycle represented all other driving [49].

For the Start cycle (Fig. 2.22), three target surfaces were developed from the database, representing three successive 80-s segments of in-use driving immediately following the initial idle. The combinations of speed and acceleration found in these distributions could largely be found in the FTP-72, but in a different sequence and with different percentages. The micro-trips that produced the best fit to these surfaces, together with an initial idle period that best matched in-use initial idles, generated a start cycle that was 258 s long [49].

The REP05, on the other hand (illustrated in Fig. 2.22, with basic parameters provided in Table 2.11), targeted speeds and accelerations, as well as micro-transient effects, not covered by the LA4. Thus, the in-use data points used in developing the REP05 target surface were those with combinations of speed and acceleration that were not represented on the LA4 cycle ('non-LA4') and, in addition, were not part of the ST01 target surfaces. These points tended to be either high-speed or high-acceleration, or both. By assembling the cycle from actual idle-to-idle driving segments, however, the cycle necessarily included some

[10]Results from a later (2004–2005) survey conducted by the EPA in Kansas City confirmed these findings. Based on instrumented vehicle studies in Kansas City and chase car in California, it was found that 28 % of driving (vehicle miles traveled) is at speeds greater than 60 mph. About 33 % of real-world driving was found outside of the FTP/HFET speed and acceleration activity region [49, 50].

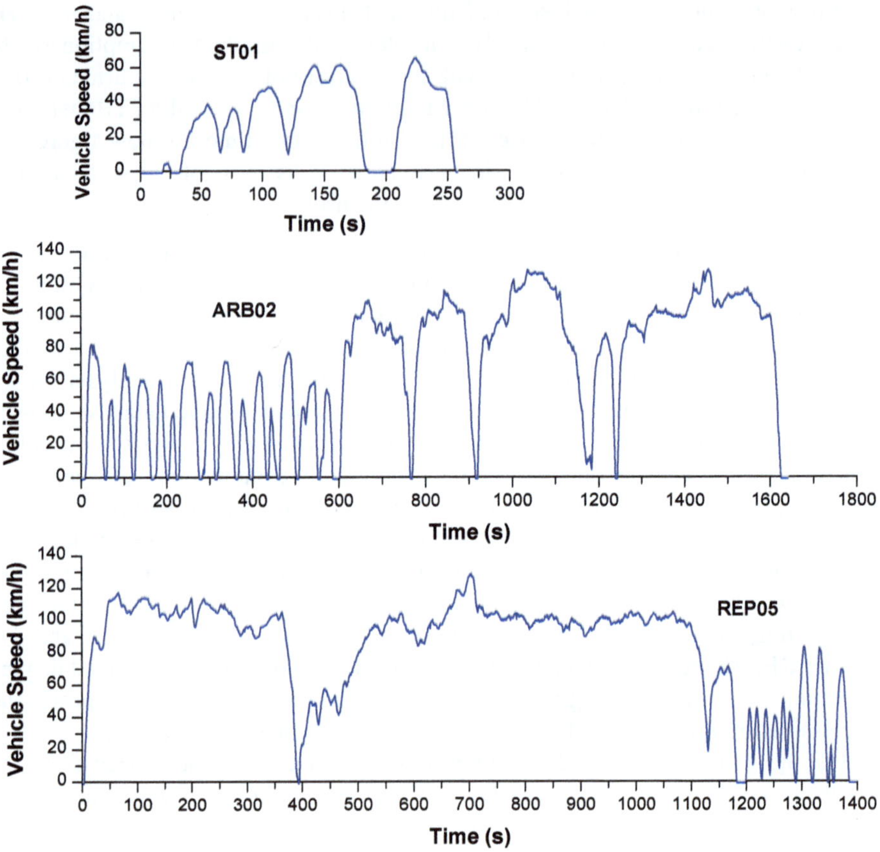

Fig. 2.22 EPA cycles REP05 and ST01, and California ARB02 from the SFTP research

Table 2.11 Summary of technical specifications of the ST01, REP05 and ARB02 cycles

Cycle	Duration (s)	Distance (m)	Maximum speed (km/h)	Average speed (km/h)	Maximum accel. (m/s^2)	Idling time (%)	RPA (m/s^2)
ST01	258	2234	66.0	31.2	2.28	16.7	0.263
REP05	1400	32,065	129.2	82.9	3.79	2.8	0.160
ARB02	1640	31,900	129.2	70.0	3.53	6.5	0.216

speed/acceleration combinations that were represented on the LA4, amounting to about 30 % of the cycle's 1400 s [49].

The Remnant/REM01 cycle was intended to represent the balance of in-use driving not already covered by ST01 or REP05 (hence, the combination of ST01, REP05 and REM01 should cover the full range of in-use driving activity). Thus, the Remnant target surface was obtained by using the remaining speed/acceleration distribution after subtracting that found in the REP05 and ST01 cycles [49].

Additionally, the ARB02 cycle was developed by the California ARB based on data from their Los Angeles chase-car study (illustrated in Fig. 2.22 too, with basic parameters given in Table 2.11). The purpose of this cycle was, again, to test vehicles over in-use operation outside the boundary of the LA4, including extreme in-use driving events. Additionally, the HL07 engineered cycle was developed by EPA in coordination with auto manufacturers. The purpose of this cycle was to test vehicles on a series of acceleration events over a range of speeds, so severe that most vehicles will go into WOT operation [49].

In order to address the vital shortcoming of the LA4, most importantly the absence of aggressive driving, the US06 cycle was the cycle that was eventually developed, illustrated in Fig. 2.23. The cycle was constructed from parts taken from the REP05 and ARB02. The US06 targets specific high emission non-FTP operation, and at the same time is based on actual segments of in-use driving. The cycle encompasses the range of non-FTP driving operation with all the most severe accelerations, incorporating also high speed cruise operation. Further to the above, the severe driving events contained in the US06 also provide for some control over the emission impact of road grade [49].

Overall, the supplemental FTP US06 addresses the need for aggressive, high-speed and/or high-acceleration driving behavior, rapid speed fluctuations and driving behavior following startup. Maximum speed is 129.2 km/h (compared to 91.2 for the LA4), whereas maximum acceleration reaches 3.75 m/s^2 or 13.5 km/h/s (1.48 m/s^2 for the LA4, and only 1.04 m/s^2 for the European NEDC). The cycle lasts 10 min and the vehicle covers a distance of 12.9 km. The structure of the cycle is as follows:

- The first two micro-trips in Fig. 2.23 (t = 0–130 s) correspond to the first 'city' part;

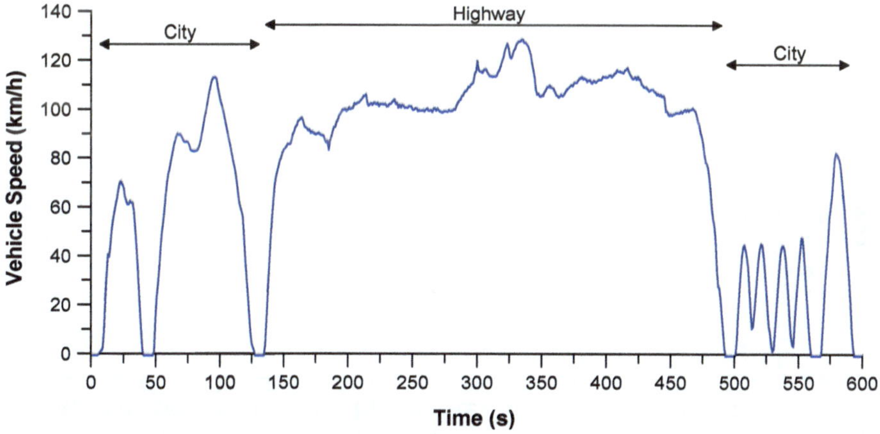

Fig. 2.23 Speed profile of the supplemental US06 (40 CFR, Pt. 86 App. I); the rule includes adjustments for low-performance vehicles

- The third micro-trip (approx. 6 min duration from t = 131–495 s) to the 'highway' part (taken from the ARB02 cycle as evidenced from Fig. 2.22); and
- The last micro-trips (t = 496–600 s) form the second 'city' part.

The US06 is run with the vehicle in the hot stabilized condition (engine and catalytic converter have reached typical operating temperatures by some type of preconditioning, e.g., running the LA4 cycle). The US06 exhaust emissions test procedure is described in 40 CFR 86.159 and 40 CFR 1066.831. Initially, exhaust sampling was a single-bag process but this changed later, as will be discussed in the next paragraphs. Indicative of the harshness of the cycle is Fig. 2.24, which compares the acceleration profiles between the FTP, HFET and the US06 during the first 300 s. Further, Fig. 2.25 supports the much broader profile of the US06 compared to the FTP by demonstrating the speed/acceleration distribution of the two cycles.

During the SFTP research, a second study was commissioned, this time focusing on the effects of the air-conditioning system on emitted pollutants. The air conditioner does not run during the FTP; instead, dynamometer load is increased by 10 % to simulate the average nationwide, year-around air-conditioning effects. Apparently, manufacturers did not optimize for low emissions with the air conditioner on, since they knew that the vehicle will not be tested in that condition. For the survey performed by an EPA contractor in Phoenix, Arizona in 1994, 20 vehicles were instrumented with data loggers for periods of 1–2 weeks, and data was collected on several air conditioner and trip parameters; data was recorded for more than 1000 trips [48]. It was concluded that engine-out NO_x emissions almost doubled with the use of air conditioners (further, fuel consumption increased by 20 %) [51].

Following the findings from this study, two cycles were considered at some point to be used as a supplemental air-condition test, namely the 568-s SC01 and

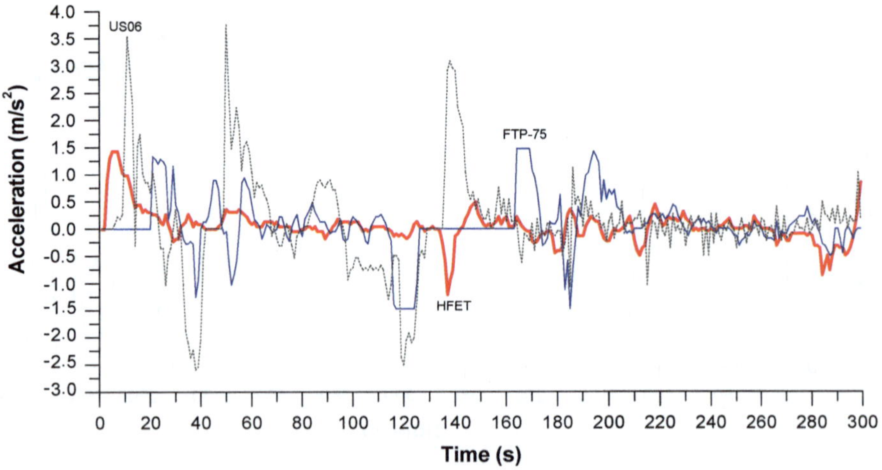

Fig. 2.24 Comparison of the acceleration profiles between the FTP, HFET and US06 cycles

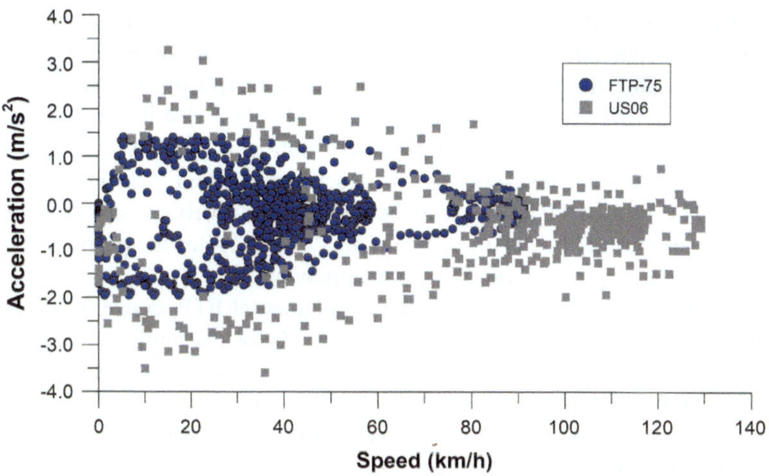

Fig. 2.25 Comparison of the speed/acceleration distribution between the FTP and the US06 (the FTP shows denser distribution at low speeds since it primarily focuses on urban driving)

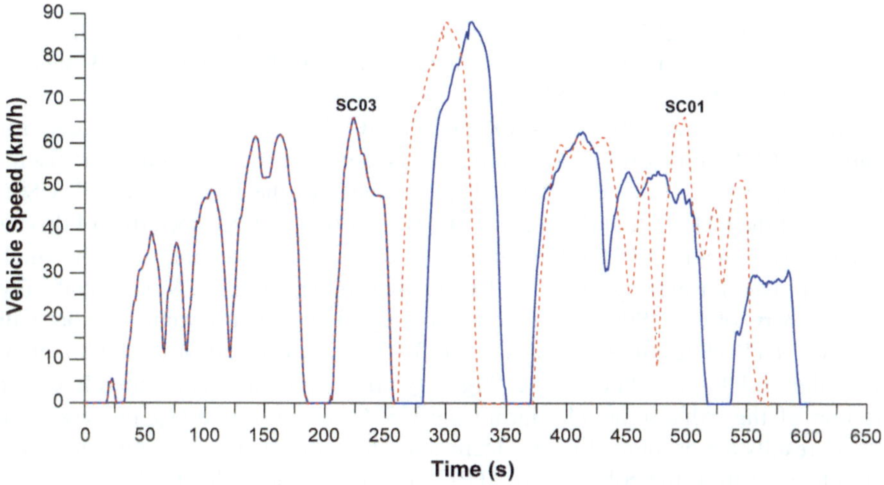

Fig. 2.26 Supplemental SC03 cycle (discontinuous line demonstrates the SC01 cycle, which was initially considered); the first 258 s of the SC0 cycles is the ST01 cycle (Fig. 2.22)

the AC866; the latter being essentially the stabilization phase (bag 2) of the LA4. In the end, the SC03 was the cycle developed to address the need to also test the engine load and emissions associated with the use of the air-conditioning system; the cycle is illustrated in Fig. 2.26.

As with the US06, the SC03 lasts 10 min, with the traveled distance 5760 m; maximum speed is 88.2 km/h (average driving speed is 42.3 km/h), while 18.3 %

of the time is spent idling. The cycle is run at 35 °C (95 °F) ambient temperature under high sun-load and high humidity. The first 258 s of the SC03 cycle is taken from the ST01 discussed previously (Fig. 2.22). The rest of the cycle was developed in such a way that the overall speed/acceleration distribution of the SC03 matched the remnant speed/acceleration distribution as closely as possible. The SC03 exhaust emissions test procedure is described in 40 CFR 86.160 and 40 CFR 1066.835. The two SFTP cycles can be run consecutively to save on preconditioning and setup time; however, separate runs of the cycles are permitted with the appropriate soak or preconditioning steps appended.

The US06 and the SC03, forming the supplemental FTP, concern all passenger cars and light-duty trucks (GVWR lower than 8500 lbs) beginning with model year 2000 and fully phased in by 2004. Beginning with the Tier 3 standard, MDPVs of Class 2b (GVWR between 8500 and 10,000 lbs), such as large SUVs, which are chassis certified, will also be tested on the US06 and the SC03, or, if their PMR is lower than 0.025 HP/lb (40 kW/t), the highway segment of the US06 and the SC03 [46]. The emission value for certification of pollutants is calculated as a weighted composite value of emissions on the FTP-75 and the two SFTPs, namely

$$0.35 \times \text{FTP-75} + 0.28 \times \text{US06} + 0.37 \times \text{SC03} \tag{2.5}$$

For the case a vehicle is not equipped with an air-condition system, the SC03 results in the above formula are substituted with the FTP-75 ones. Tier 3 standard has also legislated separate PM limit to be met on the US06 [46].

Since model year 2008, the US06 and the SC03 are also used for the determination of EPA's on-road fuel economy ratings using the 5-cycle method [replacing the 2-cycle results from Eq. (2.4)]. The 5 cycles are the FTP-75, HFET, US06, SC03 and the FTP-75 'cold' run at −7 °C. By so doing, a large spectrum of driving conditions is included in the fuel economy determination, in contrast to the single test employed for many years in the EU on the simplistic NEDC. More specifically, the city parts of the US06 contribute to the 'city' fuel economy value together with the 'hot' FTP-75 results from bags 2 and 3. The highway part of the US06, along with the HFET, contribute to the 'highway' fuel economy value (therefore, collection of the exhaust sample during the US06 is now a two-bag process). The SC03 results are accounted for by increasing the previous results. Separate values are obtained from the cold-started FTP-75 phase and the 'cold' FTP-75 at −7 °C. Weighting factors are used for the contribution of each cycle or each part of the cycle. Values obtained from the calculation are further modified to account for road roughness and grade, wind, tire pressure etc. through a 9.5 % downward adjustment [52]. Details about the exact calculation algorithm are available in 40 CFR 600.114. The results are published on the window sticker of new vehicles sold in the United States, as illustrated in Fig. 2.27. Obviously, fuel economy values, although not fully compatible with the daily driving experience, are much more cohesive compared to the ones calculated in Europe, and by extension, the respective gap between real world and certification values much smaller [16].

Fig. 2.27 U.S. EPA fuel economy window sticker for gasoline vehicles [53]

The decision to include US06 and SC03 (and 'cold' FTP-75) results on the fuel economy estimates was substantiated by studies which revealed that on average: (a) air condition operation at 35 °C reduced fuel economy by about 21 %, (b) fuel economy over the US06 cycle was almost 30 % lower than over the composite FTP and HFET, and c) fuel economy over the cold FTP-75 was about 12 % lower than over the standard FTP [52].

Tables 2.12 and 2.13 summarize some important technical specifications of the U.S. certification cycles.

It is without doubt that the suite of cycles employed for tailpipe emission and fuel economy testing in the United States, namely the FTP-75, HFET, US06 and SC03, encompass a considerable amount of real driving activity, much broader compared with the European NEDC (or the Japanese cycles, which will be

Table 2.12 Specifications of U.S. certification drive cycles

Cycle	Data collection		Engine temp.	Laboratory temp. (°C)	A/C
	Method	Year			
FTP-72/75	Instrumented vehicle on a specific route	1969	Cold	20–30	Off
HFET	Chase car	Early 70s	Hot	20–30	Off
SC03	Instrumented vehicles	1992	Hot	35	On
US06	Instrumented vehicles and chase car	1992	Hot	20–30	Off

Table 2.13 Summary of technical specifications of U.S. light-duty vehicles cycles (all are transient except for the modal California 7 mode)

Cycle	Duration (s)	Distance (m)	Maximum speed (km/h)	Average speed (km/h)	Maximum accel. (m/s^2)	Idling time (%)	RPA (m/s^2)
Cal. 7 Mode	137	1358	80.5	35.7	0.98	14.6	0.241
FTP-72	1372	11,990	91.2	31.5	1.48	17.8	0.185
HFET	765	16,507	96.4	77.7	1.43	0.5	0.071
NYCC	600	1898	44.6	11.4	2.68	32.2	0.345
SC03	600	5761	88.2	34.6	2.28	18.3	0.218
US06	600	12,888	129.2	77.3	3.76	6.5	0.222
Cal. LA-92	1735	17,706	108.1	36.7	3.08	17.5	0.250

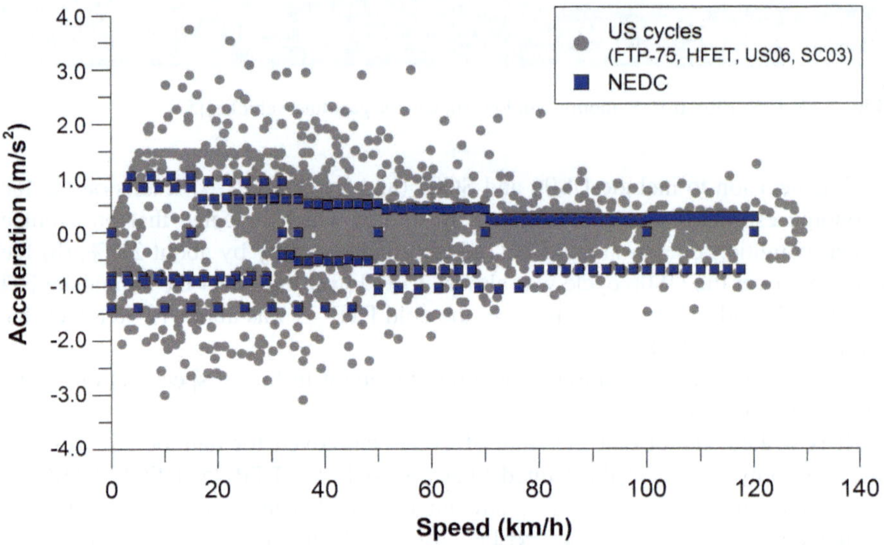

Fig. 2.28 Comparison of the speed/acceleration distribution between the U.S. and European (NEDC) cycles employed for emission certification testing of new light-duty vehicles; obviously, with the adoption of the WLTC cycle in Europe from 2017, the differences will be much smaller

presented later); Fig. 2.28 characteristically illustrates this. Although the U.S. test procedure is more costly and time lengthy, as many cycles need to be run, it is much more compatible with the daily driving activity, which is the single most important virtue a certification procedure must possess.

2.2.5 New York City Cycle—NYCC

Although the LA4 has been the primary cycle for vehicle certification in the U.S., it was acknowledged from the beginning, based on relevant observations, that it would not represent the worst-case scenario for sulfate emissions from catalyst-equipped vehicles. Under certain operating conditions, exhaust catalysts are capable of storing sulfur oxides as aluminum sulfate, which they release later on. It is therefore necessary to stabilize the vehicle and condition of the catalyst prior to emission measurement [36] (obviously, the continuous desulfurization of fuels has significantly diminished this risk). To deal with this, EPA developed in the 70s specialized chassis-dynamometer cycles.

One was the congested freeway driving schedule (CFDS), also known as sulfate emission test (SET) No. 7 or CUE (crowded urban expressway). The, now abandoned, highly transient cycle, is schematically represented in the upper sub-diagram of Fig. 2.29. It has 1398 s duration, covering a distance of 21.7 km, with 56 km/h average speed; as with the FTP cycles, maximum acceleration is 1.48 m/s^2 [54].

Another schedule was the New York City Cycle (NYCC). Oddly, very little is public knowledge regarding the construction of the NYCC. Apparently, the cycle

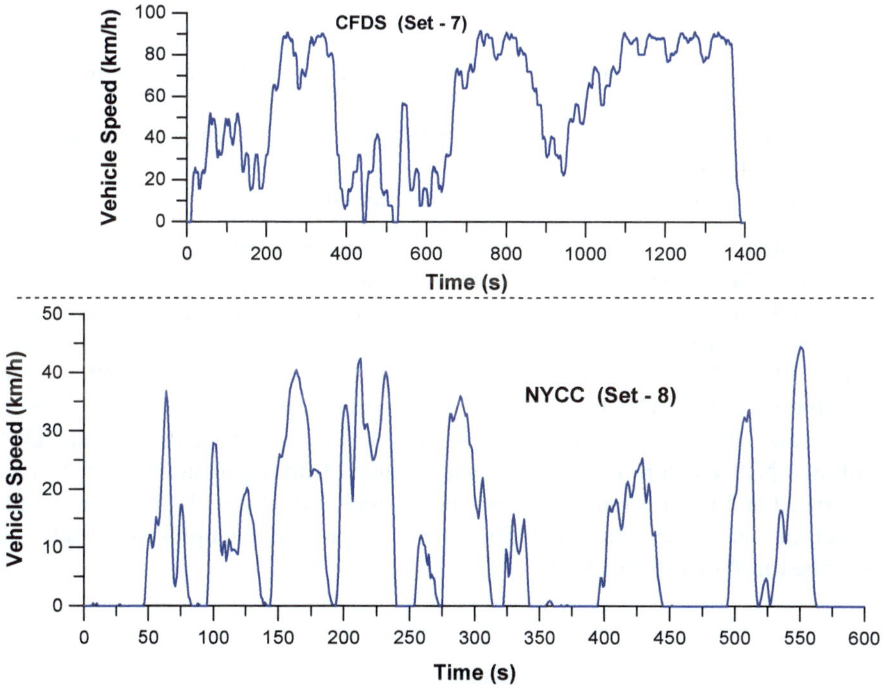

Fig. 2.29 Speed profile of the New York City cycle (40 CFR 86, App. I), also known as SET-8; the SET-7 congested freeway driving schedule—CFDS is depicted in the upper sub-diagram

was originally developed in 1975 by two employees of the New York City Department of Environmental Protection (DEP), with the aim to represent driving in highly congested urban traffic, typical of the city of New York, i.e., with frequent stops and idling. Speed/time data from midtown Manhattan was used in a stochastic model to generate speed/time tables for testing purposes. Vehicle acceleration and deceleration halves of the velocity/acceleration plane were treated separately to reflect vehicle differences implied by these mathematically inverse, but operational distinct, modes [55]. The cycle was often referred to in the 70s and 80s as SET No. 8, and is illustrated in the lower sub-diagram of Fig. 2.29. Duration of the New York City cycle is 598 s, with the traveled distance a little less than 2 km (1898 m). Maximum speed is 44.6 km/h but the average driving speed is only 16.8 km/h. If the extended (32 % of the total time[11]) idling period is accounted for, the average vehicle speed drops to 11.4 km/h. Owing to frequent and steep accelerations commencing from low speeds, this cycle exhibits a very high relative positive acceleration value of 0.345, indicative of its dynamic profile. Maximum acceleration is almost double than that of the FTP, at 2.68 m/s^2 (9.65 km/h/s).

As mentioned earlier, the New York City cycle was initially employed for (sulfate) emission testing in various research projects. Following the Clean Air Act Amendments of 1990, Section 202(k), the cycle formed an element of the HC evaporative emissions running losses test (FR Vol. 58, No. 55, March 24, 1993). Evaporative emissions, being fundamentally different from exhaust emissions, are tested:

- when the vehicle is stationary, after heating the fuel tank to simulate heating by the sun (the diurnal test);
- after the car has been driven and parked with a hot engine (the so-called hot-soak test);
- when the vehicle is driven, the so called running losses.

All these tests apply to gasoline or methanol-engined vehicles (not gaseous-fueled ones), with the whole test procedure requiring five days to be completed. For the running losses test, conducted at 35 °C, the New York City cycle is employed together with the UDDS/LA4 in the following sequence (Fig. 2.30): UDDS—2-min idle—NYCC—NYCC—2-min idle—UDDS—2-min idle. Initially, three UDDS cycle runs were considered (as proposed by General Motors and adopted by CARB [56]), but EPA decided to replace the second UDDS with two NYCCs to provide for a broader range of driving patterns, that is, more low-speed and idle operation. The test began with model year 1996 and was fully phased-in with model year 1999 [56]. The exact procedure employed nowadays is described in 40 CFR 86.134-96.

[11] 32 % of the total time is spent with the vehicle at zero velocity and acceleration; for 41 % of the time during the NYCC, vehicle speed is lower than 1 km/h.

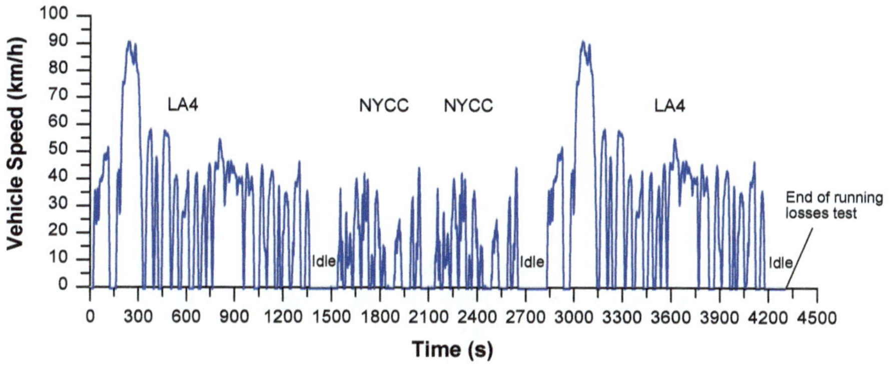

Fig. 2.30 Dynamometer schedule for HC evaporative emissions running losses measurement (40 CFR 86.134-96)

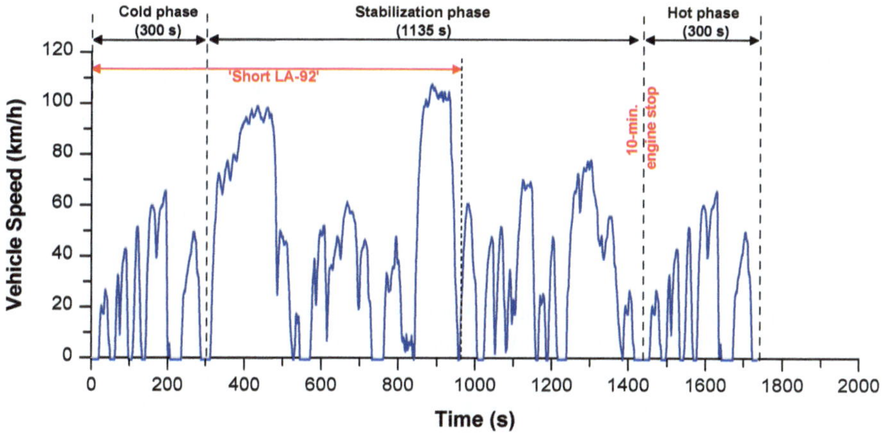

Fig. 2.31 Speed profile of the California LA-92 driving cycle (40 CFR 86, App. I)

2.2.6 *California LA-92*

The California Unified cycle, also known as Unified Cycle Driving Schedule (UCDS) or, simply, LA-92, Fig. 2.31, is a chassis-dynamometer driving schedule for light-duty vehicles developed by the California Air Resources Board (CARB) in 1992. Its construction was based on driving data—102 runs and 28 h of route-based second-by-second speed traces—recorded while following randomly selected vehicles operating on a mix of routes from the Greater Metropolitan Los Angeles area between April and May of 1992 using a refined chase car protocol [36, 42]. 'Composite' driving data, i.e., mix of chase and target vehicle data, was both used to generate the speed/acceleration frequency distribution plot and the driving cycle.

47 % of the data came from the chase car and 53 % from the target vehicle. To construct the cycle, the sample data were first divided into micro-trips. Although chase car data only accounted for about half of the sample data, it was used to generate the set of micro-trips and the target vehicle data was ultimately ignored. A total of 833 micro-trips resulted from the sampling of the 102 different routes.

For the cycle construction, a 'quasi-random' approach for micro-trip selection was utilized as discussed in Sect. 1.4.2. Firstly, 'seed' micro-trips were selected, entirely at random, to complete a seeding period forming the first 120-s start phase. Subsequent micro-trips were randomly selected in such a way that they improved the match to the sample's SAFD. Each time a micro-trip was selected, it was removed from the micro-trip set. The remaining micro-trips were scanned and subsequent micro-trips were again selected such that they improved the match of the cycle's speed/acceleration frequency distribution to the sample's SAFD. At the end of the process, 18,000 cycles had been created by repeatedly combining a subset of the micro-trips such that each cycle was approximately 20 min long. The final driving cycle, i.e., the Unified Cycle, best matched the speed/acceleration frequency distribution of the entire data set within 22 % of the sum of differences [36, 42].

The LA-92 test has a similar three-phase structure as the FTP-75. The first cold-started phase lasts 300 s, the second phase 1135 s, and the third is identical to the first but with the vehicle hot-started after a 10-min engine shutdown. The cycle was intentionally made more aggressive than the FTP-75 (maximum acceleration is 3.08 m/s^2) to account for more realistic simulation of true driving habits (cf. the supplemental driving cycle SFTP US06, which was based on driving data from the same period).

Not surprisingly, it has been found that CO and NO_x emissions during the LA-92 are higher than those produced during the FTP-75 [57]. The cycle is also characterized by higher speeds and fewer stops as well as less idle time than the FTP-75. It lasts 1735 s (phases 1, 2 and 3) for a traveled distance of 17.70 km. Maximum vehicle speed is 108.1 km/h and the average driving speed is 44.5 km/h; 17.5 % of the time is spent idling. A shorter version also exists that is limited to the first 969 s. Cycle emissions are calculated in the same manner as the weighted, overall FTP-75 formula from Eq. (2.2).

The LA-92 cycle was initially developed as a tool for future emissions inventory improvement efforts [36, 57]; for California's mobile emissions model EMFAC, emission rates are defined by testing a vehicle on the LA-92 for the base emission rate. Beginning with the Tier 3 standard, the 1735-s version of the LA-92 is the supplemental test cycle for Class 3 HDVs (GVWR 10–14,000 lbs) that are chassis tested [46].

Having concluded the discussion of the federal EPA and California CARB driving cycles employed for certification purposes in the United States, concerning passenger cars and light-duty vans, Table 2.13 (earlier in the text) summarizes some of their major technical specifications. More detailed and specialized vehicular data is available in the Appendix.

2.2.7 *Special Purpose Cycles*

Inspection and Maintenance—IM240

Various short cycles have been developed over the years in the U.S. for road-side vehicle testing. The aim of such tests is to ensure that the after-treatment devices function properly, so vehicles retain their low emission profiles in actual use. Originally, some of these tests were carried out with the vehicle idling, or on a dynamometer but the vehicle operating at one or two speeds. For example, the Federal 3-mode test was developed in the 70s as a possible short procedure for evaluating emissions from gasoline-engined cars in inspection and maintenance (I/M) programs. The car was placed on a dynamometer without a flywheel. The test involved two different speeds at 48 and 80 km/h (30 and 50 mph) with load points according to the vehicle's inertia weight. The applied dynamometer load simulated the average power which occurred at the respective speed of the UDDS. A low idle (unloaded) point was tested too. The Clayton Key Mode was another test, similar in logic to the Federal 3 mode. Obviously, such tests were only acceptable for older-technology vehicles with no electronic control.

The IM240 cycle, schematically represented in Fig. 2.32, is the current short-cycle employed in the United States to ensure that vehicles retain their low pollution profiles in actual use (purge flow test of the evaporative canister as well as pressure test of the evaporative system are two other tests applied in this regard). The abbreviation IM corresponds to Inspection and Maintenance, whereas 240 designates the (maximum) duration of the cycle in seconds. The cycle was developed in the early 90s following the 1990 CAAA. Maximum traveled distance is 3.15 km at an average driving speed of 49.1 km//h. Maximum speed is 91.2 km/h (as is the case

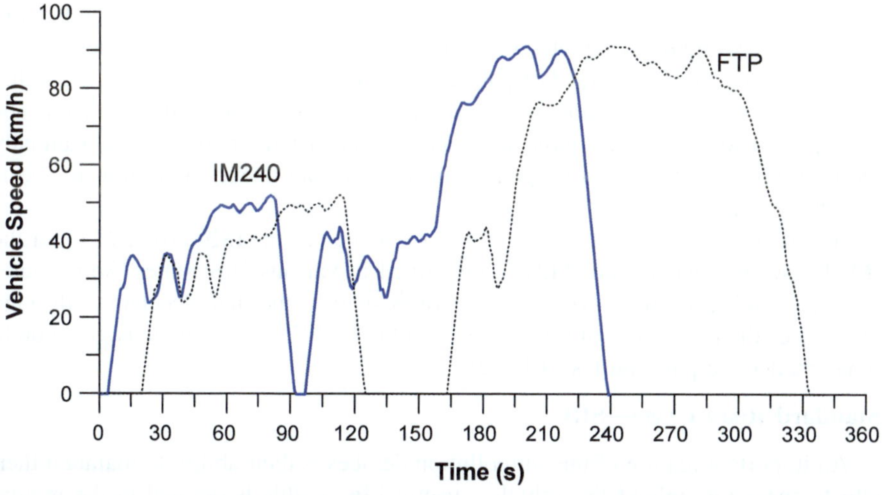

Fig. 2.32 Speed profile of the U.S. Inspection and Maintenance IM240 driving cycle (*dotted line* corresponds to the first 340 s of the FTP cycle)

with the LA4); on the other hand, only 3.8 % of the time is spent idling. Apart from CO and HC, the IM240 included from the beginning nitrogen oxides testing, and employed the CVS system, providing the emissions on a mass basis.

As is obvious from Fig. 2.32, the IM240 was patterned closely on the first two micro-trips of the FTP-72, using actual segments of this cycle. Testing over the entire range of speeds was considered important to detect malfunctioning vehicles given the discontinuous operating characteristics of electronically-controlled vehicles. It was considered important to utilize actual parts of the FTP-72 to help improve correlation and minimize errors of commission (when vehicles fail an I/M test but pass the FTP) and errors of omission (when vehicles pass the I/M test but fail the FTP). The two large decelerations from short-trips 1 and 2 are the only segments that were not taken directly from the FTP-72. The deceleration rate for both micro-trips was set at 1.56 m/s^2, whereas the maximum deceleration rate from the FTP is 1.48 m/s^2. The higher deceleration rate prevents the IM240 from exceeding 4 min, which was taken somewhat arbitrarily to be a measurable upper limit for a test time that would allow an adequate rate of vehicle processing. The 1.56 m/s^2 rate also allows time for an idle and an additional transient portion on the second short-trip (between 140 and 158 s). The IM240 test is run in two segments. The shorter segment is 93 s in duration, which is an informed guess as to the minimum amount of time needed to realize significant improvements in FTP correlation. Exhaust sampling begins simultaneously with the start of the driving schedule [58].

To determine emission levels, second-by-second instantaneous emission measurements are taken and integrated by a computer. The computer uses pass/fail algorithms to identify exceptionally clean or dirty vehicles. As soon as the emission rates indicate that a car is exceptionally clean or dirty, the computer automatically notifies the inspector to stop testing (e.g., after only 30 s). For vehicles that are close to maximum allowable emission levels, the test may continue for the full 240 s. Thus, while the complete driving cycle is 240 s long, the average test time is usually shorter. The existing fast-pass procedure is based on a specific algorithm and a large (3718 tests) database derived from an Arizona IM240 data set [59]. The principal advantage of fast-pass (and fast-fail) is that their use can substantially reduce both the time motorists spend waiting in testing queues as well as the costs of vehicle emission testing itself [60].

Alternative versions of the IM240 also exist such as the IM93 (first 93 s), or the IM147 (second part of the IM240), employed in various U.S. states. Some states use the BAR31, where the vehicle accelerates sharply and then decelerates through the 31 s of the test. For both the BAR31 and the IM147, each vehicle is given up to three chances to pass the test [61, 62].

Standard Road Cycle—SRC

An important feature of the antipollution devices is their ability to maintain their effectiveness throughout the vehicle's useful life, as this is defined in the regulations. Unfortunately, the after-treatment effectiveness is expected to deteriorate over time owing to various reasons such as poisoning, engine mechanical wear, carbon deposits, and, most importantly, thermal ageing.

In order to prove full useful life of vehicles and emission components durability in the United States, a special cycle is employed, applicable in the EU too, the standard road cycle—SRC. The SRC, as well as the similar-logic SBC (standard bench cycle—next paragraph), were introduced by the EPA in 2005 under the Compliance Assurance Program—CAP 2000; the final rulemaking was published on January 17, 2006, applicable to all 2008 and later model-year vehicles [63].

Prior to CAP 2000, EPA's regulations (40 CFR 86), as well as European ones (Directive 91/441/EEC) specified a different method to demonstrate a vehicle's emission durability, namely a whole vehicle mileage accumulation cycle,

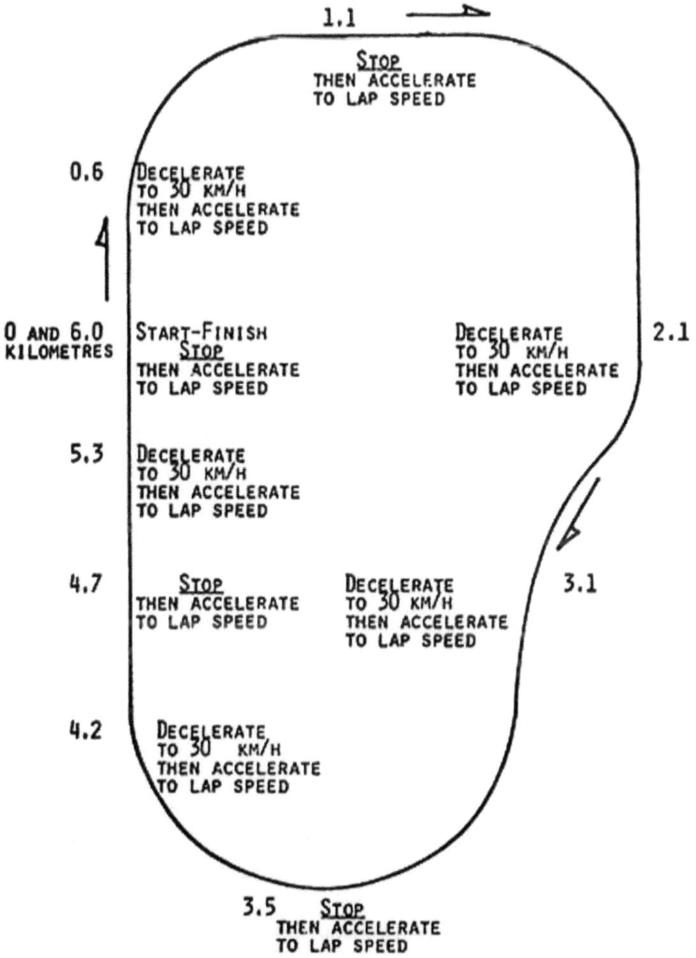

Fig. 2.33 Approved Mileage Accumulation cycle (AMA) (40 CFR 86, App. IV)

commonly referred to as the Approved Mileage Accumulation (AMA) cycle, illustrated in Fig. 2.33. This required manufacturers to accumulate mileage on a preproduction vehicle by driving it over the prescribed AMA driving cycle for the full useful life mileage. The cycle was originally defined in FR Vol. 35, No. 219, November 10, 1970.

During the AMA mileage accumulation, the car was tested in a laboratory for emissions at periodic intervals, and a linear regression of the test data was performed to calculate a multiplicative deterioration factor (DF) for each exhaust pollutant. Then, low mileage vehicles, more representative of those intended to go into production were emission-tested. The emission results from these tests were multiplied by the DFs to project the emission levels at full useful life ('certification levels'). The certification levels had to be at or below the applicable emission standards in order to obtain a certificate of conformity. EPA, however, had concerns that the AMA did not represent modern driving patterns and did not appropriately age modern-design vehicles for various reasons. First, the AMA cycle was developed (in the late 60s) before vehicles were equipped with catalytic converters. Second, the cycle contained a significant portion of low-speed driving, designed to address concerns about engine deposits. While engine deposits were a major source of emissions deterioration in pre-catalyst vehicles, the advent of catalytic converters, better fuel control, and the use of unleaded fuel shifted the causes of

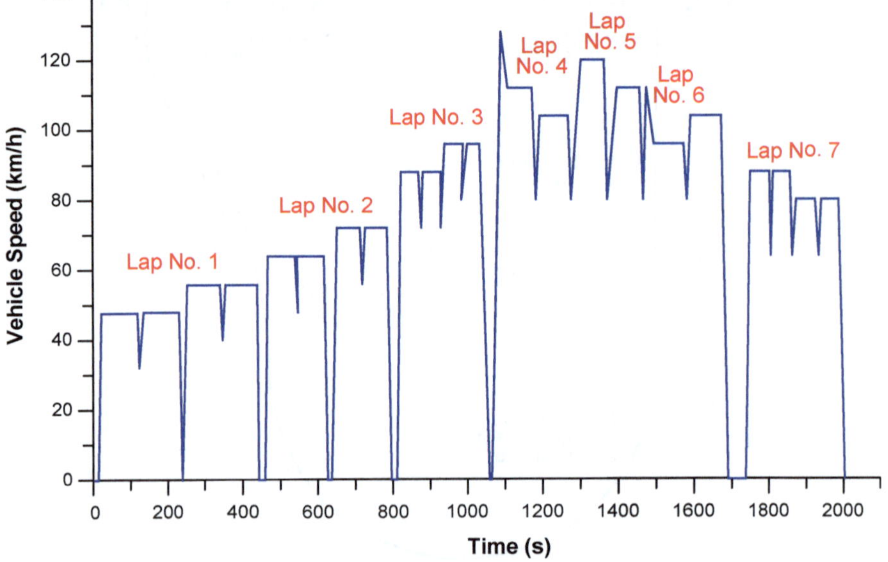

Fig. 2.34 Standard road cycle (SRC) for emission durability purposes; 40 CFR 86, App. V and European Commission Regulation 692/2008/EC provide the exact description of sequences for each lap; the cycle exhibits higher speeds and loads, being more appropriate to realistically age the antipollution devices compared to the AMA, also requiring shorter duration to be accomplished

deterioration from low-speed driving to driving modes which include higher speed/load regimes that cause elevated catalyst temperatures. In addition, manufacturers had long identified the durability process based on mileage accumulation using the AMA cycle as very costly and requiring extensive lead time for completion [63].

The new cycle developed in the 2000s for testing the durability of emission systems is the SRC, being practically a whole-vehicle aging cycle, Fig. 2.34. Manufacturers can demonstrate the emission levels of new vehicles at the end of their useful life period by running a vehicle on the SRC for the full useful life mileage of the vehicle, e.g., 120,000 miles for Tier 2 light-duty vehicles in the United States or 160,000 km in the EU (Euro 5/6 standard). The cycle, as demonstrated in Fig. 2.34, consists of seven laps of 5.95 km (3.7 miles) each [64]. As an example, lap No. 3 contains the following sequences: idle (10 s); hard acceleration to 88 km/h (1.79 m/s^2); cruise at 88 km/h for 1/4 lap; moderate deceleration to 72 km/h (−2.23 m/s^2); moderate acceleration to 88 km/h (0.89 m/s^2); cruise at 88 km/h for 1/4 lap; moderate deceleration to 72 km/h (−2.23 m/s^2); moderate acceleration to 97 km/h (0.89 m/s^2); cruise at 97 km/h for 1/4 lap; moderate deceleration to 80 km/h (−2.23 m/s^2); moderate acceleration to 97 km/h (0.89 m/s^2); cruise at 97 km/h for 1/4 lap; moderate deceleration to stop (−1.79 m/s^2).

Average speed on the SRC is 74 km/h (46.3 mph), maximum cruise speed is 120 km/h (75 mph), and the acceleration rates range from light to hard; most accelerations are moderate with no WOT ones. The SRC contains 24 fuel-cut

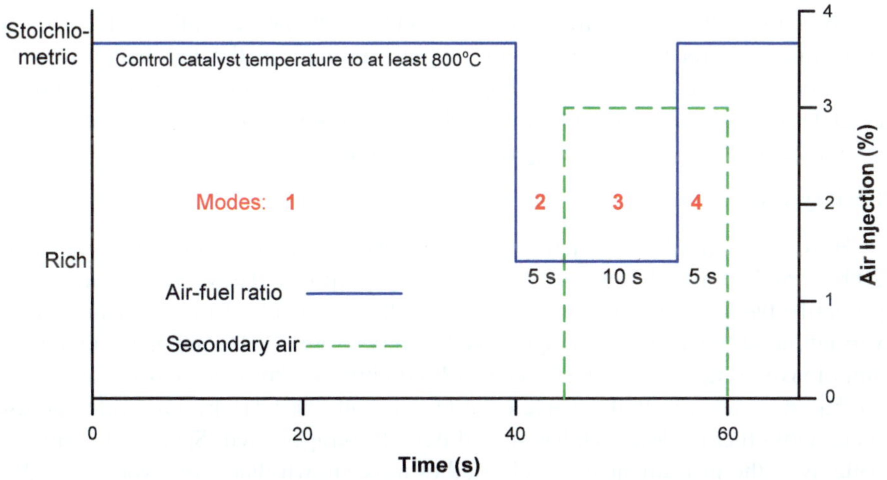

Fig. 2.35 Standard bench cycle (SBC) for catalyst aging in the U.S. and the EU (used to rapidly age emission components, primarily the catalytic converter and oxygen sensors, to the equivalent of the useful life period); 40 CFR 86, App. VII and European Commission Regulation 692/2008/EC provide details on the test sequence (for the rich operation during modes 2 and 3, catalyst temperature should be 890, or 90 °C higher than the low-control temperature)

decelerations; the deceleration rates range from coast-down (no brake force applied) to moderate [63]. Manufacturers in the United States may, with prior EPA approval, apply their own cycle or customize the SRC (or SBC) to their needs to mimic the aging of emission control components over their useful life.

Standard Bench Cycle—SBC

The standard bench cycle—SBC, Fig. 2.35, is a 60-s cycle intended for gasoline engines only, focusing on the catalyst aging, applying stoichiometric and rich operating phases. The cycle is composed of four modes, namely 40 s stoichiometric operation (mode 1), 5 s rich operation (mode 2), 10 s rich operation and air injection (mode 3), and 5 s lean operation (mode 4).

Overall, 108 days are needed to cover the 120,000 miles established in the U.S. legislation running the SRC cycle, compared to only 5–20 for the SBC, where only the engine is tested. EPA has developed a 'Bench Aging Time' (BAT) calculator that outputs the bench aging hours and temperature necessary on the SBC to replicate aging and deterioration that would occur on a road cycle (either the SRC or a manufacturer's alternative cycle). EPA's standard bench procedure specifies that the SRC be used to generate the catalyst temperature histogram needed to determine bench aging time. Specifically, catalyst temperature data is measured at a minimum rate of 1 Hz during at least two runs of the SRC. The temperatures are tabulated into a histogram with temperature bins of no larger than 25 °C [63].

Although requested by some manufacturers, a similar ageing bench cycle for diesel engines has not been defined by EPA. In Europe on the other hand, Regulation 692/2008/EC describes a standard diesel bench cycle (SDBC), which 'reproduces the engine speed and load conditions encountered in the SRC as appropriate to the period for which durability is to be determined. In order to accelerate the process of ageing, the engine settings on the test bench may be modified to reduce the system loading times. For example, the fuel injection timing or EGR strategy may be modified'. The SDBC requires use of an ageing bench with an engine as the source of feed gas for the system.

Facility Cycles

Historically, EPA has employed the MOBILE (initially) and MOVES (from 2009) simulation tools for emission inventory purposes. These models have been primarily based on testing on the FTP certification cycle. Correction factors for various conditions (e.g., average speed, temperature, fuels) are then applied to emissions measured at the FTP 'standard' conditions. The speed correction factors are based on test results for vehicles tested on both the FTP driving cycle but also on several other cycles, each having a different average speed. Specifically, driving patterns in the instrumented vehicle studies have shown that some types of facility (i.e., roadway)-specific driving contain more frequent and more extreme acceleration and deceleration than others, which reach a similar speed but remain at a steady cruise. As a result, it is important to quantify the emission differences for facility-specific speed-related traffic control measures in inventory modeling [65].

A number of specialized cycles has therefore been developed over the years to be employed in the EPA emission inventory simulation tools [66, 67]. The first set of 12 cycles (11 facility specific and one non-freeway area-wide) was developed in 1997 based on the SFTP instrumented vehicle and chase-car database from Baltimore, Spokane and Los Angeles. Sierra Research constructed the facility-specific cycles using randomly selected micro-trips to match the SAFD of all vehicle operation occurring under the conditions of interest (e.g., a particular facility type and congestion level). Moreover, a separate assessment of the highest load points (i.e., the highest combined speed/acceleration points) was conducted, to make sure the cycles had a representative sample of the high-load points. Another criterion for developing the cycles was to match the total proportion of specific power values in two groupings: between 200 and 299 mph^2/s (moderate high-load points) and higher than 300 mph^2/s (extremely high-load points). Since short-trips begin and end at idle, a modification to this methodology was required to develop cycles representative of uncongested freeway operation. More specifically, appropriate trip segments (in lieu of micro-trips) were used that were driven under the target levels of congestion on freeways [65, 66]. The main specifications of the developed cycles are provided in Table 2.14 according to roadway type and LoS (traffic amount; 'A' corresponding to uncongested conditions). As evidenced, most of the cycles have a short duration between 6 and 10 min.

Table 2.14 Summary of technical specifications of the U.S. facility cycles used in the MOBILE6 emissions inventory model (1997) [65, 66]

Cycle	Duration (s)	Distance (km)	Maximum speed (km/h)	Average speed (km/h)	Maximum acceleration (m/s^2)
Freeway, high speed	610	17.15	119.52	101.12	1.20
Freeway, LoS A–C	516	13.68	116.96	95.52	1.51
Freeway, LoS D	406	9.54	112.96	84.64	1.02
Freeway, LoS E	456	6.18	100.80	48.80	2.36
Freeway, LoS F	442	3.66	79.84	29.76	3.07
Freeway, LoS G	390	2.27	57.12	20.96	1.69
Freeway ramps	266	4.10	96.32	55.36	2.53
Arterials/Collectors LoS A, B	737	8.11	94.24	39.68	2.22
Arterials/Collectors LoS C, D	629	5.38	79.20	30.72	2.53
Arterials/Collectors LoS E, F	504	2.59	63.84	18.56	2.58
Local roadways	525	2.99	61.28	20.64	1.64
Non-freeway area-wide urban travel	1348	11.60	83.68	31.04	2.84

Since the Clean Air Act requires EPA to regularly update its mobile source emission model, EPA introduced in 2009 the successor to the MOBILE6.2, which was the MOVES2010. The newer version, which also focuses on micro-scale emission applications (e.g., from an arterial roadway corridor), is based on a set of new facility cycles [67]. Overall, 15 new cycles are incorporated in the model being specific to road types, velocity bins and traffic congestion levels.

For the cycles development, Sierra Research, which was again the subcontractor, reviewed and analyzed past driving datasets (incl. the SFTP one from 1992 used for the previous facility cycles), ultimately employing two, namely the Caltrans/CARB 2000 California route-based driving, and the CARB/Sierra 2004 Sacramento ramp driving. A total of 47 distinct link-level cycles were developed (again, on a second-by-second basis) best matching the statistical characterizations developed for each driving group. 39 cycles were constructed for individual roadway type and congestion level combinations, and 8 were specific to freeway on-ramps [67].

Initially, the speed measurements in the driving database were grouped into separate functional and LoS classes, representing the ranges of operation within each discrete driving data group for which individual cycles were to be constructed. Next, joint SAFD were generated from the data in each group; SAFD being the principal metric used to compare the speed and acceleration patterns in an individual cycle to those of the corresponding dataset. Target cycles in the range of 10–15 min were established (in contrast to 6–10 min for the MOBILE model). A computerized 'trial and error' method was used to select combinations of trip segments that best matched the population SAFD for each driving dataset for which cycles were constructed. Trip segments could be chained together only when the starting speed and acceleration of a candidate segment matched the ending speed and acceleration of the current segment within ±0.5 mph and ±0.5 mph/s respectively. The same hybrid random/best incremental cycle construction logic used for the MOBILE cycles was employed here (see Sect. 1.4.2 for more details) [67]. From the set of 47 cycles constructed, those that were ultimately incorporated in MOVES are detailed in Table 2.15.

It should be noted that apart from the above-mentioned facility cycles, MOVES incorporates 36 driving schedules for LD, medium-duty and HD vehicles, with their time length ranging from 253 to 4866 s, average speed between 2.9 and 125.2 km/h and maximum speed between 13.3 and 138.4 km/h.

Surveillance Driving Schedule—SDS

In 1971, the Surveillance Driving Schedule (SDS) was developed by the U.S. EPA to measure vehicle emissions over a variety of steady state and transient driving conditions, Fig. 2.36 [68].

The acceleration and deceleration phases represented in the SDS consist of all possible combinations of the following five speeds: 0, 15, 30, 45 and 60 mph. The average acceleration or deceleration rate observed for each mode in the Los Angeles Basin was used during the operation of 20 of the 26 transient modes. The remaining

Table 2.15 Summary of technical specifications of the 15 facility cycles used in the U.S. EPA MOVES emissions inventory software tool (2009) [67]

Cycle	Number of segments	Duration (s)	Distance (km)	Maximum speed (km/h)	Average speed (km/h)	Idling time (%)
FC01 LoS A–F	6	568	18.73	135.85	118.74	0.0
FC02 LoS D–F	7	680	14.92	117.55	78.95	5.0
FC11 LoS B	7	518	15.37	131.68	106.79	0.0
FC11 LoS C	13	904	26.02	125.81	103.64	0.0
FC11 LoS D	13	729	19.15	123.54	94.59	0.0
FC11 LoS E	13	972	20.03	115.04	74.21	0.1
FC11 LoS F	12	904	8.33	89.27	33.18	2.5
FC12 LoS C	10	886	25.21	127.74	102.45	0.0
FC12 LoS D	13	800	18.91	117.70	85.08	1.5
FC12 LoS E	12	912	17.63	114.03	69.64	0.0
FC14 LoS B	14	753	10.44	102.67	49.91	3.6
FC14 LoS C	12	512	5.81	85.42	40.85	8.0
FC14 LoS F	22	852	3.33	71.05	14.05	38.2
FC17 LoS D	18	708	5.89	80.98	29.93	16.1
FC19 LoS A–C	27	869	6.11	61.06	25.33	7.7

FC01 (interstate) and 02 (principal arterial) correspond to rural functional classes. All other FCs correspond to urban functional classes, namely *11* interstate; *12* other freeway/expressway; *14* principal arterial; *17* collector; and *19* local

6 transients were repeated using acceleration and deceleration rates higher or lower, in order to determine the effect of acceleration/deceleration rate on emissions. These were chosen to represent the full range of accelerations and decelerations observed in the CAPE-10 project [6].

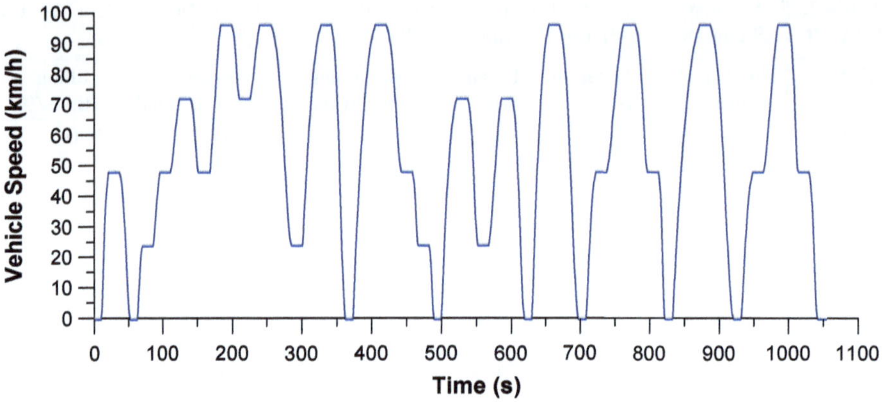

Fig. 2.36 EPA surveillance driving schedule SDS

Overall, 35 % of the total (1054 s) time during the cycle is spent cruising. However, owing to the increased frequency of accelerations, RPA is relatively high, namely 0.212 (30 % of the time is spent accelerating). Average speed is 54 km/h and, interestingly, the maximum acceleration reaches 2 m/s^2. It should be noted that the SDS is not a modal cycle, i.e., it is not described through a series of specific driving modes; instead, the cycle is defined as a vehicle speed versus time sequence on a 1-s basis [68]. Notice in Fig. 2.36 that, unlike modal cycles, the acceleration rate is not constant, but decreases as the demanded speed is approached.

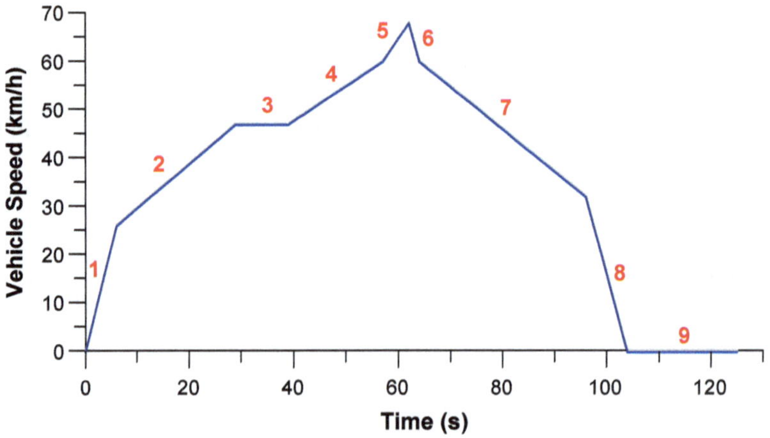

Fig. 2.37 Federal short cycle (FSC) for end-of-line testing (the 9 modes are: (*1*): acceleration from 0 to 26 km/h in 6 s; (*2*) acceleration from 26 to 47 km/h in 23 s; (*3*) cruise at 47 km/h for 10 s; (*4*) acceleration from 47 to 60 km/h in 18 s; (*5*) acceleration from 60 to 68 km/h in 4.5 s; (*6*) deceleration from 68 to 60 km/h in 2.5 s; (*7*) deceleration from 60 to 32 km/h in 32 s; (*8*) deceleration from 32 to 0 km/h in 7.5 s; and (*9*) idle for 21.5 s)

Federal Short Cycle—FSC

The FSC—federal short cycle—was developed by the EPA for fast analysis of vehicle emissions in the 70s. This was a highly transient 9-mode cycle based on the FTP-72 driving modes (specifically, the dynamometer loadings and transmission shift points follow the procedure as required for the FTP-72). Duration of the cycle was approximately 2 min (125 s), of which only 10 s spent cruising; distance traveled was 1.22 km, with 35 km/h average and 68 km/h maximum speed, for 0.151 RPA. Figure 2.37 provides a graphical illustration of the nine modes [69]. The cycle was used in various research projects during the 70s and 80s.

SAE J227a Driving Cycle for Electric Vehicles

In the 70s, SAE (Society of Automotive Engineers) developed the J227 cycle to be employed for the testing of electric vehicles (EVs) including two or three-wheelers, e.g., electric scooters. The cycle was designed to give approximately the same road-load energy demand as the FTP-72 but with lower peak road-load power. Since many EVs at the time were unable to follow the cycle, it was re-issued as J227a (now cancelled too) [70].

This is actually a suite of cycles, with four versions having different power and maximum speed requirements, named J227a-A, B, C, and D. Each version of the cycle has five phases: acceleration, cruise, coasting, braking and idle. Figure 2.38 shows the general speed profile of the J227a-C, which is the most commonly used variant.

As evidenced, the J227a is a typical modal/simplistic, short-duration and short-distance cycle, with increased cruising segment. Technical characteristics for each phase are summarized in Table 2.16 for all versions. The J227a procedure specifies only the cruise velocity and the transition time from one mode to another. The speed profile for the coasting segment (tractive force set to zero) is not

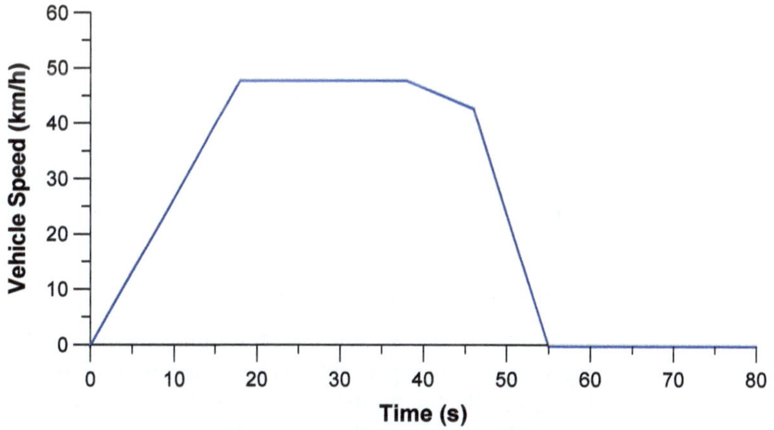

Fig. 2.38 SAE recommended J227a-C cycle for electric vehicles testing

Table 2.16 Specifications of the J227a suite of cycles [71, 72]

Cycle version	Maximum speed (km/h)	Acceleration time (s)	Cruising time (s)	Coasting time (s)	Braking time (s)	Idle time (s)	Total time (s)
A	16	4	0	2	3	30	39
B	32	19	19	4	5	25	72
C	48	18	20	8	9	25	80
D	72	28	50	10	9	25	122

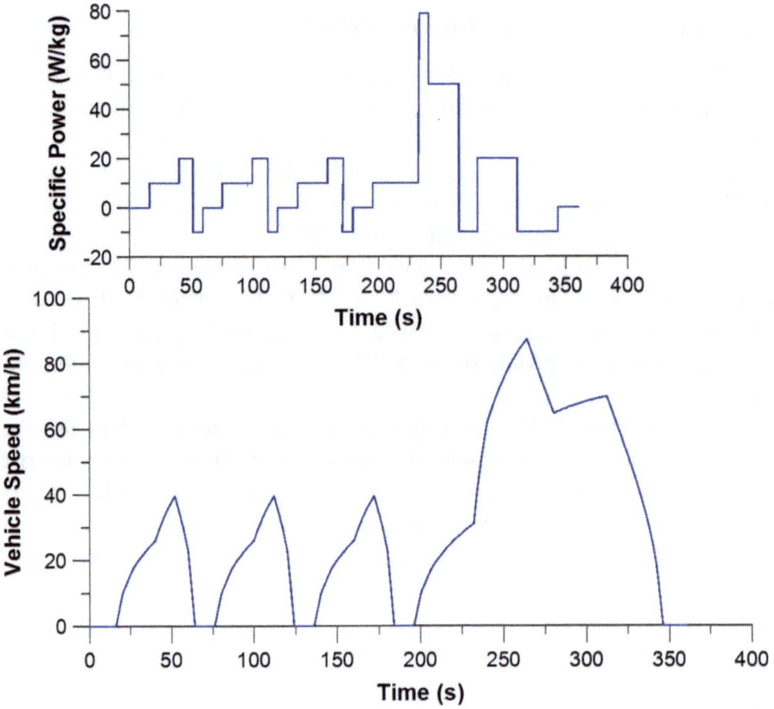

Fig. 2.39 Vehicle speed versus time schedule of the SFUDS (*lower sub-diagram*) and specific power versus time curve (*upper sub-diagram*)

explicitly defined, hence the traveled distance varies, depending on the acceleration capability of the vehicle under test [71, 72].

Simplified Federal Urban Driving Schedule—SFUDS

The prime consideration for electric vehicles is prescribing the power discharge profile of the battery and relating it to a particular vehicle design and driving cycle. Since the battery discharge profile for the FTP-72 (or FUDS) is very complex to simulate in the laboratory [73], a simplified version of the cycle, the SFUDS was developed by the U.S. Department of Energy in 1988. The cycle is illustrated in the

Table 2.17 Comparison of the technical specifications between the FTP-72 (FUDS) and the SFUDS cycle

Cycle	Duration (s)	Distance (m)	Average/Maximum speed (km/h)	Idle time (%)	Average/Maximum power (W/kg) [73]	Energy consumption (Wh/km) [73]	RPA (m/s^2)
FUDS	1372	11,990	31.5/91.2	17.8	10.1/79	225	0.185
SFUDS	360	3098	31.0/87.5	17.5	9.9/79	22.4	0.167

The specific energy (Wh/kg) of the battery primarily determines an EV's range; acceleration is determined by the battery's specific power (W/kg)

upper sub-diagram of Fig. 2.39 in terms of W/kg versus time, for testing the battery following a variable power discharge schedule.

The test comprises 20 steps and six power levels; as evidenced, the cycle also includes regeneration phases. Even though the profile was derived for a specific vehicle, it has been widely applied to other battery/vehicle combinations, irrespective of the differences in battery weight fraction, road-load parameters etc. [74, 75]. The lower sub-diagram of Fig. 2.39 demonstrates the cycle in terms of vehicle speed versus time, with the intent to test the whole electric vehicle. A comparison between the FUDS and the SFUDS in Table 2.17 indicates they produce very similar results.

A more sophisticated version of the SFUDS, termed the generic SFUDS (GSFUDS), has been also developed, to provide a test that is not vehicle specific. The GSFUDS uses the concept of average power P_{ave}, which is calculated by dividing the net energy out of the battery by the time duration of the discharge [74].

2.3 Japan

In Japan, automobile emissions were first mentioned in the 'Air Pollution Control Law' of 1966, which replaced the 'Smoke and Soot Regulation Law' of 1962. The Law required that emission standards for vehicles be determined, including CO, HC, lead, and 'all other substances which might endanger public health' [2]. The pertinent authorities responsible for emission standards and test cycles/procedures have been the Ministry of Land, Infrastructure, Transport and Tourism (MLIT), and the Ministry of Economy, Trade and Industry (METI). Japan first established emission limits for gasoline-powered, light-duty vehicles in 1966 on a concentration basis [2, 76], and in 1973 with respect to mass emissions. Mass emission limits were set in 1986 for diesel-engined passenger cars and in 1988 for light-duty vans. Nowadays, the regulated pollutants are CO, NO_x, HC (NMHC for gasoline cars

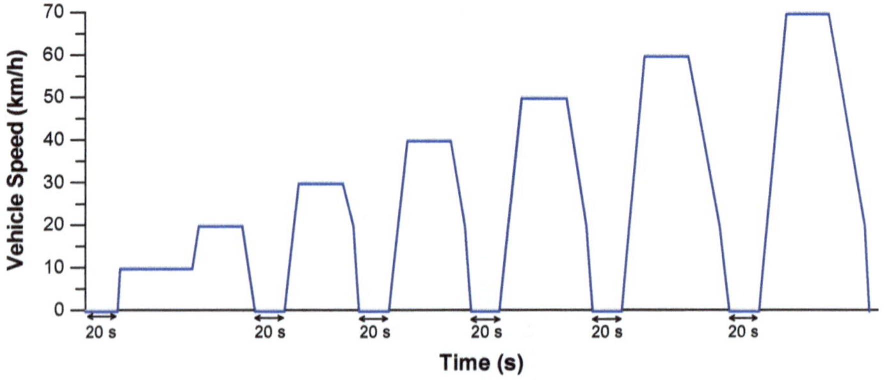

Fig. 2.40 Speed profile of the Japanese 4-mode (J4) driving cycle (only emissions during specific segments of the cycle were taken into account) [2]

from 1999) and PM. Standards apply to passenger cars and light-duty trucks with GVW lower than 3500 kg. Unlike the EU and the U.S., two types of exhaust emission standards are established, denoted as 'mean' and 'max'. The 'mean' standards are to be met as a production average and as a type approval limit. The 'max' standards are to be met generally as an individual limit in series production and as type approval limit if sales are less than 2000 per vehicle model per year [3]. In 1998, Japan joined the UNECE 1958 Agreement for mutual recognition of approvals for automotive equipment, as the first non-European country.

2.3.1 4-Mode—J4

A 4-mode cycle (J4) was initially employed in Japan, with the scope to limit the volumetric concentration of CO from gasoline and LPG fueled vehicles. The cycle, graphically illustrated in Fig. 2.40, was developed by the 'Traffic Safety and Nuisance Research Institute', after studying driving habits in Tokyo.

The cycle identifies four driving modes: idle, acceleration and deceleration at certain rates, and constant speed at specific values. On July 14, 1966, the J4 was officially adopted as the standard driving sequence in Japan, and was employed for seven years [2, 7]. As is evidenced in Fig. 2.40, the J4 consists of a simple sequence of acceleration, cruise, deceleration and idle, repeated seven times, with increasing cruise speed from 10 to 70 km/h at increments of 10 km/h. Idle time between the various sub-cycles is constant at 20 s. Since constant acceleration and deceleration rates are maintained throughout the cycle, the time length for each segment varies.

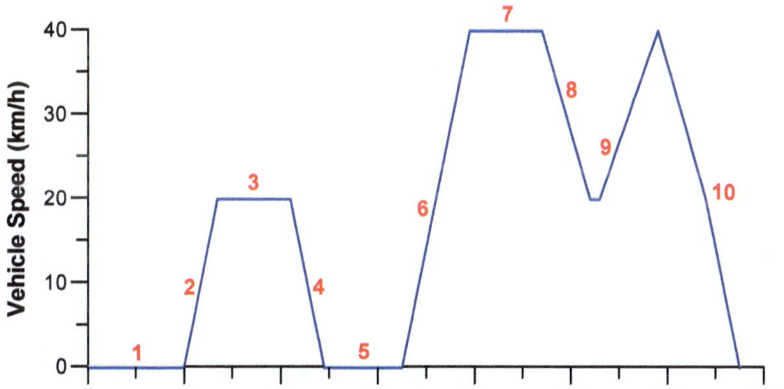

Fig. 2.41 Speed profile of the Japanese 10-mode (J10) driving cycle with identification of the various modes (see also Table 1.1 for a description of the ten modes)

2.3.2 10-Mode—J10

The 4-mode cycle was replaced by the 10 mode (J10), effective 1 April 1973. At the same time, the regulated pollutants were extended to HC and NO_x, and the test method changed from concentration measurement to the CVS technique. The J10, illustrated in Fig. 2.41 (its driving modes were detailed in Table 1.1), is similar in structure, simplicity and acceleration uniformity to the European ECE-15. It simulates urban traffic conditions in Tokyo, and was based on three earlier '8-mode' versions developed during the late 60s [2, 77, 78].

More than 25 % of the time during the J10 is spent idling compared to 31 % for the ECE-15, while the maximum vehicle speed throughout the 135 s of the cycle is limited to 40 km/h (50 km/h for the ECE-15); average driving speed is 24.1 km/h. Before sampling, there was a 15-min warm-up at 40 km/h, followed by one run of the cycle. After this warming phase, sampling commenced over five repetitions of the 10-mode cycle, i.e., for a period of 675 s [7]. As was the case with all test cycles at the time, the J10 was applicable to gasoline and LPG vehicles only. Diesel cars up to 1700 kg GVW were tested as of October 1, 1986 [2].

2.3.3 11-Mode—J11

In order to account for cold-started emissions, investigations about driving habits were conducted in 1973 that included early morning rush-hour traffic with

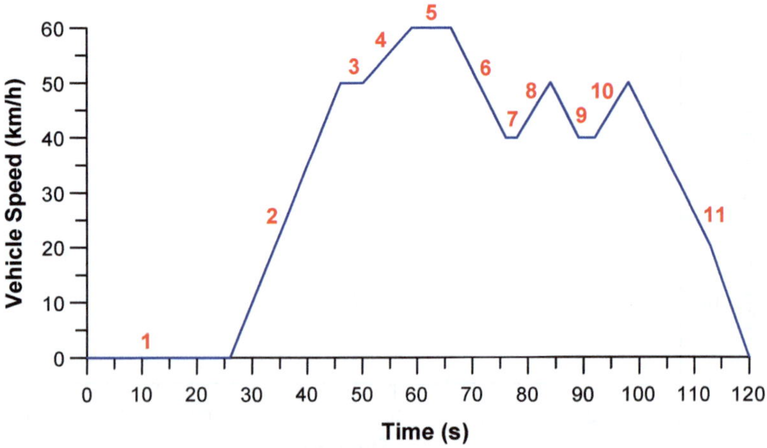

Fig. 2.42 Speed profile of the Japanese 11-mode (J11) cycle employed for emissions measurement under cold starting, with identification of the various modes

cold-started vehicles coming into Tokyo from outside the city area. The result was the development of a separate cycle, the suburban 11-mode test (J11), illustrated in Fig. 2.42.

The modal J11 lasts 120 s, with a 26-s initial idle period; it practically constitutes an expressway route after a cold start (cf. the first phase of the FTP-72) [78]. Maximum speed is 60 km/h, with the traveled distance a little over 1 km. As is evident from Fig. 2.42, the cycle is highly transient with limited constant-speed segments (only 13 %), unlike the much simpler J10; increased power demand is also established compared to the urban J10. As a result, the RPA value is rather high at 0.210.

During the certification procedure, the cycle was repeated four times, with measurements taken during all four runs [79]. Interestingly, the emission limits during the J11 were set in g/test and not g/km. The J11 came into effect as of April 1, 1975 for gasoline engined cars and was abandoned in 2008.

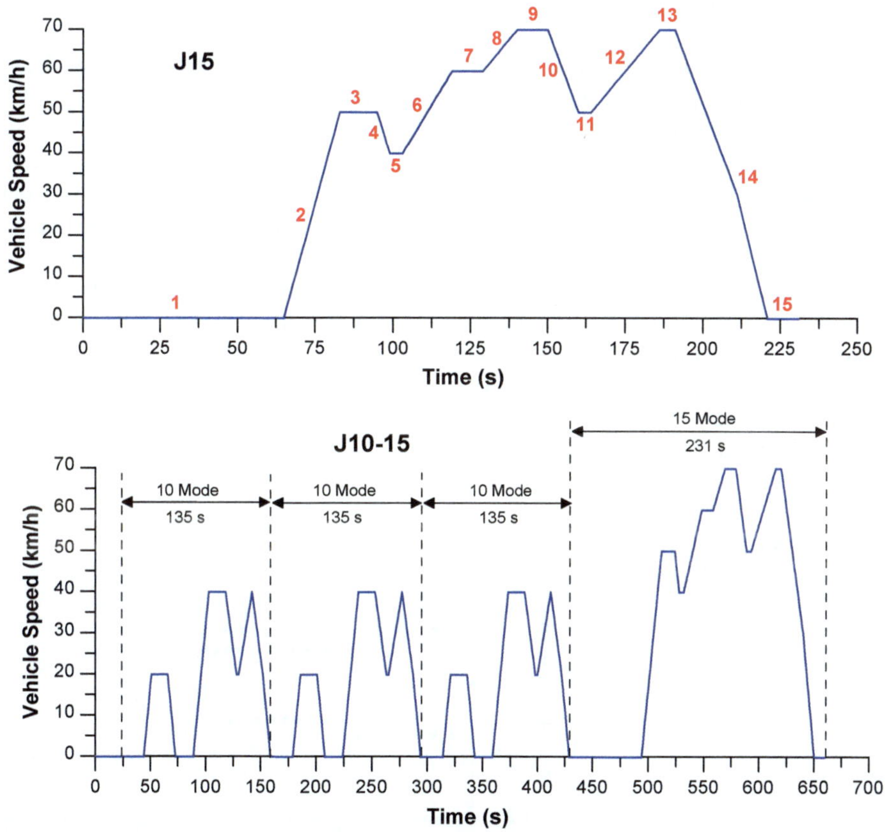

Fig. 2.43 Speed profile of the Japanese 15 mode (J15) and 10-15 mode (J10-15) driving cycles

2.3.4 10-15 Mode—J10-15

An extra-urban ('15-mode') segment was added from November 1991 (April 1993 for imported cars) to the J10 testing procedure to account for motorway driving at higher vehicle speeds [2]. The segment is illustrated in the upper sub-diagram of Fig. 2.43. Thus, the 10-15 mode (J10-15) cycle was formed comprising three repetitions of the J10 followed by one of the J15 (the latter denoted in the legislation as 'acceleration pattern'). The whole cycle is demonstrated in the lower sub-diagram of Fig. 2.43. It is specified in the Road Transport and Motor Vehicle Safety Standards (1951 Ministry of Transport Law 67), Article 31.2. It should be noted that an idling period of 24 s has been inserted prior to the first 10-mode run. The cycle reflects metropolitan driving conditions as well as highway ones, based on driving performance studies in major cities in Japan [80]. Notice the similarities between the 15-mode segment and the European EUDC in Fig. 2.2, as regards both structure and implementation date, as well as the overall similarity between the J10-15 and the European ECE+EUDC.

During the 231 s of the 15-mode segment, the maximum vehicle speed reaches 70 km/h, much lower compared to the 120 km/h of the EUDC, but fairly consistent with the driving habits of Japanese drivers at the time; average driving speed is 50.2 km/h for this motorway part. The whole 10-15 mode cycle lasts 660 s, with the traveled distance 4.16 km at an average driving speed of 33.1 km/h. As was the case with the J10, the vehicle was tested fully warmed-up; sampling was over one repetition of the whole cycle depicted in Fig. 2.43. The preconditioning included a sequence of a 15-min warm-up at 60 km/h, idle, 5-min warm-up at 60 km/h and one repetition of the 15-mode cycle [79]. Effective October 1993, the J10-15 cycle was also used for the certification of vans with GVW under 2500 kg. The cycle was used for fuel consumption measurements too, covering gasoline, LPG and diesel-engined vehicles, and for electric vehicles.

The extremely simplistic pattern of the J10 and J15 cycles is highlighted in Fig. 2.44 that demonstrates their speed/acceleration distribution.

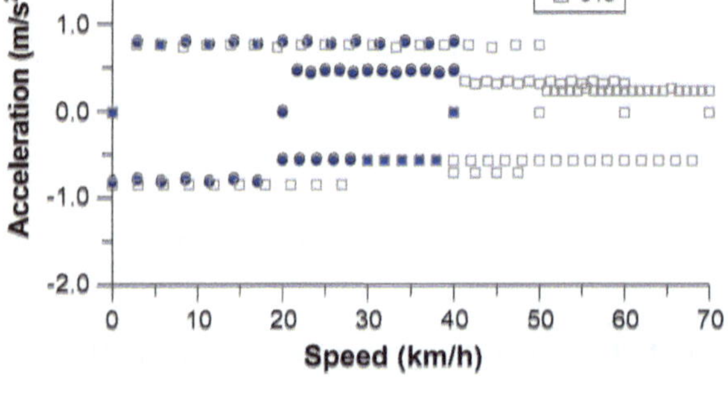

Fig. 2.44 Speed/acceleration distribution of the J10 and J15 cycles

2.3.5 JC08

An attempt to establish a driving cycle which would reflect real-world conditions and would deviate from the simplistic pattern of the J10/J11 cycles was made in 1976 by the 'Tokyo Metropolitan Research Institute for Environmental Protection' (TMRIEP). The developed cycle was composed based on 760 trips derived from Tokyo's ring artery road Meiji Dori, and took into account the frequency distribution of average vehicle speeds in relation to road characteristics and day time. Its speed versus time trace and duration (1466 s) resembled very much those of the U.S. LA4 cycle, however the average speed was only 22.5 km/h and its maximum one only 57.8 km/h. In any case, the cycle was never legislated [2].

Almost thirty years later, in 2005, the Japanese authorities finally introduced a realistic/transient test cycle, the JC08, for emission certification of passenger cars and light-duty trucks (GVW < 3500 kg). Graphical illustration of the JC08 is provided in Fig. 2.45. The cycle lasts 1204 s, with a covered distance of 8.16 km. Unlike its repetitive-type predecessors, the JC08 is highly transient with minimum cruising time; nonetheless, a long idling period (29 %), indicative of congested traffic in Japanese cities, is still maintained. Maximum speed is 81.6 km/h with an average driving speed of 34.2 km/h. The motorway segment is equal in duration

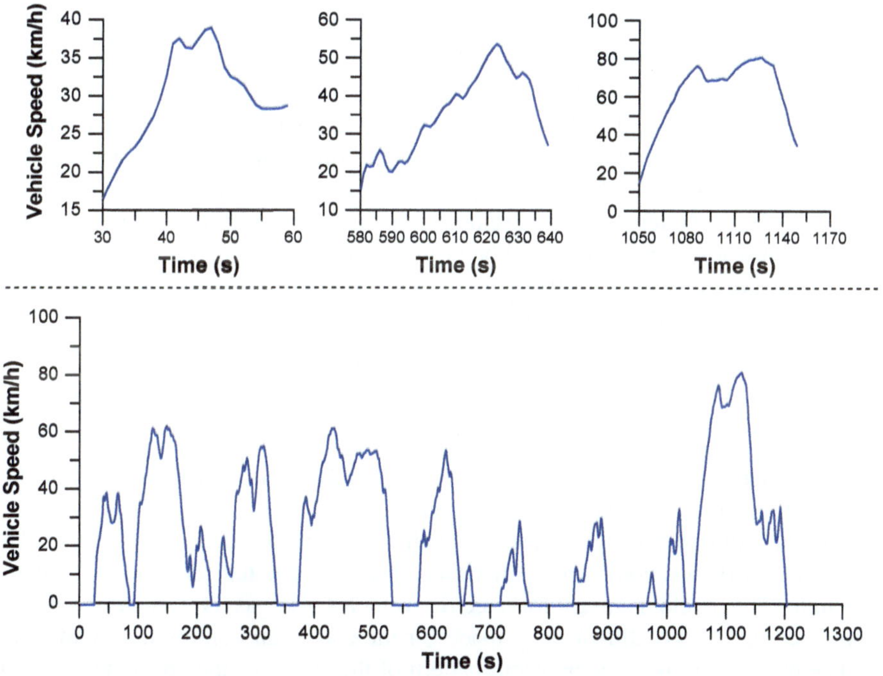

Fig. 2.45 Speed profile of the Japanese JC08 driving cycle (the *upper three sub-diagrams* illustrate specific parts of the cycle in more detail)

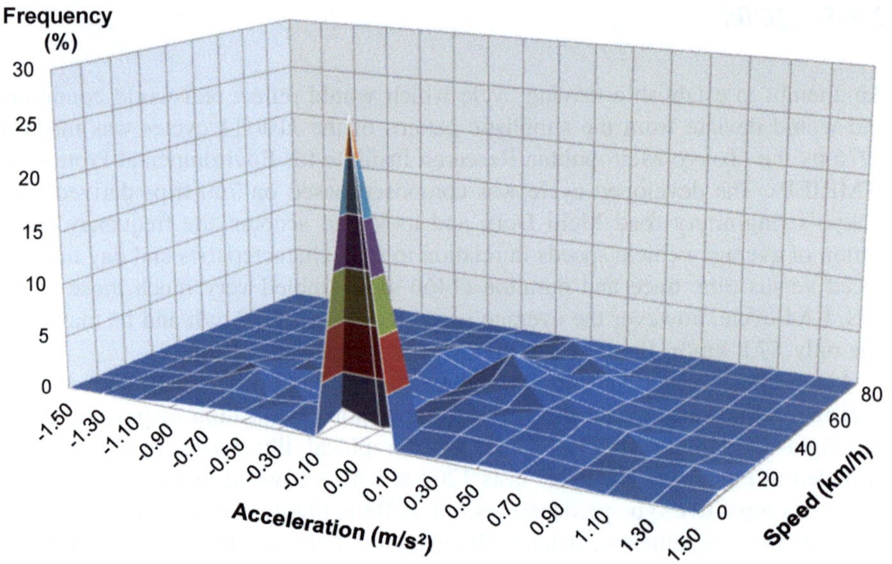

Fig. 2.46 3D speed/acceleration frequency distribution of the JC08 driving cycle (notice the high frequency at zero acceleration and speed, owing to increased idling period)

(short in any case) to the J15, with higher maximum speed (81.6 compared to 70 km/h for the J10-15). The 3D speed/acceleration frequency distribution for this cycle is provided in Fig. 2.46.

In November 2006, the 'Notice of the safety standard details for road trucking vehicles (Notice No. 619 of the Ministry of Land, Infrastructure and Transport, July 15, 2002)' was revised. For the purpose of evaluating exhaust emission performance more accurately, it was decided that driving modes of the new exhaust emission measurement method will be switched from the 11 mode to the JC08 cold started starting in fiscal year 2008, and from the 10-15 mode to the JC08 hot-started in 2011 [81], using the following weighting factors

- 2005: 12 % of J11 cold start/88 % of J10-15 hot start;
- 2008: 25 % of JC08 cold start/75 % of J10-15 hot start;
- 2011: 25 % of JC08 cold start/75 % of JC08 hot start.

The above dates were valid for domestic manufacturers. For imported vehicles, the transition phases began two years later.

For the cold-started JC08 test, the vehicle soaks for 6–36 h at 25 °C. The hot-started JC08 test commences after the vehicle is operated for at least 15 min at 60 km/h. The cycle is used for electric vehicles too; as is the case with the European NEDC, no allowance is made for the use of air-condition systems.

The much more realistic/transient pattern of the JC08 compared to the J10-15 is highlighted in Fig. 2.47 that illustrates the acceleration profiles of the two cycles during the first 300 s. Further, Fig. 2.48, comparing the speed/acceleration

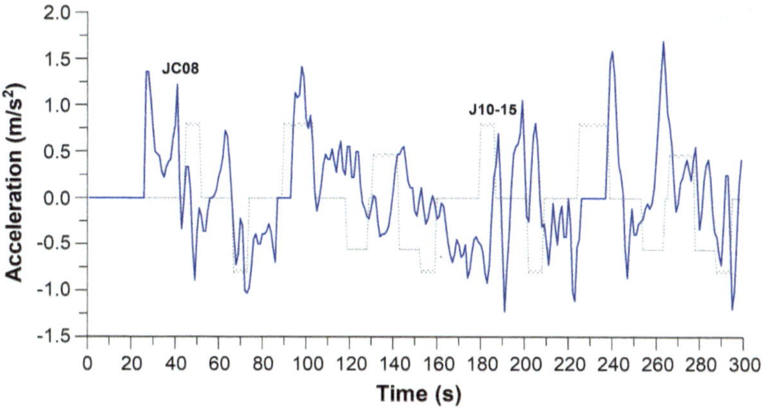

Fig. 2.47 Comparison of the acceleration profiles between the J10-15 and the JC08 driving cycles during the first 300 s (the JC08 exhibits higher, more frequent and non-steady accelerations)

Fig. 2.48 Comparison of the speed/acceleration distribution between the J10-15 and the JC08

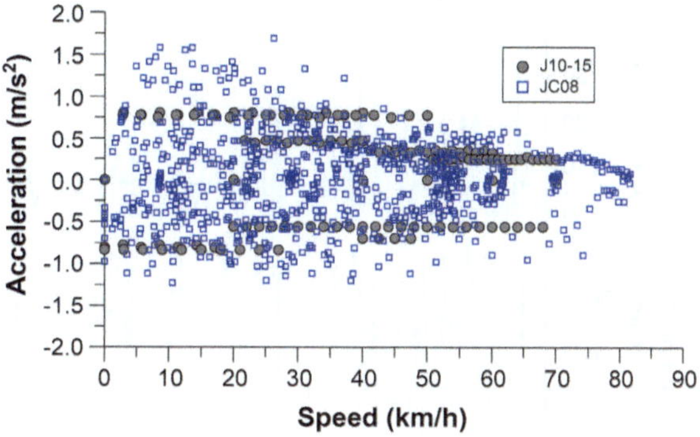

distribution between the J10-15 and the JC08, draws a more complete picture by shedding light into the latter's much broader driving activity.

For fuel economy purposes (km/L), the following formula is applicable, taking into account hot and cold JC08 results appropriately weighted (as is also the case with exhaust emissions)

$$\text{Fuel Economy} = \frac{1}{\frac{0.25}{\text{JC08C}} + \frac{0.75}{\text{JC08H}}} \tag{2.6}$$

Covered vehicles are gasoline or diesel-fueled passenger cars with a capacity of 10 passengers or less, passenger vehicles with a capacity of 11 or more passengers (GVW less than 3500 kg), and freight vehicles with GVW of less than 3500 kg that have received type designation under Article 75.1 of the Road Trucking Vehicle Law (1951, Law No. 185) [81]. Japan has expressed its intention to adopt the WLTC (Sect. 2.5), in the development of which it participated actively.

Table 2.18 Summary of technical specifications of Japanese passenger-car cycles (all are modal except for the JC08)

Cycle	Road category	Duration (s)	Distance (m)	Maximum speed (km/h)	Average speed (km/h)	Maximum accel. (m/s^2)	Idling time (%)	RPA (m/s^2)
J10	Urban	135	664	40.0	17.7	0.81	26.7	0.198
J11	Suburban	120	1021	60.0	30.6	0.69	21.7	0.210
J15	Motorway	231	2174	70.0	33.9	0.78	32.5	0.149
J10-15	Urban-motorway	660	4165	70.0	22.7	0.81	31.4	0.172
JC08	Urban-motorway	1204	8159	81.6	24.4	1.69	28.7	0.186

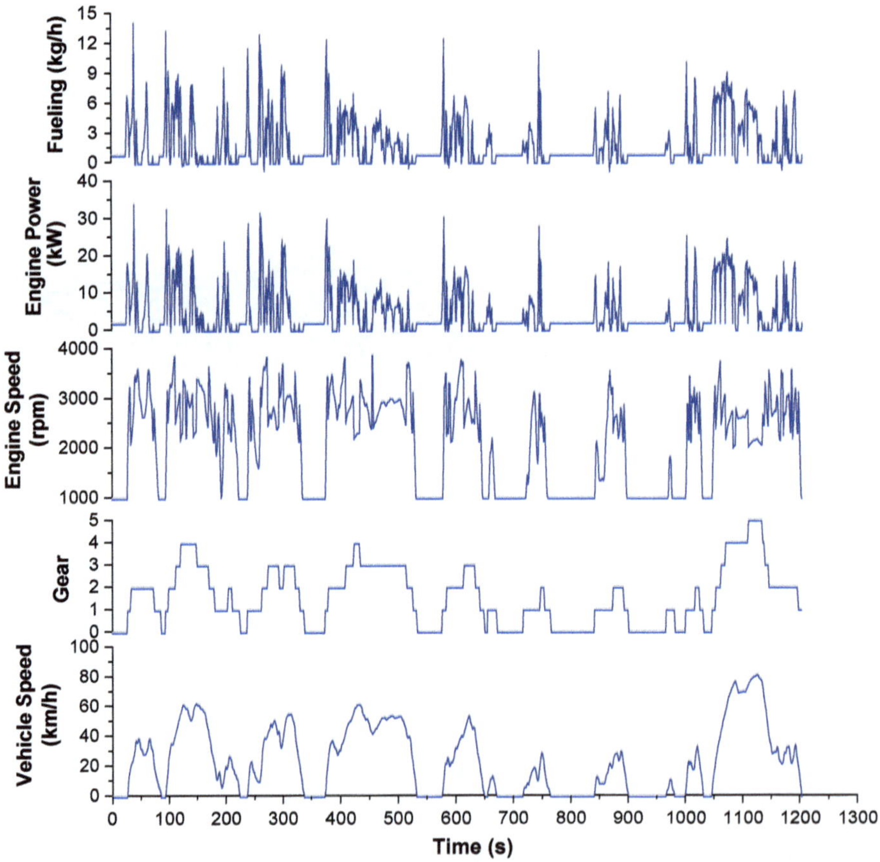

Fig. 2.49 Development of engine and vehicle properties during the JC08 for a SI-engined passenger car (notice the abrupt increase in fueling at each steep acceleration in the cycle)

Basic technical specifications of the legislated Japanese test cycles are summarized in Table 2.18. Increased idling period and rather low maximum vehicle speeds are common features in all the cycles.

Lastly, Fig. 2.49 illustrates engine and vehicle properties development during the JC08 for a gasoline passenger car.

2.4 Australia

In Australia, the Australian Design Rules (ADRs) apply, i.e., national standards for vehicle safety, anti-theft and emissions. The current (2016) standards, the third edition ADRs, are administered by the Australian Government under the Motor Vehicle Standards Act of 1989. Australia follows the European emission regulations and, from 2005, the NEDC is employed for the certification of light-duty vehicles. Nonetheless, a number of cycles have been developed in the 70s and 80s for research and inventory purposes, specifically the Sydney cycle [82], Melbourne cycle [83] and Perth cycle [84].

In the late 90s, a research project was undertaken with the goal to develop a national cycle for diesel-engined vehicles of various classes. The actual on-road driving patterns of 17 vehicles, ranging from off-road passenger to heavy-goods vehicles, were logged during normal use in Sydney in 1998. Using mathematical analytical tools, the data was analyzed according to the characteristics of micro-trips. For each vehicle, each micro-trip was allocated to a road flow category ('freeway/highway', 'arterial', 'residential/minor' or 'congested'). The most representative micro-trips in each road flow category were then combined to form an urban emissions drive cycle (UEDC) of approximately 60 min duration, for each vehicle/ADR category. The categories were:

- passenger cars (MA), forward passenger vehicles (MB) and off-road passenger vehicles (MC);
- light-goods vehicles (NA), and light buses (MD) below 3500 kg GVW;
- medium-goods vehicles (NB), and light buses (MD) above 3500 kg GVW;
- heavy buses (ME);
- heavy-goods vehicles (NC) below 25,000 kg GVW;
- heavy goods vehicles (NCH) greater than 25,000 kg GVW.

These UEDCs were recommended as 'reference' drive cycles. A CUEDC (composite UEDC) of approximately 30 min duration, and thus more suitable for test purposes, was then derived from each UEDC. The complex CUEDCs were recommended for in-service testing for inventory purposes. A 'straight line' 'simplified CUEDC' was also constructed from each CUEDC, in order to reduce the number and frequency of transients to allow greater ease of testing on less sophisticated (and less costly) chassis dynamometers [85]. Figure 2.50 illustrates the speed/time trace of the two LDV cycles and Table 2.19 provides some

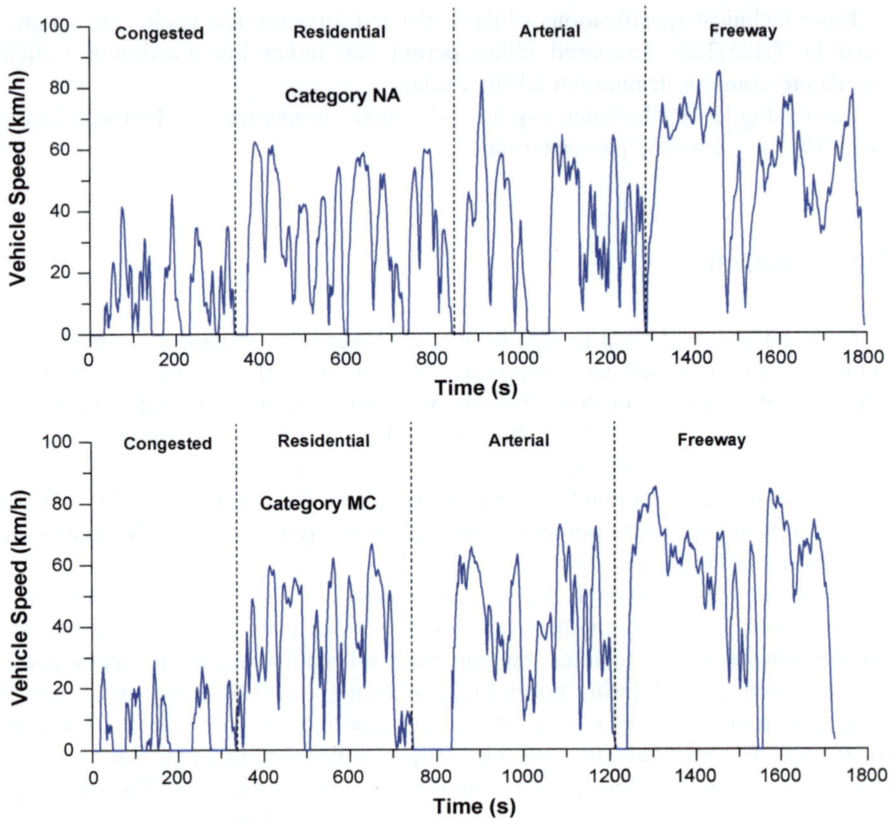

Fig. 2.50 Australian CUEDC cycles for diesel-engined light-duty vehicles [85] (see text for definition of MC and NA)

Table 2.19 Technical specifications of the Australian CUEDC cycle for LD vehicles

Cycle/Segment	Duration (s)	Distance (m)	Maximum speed (km/h)	Average speed (km/h)	Maximum accel. (m/s^2)	Idling time (%)	RPA (m/s^2)
Diesel—MC	1722	16,904	85.3	35.3	5.80	15.7	0.269
Diesel—NA	1794	17,477	85.0	35.1	5.22	11.4	0.281
Gasoline	1797	19,442	94.0	39.0	3.61	22.3	0.171
Gasoline cycle individual segments							
Residential	498	3839	58.0	27.8	3.06	21.1	–
Arterial	401	2793	67.0	25.1	3.61	13.9	–
Freeway	542	10,783	94.0	71.6	3.61	14.6	–
Congested	356	2027	59.0	20.5	2.50	28.2	–

important technical specifications. The four HDV schedules are demonstrated in Fig. 4.56.

For gasoline vehicles, a similar approach was followed in 2005 (NISE2) [86]. Sixty vehicles were monitored for a total of 431 h. Global positioning system (GPS) technology was used to collect driving data from the vehicles traveling AustRoads routes in each major Australian capital city (Sydney, Brisbane, Adelaide, Melbourne and Perth). As the AustRoads routes are arterial/freeway road types, data logging was initiated from trip origin (home/work) through to the final destination (work/home) so as to allow collection of residential road data at the start and end of each trip. Data was collected during three time periods of the day, namely AM peak, PM peak and inter-peak. Using specialized analysis techniques, each trip was broken down firstly into road flow category, and then into basic driving modes (idle, acceleration from idle, deceleration to idle, intermediate acceleration, etc.). These were then combined using VKT (vehicle kilometers traveled) weightings to account for state to state differences, to produce the average national drive profile for each road flow category.

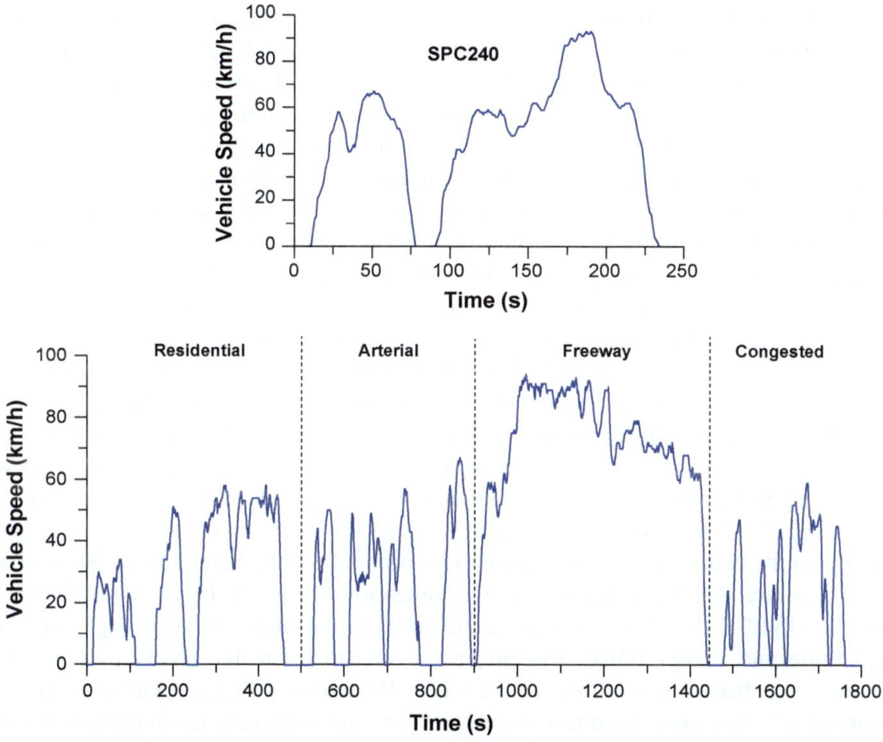

Fig. 2.51 The Australian CUEDC (composite urban emissions drive cycle), and inspection and maintenance SPC240 for gasoline vehicles

From the basic driving modes, 35 descriptive statistics were derived for each of the four road flow categories. Correlation analysis was conducted to identify individual 'trip segments', which had the highest correlation to the national average descriptive statistics for each road flow category. The corresponding composite cycle developed by the Transport Systems Centre of the University of South Australia is demonstrated in Fig. 2.51 [86].

Table 2.19 provides some technical specifications of the gasoline cycle and its intermediate phases. Notice in Table 2.19 the rather low maximum speed and the high maximum acceleration.

A shorter, 240-s version, titled SPC240 was also developed for inspection and maintenance purposes of gasoline cars, based on the last 2 min of the arterial phase and the first 2 min of the freeway phase (cf. the U.S. IM240 cycle in Sect. 2.2.7); this cycle is demonstrated in the upper sub-diagram of Fig. 2.51.

2.5 Worldwide—WLTC

Most manufacturers produce vehicles on a global scale or at least for several regions in the world. Nonetheless, since vehicle types and models tend to cater to local habits and living conditions, the vehicles produced worldwide are not identical. The compliance with different emission standards in each region, however, creates high burdens from an administrative and vehicle design point of view. Manufacturers, therefore, have a strong interest in harmonizing performance requirements and emission type approval procedures as much as possible on a global scale. Regulators have also an interest in global harmonization since it offers more efficient development and adaptation to technical progress, and facilitates the exchange of information between authorities. Harmonization doesn't necessarily mean having identical requirements/emission targets (as already mentioned, the needs of different countries vary, often by a lot). It does mean, however, that unnecessary differences can be eliminated, hence regulations can be brought closer. In this way, where possible and practical, a single vehicle specification can be built to satisfy all requirements instead of being country-specific. Common specifications of vehicle parts can reduce the cost of development and production and the retail price of vehicles. They can also make approving procedures easier, make the market larger, and give the consumers a wider range of choice. As a consequence, at its November 2007 session, the World Forum for Harmonization of Vehicle Regulations (WP.29) of the UNECE decided to set up an informal group under the working party on pollution and energy (GRPE) to prepare a road map for the development of the Worldwide harmonized Light vehicle Test Procedure (WLTP), with the aim to harmonize emission related test procedures for light-duty vehicles to the extent this was possible. The project was divided into 3 phases, as follows [87–89]:

- Phase 1 (2009–2014): Development of the worldwide harmonized light-duty driving cycle (WLTC) to reflect the actual driving conditions in real-world, as well as the associated test procedure for the measurement of pollutants, CO_2, fuel and energy consumption.

- Phase 2 (2014–2018): Low-temperature/high-altitude test procedure, durability, in-service conformity, technical requirements for on-board diagnostics (OBD), mobile air-conditioning system energy efficiency, off-cycle/real driving emissions.
- Phase 3 (2018–...): Emission limit values and OBD threshold limits, definition of reference fuels, comparison with regional requirements.

Work on the new cycle started in September 2009, the collection of driving data was launched in 2010, and the first version of the cycle was proposed by mid-2011; this was revised a number of times to take into consideration technical issues such as drivability and better representativeness of driving conditions after a first validation. The development of the WLTP was closely followed by various stakeholders (governments, industry, non-governmental organizations), as evidenced by the increased number of attendees to the UN GRPE's meetings [90]. In 2010, the U.S. EPA decided to withdraw its active participation in the WLTP, given the resource-intensive preparations for the 2012–2016 and 2017–2025 U.S. greenhouse gas standards. After the United States withdrew, the process was driven forward by the EU, South Korea, India and Japan [90].

For constructing the WLTC, driving data from all participating countries was collected and weighted according to the relative contribution of regions to the globally driven mileage and data collected for the WLTP purpose. The in-use data collected for the cycle development consisted of:

- 462,000 km from Europe (the involved countries were Belgium, France, Germany, Italy, Poland, Slovenia, Spain, Sweden, Switzerland and the UK);
- 56,000 km from India;
- 53,000 from Japan;
- 34,000 from South Korea; and
- 160,000 km from the U.S. (in total, 766,000 km).

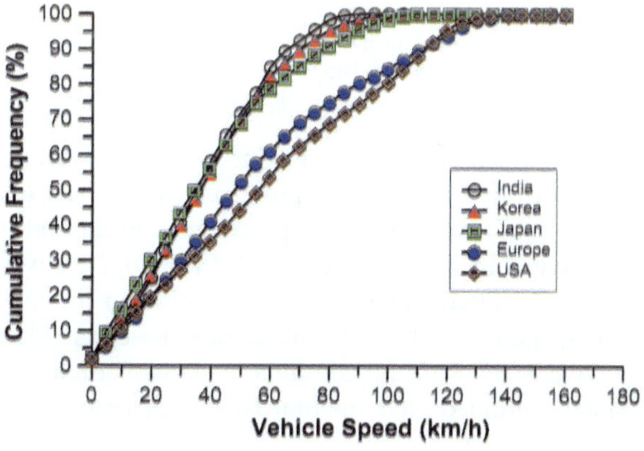

Fig. 2.52 Vehicle speed cumulative frequency distribution from the various regions in the WLTP database (data from [91])

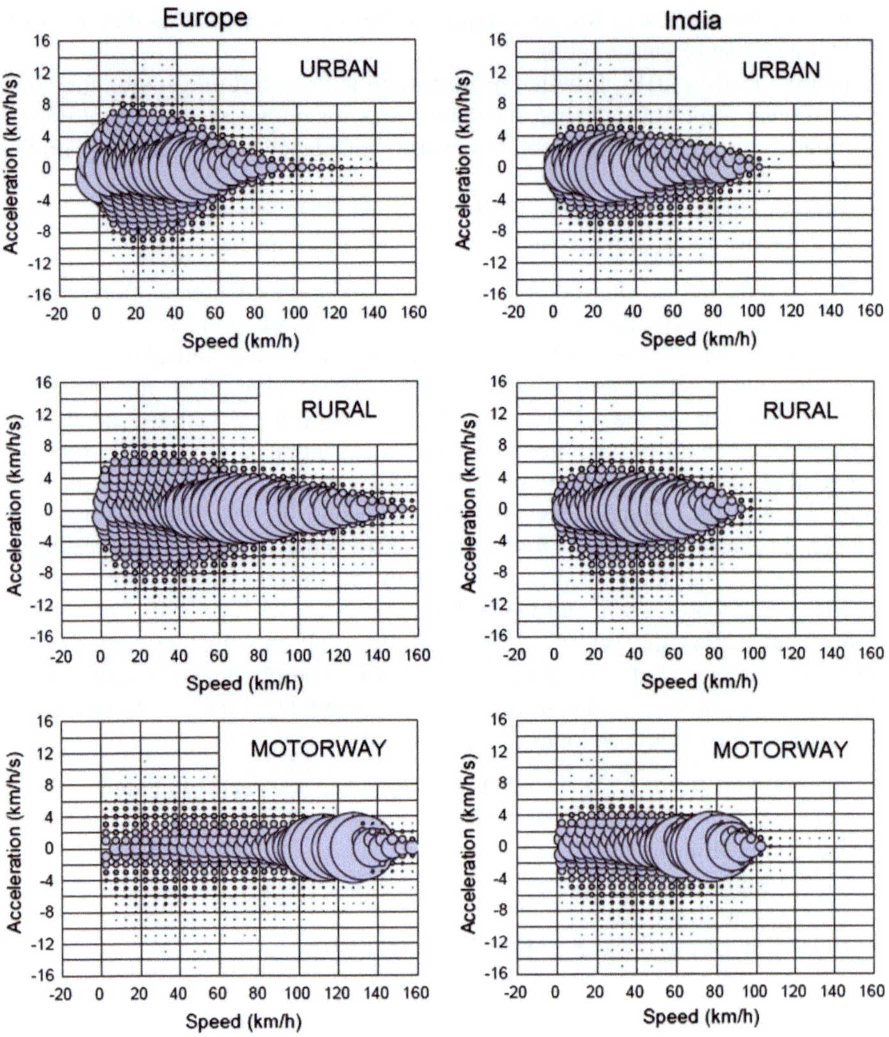

Fig. 2.53 Differences in the speed/acceleration distribution between the European (*left column*) and Indian (*right column*) data in the WLTP database [88]

394 vehicles of various engine capacities were used, with power to mass ratio ranging from 9 to127 kW/t, produced by a variety of manufacturers. The major part of this data corresponded to passenger car data but LD vans and mini-buses were represented in the database as well [88–90]. Data was collected from a combination of instrumented vehicles (Europe, India and the U.S.), 'instructed' drivers (Japan, Korea, India and the U.S.), and pure chase car (U.S. only).

Consistent with the common cycle construction practice, the data (after appropriately 'thinned' and smoothed) was separated into micro-trips and idling phases.

Unlike previous cycle development approaches, however, which were based on a road categorization, i.e., urban, rural and motorway, a different approach was followed during the WLTC development owing to the very different road categorization and speed limits existing throughout the world. The latter holds in particular true when comparing the European (or the U.S.) and Asian databases, as Figs. 2.52 and 2.53 confirm; the only exception being the urban road type, where a fairly good degree of similarity was found to exist worldwide. For the WLTC construction, the short-trips were binned according to their maximum speed, initially into three phases, i.e., low, medium and high. Subsequently, the high-speed phase was split into two segments: one high-speed phase with maximum speed representative of driving in Asia, and one extra high speed, more typical of motorway driving in Europe and the United States. These four speed-related phases are

- Low (up to 60 km/h), consisting of 97.1 % urban mileage;
- Medium or middle (between 60 and 80 km/h), consisting of 80.7 % urban and 19.2 % rural mileage;
- High (80–110 km/h), comprising 28.3 % urban and 71.5 % rural; and
- Extra high (>110 km/h), consisting of 55 % rural, 32.9 % motorway and 12 % urban mileage [87, 91, 92].

Table 2.20 Technical specifications of the regional databases [88]

	Japan	Europe	USA	S. Korea	India	Unified
Relative positive acceleration (m/s^2)						
Low	0.177	0.200	0.245	0.192	0.134	0.192
Medium	0.142	0.176	0.225	0.174	0.142	0.188
High	0.117	0.144	0.164	0.139	0.162	0.156
Extra high	0.086	0.114	0.103	0.155	–	0.108
Average speed (km/h)						
Low	19.8	20.0	18.8	17.2	21.1	19.8
Medium	40.1	39.9	37.0	34.1	39.5	38.4
High	62.9	55.6	59.7	53.9	56.1	58.0
Extra high	86.2	83.1	90.1	67.6	–	86.8
Average idle duration (s)						
Low	26.5	15.6	24.9	29.2	23.1	21.9
Medium	25.6	16.6	22.3	39.8	24.6	22.4
High	21.5	18.4	20.1	34.2	46.2	22.8
Extra high	15.5	17.1	12.2	22.1	–	14.5
Average short-trip duration (s)						
Low	66	68	63	64	148	84
Medium	161	221	125	201	642	238
High	458	473	284	691	1157	446
Extra high	1158	1082	601	1621	–	824

The disadvantage of this procedure (i.e., speed-related rather than road category-related segments) is that the phases consist of contributions from different road categories as detailed above. The threshold vehicle speeds between the various phases were chosen after comparative study of different candidate thresholds [88]. Table 2.20 presents data from the regional databases and the unified one regarding some important technical specifications for each speed segment considered.

The short-trips for the WLTC had to be selected from the unified database. The selection criteria were based on the concept that the selected short-trips must provide similar distributions of speed, acceleration, etc. to those of the unified database. Given the large number of different/possible short-trip combinations, several selection criteria (average vehicle speed, acceleration duration ratio, deceleration duration ratio) were applied. This selection was necessary to reduce the number of possible combinations and to keep computation time to a reasonable limit. The combination of the short-trips with the smallest chi-squared value was selected for the WLTC driving cycle construction [88].

The drivability of the WLTC was assessed extensively during the development process. In particular, specific cycle versions for certain vehicles with limited driving capabilities due to low power to mass ratio or limited maximum vehicle speed were introduced. As a result, three classes of the cycle were developed with respect to the vehicle power to (unladen) mass ratio PMR as follows (duration of each intermediate segment is fixed between the classes) [87, 88]:

- Within WLTC Class 3 (PMR > 34 kW/t), there are two versions of the cycle: version 3-1 (or 3a) for vehicles with maximum speed less than 120 km/h and version 3-2 (or 3b) applicable to vehicles with a maximum speed higher than 120 km/h. At the option of the country, the extra-high phase may be excluded, e.g., in India or China. Class 3-1 was developed in order to account for the special vehicle class of k-cars in Japan [91].
- WLTC Class 2 (22 kW/t < PMR $\leq$ 34 kW/t) is designed for lower powered vehicles; it has four speed phases like the WLTC Class 3 but with softer accelerations and top speeds in each phase. At the option of the country, the extra high phase may be excluded in this case too.
- WLTC Class 1 (PMR $\leq$ 22 kW/t) is to be applied to vehicles with the lowest PMR, and is designed to have only the low and medium-speed phases. The complete test cycle for type approval includes phases low, medium and again low.

For off-vehicle charging HEVs and pure electric vehicles (PEVs), the applicable cycle includes phases low and medium of Class 3 ('WLTC city').

Maximum duration of the cycle is 1800 s, similar to the earlier developed worldwide heavy-duty vehicles WTVC (Sect. 4.5.1) and motorcycles WMTC (Sect. 3.5). This cycle duration was believed to represent an acceptable compromise between statistical representativeness on the one hand and test feasibility in the laboratory on the other. The duration of each cycle segment was set in a way that reflected the mileage distribution among the phases, thus no weighting factors were

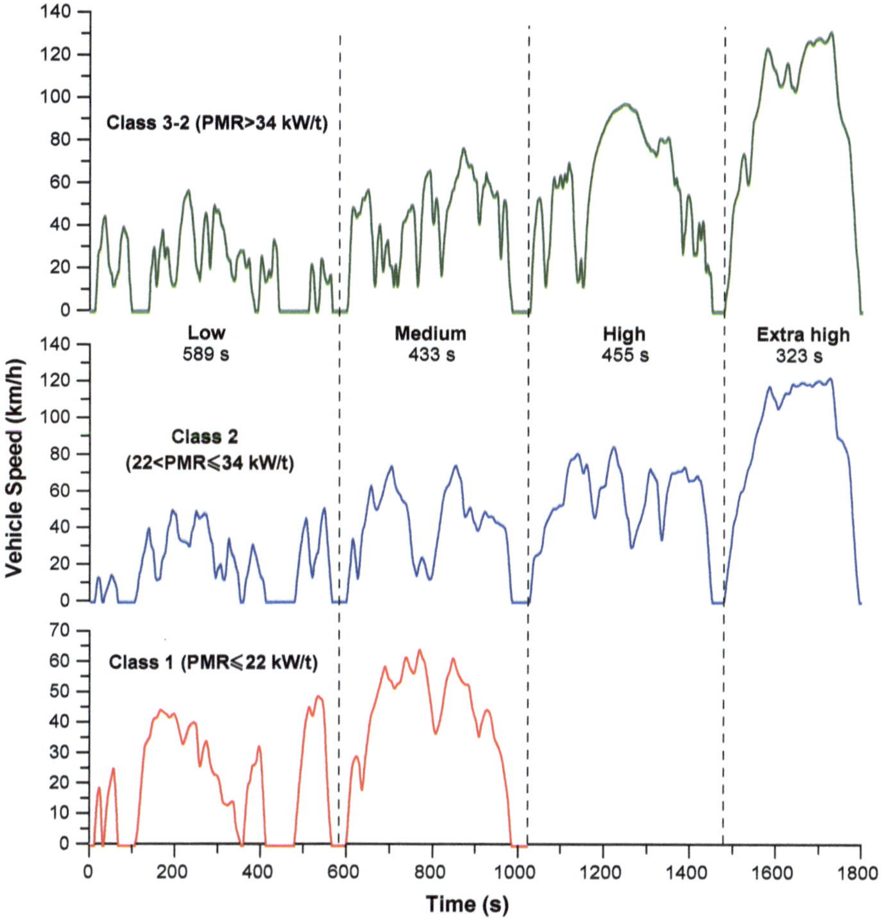

Fig. 2.54 Speed profile of the WLTC driving cycle; a downscaling procedure is applied to improve drivability for the cases of vehicles with PMR close to the borderlines between Class 2 and 3 or very low powered in Class 1, as detailed in [87]

necessary for the final result. Specifically, the low segment lasts 589 s (5 micro-trips and 6 stops), the medium 433 s (1 micro-trip, 2 stops), the high 455 s (1 micro-trip, 2 stops), and the extra high 323 s (1 micro-trip, 2 stops[12]). The final (2015) version of the WLTC suite of cycles is illustrated in Fig. 2.54, whereas Figs. 2.55 and 2.56 demonstrate the speed/acceleration distribution for Class 3-2.

Further, the corresponding UNECE Global Technical Regulation GTR No. 15 provides gear selection and shift point determination (the gear shift strategy

[12]The extra-high segment was developed based on a combination of different sub-segments, defined as take-off, cruise and slow-down.

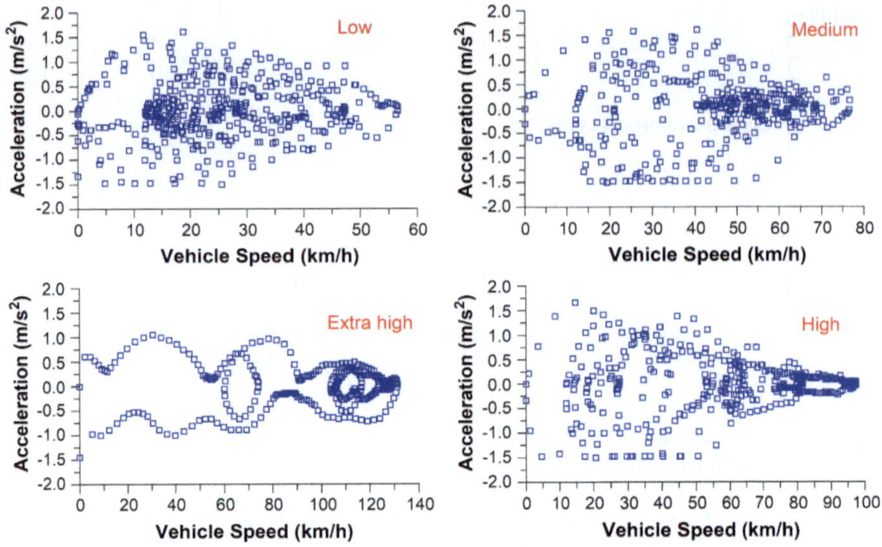

Fig. 2.55 Speed/acceleration distribution for the various segments of the WLTC Class 3-2

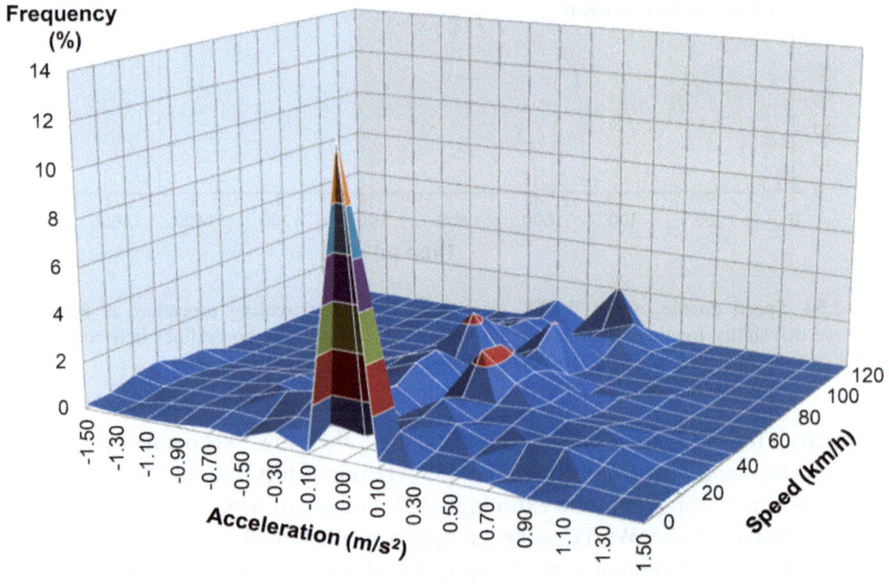

Fig. 2.56 3D speed/acceleration frequency distribution of the WLTC Class 3-2 cycle

being vehicle dependent and not fixed as is the case with the European NEDC), examples of reference fuels, road-load and dynamometer setting, test equipment such as analyzers and their calibrations, and test procedure and conditions to be

Table 2.21 Summary of technical specifications of the WLTC classes

Class	Duration (s)	Distance (m)	Maximum speed (km/h)	Average speed (km/h)	Maximum accel. (m/s^2)	Idling time (%)	RPA (m/s^2)
1	1022	8098	64.4	28.5	0.81	18.8	0.083
2	1800	22,649	123.1	45.3	0.97	12.8	0.119
3-1	1800	23,194	131.3	46.4	1.67	12.6	0.154
3-2	1800	23,266	131.3	46.5	1.67	12.6	0.159

Table 2.22 Summary of technical specifications of the WLTC Class 3-2

Segment	Duration (s)	Distance (m)	Maximum speed (km/h)	Average speed (km/h)	Maximum accel. (m/s^2)	Idling time (%)	RPA (m/s^2)
Low	589	3094	56.5	18.9	1.61	24.4	0.219
Medium	433	4756	76.6	39.5	1.61	10.9	0.206
High	455	7162	97.4	56.7	1.67	6.4	0.138
Extra high	323	8254	131.3	92.0	1.06	1.9	0.127

followed. These also cover vehicles with periodic regeneration, as well as hybrid-electric and pure electric vehicles [87].

Table 2.21 summarizes the main technical specifications of all WLTC classes, with Table 2.22 focusing on Class 3-2, providing data on all four segments

Lastly, development of various engine/vehicle properties, including CO_2 and engine-out NO_x during the WLTC is demonstrated in Fig. 2.57 for a gasoline-engined passenger car.

2.6 Other Countries

As mentioned earlier, the EU, the United States and Japan have been the pioneering regions in the world in developing emission standards and legislating the relevant test procedures for light (and heavy) duty vehicles. Not surprisingly, most of the other countries in the world follow either the European—owing to the close inter-relation with UNECE regulations—or the U.S. or Japanese legislation, often with a delay of some years. More specifically [3, 93–95]:

- Argentina follows the European emission regulations and test procedures;
- Brazil: the official emissions program 'PROCONVE' was published in May 1988. The U.S. FTP-75 cycle is employed (referred to as NBR6601);

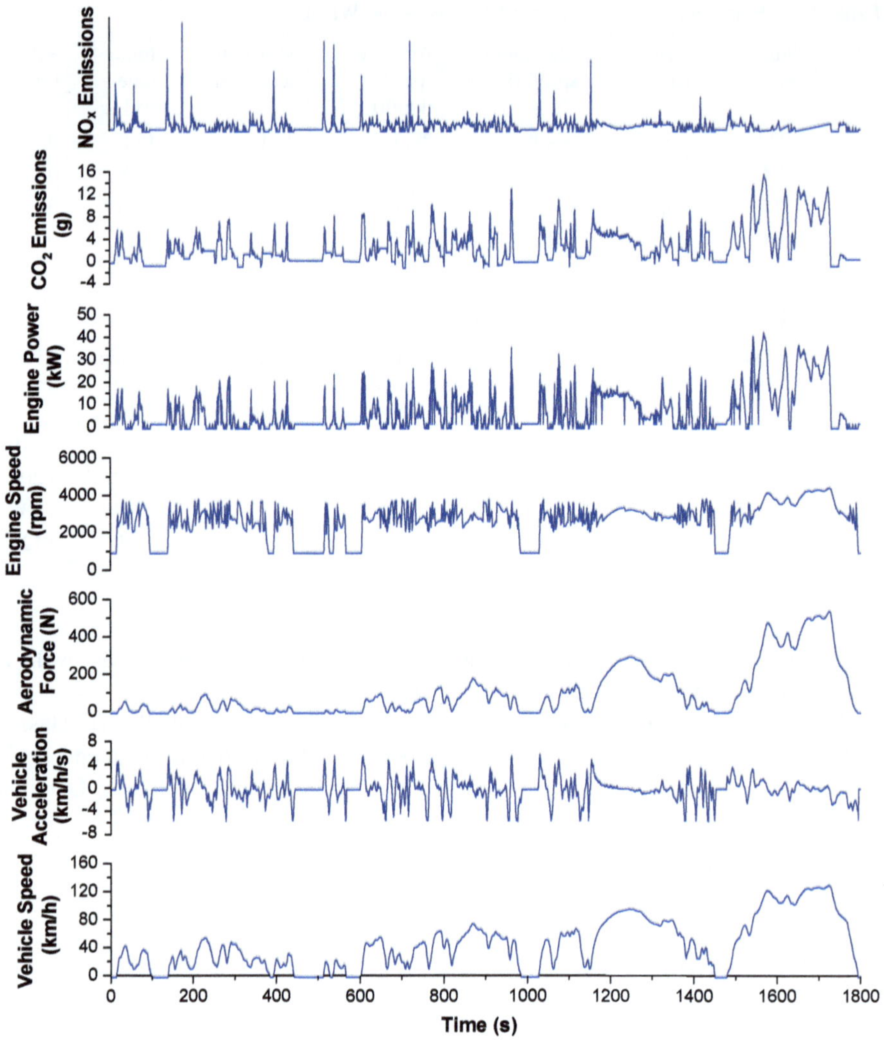

Fig. 2.57 Development of engine properties and emissions from a gasoline passenger car during the WLTC Class 3-2

- Canada: in January 2003, Environment Canada issued regulations for new on-road vehicle emissions, harmonizing its regulations with U.S. EPA standards from MY 2004 on;
- China follows the European regulations and cycles;
- Hong Kong: at the discretion of the manufacturer (EU, U.S. and Japanese tests are accepted);

- India follows the European regulations (the low-powered version of the NEDC was used for certification up to 2010 termed MIDC—modified Indian driving cycle [96]);
- Mexico: Standards that apply to model year 2004 and later are based on U.S. Tier 1 and Tier 2 standards and Euro 3 and Euro 4 limits;
- Russia has adopted the European standards;
- South Korea: the U.S. FTP-75 cycle has been in use for light-duty gasoline and diesel-powered vehicles; for small-sized diesel-engined vehicles, the ECE +EUDC has been in use since 2005;
- Taiwan: the U.S. FTP-75 cycle is employed for emissions certification; FTP-75 or NEDC for fuel consumption from 2011.
- Turkey follows the European legislation.

Apart from the schedules discussed in the previous sections, many more non-legislated cycles have been developed in many areas of the world (e.g., Australia and Asia), aiming at simulating local traffic conditions in specific cities, e.g., Melbourne, Perth and Sydney in Australia, Hong Kong, Pune (India), Bangkok etc., often for the purposes of emissions/fuel consumption modeling studies. For many of these cycles, a representative 'best' route was initially selected that was used for the collection of the driving data.

2.7 Comparative Data

Figure 2.58 summarizes the primary test cycles used for certification purposes in Europe, the United States and Japan over the years, whereas Fig. 2.59 illustrates the main modal cycles that have been at some point legislated or employed for the testing of passenger cars.

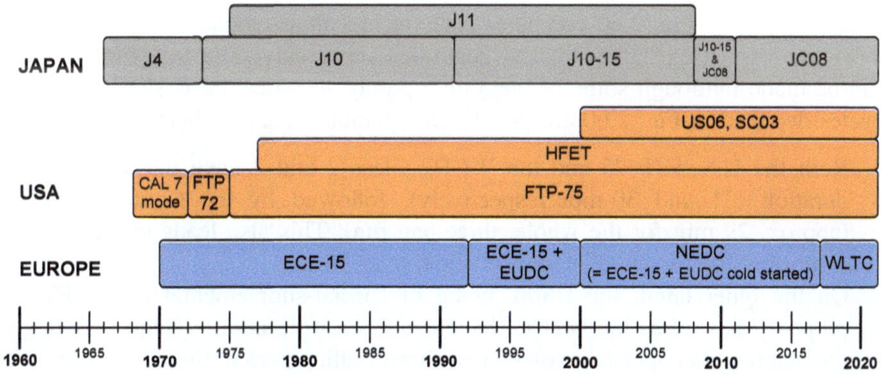

Fig. 2.58 Timeline of test cycles employed for (tailpipe) emission certification purposes in Europe, the U.S. (federal) and Japan

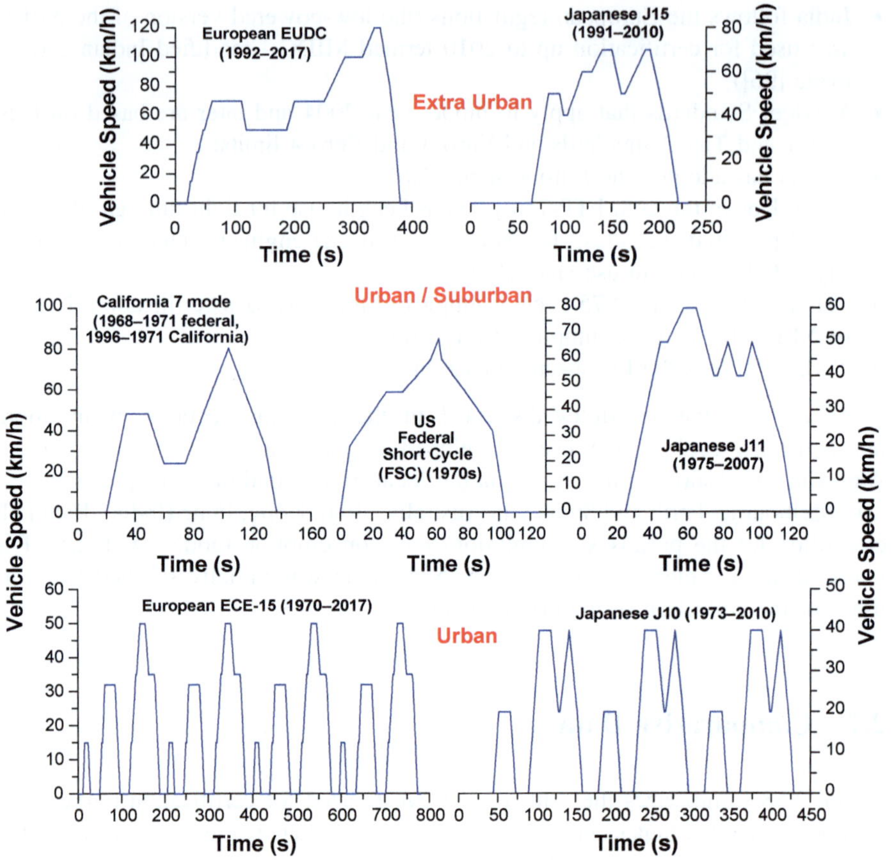

Fig. 2.59 Main modal cycles used for light-duty vehicles in the world together with application periods

The two-page Fig. 2.60 compares basic technical specifications of some of the most important (legislated) cycles mentioned previously. The following remarks can be made (although some of them on a purely statistical basis since not all the cycles depicted in Fig. 2.60 are directly comparable to each other):

- Both the U.S. FTP-75 and the WLTC Class 2 and 3 have the longest cycle duration (31 and 30 min respectively), followed by the California LA-92 (approx. 29 min for the whole, three-bag run). This also leads to these cycles exhibiting the longest distance traveled, over 16 km.
- On the other hand, the US06, being of special-supplemental to the FTP-75 purpose, as well as the NYCC, have the shortest duration (10 min). Owing to the much higher speeds involved and shorter idling period, the distance covered during the US06 is much longer than the urban NYCC.

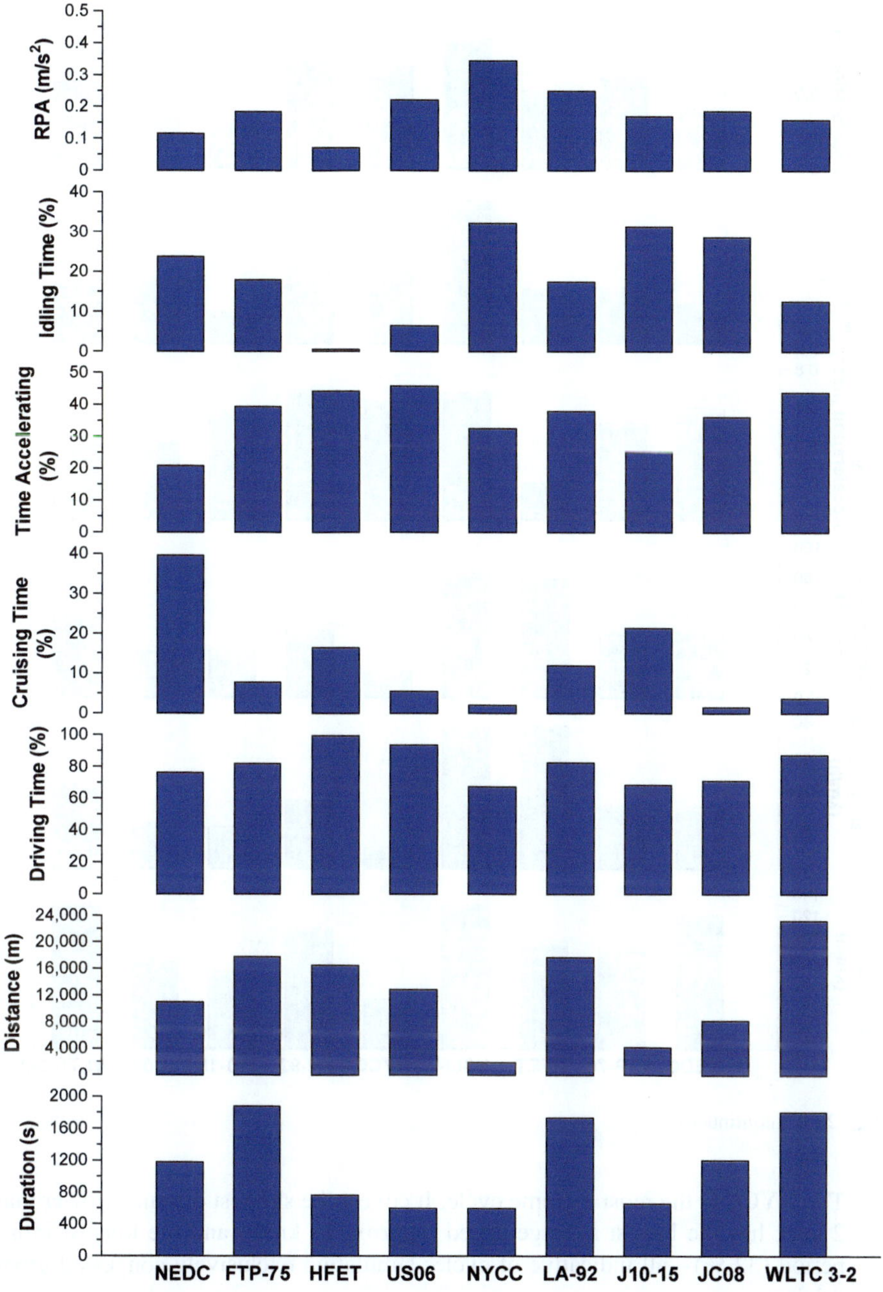

Fig. 2.60 Comparison of some important technical specifications between various light-duty chassis-dynamometer cycles

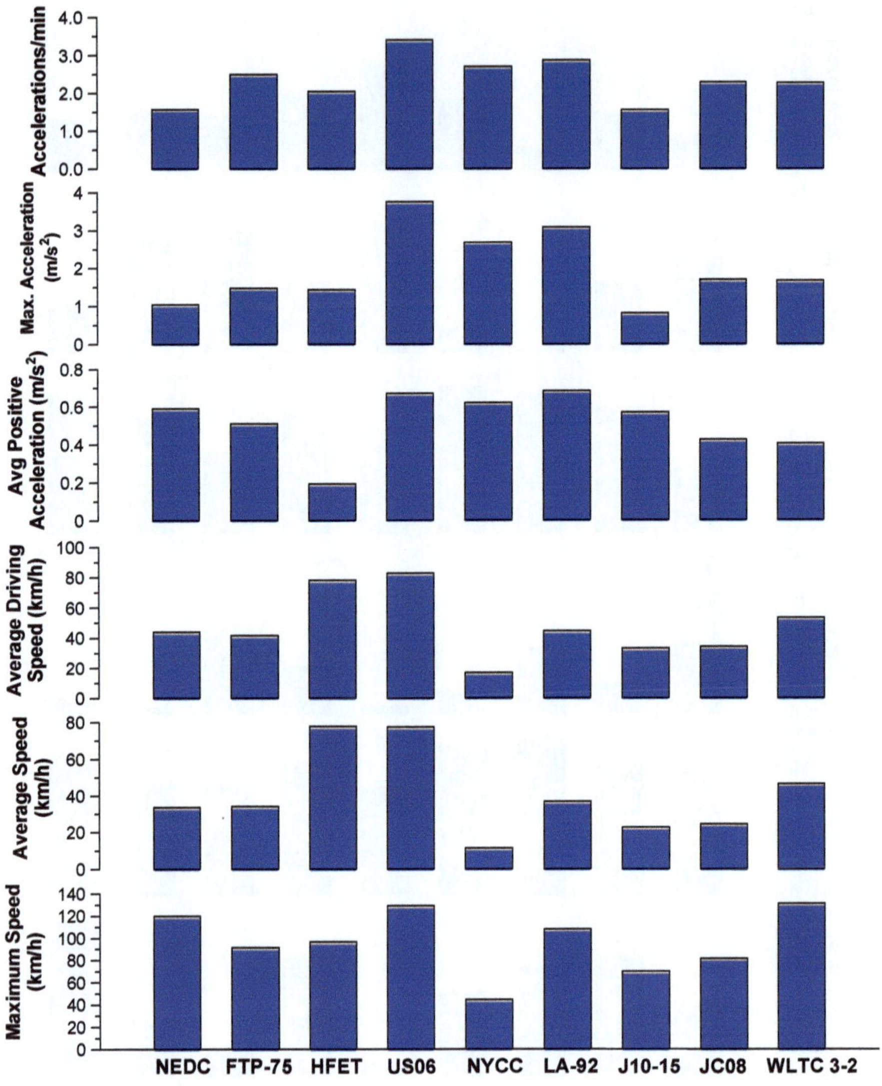

Fig. 2.60 (continued)

- The NYCC is the most extreme cycle. It covers the shortest distance (lower than 2 km), has the lowest average speed (approx. 11 km/h) and the longest idling period (32 %)—all indicative of cycles simulating exclusively congested urban driving.
- Both Japanese cycles, the earlier J10-15 and the subsequent JC08, exhibit idling values of more than 30 %, indicative of the heavily congested traffic conditions in a big Japanese city.

- The Japanese cycles have also rather low maximum, average and driving speeds, in contrast to the European NEDC and the WLTC, as driving habits are much different in Japan compared to Europe.
- In contrast to the NYCC, the US06 and HFET manifest the highest values of driving time (hence shortest idling period), of time spent accelerating, as well as the highest values of average vehicle speed; this is not surprising since both cycles simulate highway driving. The third ranking cycle in this regard is the WLTC 3-2, closely followed by the California LA-92.
- Oddly, the European NEDC is characterized by a very long cruising period (39 %), not compatible with the daily driving experience. It also shows the lowest acceleration time and second lowest maximum acceleration (after the Japanese J10-15).
- On the contrary, the NYCC, US06 and California LA-92 have the maximum number of accelerations/min and also the maximum acceleration and average acceleration; characteristics that strongly affect the amount of (engine-out) emissions particularly from turbocharged vehicles [14].
- RPA is highest for the mostly transient, exclusively urban NYCC, and lowest for the exclusively motorway HFET. Particularly low is the value for the NEDC. It is somewhat surprising that the 'stylized' J10-15 exhibits a RPA value of 0.17 (compared to 0.116 for the NEDC). This is due to the Japanese cycle's more transient profile.

Since many of the cycles presented and compared earlier are either 'special purpose' (HFET, NYCC) or even non-valid any more (J10-15), in the next paragraphs the focus will be on the four 'primary' light-duty cycles in the world, namely the European NEDC, the U.S. FTP-75, the Japanese JC08 and the soon to be implemented WLTC (Class 3-2), comparing some of their specifications and pinpointing their differences and shortcomings; their main technical specifications have been summarized in Table 2.23.

Table 2.23 Comparison of the technical specifications between the NEDC, FTP-75, JC08 and WLTC Class 3-2

Cycle	Test run	Duration (s)	Distance (m)	Maximum speed (km/h)	Average speed (km/h)	Maximum accel. (m/s^2)	Idling time (%)	RPA (m/s^2)
NEDC	Cold	1180	11,000	120.0	33.6	1.04	23.7	0.116
FTP-75	Cold + hot phase	1877	17,769	91.2	34.1	1.48	18.0	0.184
JC08	Cold run + hot run	1204	8159	81.6	24.4	1.69	28.7	0.186
WLTC	Cold	1800	23,266	131.3	46.5	1.67	12.6	0.159

Each of the four driving cycles has advantages and drawbacks. The NEDC is quite simple to drive and thus easily repeatable. However, as already argued, it does not account for real driving behavior in actual traffic, containing many constant-speed and constant-acceleration segments. In fact, from several observations it has been shown that in Europe the gap between fuel consumption experienced by the vehicle users and that measured at type approval is higher compared to other areas of the world, as already discussed in Sect. 2.1.1. Its overall simplistic pattern and the exact gear-shift schedule make it easy for the manufacturers to implement cycle-beating techniques. Moreover, since it is only run once, cold started, its short distance might over-emphasize cold-starting emission effects. Oddly for its outdated structure, the encountered speeds are relatively high, at least compared to its Japanese and U.S. counterparts.

The JC08 is a truly transient cycle but the focus is to a large extent on congested-city traffic conditions. The idle period is long, the vehicle speeds relatively low, and the motorway part rather underestimated. This cycle exhibits the longest single stop phase (76 s, followed by the WLTC with 66 s) but has the highest RPA value, in part owing to the short distance covered and the small motorway segment. The cycle is run both cold and hot, with 75–25 % weighting factors.

The FTP-75 has a real transient pattern and covers a relatively wide range of driving conditions, albeit based on data from vehicles in the late 60s. It is run cold but also incorporates a hot-started segment after a 10-min engine shut-down. It is certainly not complete, in terms of both achieved vehicle speeds and accelerations. Oddly, its limited-duration high-speed part is at the beginning, a fact that influences decisively catalyst operation. As a result, it cannot be considered fully representative, hence it is supplemented in the United States by three additional cycles that address its shortcomings, the HFET, US06 and SC03. Collectively, these four cycles constitute a very good mix of daily driving activity, better than that from the NEDC, JC08 and WLTC, see for example Fig. 2.28.

The most recent WLTC was developed with the ambitious goal to form a universally accepted and applicable test cycle. Its long duration allows for simulation of all possible driving conditions, and has acceptable harshness. However, as the data in Table 2.23 suggests, the maximum speed of 131 km/h might not be considered high enough for the majority of today's passenger cars in Europe (recall that the motorway part of the NEDC reaches 120 km/h, and this is a cycle originally developed in the late 80s), whereas the maximum acceleration of the order of 1.67 m/s^2 or 6 km/h/s suggests moderate (certainly not aggressive) driving behavior. Other features of the WLTC are its highest maximum and driving speed, its fewer stops (only 7 instead of 19 for the FTP-75 lasting at least 4 s), and the shortest total idling period compared to the other three cycles.

One of the most indicative features of the harshness of a cycle is the frequency and degree of the encountered accelerations. The upper two sub-diagrams of Fig. 2.61 present the frequency distribution during the cycles for various acceleration bins; further, Fig. 2.62 compares the acceleration profiles from the NEDC,

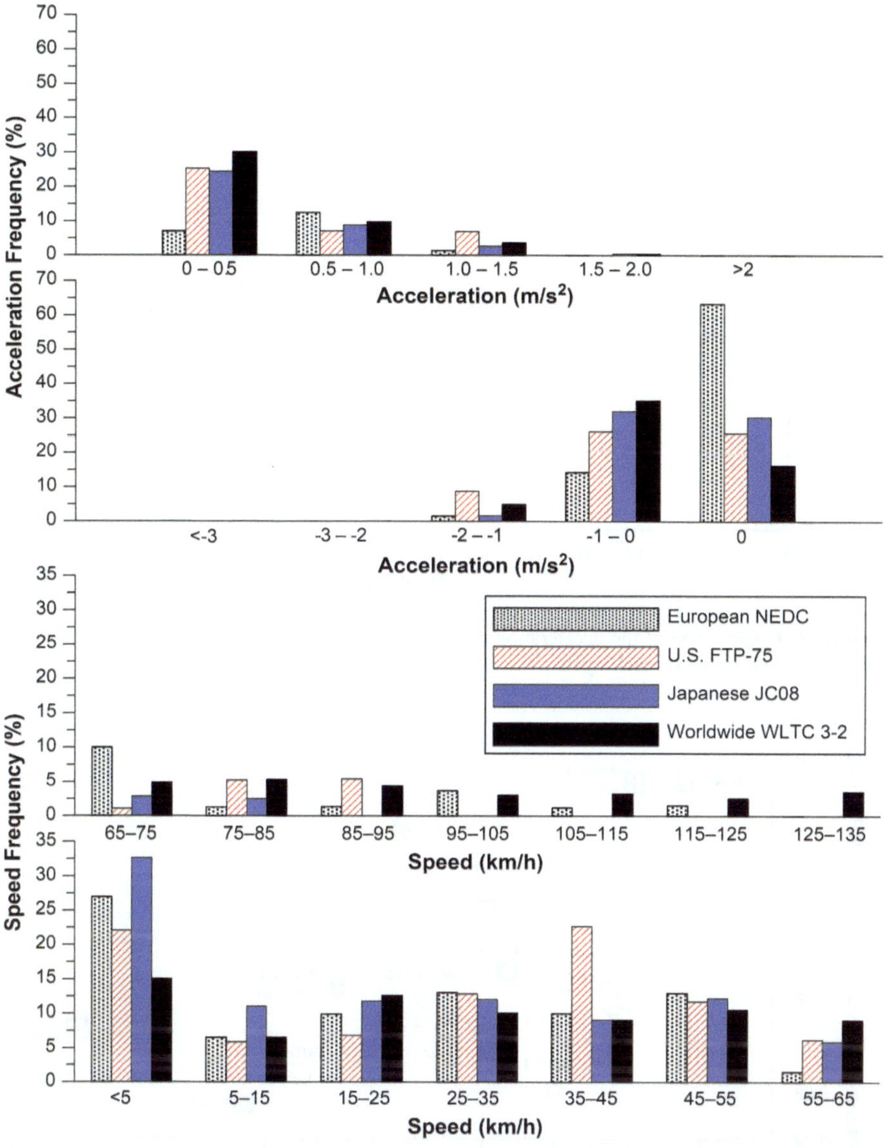

Fig. 2.61 Comparison of the speed and acceleration frequencies for the NEDC, FTP-75, JC08 and WLTC Class 3-2 (notice the increased frequency of the NEDC at zero acceleration owing to extended idling and cruise segments)

FTP-75, JC08 and the WLTC during the first 300 s (urban part). The most obvious remark is (again) the simplistic pattern of the accelerations during the NEDC (ECE-15 segment in particular). It is not only the rather low, by today's standards, maximum accelerations encountered, but also the fact that these accelerations are

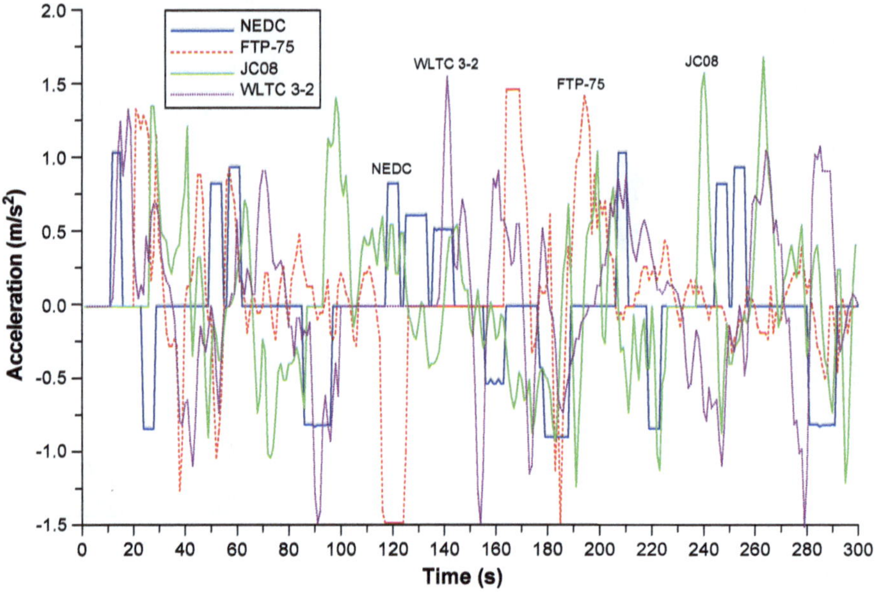

Fig. 2.62 Comparison of the acceleration profiles between various passenger cars driving cycles during the first 300 s of each cycle

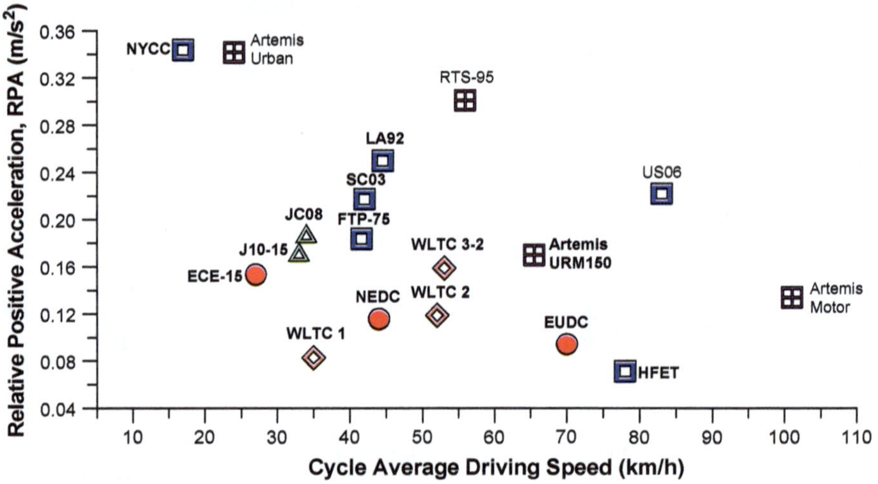

Fig. 2.63 RPA of various passenger cars cycles with respect to their average driving speed (the particularly harsh European non-legislated RTS-95 and the U.S. SFTP US06 seem to differentiate from the general trend of decreasing RPA with increasing driving speed)

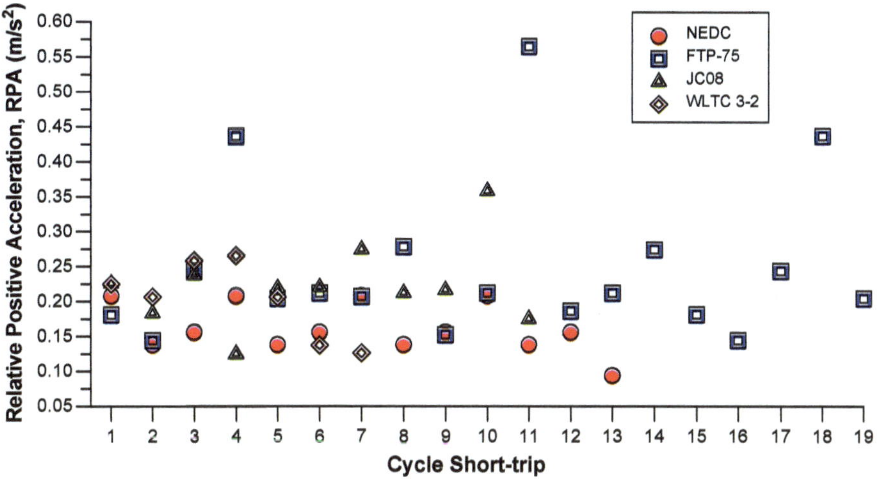

Fig. 2.64 RPA for each short-trip of the NEDC, FTP-75, JC08 and WLTC Class 3-2

for a considerable amount of time constant; for approximately two thirds during the NEDC, acceleration is zero (composed of 40 % cruising and 24 % idling time). Interestingly, the similarly structured Japanese J10-15 (Fig. 2.43) did not exhibit these non-realistic features.

The other three cycles possess a much more acceptable acceleration profile, with both higher values and greater frequency of accelerations, compatible with urban driving. Nonetheless, never is the highest acceleration encountered more than 1.7 m/s^2 (or 6.12 km/h/s, meaning that, at the minimum, 16 s are needed to reach 100 km/h from standstill). Interestingly, the maximum value of 1.7 m/s^2 is encountered in the Japanese cycle, which is, in general, more moderate, at least in terms of speeds, compared to the FTP-75 or the WLTC. From this point of view, even the newly developed WLTC does not seem to be a significant step towards acknowledging the hard driving behavior of many drivers or the increased capabilities of modern cars (cf. the acceleration profile of the FTP US06 in Fig. 2.24).

Figure 2.63 demonstrates RPA values for the four cycles, as well as from others discussed earlier, with respect to the average driving speed (excl. idling) during the cycle. The newly developed WLTC 3-2 lies almost at the same level as the U.S. FTP-75 and the Japanese JC08, being much more dynamic than the European NEDC. In general, urban cycles such as the NYCC or the Artemis Urban, comprising high frequency of hard accelerations from low speeds, are much more demanding, whereas highway cycles, such as the HFET or the EUDC, are usually less dynamic. RPA values for many important cycles are provided in the Appendix. An almost similar picture, as the one presented in Fig. 2.63 for the RPA, is drawn when the positive kinetic energy (Eq. 1.3) is calculated.

Figure 2.64 expands on the previous results by highlighting RPA values for each individual short-trip of the four cycles under comparison. It is the FTP-75 that

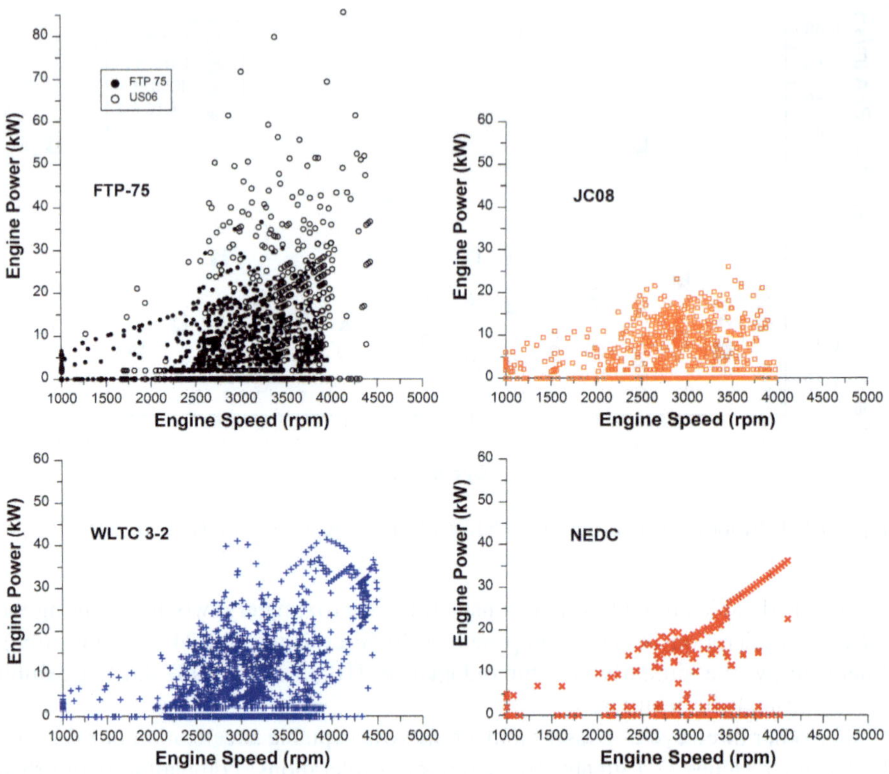

Fig. 2.65 Comparison of the engine speed/power distribution between the WLTC Class 3-2, FTP-75, NEDC and JC08 for a SI-engined passenger car; notice the considerable expansion in the U.S. cycle speed/power range when the tested points during the supplemental US06 are incorporated

differentiates now by a lot from the other three cycles, with three micro-trips having RPA higher than 0.40. Not surprisingly, all micro-trips of the (repetitive-type) NEDC are equally 'soft'.

Further comparative results between the NEDC, FTP-75, JC08 and the WLTC are presented in Fig. 2.65. For the analysis, the engine map coverage was calculated for all four cycles in terms of engine speed and power regarding a small-to-medium, SI-engined passenger car, applying the same gear-shift strategy in order for the results to be comparable. Predictably, the WLTC shows the broadest engine speed range (the JC08 the narrowest) as this correlates with the maximum cycle speed. Further, the WLTC proves to be the most demanding in terms of engine power. Notice the much narrower distribution of the modal NEDC compared with the broader ones from all transient cycles.

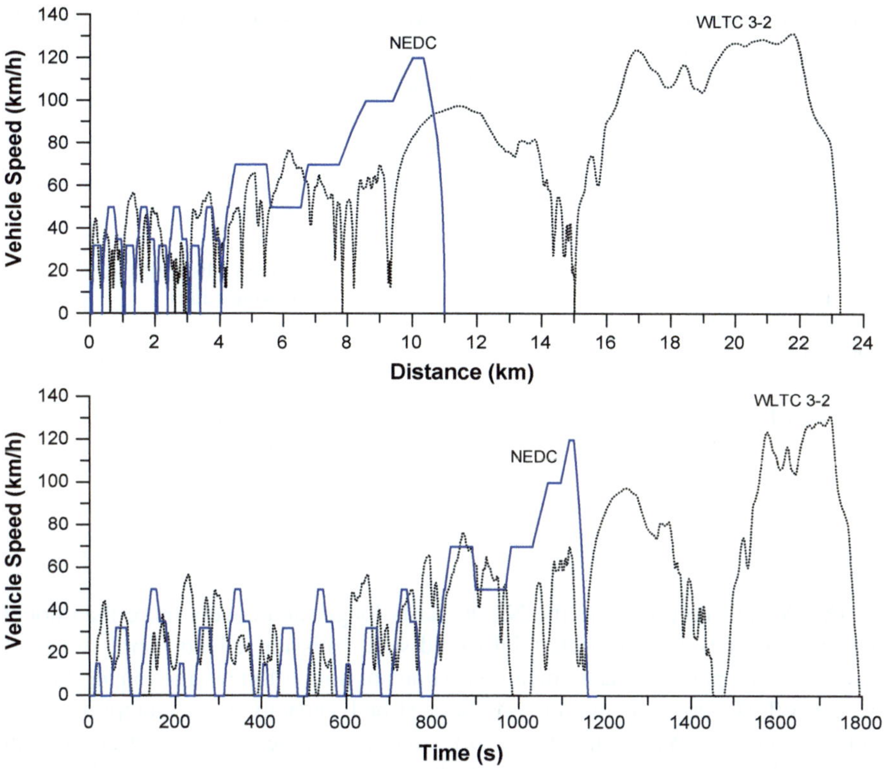

Fig. 2.66 Comparison of the speed versus time (*lower sub-diagram*) and speed versus traveled distance (*upper sub-diagram*) profile between the NEDC and the WLTC Class 3-2

Comparison between the WLTC and the NEDC

Since the beginning of the WLTP process, the European Union had already set a strong political objective by its own legislation (Regulations 443/2009/EC and 510/2011/EU) to modernize the employed test cycle by 2014. This became a major political motivation for the development of the WLTC test cycle and its time frame. The plan of the European Commission is to introduce the new test procedure together with the step 'c' of the Euro 6 standards. This means that starting on September 1, 2017, all new types of M_1 and N_1-class I vehicles should be tested according to the new procedure, with all M_1 and N_1-class I vehicles from September 1, 2018; for the N_1-class II and III vehicles applies the same with a delay of one year [92].

The following paragraphs provide a more detailed comparison between the NEDC and the WLTC. Initially, Fig. 2.66 compares the speed profiles between the NEDC and the WLTC Class 3-2; the latter is the class primarily suited to the majority of European cars. Further, Table 2.24 quantifies the differences between major technical specifications of the two cycles.

Table 2.24 Differences in major technical specifications between the WLTC 3-2 and the NEDC

Specification	NEDC	WLTC 3-2	Difference (from NEDC data) (%)	Effect
Duration (s)	1180	1800	+53	Lower influence of cold-start emissions
Distance (m)	11,000	23,266	+112	
Average speed (km/h)	33.6	46.5	+38	Most probably, better fuel efficiency
Maximum speed (km/h)	120	131.3	+9	More realistic of today's driving habits
Idling time (%)	23.7	12.6	−47	Lower influence of start-stop systems
Cruising (%)	39.6	3.7	−91	More transient, hence higher pollutant and CO_2 emissions
Transient time (%)	36.7	83.7	+128	
RPA (m/s^2)	0.116	0.159	+37	

The WLTC, compared to the NEDC, lasts longer and covers more than double distance. This is then reflected into cold-start emission effects being relatively lower. From a purely measurement/experimental point of view, the longer duration of the WLTC poses a burden on the test-bed capacity (e.g., sampling bags). Furthermore, the WLTC has higher maximum, average and driving speeds, and almost half the idling period (although the single stop with the longest duration is 66 s for the WLTC and only 27 for the NEDC). For approximately 84 % of the time during the WLTC the vehicle operates in transient conditions (accelerating or decelerating), in contrast to only 37 % during the NEDC. Cruising time is almost absent during the WLTC, and RPA is much higher. These are expected to result in higher engine-out pollutant emissions, and higher CO_2.

In Fig. 2.62 both cycles were compared with regard to vehicle acceleration; Fig. 2.67 expands these results by illustrating the respective acceleration rate (m/s^3) during the urban and motorway parts. Notice in this figure the continuous 'microscopic' changes in the WLTC profile compared to the more 'constant' development during the NEDC. Further, Fig. 2.68 compares the speed/acceleration distribution between the two cycles. The much broader speed/acceleration distribution of the WLTC confirms the considerable progress over the current NEDC in terms of simulated real-world driving behavior.

In addition, more realistic and demanding conditions of the WLTC compared to the NEDC result in increase in total energy demand when tested on a chassis dynamometer in accordance with the new WLTP requirements, such as new/higher test mass, road-load calculations, gearshift strategy, etc. For example, both kinetic energy and active resistance (rolling resistance and air drag) have been found significantly higher over the WLTC compared to the NEDC for the same gasoline vehicle [92]. Figure 2.69 regarding a diesel-engined, light-duty van and Fig. 2.70

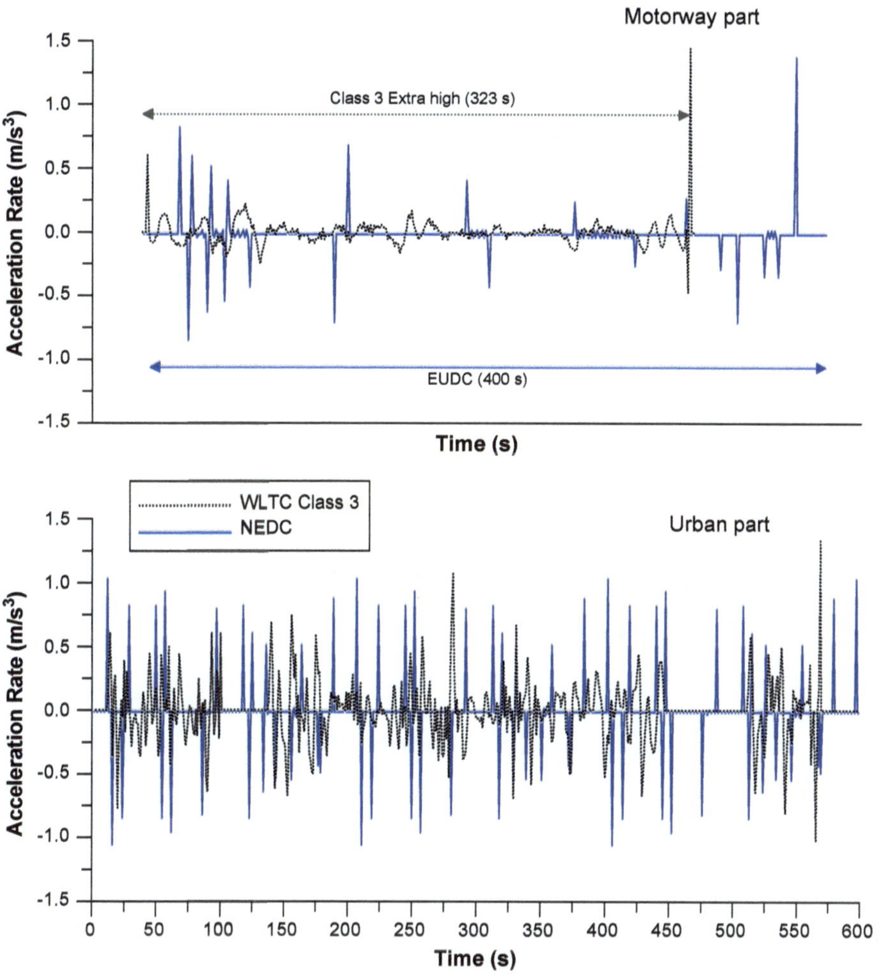

Fig. 2.67 Comparison of the rate of change of acceleration ('jerk') between the NEDC and the WLTC Class 3 for urban and motorway (extra high for the WLTC) driving segments

for a SI-engined passenger car support this argument by illustrating engine speed/power distribution during a whole run for both cycles. Higher values of power and speed as well as broader distribution are evidenced for the WLTC over its NEDC counterpart, a fact that is expected to affect accordingly the emitted pollutants [14]. Overall, 78 % of the operating points during the NEDC correspond to lower than 3000 rpm engine speed and 82 % to lower than 20 kW power in Fig. 2.69; for the WLTC, the respective numbers are much lower, 55 % and 68 %.

The transition from the NEDC to the WLTC in the EU will not be limited to the change of the test cycle; instead, several modifications in the test procedure will take place as well, namely

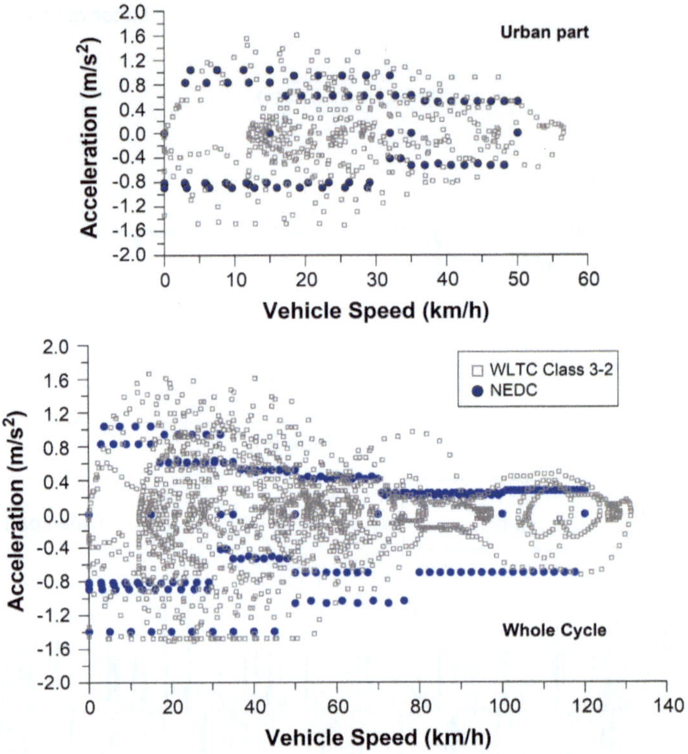

Fig. 2.68 Comparison of the speed/acceleration distribution between the NEDC and the WLTC Class 3-2 (the WLTC has 1800 operating points and the NEDC 1180)

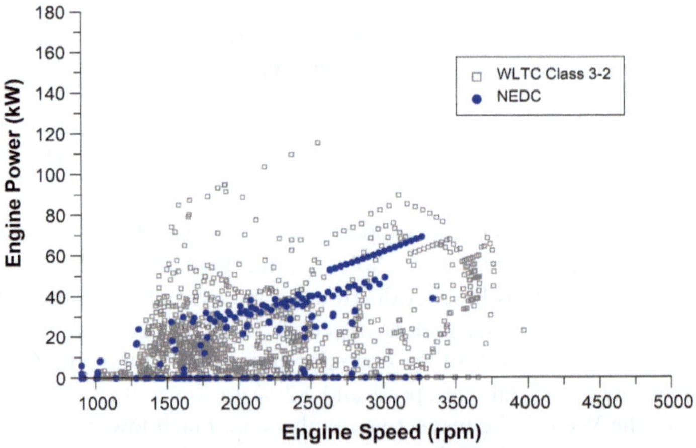

Fig. 2.69 Comparison of the engine speed/power distribution between the NEDC and the WLTC Class 3-2 for a diesel-engined van

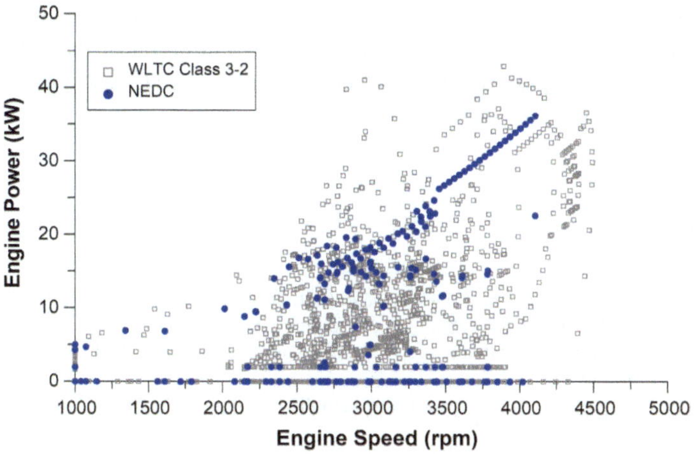

Fig. 2.70 Comparison of the engine speed/power distribution between the NEDC and the WLTC Class 3-2 for a gasoline passenger car

- more stringent requirements will be imposed in determining the representative road load for the dynamometer inertia;
- the test temperature will be somewhat lower, 23 °C instead of 20–30 °C, with a possibility to adopt 14 °C in Europe;
- instead of using discrete classes, the dynamometer inertia will correspond exactly to the vehicle reference mass;
- hybrid-electric and pure electric vehicles will be tested on dedicated to these vehicles procedures;
- one set of sampling bags for each phase of the cycle, with bag analysis conducted in parallel to the test run, etc. [23, 90].

With the introduction of the WLTP although the regulated pollutant emission limits will not be affected, the CO_2 targets (NEDC-based fleet-average value of 95 g/km for 2020–21) will, most probably, need to be adapted. The longer distance of the WLTC over the NEDC, its higher velocities and accelerations and its lower idle phase are significant parameters that will influence these new CO_2 limits (see also Table 2.24). Consequently, correlation parameters have to be determined in order to reflect any change in the regulatory test procedure for the measurement of specific CO_2 emissions referred to in Regulations 715/2007/EC and 692/2008/EC [90, 97]. As has been reported in [90], the impacts of cold start are lower on the WLTC (due to the longer duration) compared to the NEDC, the higher speeds and loads of the WLTC have the counter effect on engine efficiency, typically increasing with load, the new gearshift schedule on the WLTC can help reduce engine-speed (hence friction) effects, and the lower idle percentage on the WLTC might actually penalize enhanced systems such as start/stop. From the same analysis, it has been estimated that the NEDC-based CO_2 target for 2020 (95 g/km) might increase to 100 or even 102 g/km owing to the transition to the WLTP test procedure.

This increase is composed from: (a) 2 g/km due to the change of the certification driving cycle, (b) 3 g/km owing to changes in the definition of the vehicle's mass for the test, and (c) 2 g/km owing to lower test temperature (2 g/km might be added if the test temperature is further lowered to 14 °C).

References

1. ACEA. Official website. http://www.acea.be/statistics. Accessed 28 Aug 2016.
2. Berg W. Legislation for the reduction of exhaust gas emissions. In: Gruden D, editor. The handbook of environmental chemistry, traffic and environment, Vol. 3. Berlin: Springer; 2003. p. 175–253.
3. TransportPolicy.net. http://www.transportpolicy.net/index.php?title=Category:Light-duty_Vehicles. Accessed 12 Aug 2016.
4. Wurzel RKW. Environmental policy-making in Britain, Germany and the European Union. Manchester, UK: Manchester University Press; 2006.
5. Chapoux E, Delpeyroux D. French studies of pollution by motor car exhaust gases. In: Proceedings Institute of clean air conference, Paper V1/2. 1966, pp. 157–163.
6. Watson HC. Vehicle driving patterns and measurement methods for energy and emissions assessment. Canberra: Australian Government Publishing Service; 1978.
7. Degobert P. Automobiles and pollution. Paris: Editions Technip; 1995.
8. van Beckhoven L. An appraisal of the proposals for the further development of EEC—legislation on automotive emissions. SAE Paper No. 871077, 1987.
9. Armas O, García-Contreras R, Ramos Á, López A. Impact of animal fat biodiesel, GTL and HVO fuels on combustion, performance and pollutant emissions of a light-duty diesel vehicle tested under the NEDC. ASCE J Energy Eng, Article No. C4014009, 2014.
10. Giakoumis EG, Lioutas SC. Diesel-engined vehicle nitric oxide and soot emissions during the European light-duty driving cycle using a transient mapping approach. Transp Res Pt. D 2010;15:134–43.
11. Kühlwein J. The impact of official versus real-world road loads on CO_2 emissions and fuel consumption of European passenger cars. ICCT White Paper, May 2016.
12. Joumard R, André M, Vidon R, Tassel P, Pruvost C. Influence of driving cycles on unit emissions from passenger cars. Atmos Environ. 2000;34:4621–8.
13. Pelkmans L, Debal P. Comparison of on-road emissions with emissions measured on chassis dynamometer test cycles. Transport Res, Pt. D 2006;11:233–41.
14. Rakopoulos CD, Giakoumis EG. Diesel engine transient operation. London: Springer; 2009.
15. Tzirakis E, Pitsas K, Zannikos F, Stournas S. Vehicle emissions and driving cycles: comparison of the Athens driving cycle (ADC) with ECE-15 and European driving cycle (EDC). Global NEST J. 2006;8:282–90.
16. Mock P, German J, Bandivadekar A, Riemersma I, Ligterink N, Lambrecht U. A comparison of official and 'real-world' fuel consumption and CO_2 values for cars in Europe and the United States. ICCT White Paper, May 2013.
17. Kadijk G, van Mensch P, Spreen J. Detailed investigations and real-world emission performance of Euro 6 diesel passenger cars. TNO Report 2015 R10702, May 2015.
18. Tietge U, Zacharof N, Mock P, Franco V, German J, Bandivadekar A, et al. A 2015 update of official and 'real-world' fuel consumption and CO_2 values for passenger cars in Europe. ICCT White Paper, Sept 2015.
19. Ntziachristos L, Mellios G, Tsokolis D, Keller M, Hausberger S, Ligterink NE, et al. In-use vs. type-approval fuel consumption of current passenger cars in Europe. Energy Policy. 2014;67:403–11.

20. Franco V, Sánchez FP, German J, Mock P. Real-world exhaust emissions from modern diesel cars—part 1: aggregated results. ICCT White Paper, Oct 2014.
21. Kadijk G, Ligterink N, Spreen J. On-road NO_x and CO_2 investigations of Euro 5 light commercial vehicles. TNO Report 2015 R10192, Mar 2015.
22. Reducing NO_x emissions on the road. European conference of ministers of transport. Report prepared by the ECMT group on transport and the environment in co-operation with the OECD environment policy committee's working group on transport. 2006. http://www.itf-oecd.org/sites/default/files/docs/06nox.pdf. Accessed 5 July 2016.
23. Engeljehringer K. Emission legislation timeline: overview about emission regulation with focus on RDE. AVL-Italy RDE Roadshow, AVL List GmbH, 18 July 2015.
24. Dieselnet website. https://www.dieselnet.com//standards/cycles/. Accessed 2 Sept 2016.
25. Personal communication with Mr. Kurt Engeljehringer, AVL, Sept 2016.
26. André M. The ARTEMIS European driving cycles for measuring car pollutant emissions. Sci Total Environ. 2004;334–335:73–84.
27. André M, Joumard R, Hickmam AJ, Hassel D. Actual car use and operating conditions as emission parameters: derived urban driving cycles. Sci Total Environ. 1994;146/147:225–33.
28. Favre C, Bosteels D, May J. Exhaust emissions from European market-available passenger cars evaluated on various drive cycles. SAE Paper No. 2013-24-0154, 2013.
29. Barlow TJ, Latham S, McCrae IS, Boulter PG. A reference book of driving cycles for use in the measurement of road vehicle emissions. Published Project Report PPR354, TRL, June 2009.
30. EcoTest-testing and assessment protocol. ADAC e.V., January 2015. www.adac.de/ecotest. Accessed 15 July 2016.
31. Reitze AW Jr. Air pollution control law: compliance and enforcement. Washington DC: Environmental Law Institute; 2001.
32. Morriss AP, Yandle B, Dorchak A. Regulating by litigation: the EPA's regulation of heavy-duty diesel engines. Admin L Rev 403; 2004.
33. Stonex KA. Survey of Los Angeles traffic characteristics. In: 36th annual meeting, U.S. Highway Research Board; 1957. pp. 509–38.
34. Hass GC, Brubacher ML. A test procedure for motor vehicle exhaust emissions. J Air Pollut Control Assoc. 1962;12:505–509.
35. Clarkson D, Middleton JT. The California control program for motor vehicle created air pollution. J Air Pollut Control Assoc. 1962;12:22–8.
36. Austin TC, di Genova FJ, Carlson TR, Joy RW, Gianolini KA, Lee JM. Characterization of driving patterns and emissions from light-duty vehicles in California. California Air Resources Board, Final Report, Nov 1993.
37. Hass GC, Sweeney MP, Pattison JN. Laboratory simulation of driving conditions in the Los Angeles area. SAE Paper No. 660546, 1966.
38. Kruse RE, Huls TA. Development of the federal urban driving schedule. SAE Paper No. 730553, 1973.
39. Kearin DH, Lamoureux RL. Survey of average driving patterns in the Los Angeles urban area. TM-(L)-4119/000/01, Feb 1969.
40. SAE J1634. Battery electric vehicle energy consumption and range test procedure. Society of Automotive Engineers, Oct 2012.
41. Federal test procedure review project preliminary technical report. U.S. EPA 420-R-93-007, May 1993.
42. Lin J, Niemeier DA. An exploratory analysis comparing a stochastic driving cycle to California's regulatory cycle. Atmos Environ. 2002;36:5759–70.
43. Kruse RE, Paulsell CD. Development of a highway driving cycle for fuel economy measurements. EPA, Mar 1974.
44. Austin TC, Hellman KH, Paulsell CD. Passenger car fuel economy during non-urban driving. SAE Paper No. 740592, 1974.

45. Berry IM. The effects of driving style and vehicle performance on the real-world fuel consumption of U.S. light-duty vehicles. MSc Thesis, Virginia Polytechnic Institute and State University, 2007.
46. U.S. Environmental Protection Agency. 40 CFR Parts 79, 80, 85, et al. Control of air pollution from motor vehicles: Tier 3 motor vehicle emission and fuel standards; Final Rule, Federal Register, Vol. 79, No. 81, 28 Apr 2014.
47. Filipi Z, Wang Y, Assanis D. Effect of variable geometry turbine (VGT) on diesel engine and vehicle system transient response. SAE Paper No. 2001-01-1247, 2001.
48. U.S. Environmental Protection Agency SFTP. https://www3.epa.gov/otaq/sftp.htm. Accessed 22 July 2016.
49. U.S. Environmental Protection Agency. Final technical report on aggressive driving behavior for the revised federal test procedure notice of proposed rulemaking, 31 Jan 1995.
50. U.S. Environmental Protection Agency. Fuel economy labeling of motor vehicles revisions to improve calculation of fuel economy estimates: final technical support document, EPA420-R-06-017, Dec 2006.
51. U.S. Environmental Protection Agency. Final technical report on air conditioning for the federal test procedure revisions notice of proposed rulemaking, Jan 1995.
52. U.S. Environmental Protection Agency. 40 CFR Parts 86 and 600. Fuel economy labeling of motor vehicles: revisions to improve calculation of fuel economy estimates; Proposed rule. Federal Register, Vol. 71. 1 Feb 2006.
53. EPA fuel economy and environment labels. https://www3.epa.gov/carlabel/. Accessed 27 Aug 2016.
54. U.S. Environmental Protection Agency. Recommended practice for measurement of exhaust sulfate emissions from light-duty vehicles and trucks, Sept 1977.
55. Gibbs R, Hyde JD, Whitby RA, Choudhury DR. Characterization of emissions and fuel economy of in-use diesel automobiles. Final Report, EPA-600/3-83-087, Sept 1983.
56. U.S. Environmental Protection Agency. Final regulatory impact analysis and summary and analysis of comments—control of vehicular evaporative emissions, Feb 1993.
57. California Air Resources Board. http://www.arb.ca.gov/research/resnotes/notes/96-11.htm. Accessed 24 May 2015.
58. Pidgeon WM, Dobie N. The IM240 transient I/M dynamometer driving schedule and the composite I/M test procedure. U.S. EPA, EPA-AA-TSS-91-1, Jan 1991.
59. U.S. Environmental Protection Agency. High-Tech I/M test procedures, emission standards, quality control requirements, and equipment specifications: IM240 and functional evaporative system tests—revised technical guidance. EPA-AA-RSPD-IM-96-1, June 1996.
60. Ando AW, Harrington W, McConnell V. Estimating full IM240 emissions from partial test results: evidence from Arizona. Discussion Paper 98-24, Mar 1998.
61. Committee on Vehicle Emission Inspection and Maintenance Programs. Evaluating vehicle emissions inspection and maintenance programs. Washington DC: National Academy press; 2001.
62. St. Denis M, Lindner J. Review of light-duty diesel and heavy-duty diesel gasoline inspection programs. J Air Waste Manage Assoc 2005;55:1876–84.
63. U.S. Environmental Protection Agency. 40 CFR Part 86, Emission durability procedures and component durability procedures for new light-duty vehicles, light-duty trucks and heavy-duty vehicles; Final rule and proposed rule, Federal Register, Vol. 71. 17 Jan 2006.
64. Cuerden R, Nathanson AA, Goodacre O, McCarth M, Knight I, Muirhead M, et al. Durability of pollution control measures for L-category vehicles. Final Report, PPR627, Dec 2012.
65. Brzezinski D, Hart C, Enns P. Final facility specific speed correction factors. U.S. EPA Final Report, EPA420-R-01-060, Nov 2001.
66. Carlson TR, Austin TC. Development of speed correction cycles. Sierra Research Report SR97-04-01 prepared for the U.S. EPA, Apr 1997.
67. Carlson TR, Austin TC, McClement D, Yoon S-H. Development of generic link-level driving cycles. Sierra Research Report SR2009-05-02, prepared for the U.S. EPA, May 2009.

68. Kunselman P, McAdams HT, Domke CJ, Williams M. Automobile exhaust emission modal analysis model. U.S. EPA-460/3-74-005, Jan 1974.
69. Spindt RS, Dizak RE, Stewart RM, Meyer WAP. Effect of ambient temperature on vehicle emissions and performance factors. U.S. EPA-460/3-79-006A, Sept 1979.
70. Chan CC, Chau KT. Modern electric vehicle technology. New York: Oxford University Press; 2001.
71. Husain I. Electric and hybrid vehicles: design fundamentals. 2nd ed. Boca Raton: CRC Press; 2011.
72. Larminie J, Lowry J. Electric vehicle technology explained. Chichester: Wiley; 2003.
73. Brandt DD. Driving cycle testing of electric vehicle batteries and systems. J Power Sour. 1992;40:73–9.
74. Burke AF, Cole GH. Application of the GSFUDS to advanced batteries and vehicles. In: 10th international electric vehicle symposium, Hong Kong, Dec 1990.
75. Cole GH. A generic SFUDS battery test cycle for electric road vehicle batteries. SAE Paper No. 891664, 1989.
76. Itagaki A. The legislative process of automobile exhaust emissions control in Japan—the 1966 regulation and the role of the ministry of transport. In: Takeda H, editor. Micro-performance during postwar Japan's high-growth era. Singapore: Springer; 2016. p. 39–68.
77. Yoshizumi K, Nakamura K, Inoue K, Ishiguro T. Automotive exhaust emissions in an urban area. SAE Paper No. 800326, 1980.
78. Japanese Industrial Safety and Health Association. Technical standard for 10-mode and 11-mode exhaust emission measurement for gasoline-fueled motor vehicles, JISHA 899, 1983.
79. Vehicle Emission Standards—Japan. MECA World Regulations, 2006.
80. Kimura T. General resources energy investigation committee, Energy savings standards section, Automobile standards judging subcommittee, LPG car fuel efficiency standards evaluation group, Final Report, Feb 2003. https://www.eccj.or.jp/top_runner/pdf/tr_lpg_passenger.pdf. Accessed 7 July 2016.
81. Final report of joint meeting between the automobile evaluation standards subcommittee, energy efficiency standards subcommittee of the advisory committee for natural resources and energy, and the automobile fuel efficiency standards subcommittee (automobile transport section, land transport division of the council for transport policy), concerning revisions of evaluation standards for manufacturers with regard to improvement of automobile energy consumption efficiency, Japan, Mar 2007.
82. Rule G, Allen GH, Kent JH. A driving cycle for Sydney. Charles Kolling Research Laboratory Technical Note ER-15, Department of Mechanical Engineering, University of Sydney; 1976.
83. Watson HC, Milkins EE, Braunsteins J. Development of the Melbourne peak cycle. In: Joint SAE-A/ARRB 2nd conference on traffic energy and emissions, 19–21 May 1982, Melbourne, SAE Paper 82148.
84. Kenworthy JR, Newman PWG, Lyons TJ. A driving cycle for Perth. Report to National Energy Research Development and Demonstration Council, NERDDP/EG/83/120, 1983.
85. Brown S, Bryett C, Mowle M. In-service emissions performance—drive cycles, Vol. 1, NSW Environment Protection Authority, Mar 1999.
86. NISE 2—drive cycle and short test development. Final Report, Department of the Environment and Heritage, Sept 2005.
87. Global Technical Regulation No. 15. Worldwide harmonized light vehicles test procedure. Established in the Global Registry on 12 Mar 2014.
88. Tutuianu M, Marotta A, Steven H, Ericsson E, Haniu T, Ichikawa N, et al. Development of the worldwide harmonized light-duty driving test cycle (WLTC). Technical report, UNECE/WP.29/GRPE/WLTP-IG DHC subgroup, Dec 2013.

89. Tutuianu M, Bonnel P, Ciuffo B, Haniu T, Ichikawa N, Marotta, A, Pavlovic J, et al. Development of the worldwide harmonized light duty test cycle (WLTC) and a possible pathway for its introduction in the European legislation. Transport Res, Pt. D 2015;40:61–75.
90. Mock P, Kühlwein J, Tietge U, Franco V, Bandivadekar A, German J. The WLTP: how a new test procedure for cars will affect fuel consumption values in the EU. ICCT Working Paper 2014-9, Oct 2014.
91. Steven H. Homologation test cycles worldwide—status of the WLTP. Green global NCAP labelling/green scoring Workshop, Apr, 2013.
92. Ciuffo B, Marotta A, Tutuianu M, Anagnostopoulos K, Fontaras G. Pavlovic J, et al. The development of the world-wide harmonized test procedure for light duty vehicles (WLTP) and the pathway for its implementation into the EU legislation. Transportation Research Board, 94th Annual Meeting, Washington DC, United States, 2015.
93. Worldwide emission standards—Passenger cars and light-duty vehicles. Delphi, 2015/16.
94. Worldwide emission standards and related regulations—Passenger cars, light and medium duty vehicles. Continental, Feb 2009.
95. Faiz A, Weaver CS, Walsh MP. Air pollution from motor vehicles—standards and technologies for controlling emissions. Washington DC: The International Bank for Reconstruction and Development/World Bank; 1996.
96. Document on test method, testing equipment and related procedures for testing type approval and conformity of production (COP) of vehicles for emission as per CMV Rules 115, 116 and 126. Ministry of Road Transport and Highways, Government of India, Mar 2010.
97. Marotta A, Pavlovic J, Ciuffo B, Serra S, Fontaras G. Gaseous emissions from light-duty vehicles: moving from NEDC to the new WLTP procedure. Environ Sci Technol. 2015;49:8315–22.

Chapter 3
Motorcycles

Abstract This chapter presents the driving cycles used for the certification of motorcycles, which are all of the chassis-dynamometer type. European/UNECE (R40 and R47), American (FTP-72), Indian (IDC) and worldwide (WMTC) cycles are presented, with graphic illustration of the speed/acceleration profiles together with a historical overview. A summary of the main technical specifications is provided as well, along with detailed comparisons between the cycles, and results from real light-vehicles operation.

Motorcycles, tricycles, quadricycles and mopeds form a significant part of the worldwide motor vehicle population, particularly in Asia.[1] Their lower cost compared to passenger cars and much easier navigation in heavy traffic make them appealing to many customers, particularly young ones. Although their contribution to emissions is smaller than that of passenger cars and heavy-duty vehicles, it is by no means negligible, mainly with regard to carbon monoxide and unburned hydrocarbons. This holds particularly true when taking into account the decreasing trend in the amount of emitted pollutants from passenger cars and heavy-duty vehicles (engines), following the constantly increasing stringency in their emission limits. Not surprisingly, emission targets and drive cycles for motorcycles have been legislated in many countries around the world (at a later stage compared to passenger cars); the pioneering country was the United States, enforcing emission standards in 1978. Motorcycles are powered by a variety of engines, e.g., two-stroke ones with carburetors (installed in mopeds and preferred over four-stroke ones owing to higher power to mass ratio and lower maintenance cost); four-stroke SI with sophisticated electronic fuel injection control and three-way catalysts; simple (not electronically-controlled) or modern-technology diesel engines; hybrid or pure electric power-trains are also available [1, 2].

[1]Estimated 65 % of the total vehicle population in the 10 largest Asian vehicle markets in 2010 [1].

E.G. Giakoumis, *Driving and Engine Cycles*,
DOI 10.1007/978-3-319-49034-2_3

3.1 ECE R47

In Europe, the first legislation concerning emission limits and test cycles for motorcycles was published on June 17, 1997 with Directive 97/24/EC regarding Euro 1 and Euro 2 standards (implementation dates of the emission regulations for motorcycles do not temporally coincide with those of the passenger cars/light vans). Motorcycles, tricycles and mopeds were covered, with mopeds defined in Directive 92/61/EEC as light vehicles with engine capacity lower than 50 cm^3 and maximum speed lower than 45 km/h. For the latter, CO and HC + NO_x were the controlled pollutants.

For mopeds and scooters, the ECE R47 has been the European test cycle for emission standards Euro 1 and 2, Euro 3 (Directive 2013/60/EU) and Euro 4 (Regulation 134/2014/EU); the same cycle applies to quadricycles for the Euro 4 standard. It should be noted that during the Euro 1, 2 and 3 standards, motorcycle classes were defined on the basis of engine capacity and vehicle speed; Euro 4 and 5 standards, on the other hand, define light vehicles (category L) more analytically[2] (see also Table 3.2 later in the text).

The modal ECE R47 cycle was first established in UNECE Regulation No. 47 'Uniform provisions concerning the approval of mopeds equipped with a positive-ignition engine with regard to the emission of gaseous pollutants by the engine', published August 28, 1981 [3]. In this regulation, moped was defined as a two-wheeled or three-wheeled vehicle with an unladen weight of less than 400 kg, a maximum design speed not exceeding 50 km/h and a cylinder capacity not exceeding 50 cm^3 (i.e., slightly different from the definition adopted in Europe).

Description of the various R47 moped-cycle modes is provided in Table 3.1. They are seven in total, two of which (one at the beginning and one at the end) are at idle conditions; the duration of one run of the cycle is 112 s. The single acceleration in the cycle, as well as the subsequent constant-speed driving at the maximum speed are both accomplished with wide open throttle. Maximum speed is not fixed but depends on the tested moped; the same holds true for the rate of speed change during the single acceleration of the cycle. Notice in Table 3.1 that the constant velocity segment at full speed is not explicitly defined either in terms of duration since it is influenced by the preceding acceleration. Hence, low-PMR mopeds will spend more time accelerating, while for higher-PMR ones, the constant-speed segment will be longer. For example, a moped with 42 km/h maximum speed and 0.56 m/s^2 (=2 km/h/s) full throttle acceleration, spends 54 % of the cycle time cruising, whereas if the acceleration is 0.97 m/s^2 (=3.5 km/h/s), the cruising segment increases to 63 % of the total time.

[2]L category is the family name of light vehicles, such as powered cycles, two and three-wheeled mopeds, two and three-wheeled motorcycles, motorcycles with sidecars, light and heavy on-road quadricycles, and light and heavy quadrimobiles.

Table 3.1 Description of the R47 cycle seven modes [3]

Mode	Operation	Acceleration (m/s^2)	Speed (km/h)	Duration (s)	Cumulative time (s)
1	Idling	0	0	8	8
2	Acceleration	Full throttle	0–max.	57	65
3	Steady speed	Full throttle	max.		
4	Deceleration	−0.56	max.–20		
5	Steady speed	0	20	36	101
6	Deceleration	−0.93	20–0	6	107
7	Idling	0	0	5	112

Graphic representation of the modes described in Table 3.1 is provided in Fig. 3.1 concerning a moped with 42 km/h maximum speed. As evidenced, the R47 is a typical repetitive, simply-structured/stylized cycle. Unlike the passenger cars ECE-15 (Fig. 2.1) however, idling is considerably lower, namely 11.6 % instead of 31 % during the ECE-15, as one would expect when driving a motorcycle (see Table 3.4 later in the text for more technical data). In the European regulation, the R47 test has been designated the Type I test (tailpipe emissions), with the Type II one concerning measurement of HC and CO emissions at idling conditions (conducted immediately after its Type I counterpart).

For the Euro 1 and 2 emission standards, the test lasted 448 s (four repetitions), with the sampling beginning after four warming-up cycle runs; temperature at the

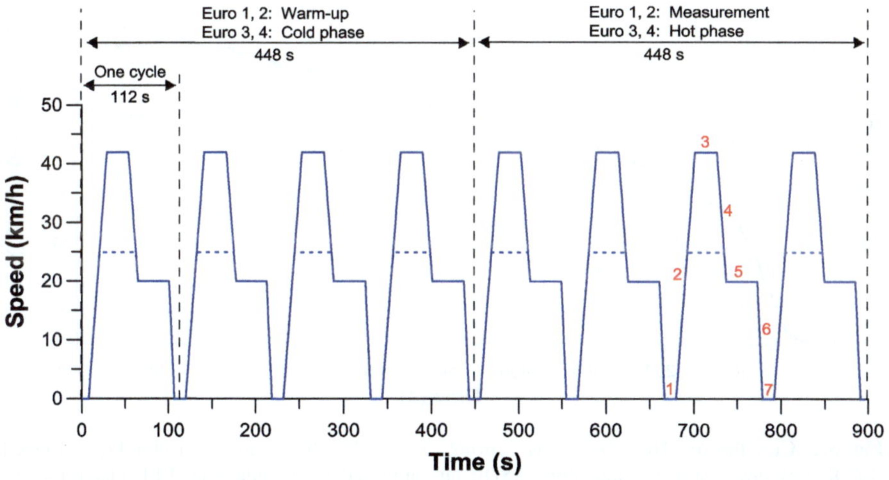

Fig. 3.1 Speed profile of the R47 cycle identifying the seven driving modes from Table 3.1; the truncated vehicle speed trace at 25 km/h corresponds to categories L1e-A (powered cycles) and L1e-B (two-wheeled mopeds) with a maximum vehicle speed of 25 km/h

place of the test was 20–30 °C. On the contrary, for the Euro 3 and 4 certification, sampling begins with the engine cold started; the weighting factors in this case are 30 % for the cold phase (four repetitions = 448 s) and 70 % for the hot one (448 s). The reason for the lower weighting cold-phase factor is that under real-world conditions, the engine will operate for a higher percentage of time when it is warm. In any case, the moped manufacturer must ensure from the Euro 3 standard (2014) onwards, that emissions are controlled either by an exhaust after-treatment system that also operates efficiently under cold engine conditions (e.g., lower light-off or operating temperature) and/or by optimizing the engine calibration to take account of cold-start conditions and warming-up of the engine.

The R47 has been employed in many Asian countries such as China, South Korea and Indonesia, which have a large population of low-powered motorcycles [4, 5].

Figure 3.2 demonstrates cumulative hydrocarbon emissions from various technology mopeds running on the R47 certification driving cycle. Unsurprisingly, the two-stroke carbureted engine exhibits the highest amount of HC emissions (and the lowest amount of NO_x owing to the inherent internal EGR of two stroke engines) [6].

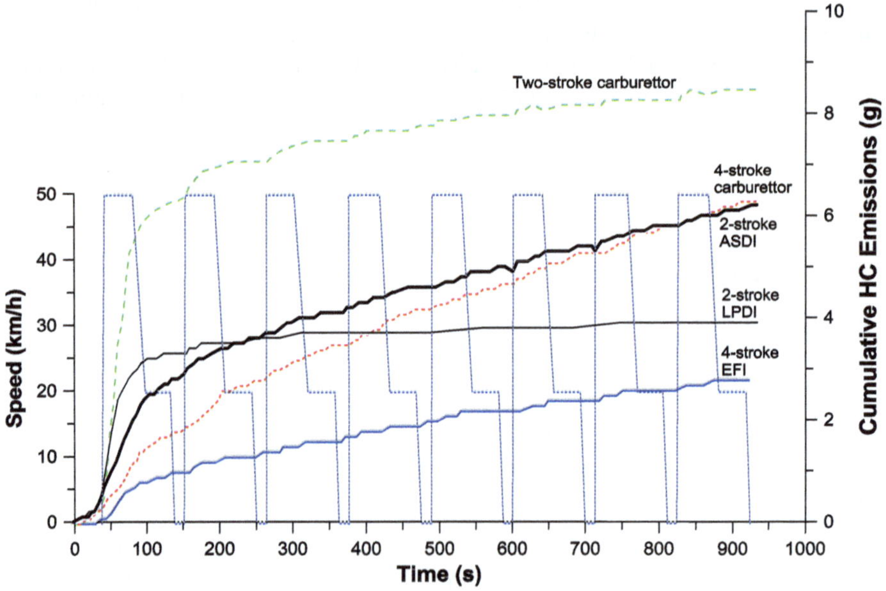

Fig. 3.2 Cumulative HC emissions during the R47 for various technology mopeds (LPDI: low-pressure direct injection, ASDI: air-supported direct injection, EFI: electronic fuel injection) adapted from [6]

3.2 ECE R40

The ECE R40 has been employed for the testing of motorcycles (excluding mopeds, i.e., for engine capacity higher than 50 cm^3 and maximum speed higher than 45 km/h) in the EU, and is graphically presented in Fig. 3.3.[3] The cycle was originally defined in UNECE Regulation No. 40, 'Uniform provisions concerning the approval of motor cycles equipped with a positive-ignition engine with regard to the emission of gaseous pollutants by the engine', published May 7, 1979. The regulation applies to the emission of gaseous pollutants from positive-ignition engines of two-wheeled or three-wheeled motorcycles with an unladen weight of less than 400 kg, having a maximum design speed exceeding 50 km/h and/or cylinder capacity exceeding 50 cm^3 [7].

The cycle is similar in logic to the moped R47. It has its origins in passenger car driving, however, having been derived from the European ECE-15. More specifically, it is comprised of six repetitions of the elementary urban cycle UDC depicted in Fig. 2.1 (cf. the NEDC consisting of four) for a total duration of 1170 s (=6 × 195 s). In any case, it is questionable whether a typical motorcycle urban route can be considered identical to that of a car, particularly as regards percentage of time spent idling and rate of speed change.

In Europe, the test has been conducted as follows according to the emission standard and engine displacement (Fig. 3.3) [8, 9]:

- For the Euro 1 and 2 standards, the motorcycle (run-in and driven for at least 1000 km) was tested with the engine warmed-up; the preconditioning consisted of two repetitions of the elementary urban cycle (UDC) and the sampling took place during the last four repetitions.
- For the Euro 3 standard, sampling began with the engine cold-started (Directive 2002/51/EC). For motorcycles with engine capacity greater than 150 cm^3, apart from the six UDC repetitions, there was an extra-urban (motorway) part, identical to the EUDC segment of the passenger cars NEDC (Fig. 2.2); the total time length of this 'extended' cycle is 1570 s.
- For the Euro 4 emission standard (including for the first time PM limits), the motorway part is excluded, and the light vehicle is tested with the engine cold-started on the urban segments only; weighting factors for the cold and hot phases are in this case 30–70 % (Regulation 134/2014/EU). For all the above cases, the various sub-parts of the cycle are run consecutively.

During the 1170 s of the R40, the traveled distance is approximately 6 km; maximum speed is 50 km/h, average driving speed 27 km/h, while 31 % of the time is spent idling (compared to 11.6 % during the R47). If the EUDC segment is accounted for (Euro 3), then the traveled distance increases to 13 km, with 120 km/h maximum and 40 km/h average driving speed; the idling period drops to

[3]The relevant EU documents for the R40 cycle are Directives 97/24/EC, 2002/51/EC, 2003/77/EC and Regulation 134/2014/EU.

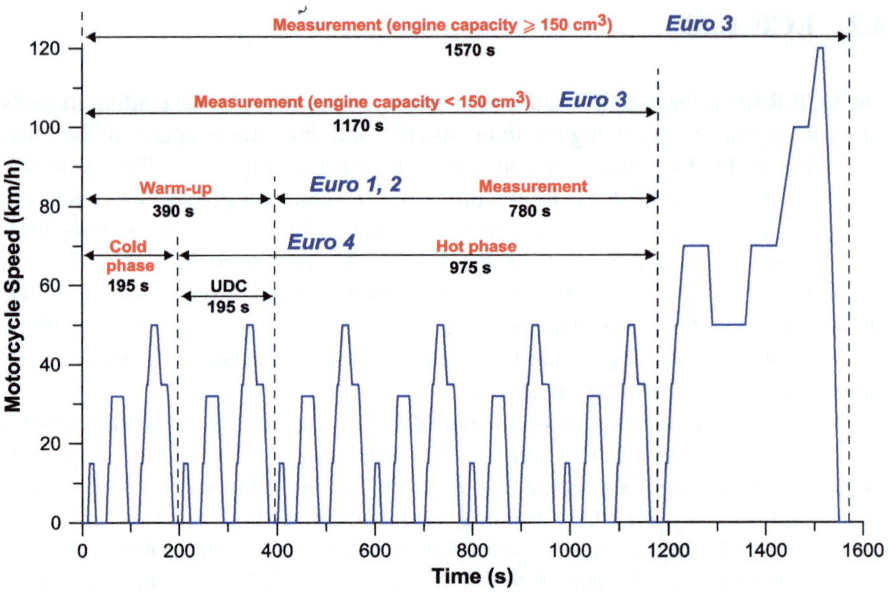

Fig. 3.3 Speed profile of the various versions of the R40/R40 + EUDC driving cycle for emission certification of motorcycles in Europe; a tolerance of ±2 km/h is allowed [7]

25.5 % of the total (1570 s) time; see also Table 3.4 later in the text for more technical specifications.

Figure 3.4 illustrates NO_x emission results from a modern light vehicle running on the R40 + EUDC cycle. Clearly, the first elementary urban cycle is responsible for the biggest amount of NO_x, owing to the catalyst not having reached its operating (light-off) temperature yet.

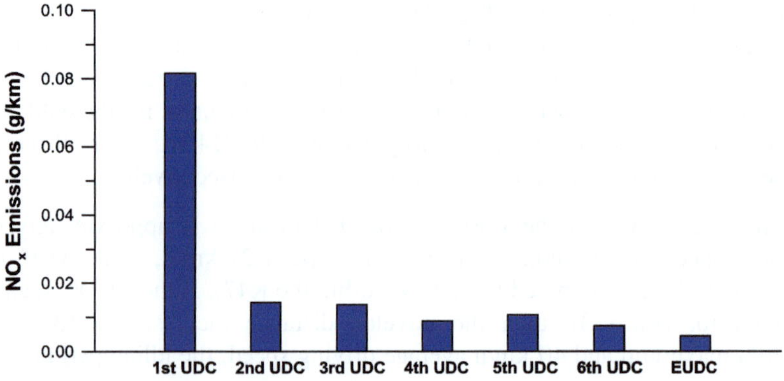

Fig. 3.4 NO_x emissions during the various parts of the R40 + EUDC cycle for a light-duty SI vehicle with three-way catalyst (data from [10])

Table 3.2 EU-valid test cycles for mopeds/motorcycles (Category L) over the years with respect to the emissions standard

Emission standard	Directive or Regulation (Date)	Category	Cycle
Euro 1	97/24/EC	Mopeds (<50 cm^3)	R47 hot
		Motorcycles ($\geq$ 50 cm^3)	R40 hot
Euro 2	97/24/EC	Mopeds (<50 cm^3)	R47 hot
	2002/51/EC	Two wheeled ($\geq$ 50 cm^3)	R40 hot
		Three wheeled ($\geq$ 50 cm^3)	
Euro 3	2013/60/EU	Mopeds	R47 cold
	2002/51/EC	Two wheeled	R40 cold (<150 cm^3)
			R40 + EUDC cold ($\geq$ 150 cm^3)
	2006/72/EC		WMTC Stage 1 cold ($\geq$ 150 cm^3)
Euro 4	134/2014/EU 168/2013/EU (1/1/2016–31/12/2020 depending on the specific sub-category)	L1e-A (powered cycles) L1e-B (two-wheeled mopeds) L2e (three-wheeled mopeds)	R47 cold
		L3e (two-wheeled motorcycles) L4e (two-wheeled motorcycles with side-car) L5e-A (powered tricycles for passenger transport) L7e-A (heavy on-road quadricycles)	WMTC Stage 2 cold
		L5e-B (powered tricycles exclusively designed for the carriage of goods)	R40 cold
		L6e-A (light on-road quadricycles) L6e-B (light quadrimobiles)	R47 cold
		L7e-B (heavy all-terrain quadricycles) L7e-C (heavy quadrimobiles)	R40 cold
Euro 5	134/2014/EU 168/2013/EU (1/1/2020–21)	All categories	Revised WMTC cold

Table 3.2 summarizes the test valid in the EU over the years for all types of mopeds/motorcycles or L-category light vehicles (Euro 1–5).

The R40/R40 + EUDC cycle has been employed in Japan as well (ISO 6460) following the Japanese Regulation TRIAS: Road vehicle Act, Article 41 'Systems and Devices of Motor Vehicles' and Safety Regulations for Road Vehicles, Article 31 'Emission Control Devices'. For the first-time Japanese emission standards in 1998, motorcycles were tested on the R40 hot started. For certification under the emission standard of 2006/7, motorcycles were tested with the engine cold started [11]. Motorcycles with engine capacity lower than 125 cm^3 were tested on the R40, whereas larger capacity ones on the R40 + EUDC (both cold started) [4]. For the emission standards established by the Japanese Ministry of Environment and the Ministry of Land, Infrastructure and Transport in September 2013, the test cycle for motorcycles and larger mopeds is the worldwide WMTC (Sect. 3.5). The R40 (or R40 + EUDC) has been used for emission testing of motorcycles in many other countries as well such as China, South Korea, Indonesia, Taiwan, Philippines, Thailand, Malaysia, Singapore, Turkey and Brazil [2, 4, 5].

European Durability Cycles

Further to the above, Regulation 134/2014/EU defines two cycles for durability purposes (Type V test) applicable to Euro 4 and 5 standards. One choice is to use a modified version of the SRC (standard road cycle) depicted in Fig. 2.34. The SRC for L-category vehicles (SRC-LeCV) is a representative kilometer accumulation cycle to age L-category vehicles and in particular their antipollution devices in a defined, repeatable and representative way [12]. The test vehicles may run the SRC-LeCV on the road, on a test track or on a kilometer accumulation chassis dynamometer. The cycle consists of five laps of a 6 km course and four different vehicle speed profiles. The length of the lap may be changed to accommodate the length of the kilometer accumulation test track or test road. The manufacturer may request to be allowed alternatively to perform the next higher numbered test cycle, with the agreement of the approval authority, if he considers that this better represents the real-world use of the vehicle. The total distance traveled is the applicable durability mileage set out in Part A of Annex VII to Regulation 168/2013/EU (ranging from 5500 to 35,000 km according to the L-category vehicle), plus one complete SRC-LeCV sub-cycle (30 km).

A second option for the durability test is to employ the U.S. AMA (accumulation mileage cycle), Fig. 2.33; last regulation is 2016/1824/EU.

3.3 U.S. EPA Urban Dynamometer Driving Schedule

U.S. federal regulations define motorcycle in 40 CFR 86.402-98 as 'any motor vehicle with a headlight, taillight, and stoplight and having: two wheels, or three wheels and a curb mass less than or equal to 793 kg (1749 lb)'. Federal standards for motorcycles were first established for the 1978 model year with limits set for CO and HC only. Beginning with model year 2006, U.S. federal emission standards for motorcycles were harmonized with those of the California ARB. At the same time, the light vehicles classification was slightly changed. Motorcycles are classified into three categories based on the engine capacity. These are: Class I (<170 cm^3), Class II (170–279 cm^3) and Class III ($\geq$ 280 cm^3; this class represents more than 90 % of annual U.S. sales [13]). Further, Class I is divided into sub-class I-A (mopeds and scooters with engine capacity lower than 50 cm^3) and sub-class I-B (small motorcycles with engine capacity between 50 and 169 cm^3) [13]. Prior to

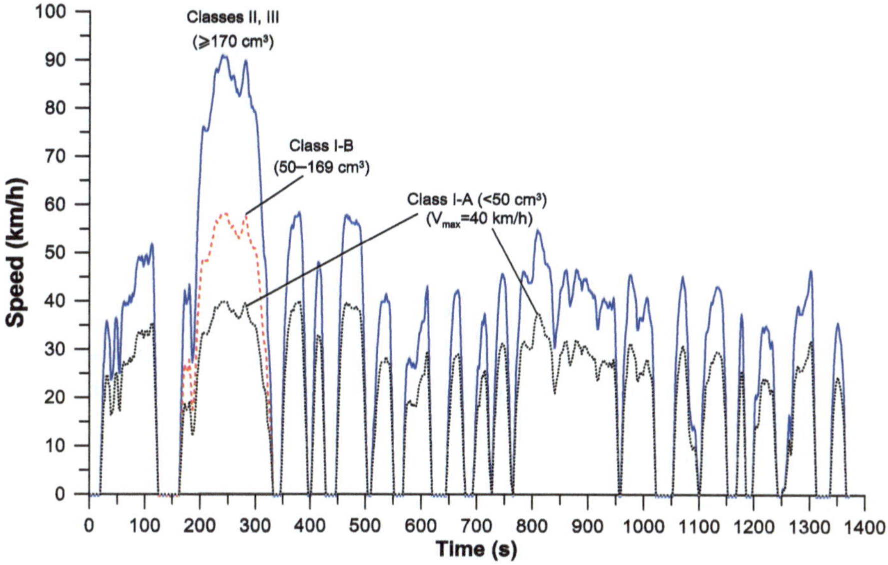

Fig. 3.5 Speed profiles of the EPA UDDS driving cycles for motorcycles (40 CFR 86, App. I)

this classification, mopeds with engine capacity lower than 50 cm^3 were not regulated. The relevant citation for motorcycles is 40 CFR 86 Subparts E and F.

The cycle valid for emission certification of motorcycles with capacity greater than 170 cm^3 (Classes II and III) is the Urban Dynamometer Driving Schedule (UDDS or LA4) valid for passenger cars and light-duty trucks (Sect. 2.2.2). As is the case with the R40, the FTP-72 has been developed based on measurements on passenger cars; EPA believes it adequately represents motorcycles too. For motorcycles belonging to Class I-B (engine capacity 50–169 cm^3), a slight variation of the UDDS is employed, illustrated in Fig. 3.5. The difference is located in the structure of the second micro-trip of the cycle (t = 164–332 s), where the maximum velocity is limited to 58.7 km/h (36.5 mph) instead of 91.2 km/h for the LA4.

Beginning with model year 2006, mopeds/scooters (Class I-A) are tested on the 'reduced' FTP-72 further adapted to their maximum speed. This is accomplished by 'adjusting each speed point of the cycle by the ratio of the top speed of the moped to 58.7 km/h (the top speed of the existing Class-I drive cycle)' [13]. Figure 3.5 illustrates such a version of the cycle for a moped with a maximum speed of 40 km/h. Table 3.3 summarizes some technical specifications of the driving cycles depicted in Fig. 3.5. Predictably, the moped version has the lowest RPA owing to both lower speeds and accelerations throughout the cycle.

As is the case with passenger cars, the dynamometer test also includes a hot-start run of the first 505-s segment after a 10-min engine shut down, for a total time length of 1877 s. Equation (2.2), providing the calculation formula for total emitted

Table 3.3 Summary of technical specifications of the U.S. EPA test cycles for motorcycles

Cycle	Duration (s)	Distance (m)	Max. Speed (km/h)	Average Speed (km/h)	Driving Speed (km/h)	Max. Accel. (m/s^2)	RPA (m/s^2)
Class I-A (V_{max} = 40 km/h)	1372	7403	40.0	19.4	23.6	1.02	0.123
Class I-B	1372	10,864	58.7	28.5	34.7	1.48	0.180
Classes II, III	1372	11,990	91.2	31.5	38.3	1.48	0.185

idling time for all versions is 17.8 %

pollutant values from the three distinct cycle phases (505-s cold start, 867-s stabilization segment, and 505-s hot start), is valid for motorcycles as well.

Apart from the United States, the U.S. legislation on motorcycles is also followed by Canadian rulemaking.

U.S. Durability Cycles

For durability purposes of motorcycles, the AMA cycle illustrated in Fig. 2.33 can be applied (40 CFR 86, Appendix IV). The entire test, Fig. 3.6, consists of 11 laps of 6 km distance each, with different 'basic' speeds per lap according to the motorcycle class, ranging from 45 to 70 km/h for Class I-B up to 45 to 110 km/h for Class III vehicles. During each of the first nine laps, there are four stops with 15-second idle. Normal accelerations and decelerations are used; in addition, there are five light decelerations in each lap from the base speed to 30 km/h followed by light accelerations to the base speed. The 10th lap is run at a constant speed. The 11th lap begins with a wide open throttle acceleration from stop. A normal

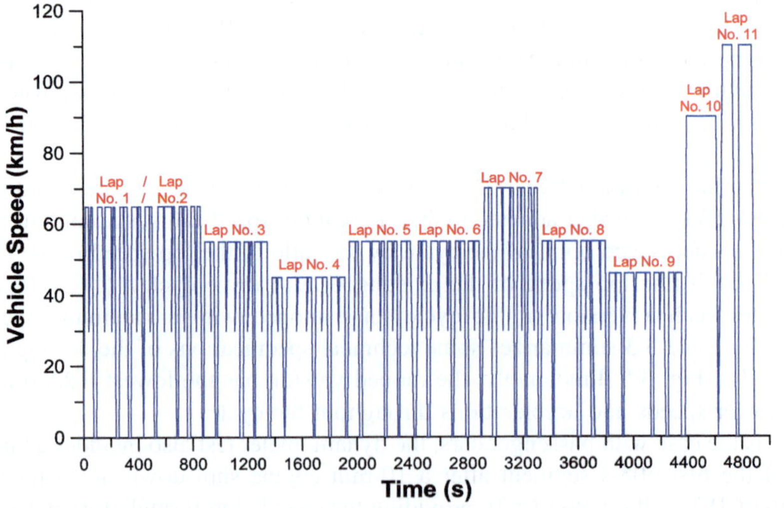

Fig. 3.6 Speed-time trace during the U.S. AMA durability cycle for Class III motorcycles

deceleration to idle followed by a second wide open throttle acceleration occurs at the midpoint of the lap. This schedule may be modified with the advance approval of the Administrator if it results in unsafe operation of the vehicle. Apart from the United States, the AMA cycle is employed in China, Japan, Taiwan, Thailand, and Brazil [14].

3.4 Indian Driving Cycle—IDC

In India, a special drive cycle has been employed for many years, demonstrated in Fig. 3.7. The cycle was developed in 1985 by the Automotive Research Association of India (ARAI) in Pune, Maharashtra, based on actual motorcycle measurements; it consists of 19 operating phases. Although it is of the simple-structure type, it is believed to suit the specifications of the low-powered, low-capacity light vehicles used in the country [15].

Despite its modal form, the IDC exhibits some marked dissimilarities compared to other synthesized urban cycles like the European ECE-15/R40 and R47 or the Japanese J10, namely very small idling period (16 s at the beginning) and equally limited cruising time (12 %). On the contrary, it is quite transient, with the accelerations during the cycle covering 39 % of the time and the decelerations 34 % (in contrast, during the ECE R47, the combined acceleration and deceleration time is approximately 35 %). Accordingly, the IDC's RPA is quite high at 0.195 m/s^2, although the maximum acceleration during the cycle is 'only' 0.65 m/s^2 (2.34 km/h/s). The maximum cycle speed is set at 42 km/h and the time length is 108 s, with the average speed at 22 km/h. For certification purposes, the cycle is run six times for a total duration of 648 s and approximately 4 km traveled

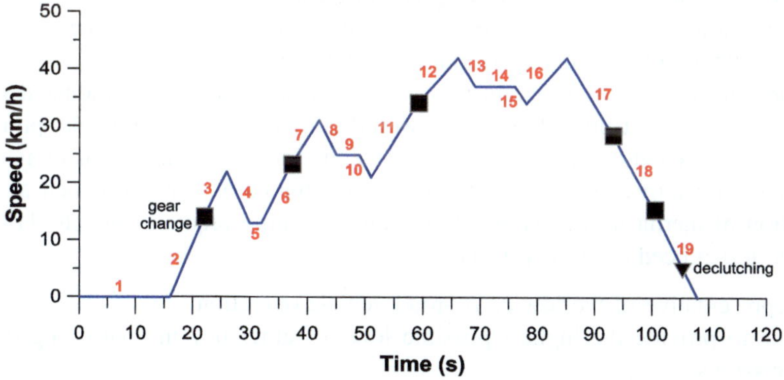

Fig. 3.7 Speed profile of the IDC, showing also recommended gear-shift positions; the cycle consists of 19 operating phases (as is the case with every modal cycle, the IDC comprises a series of driving modes, the graphical illustration of which is provided in this figure) [15]

distance. Comparing the IDC to the passenger car modal cycles from Chap. 2 (Fig. 2.59), a reasonable conclusion seems to be that it resembles the Japanese J11.

The test, designated as Type I, is carried out with the motorcycle warmed-up (vehicle running on four IDC cycles) for SI, and cold-started for diesel-engined vehicles (PM limits included too). If the maximum speed of the vehicle is less than the maximum speed of the cycle, for that part of the cycle where the speed is exceeding the vehicle's maximum speed, the vehicle is driven with the throttle fully open. Type II test is CO and HC measurement at idling conditions for SI engines, or free acceleration smoke test for diesel-engined three-wheelers [16, 17].

3.5 Worldwide Motorcycle Test Cycle—WMTC

The worldwide harmonized motorcycle test cycle has been developed by a special UN working group with the aim to provide a universal and realistic, in terms of daily driving behavior, test cycle for motorcycles. The WMTC project began in 1999 as a tripartite research effort between the Netherlands Ministry of the Environment (VROM), TNO Automotive and the International Motorcycle Manufacturers Association (IMMA). VROM dealt with the political aspects of constructing a worldwide test cycle, TNO Automotive did the development work and the technical management of the project, while IMMA contributed the collection of in-use worldwide driving behavior data. One year later, in May 2000, the project was brought under the auspices of the UNECE WP.29. Under the guidance of WP.29, the Group of Experts on Pollution and Energy (GRPE) mandated the ad hoc group WMTC with the development of a 'worldwide harmonized motorcycle emissions certification procedure' and its establishment in the framework of the 1998 Agreement on Global Technical Regulations (GTR). The WMTC group was formed by members of the following countries/organizations: AECC (Association for Emission Control by Catalyst), European Commission, Germany, IMMA, Japan, Netherlands, Spain, Switzerland, the UK and the U.S. [18].

The objective of the program was to develop a worldwide harmonized motorcycle emissions test procedure which would comprise: a driving cycle, procedure for gearshift schedule, as well as exhaust emissions sampling, measurement and analysis procedures (with support from/in collaboration with ISO) such as determination of inertia mass, road-load resistance, cooling requirements etc. [18, 19].

The test procedure needed to be:

- 'representative of worldwide on-road vehicle operation;
- able to provide the highest possible level of efficiency in controlling on-road emissions;
- corresponding to state-of-the-art testing, sampling and measurement technology;
- applicable in practice to existing and foreseeable future exhaust emission abatement technologies;

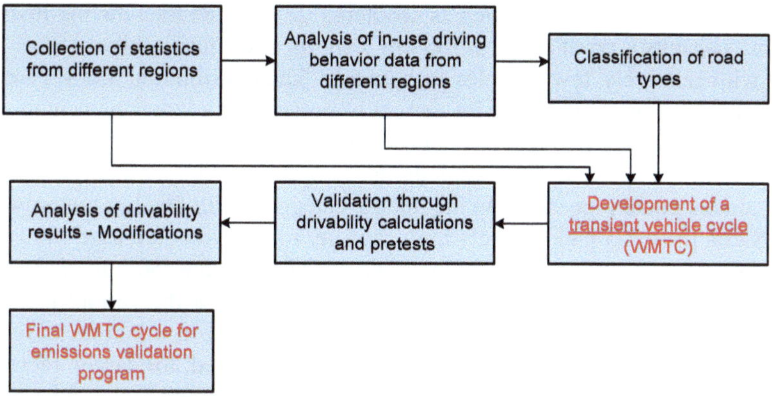

Fig. 3.8 Flow chart of the WMTC development process [18]

- capable of providing a reliable ranking of exhaust emission levels from different engine types;
- consistent with the development of appropriate emission factors, and inclusive of adequate cycle by-pass prevention provisions' [18, 19].

From October 2000, RWTÜV Fahrzeug GmbH joined the WMTC group and focused on the cycle development, its test protocol as well as the corresponding gearshift procedure. The basis of the cycle development was the collection and analysis of driving behavior data and statistical information about motorcycle use from different regions of the world (see also Fig. 3.8 illustrating the corresponding flow chart of the cycle development phases). In a second step, the in-use driving behavior data was combined with the statistics on vehicle use in order to create a reference database representative of worldwide motorcycle driving behavior. This was achieved using a classification matrix for the most influencing parameters. In the final classification matrix, three regions (Europe, Japan and USA), three vehicle classes and three road categories were included [18].

The in-use driving behavior data for the WMTC project consisted of data measured by various institutions and Universities worlwide, namely in:

- Europe; measurements carried out in 1994 and 1999 in Paris, Pisa, Amsterdam, Frankfurt, Mandeure (France), Munich, as well as in Biel (Switzerland) and Darmstadt,
- Japan; 1992, 1997 and 2000 in Tokyo,
- China; Ji Nan area, and
- the United States; Birmingham, Alabama in 1999.

The duration of the in-use data set was 518 h for a total mileage of 27,224 km. A wide range of motorcycles was used from a variety of manufacturers, with engine capacity ranging from 49–1500 cm^3, power from 5–110 kW and power to mass ratio from 30–335 kW/t. Europe was found to have a fleet mainly consisting of either small capacity vehicles such as scooters (35 % of the total having lower than 125 cm^3 capacity), or of large (>500 cm^3) sports and touring bikes (51 % of the total), with relatively few vehicles in between. The distribution was very country dependent, with northern countries generally possessing a larger percentage of big motorcycles, whereas southern countries having mainly small vehicles. Japan showed a similar share of small vehicles as Europe (31 % lower than 125 cm^3), but has an almost equal percentage in the middle category, with few big motorcycles. In the U.S., on the other hand, mainly high capacity motorcycles are used (86 % of the population concerns motorcycles with capacity higher than 450 cm^3) [18].

Predictably, different driving habits were determined for the various parts of the world. For example in Europe, scooters are mainly used in urban areas for small trips (97 %), with big motorcycles being primarily used for larger recreational country road (59 %) and motorway trips (22 %). In the United States, a larger share of urban use and a smaller of motorway (25 %) compared to Europe was established. In Japan, on the other hand, the majority of the mileage was found to be performed on rural and urban roads (94 %), with, generally, small average trip lengths and low average speeds established.

The next step was to compact the reference cycle into a test cycle of the desired length. It was decided that the cycle should consist of three parts, each representing a separate road category and lasting 600 s. Part 1 should be a low-speed part, essentially representative of urban traffic, beginning with a cold start; Part 2 a medium-speed part, representing slower country-road type of traffic; Part 3 a high-speed part representative of faster country roads and motorways. The vehicle speed modules were initially calculated as follows: below 60 km/h, between 60 and 90 km/h and above 90 km/h respectively; the lower limit was set to 50 km/h at a later stage [18]. A computer search program selected a number of micro-trips to represent, by approximation, the desired time length. The statistical characteristics of these micro-trips were then compared to those of the database on the basis of the chi-squared statistical method for the speed-acceleration matrix. The selection of micro-trips with the lowest chi-squared values was then selected as the ideal combination. After that, stops were added, with the total stop time taken from statistics. This total stop time was divided into stops on the basis of the statistical distribution of the stop lengths in the in-use database.

After several updates and modifications, the cycle that was eventually developed is illustrated in Fig. 3.9. It is divided into five classes (1, 2-1, 2-2, 3-1 and 3-2) depending on motorcycle engine capacity and maximum speed (the latter as defined in Directive 95/1/EC); details about the classes are provided in Fig. 3.10:

- Part 1 consists of eight micro-trips; the top speed is 60 km/h (50 km/h for low-powered vehicles in Classes 1 and 2-1).
- Part 2, representing driving on secondary rural roads, consists of two micro-trips with the top speed 95 km/h for Class 2-2 and 80 km/h for the Class 2-1 lower-powered vehicles.
- Part 3, representing primary rural roads and motorways, consists of one micro-trip with the top speed at 125 km/h (Class 3-2) or 110 km/h (Class 3-1);

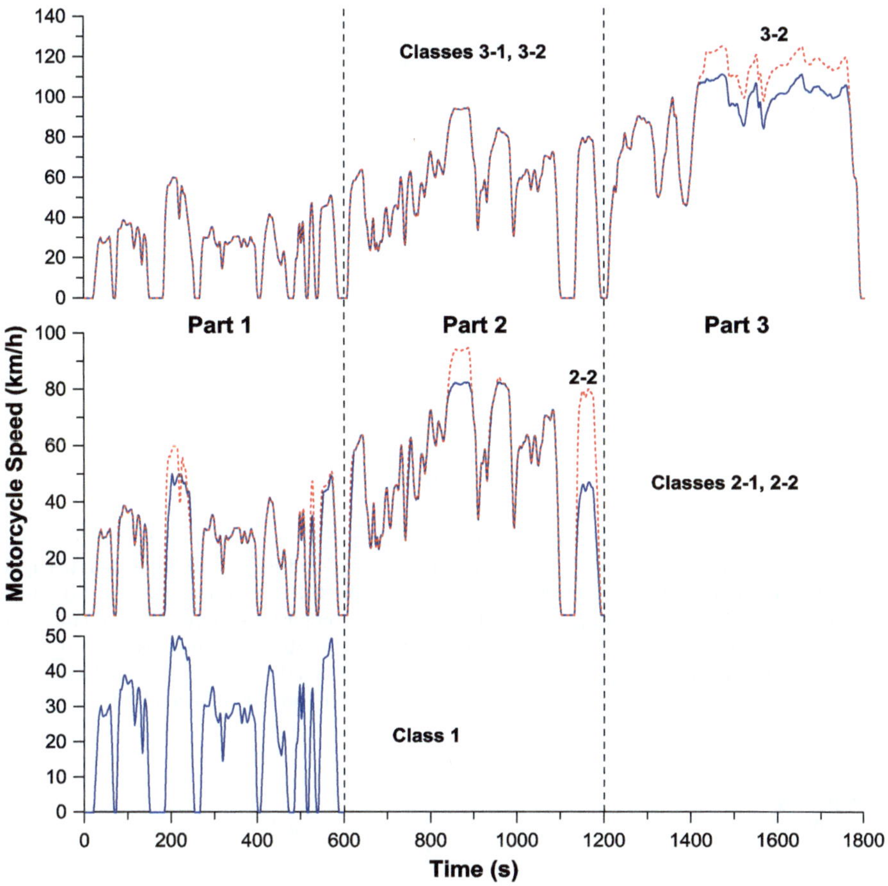

Fig. 3.9 Speed profile of the WMTC driving cycle (discontinuous line corresponds to high-speed Classes 2-2 and 3-2 [19]

notice that the speed trace during Part 3 exhibits frequent steep speed changes unlike passenger cars and HDVs highway segments, which resemble steady-state operation. Specific gear-shift strategy and laboratory conditions are defined in the corresponding UNECE GTR No. 2 [19].

Three versions of the WMTC have been developed so far. The version 'WMTC-stage 1' is the basis, adopted as GTR No. 2 in August 2005. With Amendment 1 to GTR No. 2, slight modifications to the classification (Classes 1 and 2-1) and the test cycles (alternative Part 1 and 2 traces) were introduced in 2007 (version 'WMTC-stage 2'). The current version is WMTC stage 3.

Figure 3.11 demonstrates the 3D speed/acceleration distribution of the WMTC (Class 3-2), exhibiting a rather uniform profile with no dominating bin.

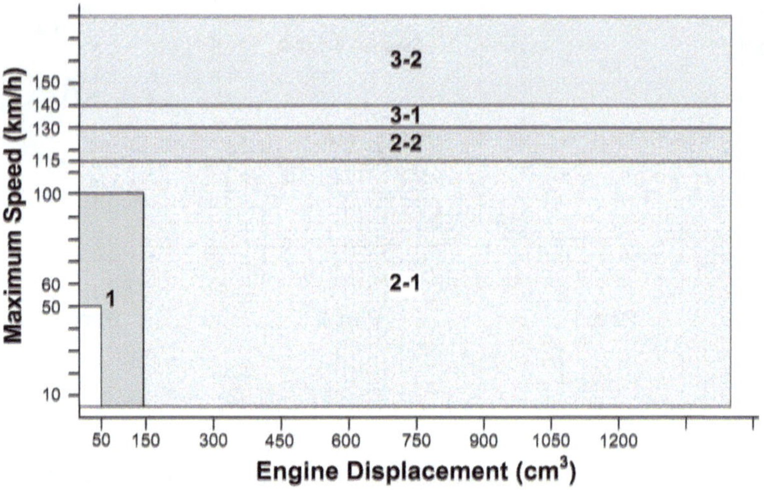

Fig. 3.10 Specifications of the WMTC motorcycle classes according to UNECE GTR No. 2 (Class 1 corresponds to: (a) engine capacity between 50 and 150 cm^3 and V_{max} < 50 km/h, and (b) engine capacity < 150 cm^3 and V_{max} between 50 and 100 km/h; Class 2 corresponds to: (a) engine capacity < 150 cm^3 and V_{max} = 100–115 km/h, or engine capacity ≥ 150 cm^3 and V_{max} < 115 km/h (sub-class 2-1), and (b) V_{max} from 115–130 km/h (sub-class 2.2); Class 3 corresponds to: (a) V_{max} from 130–140 km/h (sub-class 3-1), and (b) V_{max} ≥ 140 km/h (sub-class 3.2); mopeds (engine capacity < 50 cm^3 and V_{max} < 50 km/h) are not included [19]

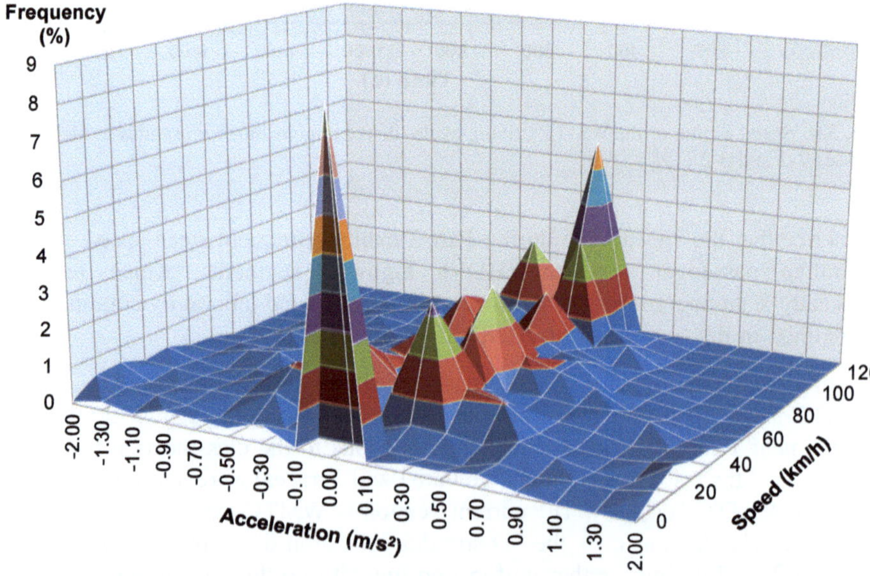

Fig. 3.11 Speed/acceleration frequency distribution of the WMTC Class 3-2 driving cycle; notice the much broader distribution compared to the frequency distribution figures of the passenger car cycles in Chap. 2, with no single bin exhibiting higher than 9 % frequency

Table 3.4 Summary of motorcycle test cycles technical specifications (duration and distance for the R40 and the IDC correspond to six runs, for the R47 to 8 runs)

Cycle	Type	Duration (s)	Distance (m)	Maximum Speed (km/h)	Avg. Speed (km/h)	Maximum Accel. (m/s^2)	Idling Time (%)	RPA (m/s^2)
WMTC 1	Transient	600	3838	50.0	23.0	1.72	17.5	0.187
WMTC 2-1	Transient	1200	12,287	82.5	36.9	1.75	12.4	0.181
WMTC 2-2	Transient	1200	13,178	94.9	39.5	2.69	12.0	0.202
WMTC 3-1	Transient	1800	27,614	111.3	55.2	2.69	8.8	0.154
WMTC 3-2	Transient	1800	28,915	125.3	57.8	2.69	8.8	0.154
R40	Modal	1170	6068	50.0	18.7	1.04	30.8	0.154
R40 + EUDC	Modal	1570	13,023	120.0	29.9	1.04	25.5	0.122
R47*	Modal	896	5804	42.0	23.3	0.56	11.6	0.098
Indian IDC	Modal	648	3948	42.0	21.9	0.65	14.8	0.195
U.S. FTP-72	Transient	1372	11,990	91.2	31.5	1.48	17.8	0.185

*V_{max} = 42 km/h, wide open throttle acceleration = 0.56 m/s^2

Table 3.5 Summary of technical specifications for the WMTC Class 3-2

Cycle Part	Duration (s)	Distance (m)	Maximum Speed (km/h)	Average Speed (km/h)	Maximum Accel. (m/s^2)	Idling Time (%)	RPA (m/s^2)
1	600	4066	60.0	24.4	2.50	16.8	0.215
2	600	9112	94.9	54.7	2.69	7.2	0.197
3	600	15,737	125.3	94.4	1.56	2.5	0.113

Main technical specifications for all WMTC classes are provided in Table 3.4, in comparison to the earlier R47, R40, R40 + EUDC, Indian IDC and American FTP-72 cycles (more data is available in the Appendix). Further, Table 3.5 details the main specifications for the three parts of Class 3-2. The most striking observation from Table 3.4 is the considerably higher acceleration of the WMTC Classes 2 and 3 compared to the 'soft' R40, being thus much more compatible with the driving habits of motorcyclists. Notice also in Table 3.4 the much higher RPA of the IDC compared to the R47. RPA values for all motorcycle tests are illustrated in Fig. 3.12 with respect to the average driving speed during the cycle.

Figure 3.13 illustrates a direct comparison between the WMTC Class 3-2 and the R40/R40 + EUDC cycle employed in earlier certification procedures. The differences are further highlighted in Fig. 3.14 that illustrates the speed/acceleration distribution between the two cycles. As it is obvious from Fig. 3.14, the WMTC has much broader speed/acceleration range, significantly higher accelerations, and is, in general, much more aggressive and transient compared to the simplistic R40.

Figure 3.15 expands on the findings from Fig. 3.14 by illustrating the engine speed/power distribution between the WMTC Class 3-2 and the R40 + EUDC for a mid-sized motorcycle. It is obvious that during the WMTC test, the sampling points

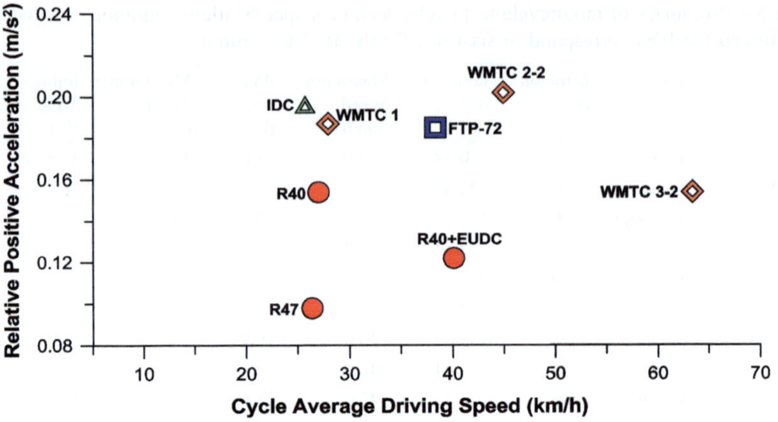

Fig. 3.12 RPA with respect to average driving speed for all motorcycle and moped test cycles

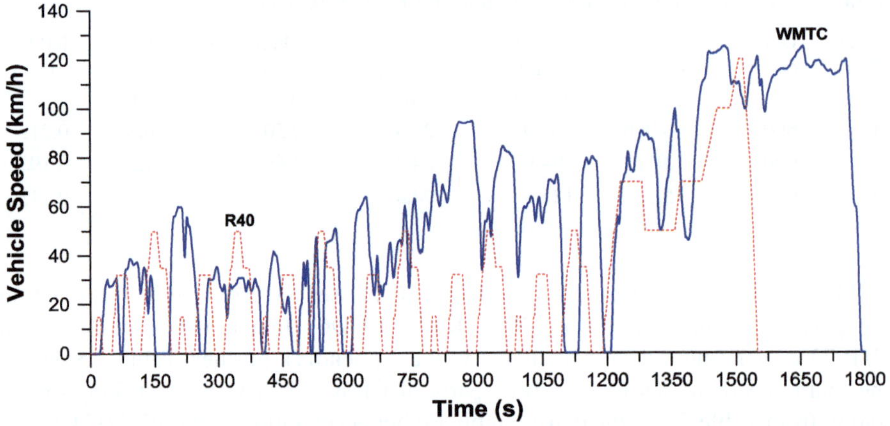

Fig. 3.13 Comparison of the speed profile between the WMTC 3-2 and the R40 + EUDC

cover a larger area compared to the R40 + EUDC. Nonetheless, a significant range of the engine load (mainly mid to high loading points) still remains uncovered as is detailed in [20] based on measurements on a large number of motorcycles. In any case, the improvement compared to the R40 is huge.

An important feature of the cycles' dynamics, namely the acceleration profile, is on display in Fig. 3.16 for all test cycles discussed earlier. Predictably, Class 3 (and 2) of the WMTC suite of transient cycles exhibits much higher and frequent accelerations compared to the modal R40. Cycles intended primarily for mopeds or, in general, low-PMR motorcycles, such as the IDC, R47 and WMTC class 1, on the other hand, manifest much lower acceleration values. The unique pattern of the

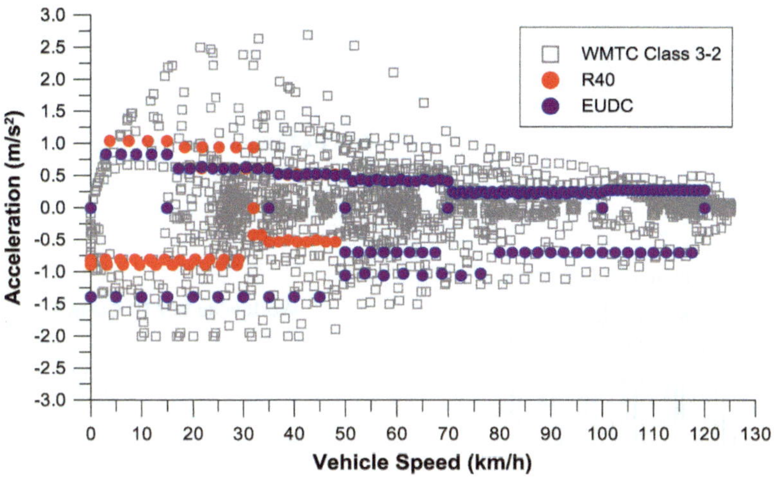

Fig. 3.14 Comparison of the speed/acceleration distribution between the WMTC Class 3-2 and the R40 + EUDC

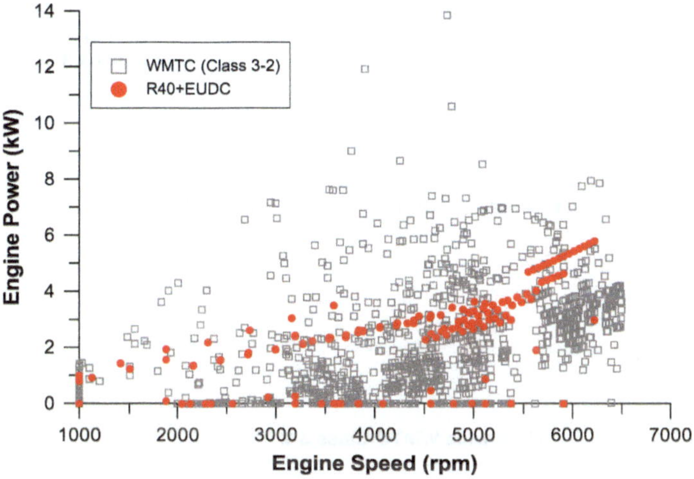

Fig. 3.15 Comparison of the speed/power distribution between the WMTC Class 3-2 and the R40 + EUDC for a mid-size motorcycle

moped R47 is also highlighted in Fig. 3.16. Despite its modal form and the low maximum acceleration, the overall acceleration profile of the IDC is quite dense.

Lastly, Fig. 3.17 compares the speed and acceleration frequencies between the worldwide motorcycle test cycle WMTC and its passenger car counterpart WLTC (both regarding the high-speed Class 3-2). As it is evidenced in Fig. 3.17, the motorcycle WMTC exhibits more abrupt accelerations and decelerations (>1.5 and <−1 m/s^2), as well as more operating points with speeds higher than 55 km/h

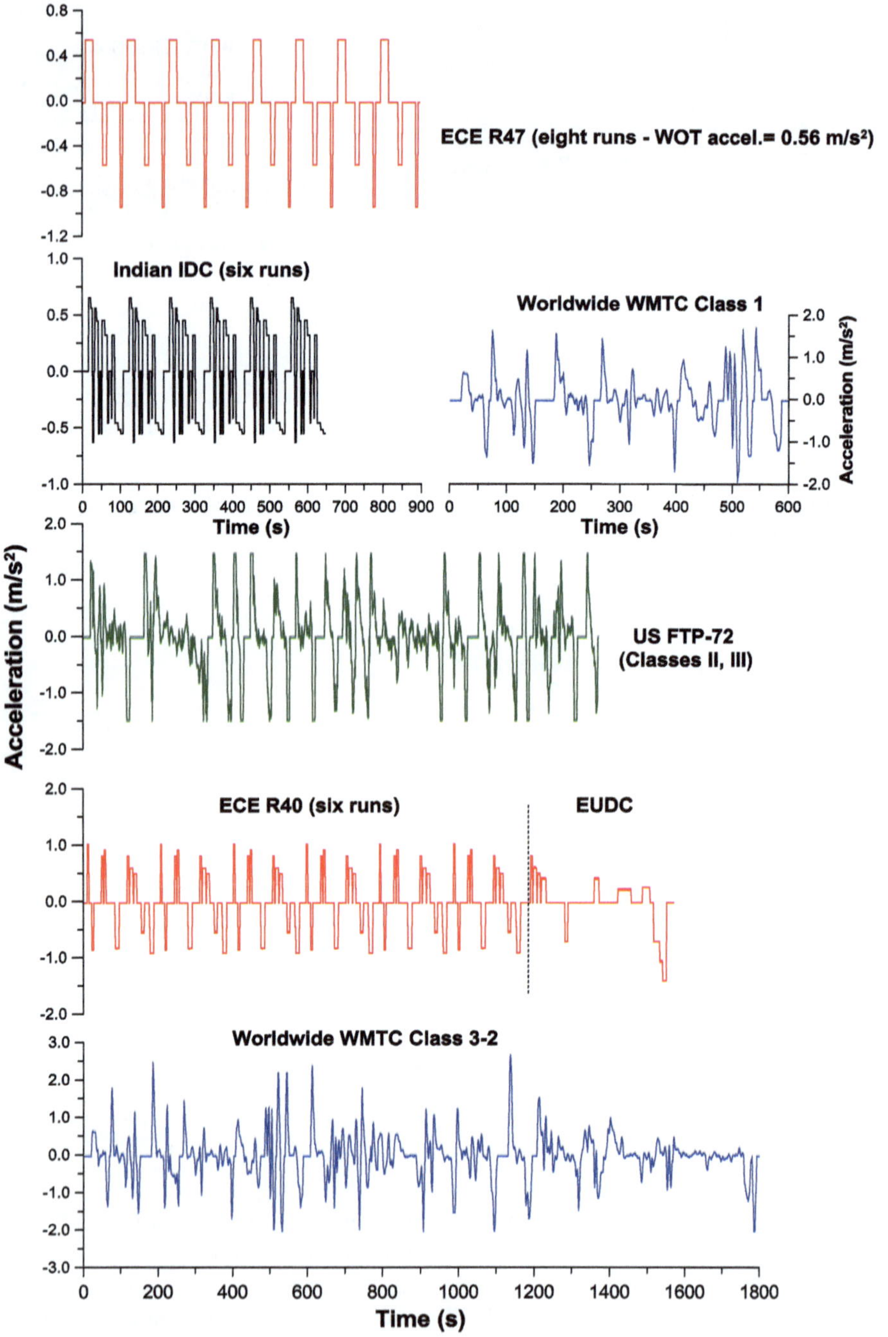

Fig. 3.16 Acceleration profiles of driving cycles employed for motorcycles testing in the world

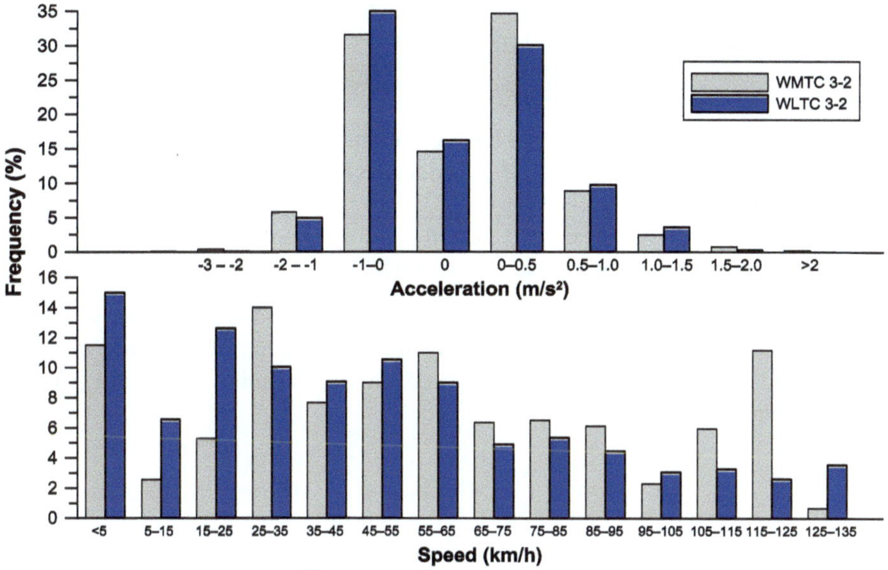

Fig. 3.17 Comparison of the speed/acceleration frequency between the WMTC Class 3-2 and the WLTC Class 3-2 for light-duty vehicles

and fewer with (urban driving) speeds lower than 25 km/h. Both remarks were rather anticipated. Motorcycles typically possess higher power-to-mass ratios, hence achieve higher accelerations during daily driving compared to LDVs.

In Europe, the worldwide motorcycle WMTC was initially adopted in 2006 with Directive 2006/72/EC. Manufacturers were given the choice to test their vehicles (only motorcycles with engine capacity greater than 150 cm^3) on the WMTC as an alternative to the R40 + EUDC for compliance with the Euro 3 emission standard; slightly higher emission levels were accepted in this case compared to the R40 test. The cycle has become mandatory for certain light-vehicle categories provided in Table 3.2 for the Euro 4 emission standard with slight modifications (WMTC stage 2). The revised WMTC (or WMTC stage 3—Fig. 3.9) will be mandatory for all motorcycle types (L-category vehicles) in Europe for compliance with the Euro 5 standard from 2020. According to Regulation 134/2014/EU, and based on the specific L-category vehicle, the cycle is run as follows with reference to Fig. 3.9:

- Class 1: Part 1 of the reduced vehicle speed in cold-started condition, followed by a repetition of the same Part 1, now under warm conditions (total duration of the test is 1200 s);
- Class 2:
  - Sub-class 2-1: Part 1, reduced vehicle speed in cold-started conditions, followed by Part 2, reduced vehicle speed in warm conditions (1200 s);

- Sub-class 2-2: Part 1, in cold-started conditions, followed by Part 2, in warm conditions (total duration 1200 s);

- Class 3:
  - Sub-class 3-1: Part 1, in cold-started conditions, followed by Part 2, in warm conditions, followed by Part 3, reduced vehicle speed in warm conditions (total duration 1800 s);
  - Sub-class 3-2: Part 1, in cold-started conditions, followed by Part 2 and Part 3, in warm conditions (total duration 1800 s).

For the certification procedure, Regulation 134/2014/EU assigns weighting factors to each segment of the previously mentioned runs of the cycle parts, as summarized in Table 3.6. The final weighted emission result of pollutant i (gaseous or PM) is calculated as follows

$$E_{wi}\,(\text{g/km}) = \sum_j wf_j \cdot E_{i-j} = \sum_j wf_j \cdot \frac{M_{i-j}}{S_j} \tag{3.1}$$

with wf_j the weighting factor of part j from Table 3.6, E_{i-j} the emission of pollutant i during part j in g/km, M_{i-j} the respective emissions in g (as described in Chap. 6.5), and S_j the distance of part j; j = 1, 2 or 3 (depending on the tested light vehicle). The same equation is used for fuel consumption calculations.

For low-speed light vehicles such as mopeds, for which UNECE GTR No. 2 is not valid (there is no class available, as evidenced in Fig. 3.10), the WMTC Class 1 will be the applicable cycle for the Euro 5 standard from 2020, replacing the R47; this WMTC version will be run tailored to the maximum vehicle speed, as demonstrated in Fig. 3.18 for the case of V_{max} = 45 km/h.

Table 3.6 Emission weighting factors for the various WMTC parts valid in the EU according to the emission standard (see Table 3.2 for information on the L category vehicles)

Emission Standard	Cycle	Category	Weighting factors (%)
Euro 4	WMTC-2 (stage 2)	L3e/L4e/L5e-A/L7e-A (V_{max} < 130 km/h)	30–70 (Class 2)
Euro 4	WMTC-2 (stage 2)	L3e/L4e/L5e-A/L7e-A (V_{max} ≥ 130 km/h)	25–50–25 (Class 3)
Euro 5	WMTC-3 (stage 3)	L1e-A, L1e-B, L2e, L6e-A, L6e-B (formerly R47) L3e/L4e/L5e-A/L7e-A (V_{max} < 130 km/h)	50–50 (Class 1) 50–50 (Class 2)
Euro 5	WMTC-3 (stage 3)	L5e-B, L7e-B, L7e-C (formerly R40)	30–70 (Class 2)
Euro 5	WMTC-3 (stage 3)	L3e/L4e/L5e-A/L7e-A (V_{max} ≥ 130 km/h)	25–50–25 (Class 3)

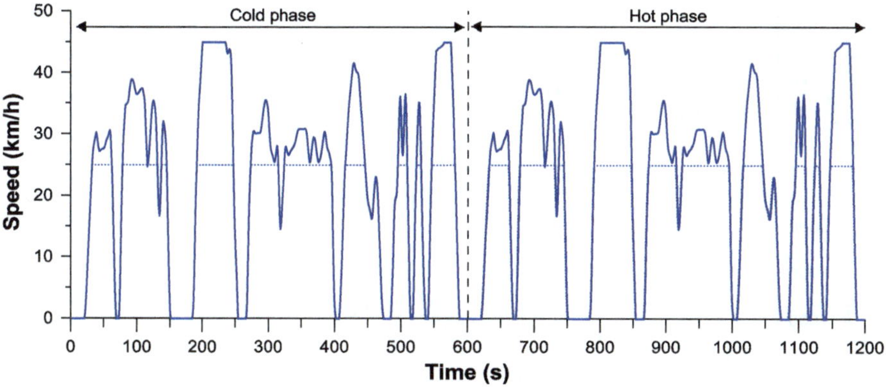

Fig. 3.18 Speed profile of the WMTC moped driving cycle with 45 km/h limit, valid in the EU from 2020 (categories L1e-A, L1e-B, L2e, L5e-B, L6e-A and L6e-B); the truncated vehicle speed trace at 25 km/h is applicable to L1e-A and L1e-B vehicles with a limited maximum design vehicle speed of 25 km/h (speed traces for the cold and hot phases are identical)

In the United States, it is expected that the WMTC will be introduced as an alternative to the UDDS cycle with equivalent limits to the present U.S. emission standards [2]. After a period of time (which will be determined through the U.S. rulemaking process), phasing out the UDDS is intended in order to ultimately rely exclusively on the WMTC for motorcycle certification purposes. The timing of U.S. regulatory action is currently not determined.

The WMTC has already been adopted by the Japanese legislation [5]. It is estimated that China (as well as other Asian countries, such as Taiwan, Vietnam, Thailand and Indonesia) and Brazil will follow the European approach in due course [5].

India, on the other hand, initially expressed concerns based on the significant percentage of motorcycles in the country falling under the lower range of Class 1. With the adoption of Bharat Stage IV regulations in 2016, however, the use of the WMTC for type approval of two-wheelers has become mandatory, employing the same weighting factors for the various parts as the ones detailed in Table 3.6 for the EU [21]. For three-wheelers, the Bharat Stage IV (2016) and Bharat Stage VI (2020) regulations employ the Indian IDC driving cycle [22]; the IDC continues to be the applicable test cycle for mopeds [21].

Concluding this chapter, Fig. 3.19 illustrates engine speed and torque development during the worldwide motorcycles WMTC Class 3-2 from a 240-kg, six-speed two-wheeled motorcycle. In particular, the acceleration at $t = 1133$–1152 s proves to be the most challenging.

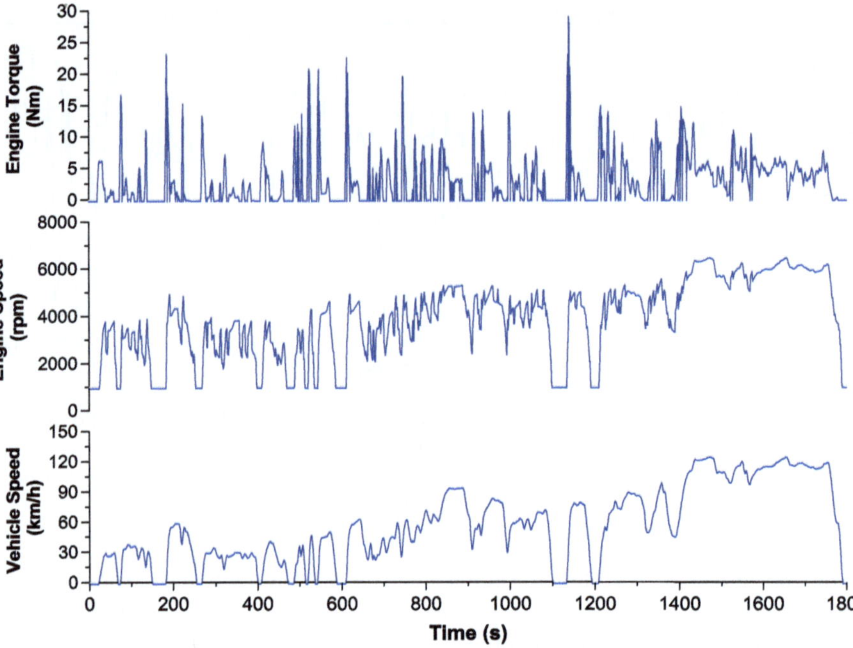

Fig. 3.19 Engine speed and torque development during the WMTC for a mid-size motorcycle

References

1. Posada F, Kamakate F, Bandivadekar A. Sustainable management of two- and three-wheelers in Asia. ICCT Working Paper 2011–13, December 2011.
2. Manufacturers of Emission Controls Association. Emission control of two- and three-wheel vehicles. White Paper, September 2014.
3. UNECE Regulation No. 47. Uniform provisions concerning the approval of mopeds equipped with a positive-ignition engine with regard to the emission of gaseous pollutants by the engine. Date of entry into force: 1 November 1981.
4. Worldwide emission standards—Passenger cars and light duty vehicles, Delphi 2015/16. http://delphi.com/docs/default-source/catalogs/delphi-worldwide-emissions-standards-pc-ldv-15-16.pdf. Accessed 11 August 2016.
5. Summary of the future emission reduction measures for motorcycles from the future policy for motor vehicle emission reduction (11th report). Japanese Ministry of the Environment, EPPR-01-05e; 2013.
6. Favre C, May J, Bosteels D, Tromayer J, Neumann G, Kirchberger R, et al. Regulated and non-regulated emissions of selected state-of-the-art European mopeds. AECC SIA 2011 Congress, Strassburg, France.
7. UNECE Regulation No. 40. Uniform provisions concerning the approval of motor cycles equipped with a positive-ignition engine with regard to the emission of gaseous pollutants by the engine. Date of entry into force: 1 September 1979.
8. Geivanidis S, Ntziachristos L, Samaras Z, Xanthopoulos A, Steven H, Bugsel B. Study on possible new measures concerning motorcycle emissions. Final report No. 08.RE.0019.V2, Thessaloniki, November 2008.

9. Giechaskiel B, Zardini A, Martini G. Particle emission measurements from L-category vehicles. SAE Paper No. 2015-24-2512; 2015.
10. Regulated emissions of a Euro 5 passenger car measured over different driving cycles. European Commission Joint Research Centre; 2010.
11. Vehicle Emission Standards—Japan. MECA World Regulations; 2006.
12. Cuerden R, Nathanson AA, Goodacre O, McCarthy M, Knight I, Muirhead M, et al. Durability of pollution control measures for L-category vehicles. Final report, PPR627, November 2012.
13. U.S. Environmental Protection Agency. 40 CFR Parts 9, 86, 90, and 1051. Control of emissions from highway motorcycles; Final rule, Federal Register, Vol. 69, No. 10, January 15, 2004.
14. Association des Constructeurs Européens de Motocycles. L-category vehicles type approval regulation—ACEM comments on draft TRL durability study, October 2012.
15. Document on test method, testing equipment and related procedures for testing type approval and conformity of production (COP) of vehicles for emissions as per CMV rules 115, 116 and 126. Ministry of Road Transport and Highways, Government of India, March, 2010.
16. The Central Motor Vehicle Rules, 1989 as amended by the Central Motor Vehicle (3rd amendment) Rules. Universal Law Publishing PVT Ltd; 2010.
17. ECMA website. http://www.ecmaindia.in. Accessed 25 July 2016.
18. UNECE WP.29 GRPE. Worldwide Harmonized Motorcycle Emissions Certification Procedure. Informal document No. 9, 45th GRPE, 13–17 Jan 2003.
19. UNECE Global Technical Regulation No. 2. Measurement procedure for two-wheeled motorcycles equipped with a positive or compression ignition engine with regard to the emission of gaseous pollutants, CO_2 emissions and fuel consumption. Established in the Global Registry on 22 June 2005.
20. Zardini A, Clairotte M, Lanappe G, Giechaskiel B, Martini G. Preparatory work for the environmental effect study on the Euro 5 step of L-category vehicles. JRC Science for Policy Report; 2016.
21. Bharat stage IV emission standards for two-wheelers in India. ICCT policy update, August 2014.
22. Transportpolicy.net http://www.transportpolicy.net/index.php?title=Category:Motorcycles. Accessed 7 August 2016.

Chapter 4
Heavy-Duty Vehicles and Engines

Abstract This chapter reviews the most important cycles used for the certification, and testing in general, of heavy-duty vehicles in the world. For trucks and buses, both chassis and engine-dynamometer cycles have been employed. The latter were initially steady-state but later evolved to a more realistic transient form. European, U.S., Japanese, Australian, Chinese and worldwide cycles are presented and discussed. A historic review regarding the development of each cycle is provided at first, together with technical specifications, 2D and 3D frequency diagrams, detailed comparative data, and examples from real heavy vehicles operation.

4.1 Introduction

Medium and heavy-duty on-road vehicles, such as trucks/lorries, tractors, trailers, buses, and coaches, Fig. 4.1, are primarily powered by diesel engines. Unlike passenger cars, heavy-duty vehicles (HDVs) are not manufactured in large scales but in much broader variety of engine/drive-train/chassis combinations. Both emission regulations and test cycles for heavy-duty vehicles were legislated at a later stage compared to light-duty ones.

Owing to the considerable diversity in HDV applications, every truck or bus application would ideally require a different test cycle. For example, the daily schedule of urban buses or pickup and delivery vans is heavily transient, in contrast to line-haul trucks which spend a significant amount of time cruising. Consequently, it is very challenging to develop a single test schedule that accurately simulates the real driving profile of all target HDVs.

Contrary to the passenger car cycles presented in Chap. 2, of which all have been of the chassis-dynamometer type, HDVs have been certified in the past either on a chassis dynamometer (for example, in Japan) or, primarily, an engine dynamometer. An important classification as regards the engine-dynamometer cycles is whether they are steady-state or transient. Figure 4.2 demonstrates typical examples of HDV and heavy-duty engine (HDE) cycle types.

E.G. Giakoumis, *Driving and Engine Cycles*,
DOI 10.1007/978-3-319-49034-2_4

Fig. 4.1 Various heavy-duty vehicles (dump trucks, tractors, car carriers, tank trucks, urban buses, coaches, trailers)

The United States, EU (often through UNECE Regulations) and Japan have been the pioneering regions that primarily developed test cycles for heavy vehicles and engines, with worldwide test procedures recently constructed and employed too [1–3]. Since HDV engine calibration has traditionally focused on low fuel consumption, it is without doubt a very demanding task to maintain this objective while at the same time dealing with increasingly stringent exhaust emission targets. This holds particularly true when NO_x emissions are concerned since these exhibit an inverse relation relation with fuel consumption.

As was the case with the light-duty vehicles, *chassis-dynamometer* cycles for HDVs involve testing of the whole vehicle (Fig. 4.2a). During the test, the vehicle is stationary, mounted on a dynamometer, with the drive wheels spinning cylindrical rolls to simulate driving at different speeds. The dynamometer simulates the varying vehicle inertia load, and rolling and aerodynamic resistance throughout the cycle. The driver follows closely the specific profile of the test schedule with the help of a computer screen showing real-time and theoretical velocity. The most significant benefit of this test method is that it effectively brings the entire

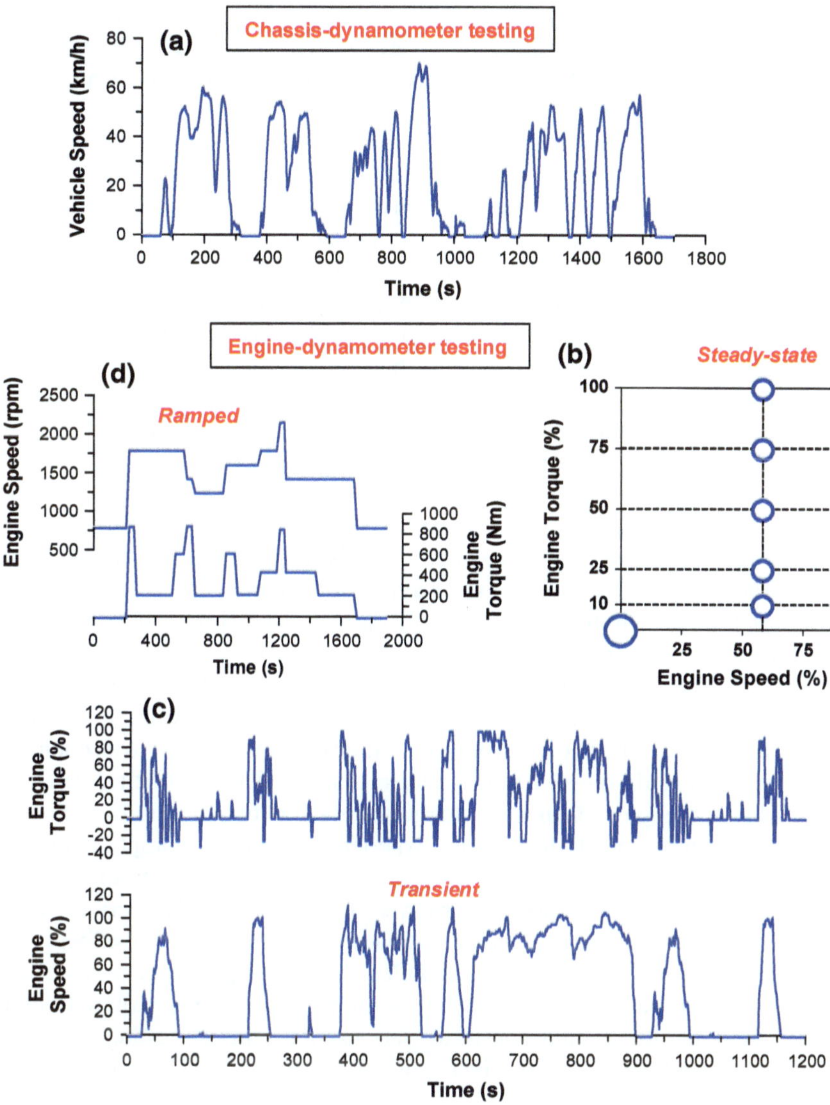

Fig. 4.2 Typical vehicle and engine cycles for heavy-vehicles testing

drive-train into test (see also Table 1.2 highlighting the differences between chassis and engine testing). As such, it can be used to provide a realistic assessment of distance-specific emissions and fuel consumption [4].

Engine-dynamometer cycles, on the other hand (Fig. 4.2b–d), are defined in the form of normalized engine speed and load operating points (%). Load usually means torque but there have been some occasions where the (SI-engine stationary) cycle was defined with respect to intake manifold vacuum. Since normalized speed/load points are used, this means that it is the cycle that is practically adapted to the engine and not the other way around (as was the case with the chassis-dynamometer cycles). Therefore, each engine to be tested on an engine dynamometer is equally challenged by the cycle. Prior to the test, and based on the specific engine's speed/load mapping, a denormalization procedure is performed that defines the exact operating points to be followed by the engine in question during the test. What varies between the different cycles is the exact form of the speed denormalization equation, with the torque denormalized always using the same expression (see Eq. (4.4) later in the text). During the test, the emissions are measured and reduced to the total work produced by the engine, usually in g/kWh or g/HPh. This way of expressing emission levels practically removes inequities between large and small engines, placing both on the same comparative scale. Testing on an engine-dynamometer cycle was initially of the stationary type, in which case reference is made of *steady-state* cycles; more recently developed cycles, on the other hand, employ a true *transient* pattern.

Steady-state engine-dynamometer cycles, Fig. 4.2b, are formed from a number of discrete, and limited in any case, engine speed/load points, referred to as modes or notches, representative of the specific vehicle's duty cycle. The engine under study undergoes the test by passing from each mode in a predefined order, allowing for adequate time to stabilize at each one. Emissions are expressed in g/kWh or g/HPh, but in the earlier versions of these cycles, concentrations in ppm were sometimes used too. The obvious advantages of stationary cycles are the relatively low cost and the real time results, which can be easily compared mode to mode. However, the critical effect of transient conditions is not taken into account. In general, and as regards diesel engines, steady-state cycles were better suited to naturally aspirated engines than turbocharged, since for the latter an overshoot of emissions, in particular PM/soot, is experienced during abrupt transients. Furthermore, they were better suited to engines that were not electronically controlled, therefore their calibration pattern was invariable. Another serious shortcoming of a steady-state test is its inability to account for regenerating effects from modern after-treatment systems such as DPFs. In addition, cold starting effects cannot be taken into account, since stabilization at each mode is required before sampling. In order to account for these serious deficiencies, testing of HDEs evolved from steady-state to transient (Fig. 4.2c), or at least pseudo-transient (Fig. 4.2d) form.

Transient engine-dynamometer cycles are expressed as a sequence of speed/torque operating points versus time; this sequence is derived from a vehicle speed versus time trace after having applied a suitable drive-train model. Both the engine speed and torque are provided in normalized form (in order for their application to be universal), and the time step is one second. For the denormalization procedure, specific equations are employed with reference to the speed/load mapping of the engine under test. Hence, the cycle is always adapted to the tested engine, which means that each engine is equally challenged by the cycle. During the test, emissions are measured continuously (see Sect. 6.5), and the final total mass of each pollutant is reduced to the total work produced by the engine during the test; the result is provided in g/kWh or g/HPh.

Interestingly, even when steady-state cycles were the primary procedure for testing the compliance of (diesel) engines with the gaseous emission standards, a simple transient test, mostly in the form of a free acceleration or load step increase, was legislated in many countries for the control of soot. This was utilized as an indirect approach to PM control (particulate emissions were unregulated in the United States until 1988 and in Europe until 1992). Such smoke tests are only able to effectively identify high emission vehicles/engines and, in certain cases, also detect tampering with an injection system, that is why they are often used for inspection and maintenance reasons. Although these smoke tests are not strictly test cycles and not necessarily HDV-oriented either, they will be discussed in this section too for the sake of completeness.

4.2 Europe

European emission regulations for new heavy-duty engines are commonly referred to as 'Euro' followed by a Roman numeral (I … VI). The regulations were originally introduced in 1988 by Directive 88/77/EEC (corresponding to UNECE R49/01), known as 'Euro 0'. Directive 88/77/EEC was applicable to 'vehicles propelled by a diesel engine, intended for use on the road, with or without bodywork, having at least four wheels and a maximum design speed exceeding 25 km/h, with the exception of vehicles of category M_1 (passenger cars) and vehicles running on rails, agricultural tractors and machines, and public works vehicles'. Subsequently, HDE directives were consolidated by Directive 2005/55/EC. As with their light-duty counterparts, Euro levels for heavy-duty engines began in 1992 (Euro I), followed by Euro II in 1996. Beginning with the Euro VI stage, regulations are used, instead of directives, being directly applicable to all Member States.

The emission standards apply to all motor vehicles of categories M_3 (buses) and N_3 (trucks), and those of categories M_1 (passenger cars), M_2 (small buses), and N_1 and N_2 (vans) having a reference mass greater than 2610 kg, equipped with compression ignition engines or positive ignition natural gas (NG) or LPG engines. At the time of writing, the Euro VI emission standard is valid, introduced by Regulation 595/2009/EC (implementing Regulations 582/2011/EU and 64/2012/EU). This emission limit is comparable in stringency to the U.S. 2010 standard (Fig. 1.4), and became effective in 2013 for new type approvals and in 2014 for all registrations [1–3].

Table 4.1 provides details concerning the European legislation on heavy-duty engines, including the applicable test schedule according to the emission level, the controlled pollutants and the corresponding directives/regulations together with the applicable dates. The test cycles will be presented in detail in the next paragraphs, in chronological order.

Table 4.1 EU regulations on HDE emission levels and test cycles (adapted from [5])

Emission level (application date)	Test cycle	Controlled pollutants	EU Directive (UNECE Regulation)
Euro '0' (October 1, 1990)	ECE R49	CO, HC, NO_x	88/77/EEC (R49/01)
Euro I (July 1, 1992)		CO, HC, NO_x, PM	91/542/EEC (R49/02)
Euro II (October 1, 1995)			
Euro III (October 1, 2000)	ESC/ELR + ETC	CO, HC (NMHC + CH_4), NO_x, PM, Smoke opacity	1999/96/EC 2005/55/EC (R49/03, 04, 05)
Euro IV (October 1, 2005)			
Euro V (October 1, 2008)			
Euro VI (January 1, 2013)	WHSC + WHTC	CO, HC (NMHC+CH_4 for gas engines), NO_x, PM, NH_3, PN	595/2009/EC (GTR No. 4, R49/06)

4.2.1 *Free Acceleration Test*

In Directive 72/306/EEC (UNECE R24), a free acceleration smoke test was legislated to be conducted on diesel engines or diesel-engined vehicles (of all sizes), initially for type approval purposes. It comprised the following sequences (with gear lever in neutral and clutch engaged): with the engine idling, the accelerator control is operated quickly so as to obtain maximum delivery from the injection pump. This position is maintained until maximum engine speed is reached and the governor comes into action. Afterwards, the accelerator is released until the engine resumes its idling speed and the opacimeter reverts to the corresponding conditions. This operation is repeated at least six times in order to clear the exhaust system and to allow for any necessary adjustment of the apparatus. The maximum opacity values read at each successive acceleration is noted until stabilized values are obtained. No account is taken of the values read while the engine is idling after each acceleration. The values read are regarded as stabilized when four consecutive readings are situated within a bandwidth of 0–25 m^{-1} and do not form a decreasing sequence. The absorption coefficient to be recorded is the arithmetic mean of these four values.

The test was replaced by the ELR (Sect. 4.2.5) as regards type approval of HDEs, but a similar free acceleration procedure was legislated in Directives 2009/40/EC and 2010/48/EU for the purposes of roadworthiness testing for motor vehicles and their trailers (both light-duty and heavy-duty diesel-engined vehicles). The certified smoke value is indicated on the engine nameplate and serves as the basis for periodic inspection in some EU countries.

4.2.2 *Steady-State R49*

The first employed European test schedule for HDEs was the stationary R49, corresponding to the ISO 8178 A cycle. Steady-state cycle ECE R49 or just R49 (or 13 mode) was introduced with UNECE Regulation No. 49, and adopted in Europe through Directive 88/77/EEC; it was valid for the Euro I and II standards during the 90s. The R49 cycle is detailed in Table 4.2 and schematically represented in Fig. 4.3; the cycle is closely patterned after the earlier California/U.S. 13-mode cycle for diesel engines discussed later in the text (Fig. 4.18).

Table 4.2 Speed/load points of the European ECE R49 cycle [6]

Mode	Engine speed	Engine load (%)	Weighting factor (%)
1	Idle	–	8.33
2	Intermediate	10	8.00
3		25	8.00
4		50	8.00
5		75	8.00
6		100	25.00
7	Idle	–	8.33
8	Rated	100	10.00
9		75	2.00
10		50	2.00
11		25	2.00
12		10	2.00
13	Idle	–	8.33

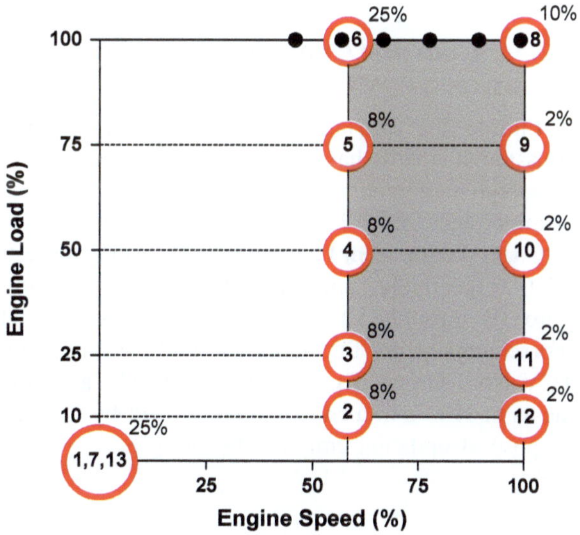

Fig. 4.3 Illustration of the European R49 engine speed/load points (*percentages* denote weighting factors of the specific modes); *small black-colored circles* at full load denote test points of the smoke opacity measurement defined in Directive 72/306/EEC (UNECE Regulation No. 24/03)

The cycle consists of three idle sample points (25 % total weighting factor), as well as intermediate speed and rated-speed modes. Rated speed is defined as the maximum full-load speed allowed by the governor as specified by the manufacturer in the sales and service literature. Intermediate speed, on the other hand, is defined as the peak-torque speed, provided it is within the range 60–75 % of rated speed; otherwise it is 60 % of the rated speed. The cycle was run with the engine fully warmed-up, and the results were expressed in g/kWh. As is obvious from Fig. 4.3, emphasis is placed on the idle and full-load sample points, which possess the highest weighting factors. Moreover, medium to high speeds are under test, representative of rural or motorway driving. On the contrary, urban driving operating points seem to be missing from the test. This means that the cycle was better suited to highway trucks rather than (urban) buses. In any case, at the lower speed range, the engine could be calibrated with regard to fuel consumption alone, without concern about emissions. Notice also in Fig. 4.3 that the loads at each speed were tested sequentially. The maximum-torque points were accounted for with an overall weighting factor 57 % and the rated-speed ones with 18 %.

Initially, only gaseous pollutants (NO_x, CO and HC) were intended to be measured on the ECE R49. For particulates, a two-part smoke test had been defined in Directive 72/306/EEC. One part of the test consisted of a free acceleration (Sect. 4.2.1), and the other was a 6-mode steady-state test. For the latter, the opacity of the exhaust gases produced by the engine was measured with the engine running at full load and at six speeds spaced out uniformly between that corresponding to maximum power and the higher of the following two engine speeds: 45 % of the engine speed corresponding to maximum power or 1000 rpm (Fig. 4.3). Directive 91/542/EEC (Euro I standard) changed the particulates measurement method by expanding the R49 test procedure to also include PM.

4.2.3 Chassis-Dynamometer FIGE

In Europe, the legislation for testing the compliance of heavy vehicles with the emission standards has always employed an engine-dynamometer cycle. Thus, no chassis-dynamometer cycle has ever been legislated. The FIGE cycle is, however, of importance, since it was used as the basis for the development of the stationary ESC and the transient ETC engine-dynamometer tests discussed in Sects. 4.2.4 and 4.2.6 respectively, applicable for many years in the EU (emission standards Euro III through V). The FIGE cycle was developed by the FIGE Institute (Forschungsinstitut für Geräusche und Erschütterungen[1]), Aachen, Germany, now TÜV Nord Group, Essen, in the early 90s, and was based on real road measurements of HDVs in a collaborative project by German and Swiss authorities for the purpose of updating emission factors [7].

[1]Research Institute for Noise and Vibration.

Table 4.3 Cycle categories of the FIGE research [9]

Categories			
Road types	Load types	Road gradients (%)	Vehicle types
Urban	No load	–6	*Trucks*
Rural	Partial load	–3	Coaches
Motorway	*Full load*	*0*	Buses
		3	Special vehicles
		6	

italics denote the categories used for the creation of the FIGE cycle

FIGE measured continuously vehicle speed, engine speed, and torque from 20 in-use vehicles (17 goods vehicles ranging from small delivery vans to 40,000 kg articulated vehicles, and 3 local public transport buses) operated in normal service for 3–4 days. The tests were carried out in Germany, and, as part of it, 5 representative vehicles from the former East Germany were also tested [8]. During the research, representative driving cycles, consisting of a series of vehicle speeds vs. time, were categorized according to road type and gradient, traffic density, vehicle type and loading. Table 4.3 demonstrates the various categories considered. Eventually, the FIGE Institute proposed a cycle by selecting and combining three representative 900-s duration driving cycles on each of the three road types, urban, rural, and motorway [7–9]. All three segments were from truck operation at zero gradient and full-load condition (80–100 % vehicle capacity).

The final version of the cycle, demonstrated in Fig. 4.4 and used for the development of the ETC, lasts however 1800 s (30 min) for a total distance of 29.5 km, with 91.1 km/h maximum speed, 59 km/h average and 61 km/h driving speed; only 3.5 % of the time is spent idling, while for two thirds of the time the speed is maintained higher than 61 km/h. Clearly, emphasis is placed on rural and

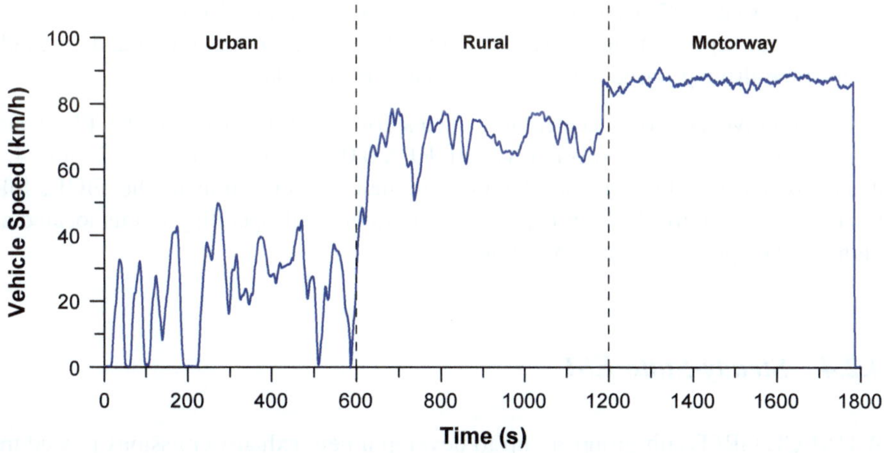

Fig. 4.4 Speed profile of the European FIGE driving cycle

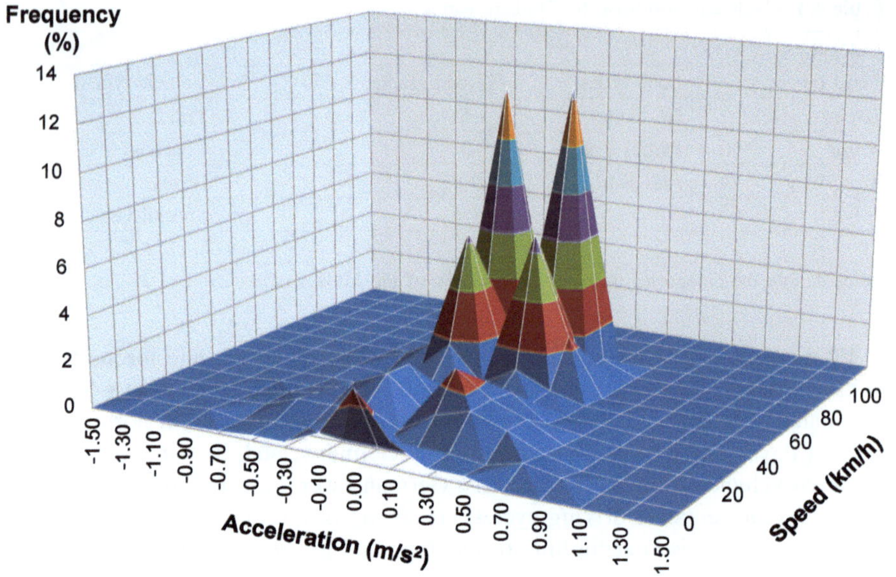

Fig. 4.5 3D speed/acceleration frequency distribution of the FIGE driving cycle

highway driving, representative of highway trucks. On the other hand, only the first phase seems to possess the kind of transient schedule expected to be encountered by an urban bus.

The cycle is divided into three equal parts of 600-s duration:

- Part one represents urban driving conditions with a maximum speed of 49.9 km/h, 8.33 % idling, and 3.87 km distance,
- Part two simulates rural driving starting with a steep initial acceleration; maximum speed is 87.5 km/h for a traveled distance of 11.55 km,
- Part three represents motorway driving with 91.1 km/h maximum and 86 km/h average driving speed for a traveled distance of 14 km.

The relative positive acceleration values for the three parts of the FIGE are 0.146, 0.088 and 0.079 respectively, with RPA 0.07 for the whole cycle. Figure 4.5 demonstrates the 3D speed/acceleration frequency distribution of the FIGE; following the speed profile from Fig. 4.4, the frequency of operating points located at high engine speeds in Fig. 4.5 is fairly high.

4.2.4 Steady-State ESC

A UNECE GRPE sub-group set up to develop a new exhaust emissions procedure for heavy-duty vehicles for implementation with the Euro III emission standard in

2000, reported its conclusions in May 1996. During the course of its work, it compared five candidate cycles with the steady-state R49 and the U.S. FTP *transient* cycle (Sect. 4.3.5) on eight modern engines at four laboratories. It proposed the adoption of two separate tests, each of about 30 min duration based on the FIGE chassis cycle. One was a new 13-mode steady-state cycle (later called ESC) with an additional dynamic load response test for smoke measurements (ELR), and the other a transient cycle (ETC—Sect. 4.2.6). In a ranking procedure, and taking into account precision, applicability to new technology and fuels, cost, and ability to control emissions, both the ESC and the transient ETC were found to offer a substantial improvement over the ECE R49 test [7].

The ESC (European Steady-state Cycle, also known as OICA/ACEA cycle), detailed in Table 4.4 and illustrated in Fig. 4.6, replaced the R49 beginning with the emission standard Euro III in 2000 (Directive 1999/96/EC, equivalent to UNECE R49/03) covering NO_x, CO, HC and PM emissions.

As was the case with the R49, the ESC consisted of 13 modes, but now the tested engine speeds were three, designated A, B and C. These test speeds were chosen following an investigation of the power curves on a number of engines, which showed that the usable speed range of an engine was between 50 % of the rated power before and 70 % beyond the peak in the power curve respectively (based on the FIGE speed profile). Equations (4.1) provide the calculation method for engine speeds A, B and C with reference to Fig. 4.7, which demonstrates a typical engine mapping

$$\begin{aligned} A &= N_{lo} + 0.25\,(N_{hi} - N_{lo}) \\ B &= N_{lo} + 0.50\,(N_{hi} - N_{lo}) \\ C &= N_{lo} + 0.75\,(N_{hi} - N_{lo}) \end{aligned} \tag{4.1}$$

Table 4.4 Speed/load points of the European ESC with weighting factors and duration of engine operation at each mode (transition must be completed within the first 20 s of the mode) [10]

Mode	Engine speed	Engine load (%)	Weighting factor (%)	Duration (min)
1	Idle	–	15	4
2	A	100	8	2
3	B	50	10	2
4	B	75	10	2
5	A	50	5	2
6	A	75	5	2
7	A	25	5	2
8	B	100	9	2
9	B	25	10	2
10	C	100	8	2
11	C	25	5	2
12	C	75	5	2
13	C	50	5	2

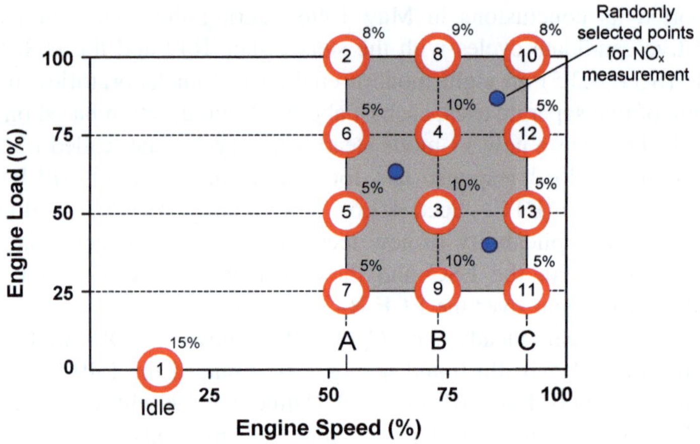

Fig. 4.6 Illustration of the European ESC engine speed/load points with reference to each mode's weighting factor (at the manufacturer's request, the test sequence can be repeated a number of times for sampling more particulate mass on the filter)

Fig. 4.7 Typical engine power/speed mapping highlighting the control area for the purposes of the ESC test

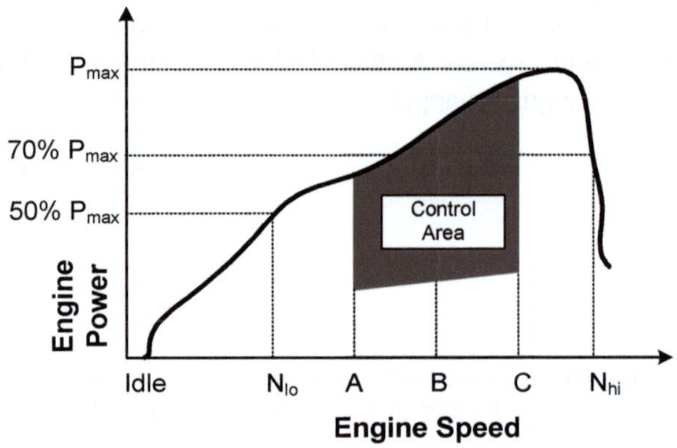

Engine speed N_{hi} corresponds to the maximum speed (above the rated N_{ref}) where 70 % of the declared by the manufacturer maximum net power is produced. Similarly, N_{lo} corresponds to the lowest speed (below the rated) where 50 % of the declared by the manufacturer maximum net power is produced (speeds A, B and C were denoted by the manufacturer and verified prior to the test procedure) [10].

As evidenced from Fig. 4.6, emphasis during the ESC is placed on the B-speed modes, with total weighting 39 %, compared to 23 % for the low and the high-speed modes. In contrast to the R49 cycle, where all sample load-points at each speed were tested sequentially (for the intermediate speed in ascending load order, and for the rated in descending), during the ESC cycle, the sampling sequence is mixed. Similar to the R49, the sampling took place with the engine fully warmed-up. For example, a dummy test cycle could be conducted for the

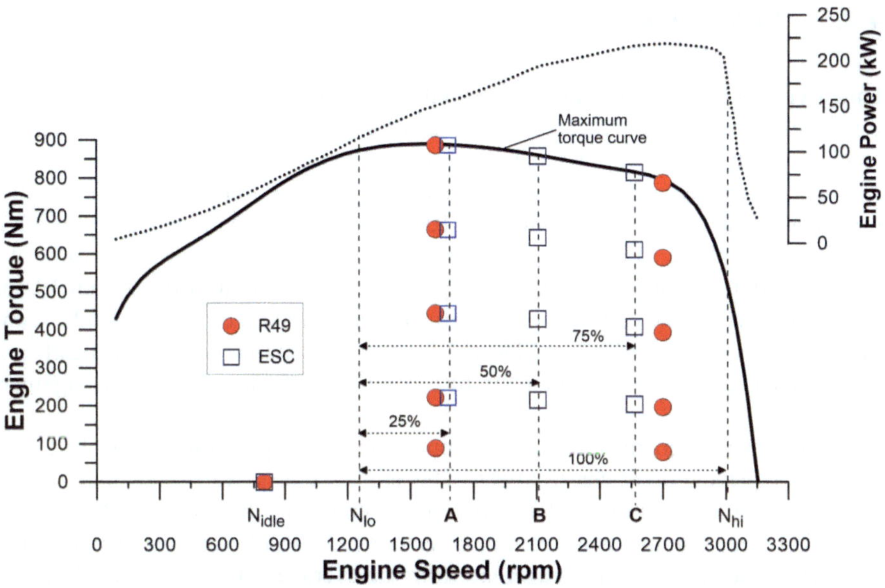

Fig. 4.8 Torque mapping of a heavy-duty turbocharged diesel engine demonstrating the points tested during the ESC compared to the R49 (owing to the specific engine's peak torque speed being lower than 60 % of the rated, the 60 % rated speed is taken as the intermediate speed for the purposes of the R49 cycle)

engine and the exhaust system to be conditioned. Figure 4.8 compares the tested engine areas for a medium/heavy-duty engine running on both the R49 and ESC cycles.

Since the tested points were known to the manufacturer prior to the test, and in order to avoid possible 'cycle beating' attempts, during the certification procedure three randomly selected modes inside the cycle control area were tested for NO_x emissions (see Fig. 4.6), upon completion of mode 13. The duration for each extra NO_x measurement was again 2 min. The emissions from these extra three modes were not allowed to exceed by more than 10 % the averaged value of the four adjacent modes [10].

Emission sampling during the ESC was based on concentration values for all gaseous pollutants using the average chart reading during the last 30 s at each mode. The average concentration for each pollutant on a wet basis was then multiplied by the exhaust gas flow in order to obtain the emission mass flow-rate E_{gi}, as detailed in Sect. 6.5.

The ESC was valid through emission standards Euro III, IV and V, and was abandoned from the Euro VI standard (1/2013) having been replaced by the steady-state WHSC, discussed later in the text.

4.2.5 ELR Smoke Test

Directive 1999/96/EC introduced a special test, the ELR (European load response), for checking the compliance of HDV diesel engines with the acceptable smoke levels, to be used as a supplemental test to the ESC cycle. The test is illustrated in Fig. 4.9 in generalized form and Fig. 4.10 for a specific engine; it consists of a sequence of three load steps from 10 to 100 % at each of the three engine speeds A (phase 1), B (phase 2) and C (phase 3). These are then followed by phase 4 at a speed within the control area, between speeds A and C, and a load between 10 and 100 % selected by the certification personnel. Total duration of the test is 420 s, and the opacity is measured in m^{-1}. During the test, the engine covers practically the whole range of possible air-fuel mixture conditions defined by the management system.

Smoke measurement values were continuously sampled during the ELR test with a frequency of at least 20 Hz. The smoke traces were then analyzed to determine the final smoke values applying a procedure based on a Bessel filter averaging algorithm that is detailed in Directive 1999/96/EC. The final smoke value was determined as a weighted average from the average values at speeds A, B and C, with weighting factors 43, 56 and 1 % respectively. The opacity measured at the free selectable speed in phase 4 was not allowed to exceed the value of the two

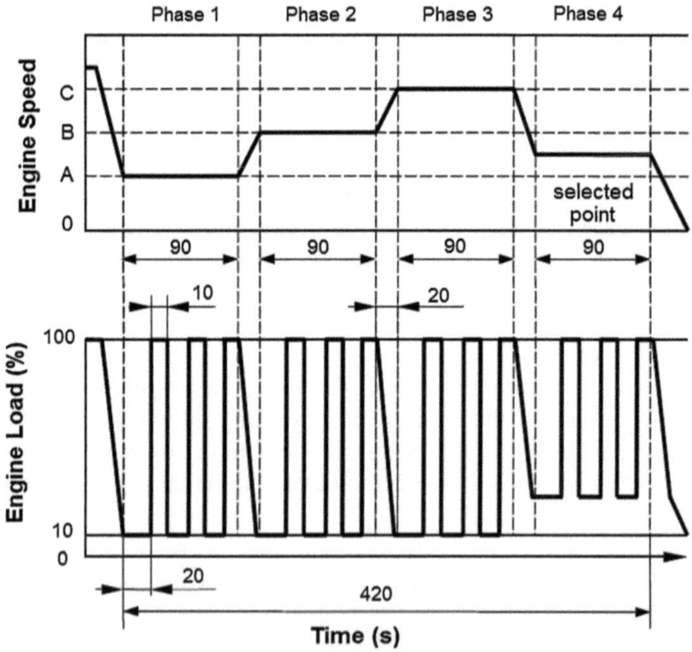

Fig. 4.9 Engine speed and load profile of the European ELR smoke test [10]

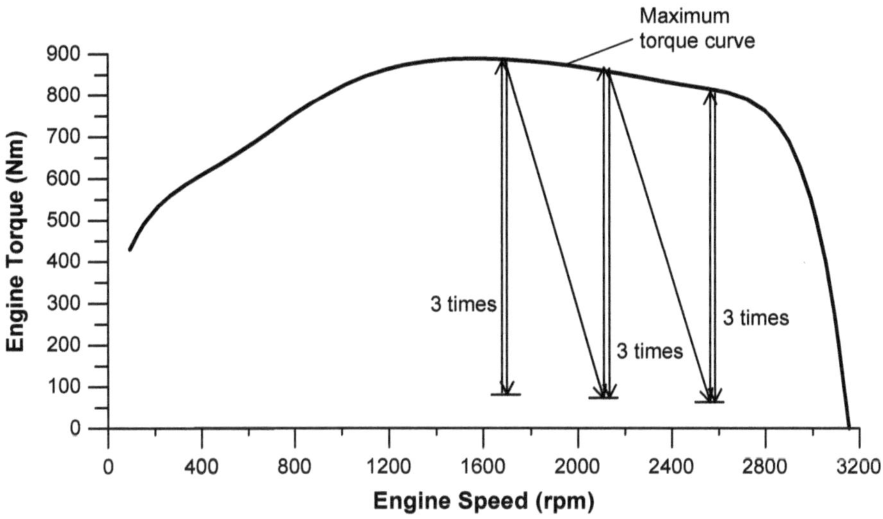

Fig. 4.10 Load steps during phases 1, 2 and 3 of the ELR test on the torque/speed map of a heavy-duty engine

adjacent speeds by more than 20 % or the limit value by more than 5 % (whichever is greater) [10].

The ELR was run with the engine fully warmed-up; the preconditioning consisted of engine operation at full power in order to stabilize the engine parameters, and then, within 20 s, the test was started. As was the case with the ESC and the ETC (next section), the ELR was valid through emission standards Euro III, IV and V.

4.2.6 Transient ETC

The transient ETC (European Transient Cycle) was introduced together with the steady-state ESC and the smoke test ELR for emission certification of heavy-duty diesel and SI engines in Europe starting in the year 2000 (Directive 1999/96/EC), being the first transient HD engine-dynamometer cycle in the EU. The cycle was based on research done by the FIGE Institute on heavy vehicles in the early 90s. The chassis-dynamometer cycle proposed by FIGE, after shortened and modified, resulted in the vehicle speed trace illustrated in the upper sub-diagram of Fig. 4.11. The time curve of this cycle was then normalized for engine speed and torque to obtain the engine-dynamometer ETC with the assumption of a 28,000 kg vehicle weight [7]. The resulting engine-dynamometer ETC is demonstrated in Fig. 4.11 in terms of normalized engine speed and torque versus time; similarly to the chassis-dynamometer FIGE cycle, it lasts 1800 s (30 min), and consists of three parts, namely urban, rural and motorway, of equal (600 s) duration.

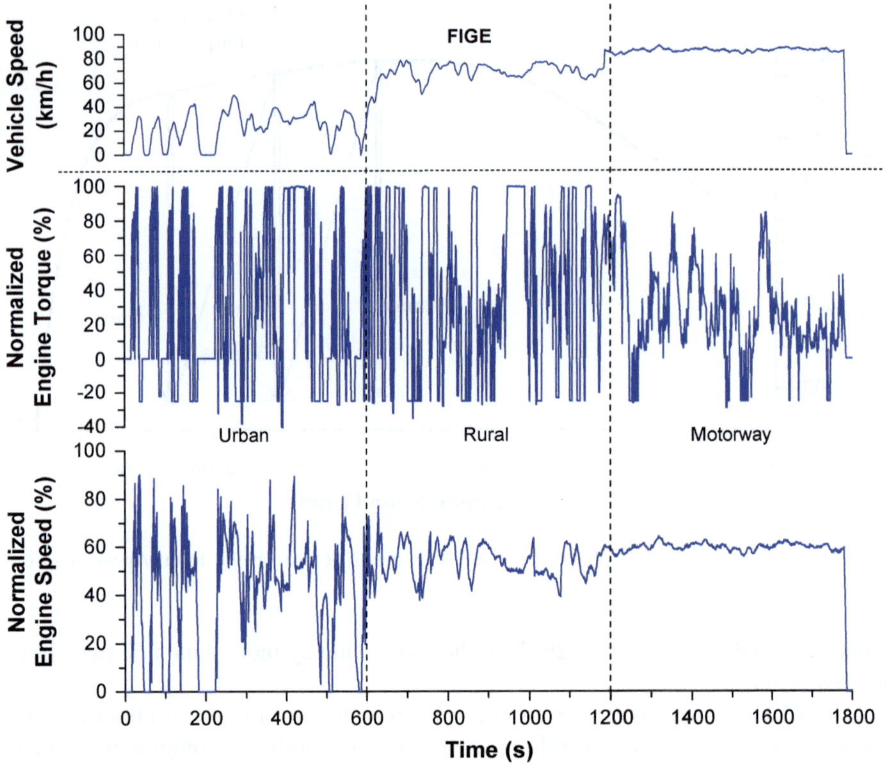

Fig. 4.11 Engine speed and torque profile of the European ETC transient cycle (the upper sub-diagram illustrates the chassis-dynamometer FIGE cycle from which the ETC was derived) [10]

For certification purposes, the engine under study was first mapped in order to determine the actual speeds N (in rpm) and torques M (in Nm) throughout the cycle from the normalized N_{norm} and M_{norm} ones. Speed was denormalized using the following equation [10]

$$N = \frac{N_{norm} \times (N_{ref} - N_{idle})}{100} + N_{idle} \tag{4.2}$$

with the reference speed N_{ref} corresponding to the 100 % speed values specified in the engine dynamometer schedule, defined as follows

$$N_{ref} = N_{lo} + 95\,\%\,(N_{hi} - N_{lo}) \tag{4.3}$$

N_{hi} is the highest engine speed where 70 % of the declared by the manufacturer maximum power occurs, and N_{lo} the lowest engine speed where 50 % of the declared maximum power occurs (Fig. 4.7). Engine torque was denormalized to the maximum torque at the corresponding speed using the following equation

$$M = \frac{M_{norm} \times M_{\max}}{100} \tag{4.4}$$

with M_{max} the maximum torque at each speed found from the respective engine mapping curve.

Negative values in the torque sub-diagram of Fig. 4.11 correspond to motoring segments, and for the purpose of reference cycle generation take denormalized values determined in Directive 1999/96/EC as follows: one choice is to use negative 40 % of the positive torque value at the associated speed point; another to map the negative torque required to motor the engine from minimum to maximum speed; the third option is to determine the negative torque required to motor the engine at idle and reference speeds, and linearly interpolate between them [10].

The ETC manifests increased frequency in the 50–70 % normalized speed region, as is evidenced in Fig. 4.12 that displays the 3D normalized speed/torque distribution of the cycle. Further, it exhibits less than 7 % idling period. The above are direct consequences of the chassis-dynamometer FIGE's extended rural and motorway segments.

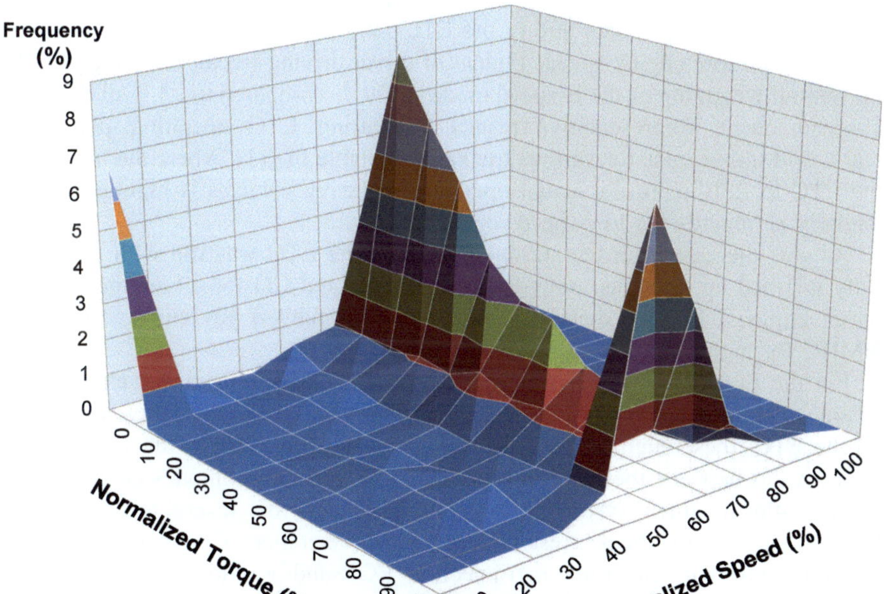

Fig. 4.12 3D normalized speed/torque frequency distribution of the European ETC (notice the increased frequency between 50 and 70 % normalized speeds)

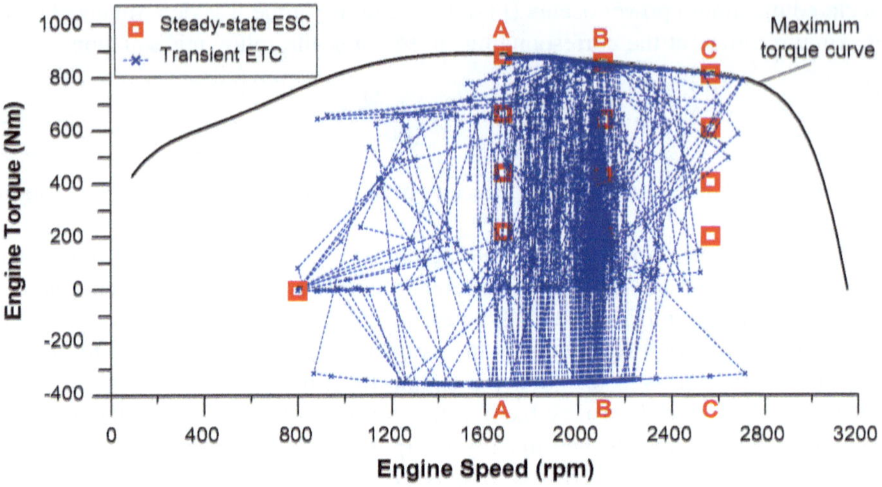

Fig. 4.13 Comparison between the transient ETC and the steady-state ESC cycles for a heavy-duty diesel engine (the area mostly covered by the ETC is the medium-speed one (between speeds A and B); low-speed operating points are represented with much lower frequency)

Figure 4.13 compares the tested speed/torque points (actual values) between the steady-state ESC and the transient ETC for a heavy-duty diesel engine. The greatest volume of operating points during the ETC falls around speed B, with relatively few points at low engine speeds typical of urban driving (recall from Fig. 4.6 that the overall weighting of the B-speed modes is 39 % compared to 23 % of the A and C-speed ones). Representative results comparing ETC operating points and real-world in-use ones for various trucks is available in [11], where the inability of the ETC to fully represent actual operation of heavy vehicles is highlighted, particularly as regards the low-speed region.

As was the case with the steady-state ESC, the ETC was run with the engine fully warmed-up. Sampled emissions were: NO_x and CO for all engines; PM for diesel engines; HC for diesel engines and LPG fuelled gas engines; CH_4 and NMHC for natural gas fueled gas engines [10].

For the Euro III emission standard, adopted in 2000, only diesel engines with advanced emission control systems, such as NO_x catalysts and/or DPFs, were tested on the ETC; these engines were also tested on the ESC/ELR, with different limit values valid between the ETC and the ESC test. For those engines equipped with exhaust after-treatment systems with periodic regeneration, several test cycles were to be conducted, and the test results averaged between them. Conventional engines, meaning those with electronic fuel injection, EGR and/or oxidation catalysts, were only tested on the ESC/ELR. For compliance with the Euro IV (adopted in 2005) and Euro V (2008) standards, all diesel engines were tested on the ETC + ESC/ELR; again, the limits between the ETC and the ESC were different, with the ETC ones being higher than those for the ESC. Regarding heavy-duty gas engines, the ETC has been the only applicable cycle from Euro III.

Comparing the steady-state ESC and its transient counterpart ETC, the UNECE sub-group concluded that although the ESC could not capture the true driving pattern of a vehicular engine, it gave better protection against an engine being operated at worse levels of emission control outside the envelop defined by the cycle (owing to the three extra tested points unknown to the manufacturers). Further, it required relatively minor changes with respect to the R49 cycle, which was the employed test cycle at the time [7].

Typical engine results (performance and engine-out soot and NO) during the ETC are demonstrated in Fig. 4.14 for a heavy-duty turbocharged diesel engine. Similar to the remarks made earlier for chassis-dynamometer cycles, abrupt accelerations and load increases throughout the cycle are responsible for spikes in both gaseous emissions and soot/PM. For this behavior, the probable cause is the instantaneous shortage of air due to turbocharger lag. For NO emissions, in particular, it is the lag between increased fueling and the response of the air-charging system that is responsible for the overshoot in emissions. For engines equipped with exhaust gas recirculation, the shutting off of the EGR valve at the onset of each transient event in the cycle, in order to limit intolerable soot, enhances the NO increase. The latter is, in any case, smaller than its soot/PM counterpart as is evidenced in Fig. 4.14 [12].

The ETC has been abandoned in Europe from the Euro VI standard (1/2013), having been replaced by the worldwide WHTC discussed later in the text; it continues, however, to be valid in other areas of the world (see Fig. 4.55). Detailed technical characteristics of the transient ETC will be presented in Sect. 4.7.2 in comparison to the other two legislated heavy-duty engine cycles (FTP and WHTC).

4.2.7 Other European Non-legislated Chassis-Dynamometer Cycles

Braunschweig Cycle

A European heavy-duty cycle, frequently employed in research projects around the world, is the Braunschweig cycle ('Stochastischer Fahrzyklus für Stadtlinien Omnibusse'). It was developed by the Braunschweig University in the mid 70s. The cycle, illustrated in Fig. 4.15, was based on the passenger cars U.S. FTP-75 (Fig. 2. 18) adding more stops, increasing the idling period and suppressing high vehicle speeds [13].

The cycle lasts 1740 s for a traveled distance of approximately 11 km, with 58.2 km/h maximum and 22.5 km/h average speed. It demonstrates typical urban bus route with many idling phases (30 in total or 2.75 per km). Excluding the initial and final idling period, there are 16 bus stops and 12 stops due to traffic conditions. The above were defined based on a survey conducted in nine German cities at the

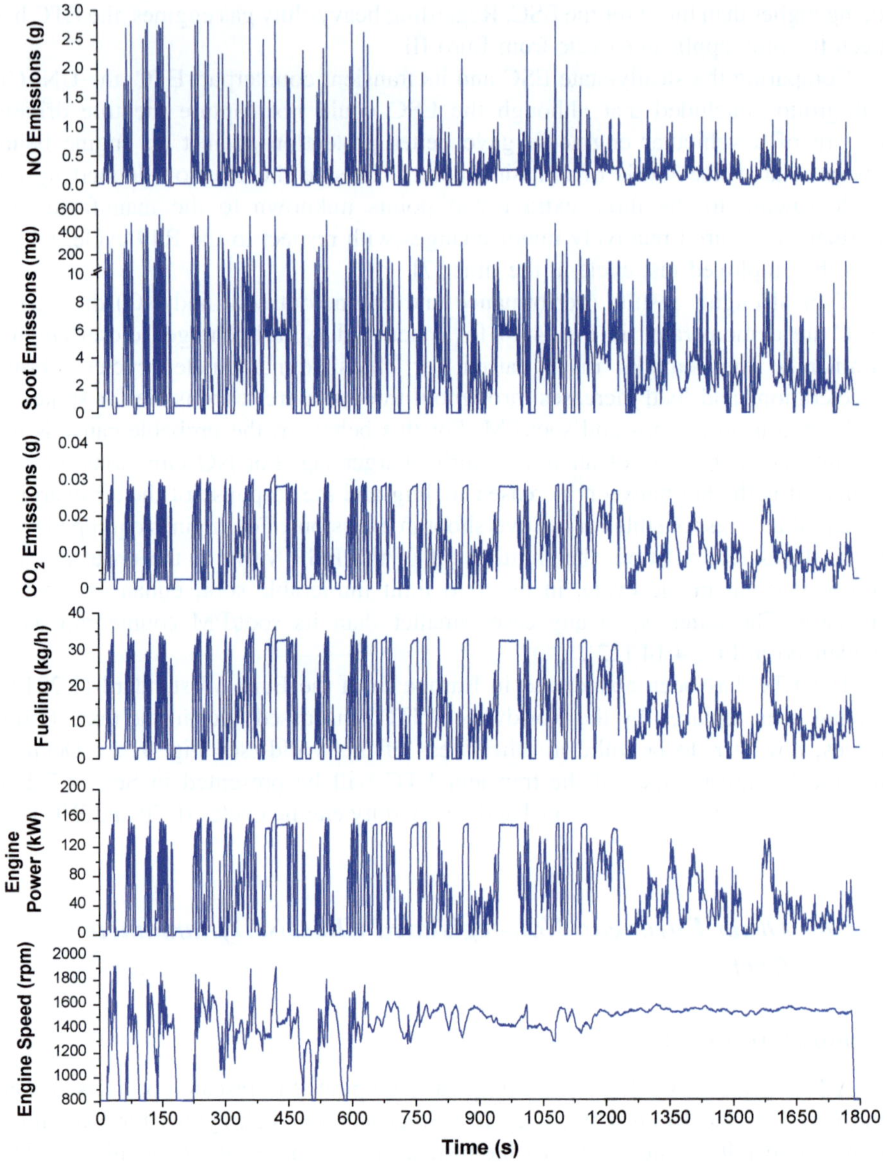

Fig. 4.14 Development of engine properties and engine-out emissions during the ETC for a turbocharged diesel engine (reprinted from [12], copyright 2010, with permission from Elsevier)

time [14]. RPA is 0.235 (typical value for exclusively urban routes with frequent and steep accelerations), with the maximum acceleration 2.42 m/s^2 (8.7 km/h/s); 23.7 % of the time is spent idling and 70 % is spent accelerating/decelerating.

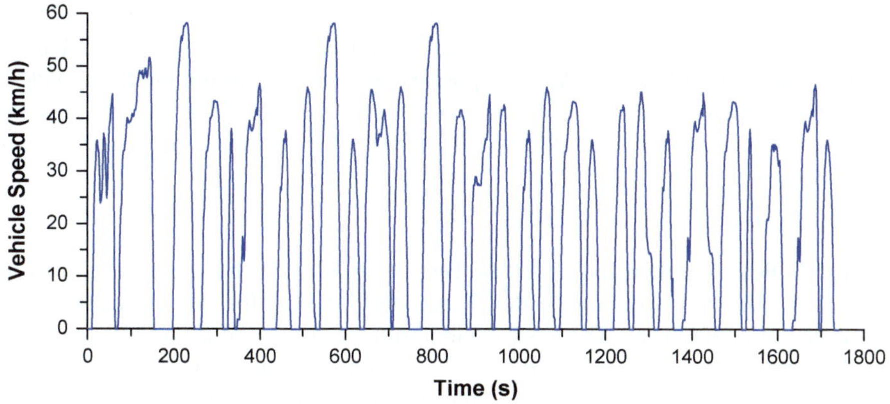

Fig. 4.15 Speed profile of the Braunschweig bus driving cycle

Paris ADEME Cycle

The French Agency of Environment and Energy Management (ADEME) engaged a comprehensive program in 1998 concerning urban buses. During the course of the program, a bus cycle was developed with the cooperation of RATP (Régie Autonome des Transports Parisiens) based on real-time data from urban line No. 21 in Paris [15]. The cycle, illustrated in Fig. 4.16, is known as the ADEME-RATP or Paris cycle and has been employed in various research studies relating to bus (or generally heavy vehicles) emissions [16].

The cycle is highly transient (68 % of the time is spent accelerating or decelerating, with less than 1 % cruising), and lasts almost 32 min, for a traveled distance of only 5.68 km. Maximum speed is 48 km/h but the average speed is 10.8 km/h, with 31 % of the time spent idling (cf. the data for the NY Bus cycle

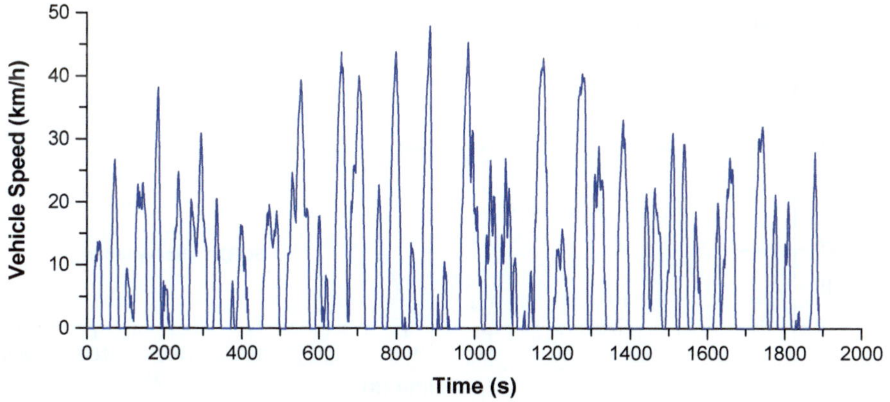

Fig. 4.16 Speed profile of the ADEME-RATP Paris bus driving cycle

later in the text). The cycle is more dynamic than the Braunschweig, with the RPA at 0.276, although the maximum acceleration is somewhat lower, at 2.1 m/s^2; overall there are forty stops. Average positive acceleration during the Paris cycle is 0.51 m/s^2, almost equal to the Braunschweig (0.54 m/s^2), and much higher than the extra-urban oriented FIGE (0.19 m/s^2) or even the latter's initial urban phase (0.30 m/s^2).

SORT Cycles

The SORT (Standardized On-Road Test) suite of cycles has been developed by the UITP (Union Internationale des Transports Publics) for measuring fuel consumption of urban buses. The basic cycle design has been defined as a combination of three trapezes, each composed of three sections: acceleration, constant (target) speed, and deceleration as follows, with reference to Fig. 4.17 [17, 18]:

- The accelerations must be adapted in a way that allows them to meet the target speed at the traffic cone No. 2. The accelerations cannot be inferior to the values provided in Table 4.5 for each trapeze.
- The target speed must be maintained at least until traffic cone No. 3.
- The braking phase starts at cone No. 3 and finishes at traffic cone No. 4; it must be as uniform as possible.

Based on 5 basic trapezes (20, 30, 40, 50 and 60 km/h target speed), three different base cycles have been defined, as illustrated in Fig. 4.17, to represent:

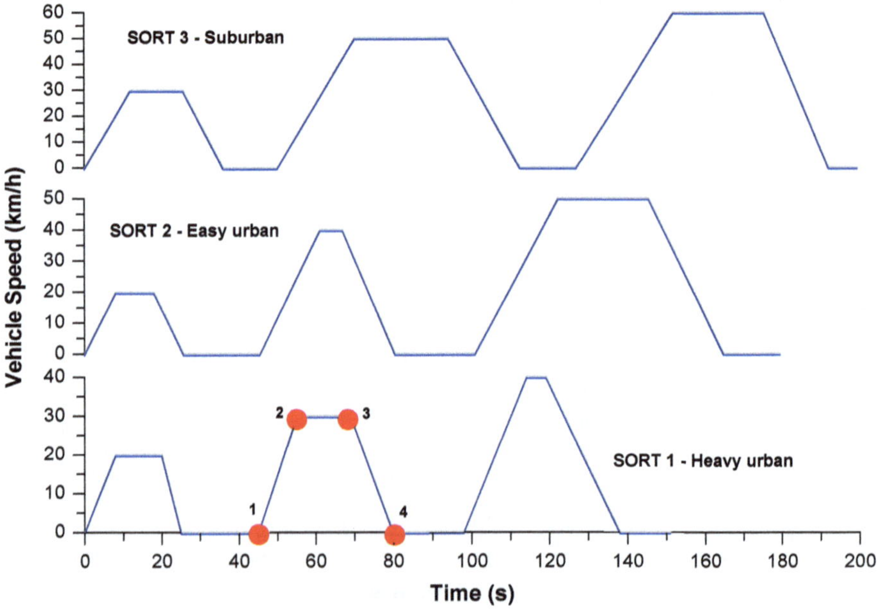

Fig. 4.17 Speed profile of the SORT bus driving cycles [18]

Table 4.5 Main technical specifications of the various versions of the SORT cycles [17, 18]

Cycle	Distance (m)	Maximum speed (km/h)	Rated average speed (km/h)	Idling time (%)	Stop length (s)	Acceleration (m/s^2) (for each trapeze)[a]
SORT 1	520	40	12.1	39.7	20/20/20	1.03/0.77/0.62
SORT 2	920	50	18.0	33.4	20/20/20	1.03/0.62/0.57
SORT 3	1450	60	25.3	20.1	20/10/10	0.77/0.57/0.46

[a]deceleration is 0.8 m/s^2 for all versions and trapezes

(a) heavy urban traffic ('SORT 1', representative of Paris or London), (b) easy urban traffic, also called mixed traffic ('SORT 2', e.g., in Madrid or Munich), and (c) easy suburban traffic ('SORT 3', e.g., Paris suburbs). Each cycle includes a certain time for standstill, split among the different stops. The standstill time was adapted to the average speed of each cycle corresponding to practical experience. Table 4.5 summarizes the main technical specifications of the three cycle versions from Fig. 4.17; the values have been compared to actual in-use data from buses and have been found to correlate satisfactorily [18].

4.3 United States of America

In the United States, The Clean Air Act initially defined heavy duty vehicle as 'a truck, bus, or other vehicle manufactured primarily for use on the public streets, roads, and highways (not including any vehicle operated exclusively on a rail or rails) which has a gross vehicle weight (GVW) in excess of six thousand pounds. Such term includes any such vehicle which has special features enabling off-street or off-highway operation and use'. Effective 1979, EPA defines HDVs in the federal jurisdiction as vehicles of GVWR of above 8500 lbs (3850 kg); in California, HDVs are those with GVWR above 14,000 lbs (6340 kg) for model year 1995 and later [19]. Federal emission standards have been in place for highway diesel and gasoline-fueled heavy-duty engines since the early 70s (in California, from model year 1969). For diesel engines in particular, the first regulations focused only on control of smoke emissions. Subsequent regulations broadened emission control requirements to include gaseous emissions in 1974 and particulate emissions in 1988 (six years later compared to LDVs). More stringent regulations were phased-in during the 2000s. Current limits are defined in the '2007 Heavy-duty Highway Rule', and were fully phased-in as of 2010 [1–3]. The relevant citation for HD engines is: 40 CFR 1065 for emissions measurement, 40 CFR 86 subparts A and N for criteria pollutants, and 40 CFR 1036 for greenhouse gases. Complete medium-duty passenger vehicles (MDPVs), i.e., those with GVWR between 8500 and 10,000 lbs (Class 2b) are chassis tested, and, generally, follow the LDV test procedures (Sect. 2.2), beginning with model year 2004 (Tier 2 standard).

Table 4.6 U.S.-valid test cycles for HDV engines

Model Year	Test cycle	
	Otto engines	Diesel engines
1972	9 mode (1st version) (emission levels in ppm)	–
1974	9 mode (1st version) (emission levels in g/bhp-h)	13 mode
1979	9 mode (2nd version)	
1984	9 mode (2nd version)/FTP transient ('EPA version')	13 mode/FTP transient
1985/86	FTP transient ('EPA version')/FTP transient ('MVMA version')	FTP transient
1987	FTP transient ('MVMA version', 40 CFR 86 Appendix I § (f) (1))	
2007–2009		FTP transient + SET (ESC steady-state or ramped)
2010		FTP transient + SET (ESC ramped)

Beginning with the Tier 3 standard, HDVs with GVWR between 10,000 and 14,000 lbs (Class 3) are also chassis certified.[2]

A thorough review of EPA's regulation regarding HD diesel engines is available in [20]. In this reference, EPA's regulation of heavy-duty diesel engines is divided into five phases. 'The first phase was the pre-Air Quality Act of 1967 period, during which diesel emissions were left unregulated. The second was the period of regulation-by-rulemaking from the Air Quality Act of 1967 through the 1990 Clean Air Act Amendments. The third phase was a brief flurry of formal and informal regulation-by-negotiation after the 1990 Amendments. The fourth, involving regulation-by-litigation, began with the EPA's suit against the heavy-duty engine manufacturers in 1998 (see Sect. 1.3.2). The fifth is the post-1998-consent decree period, a return to regulation-by-rulemaking' [20]. Table 4.6 summarizes the applicable test cycle valid in the U.S. over the years for both Otto-cycle and diesel HD engines.

4.3.1 *Steady-State Gasoline 9 Mode*

For heavy-duty gasoline-powered vehicles, a 9-mode cycle was applicable in the United States between 1972 and 1983 [19, 21]; the cycle was patterned after a similar test schedule already in use in California. The cycle was developed by the

[2]The HDV weight classification used in the U.S. is (lbs): Class 2b (8,501–10,000); Class 3 (10,001–14,000); Class 4 (14,001–16,000); Class 5 (16,001–19,500); Class 6 (19,501–26,000); Class 7 (26,001–33,000); and Class 8 (>33,000)—1 lb equals 0.45 kg.

Table 4.7 The U.S. 9-mode cycle for gasoline heavy-duty engines

Mode (1972)	Mode description	Manifold vacuum (kPa/in Hg)	Time in mode (s)	Weighting factor (%)
1	Idle	–	70	3.6
2	Cruise	54/16	23	8.9
3	Part-throttle acceleration	34/10	44	25.7
4	Cruise	54/16	23	8.9
5	Part-throttle deceleration	64/19	17	4.7
6	Cruise	54/16	23	8.9
7	Full load	10/3	34	28.3
8	Cruise	54/16	23	8.9
9	Closed throttle	–	43	2.1

Mode (1979)	Mode description (each mode lasts 60 s)	Engine torque (%)	Weighting factor (%)
1	Idle	–	23.2
2	Cruise	25	7.7
3	Part-throttle acceleration	55	14.7
4	Cruise	25	7.7
5	Part-throttle deceleration	10	5.7
6	Cruise	25	7.7
7	Full throttle	90	11.3
8	Cruise	25	7.7
9	Closed throttle	–	14.3

For the 1979 version, the mode sequence during the 2nd run was: 2–3–4–5–6–7–8–9–1

Motor Vehicle Manufacturers Association (MVMA) [22]. Table 4.7 lists the modes for the 1972 and the 1979 model year versions.[3]

The 9-mode schedule was intended to simulate truck driving in a metropolitan area, as most of the SI HDVs were of urban-oriented use. Except for the idling phases, all other modes were operated at 2000 rpm. Initially (FR Vol. 33, No. 2, January 4, 1968), the cycle was intended to be run either on a chassis or engine dynamometer. Subsequently, it was regulated as an engine test schedule only (FR Vol. 33, No. 108, June 4, 1968, 'Control of air-pollution from new motor vehicles and new motor vehicle engines', issued by the Dept. of Health, Education and Welfare (HEW), the predecessor or EPA). For the 1972 version, the cycle was run four times, for a total time length of 20 min, preceded by a 5-min idle. For the final reported emission value, the average concentrations of the two warm-up runs were weighted 35 % and of the two hot 65 %. Notice in Table 4.7 that modes 3 and 7

[3]The population of SI-engined HDVs in the U.S. in the late 60s was considerably larger compared to the diesel-fueled ones and so was their emissions contribution. Thus, both emission control and the applicable test cycle for SI engines were implemented earlier [19, 20].

accounted for 54 % of the schedule, with all the other modes corresponding to very low engine load. Emission limits for HC and CO were initially set at ppm and % respectively. Emission limits were set in g/bhp-h beginning with model year 1974, and covered NO_x as well [19].

With the advent of supercharged engines and exhaust gas recirculation, the use of manifold vacuum made no sense anymore as the sole parameter for engine load characterization. Based on data from 19 engines already tested in various EPA contracts, a conversion from manifold vacuum to % torque was performed [23]. Thus, the second/modified version of the 9-mode cycle was defined, applicable to 1979 and later model year HD gasoline engines. In this version of the cycle, each mode lasted 60 s, and the test consisted of one warm-up and one hot cycle run, again preceded by a 5-min warm-up phase (40 CFR 86.335-79, defined in FR Vol. 42, No. 174, September 8, 1977). At the same time point, the definition of HD vehicles changed, covering now those with GVW higher than 8500 lbs. In essence, the 1979 version's factors, derived from California ARB, gave substantially more weight to closed throttle and idle modes and less to the part throttle acceleration and full load ones. Since neither the closed throttle nor the idle are power producing modes, the cycle power was less when weighted, making the denominator smaller and the quotient (g/bhp-h) larger [24].

4.3.2 Steady-State Diesel 13 Mode

For diesel engines, a 21-mode cycle was initially proposed by EPA in 1971 (FR Vol. 36, No. 193, October 8, 1971). The cycle consisted of 3 idle modes and 9 loads (2, 12.5, 25, 37.5, 50, 62.5, 75, 87.5 and 100 %) at intermediate and rated speed. Since the cycle was different from the one already adopted in California, the manufacturers objected on the grounds of increased costs for separate certifications. One year later (FR Vol. 37, No. 175, September 8, 1972), EPA adopted what was essentially the California HD diesel engine test procedure, subsequently known as the U.S. 13-mode cycle. This was the steady-state cycle employed for testing heavy-duty diesel engines in the United States from 1974 to 1983/84.

The cycle, detailed in Table 4.8 and Fig. 4.18, was developed by EMA (Engine Manufacturers Association) [22], and is very similar (but not identical) to the European/UNECE R49. Two speeds are tested, the peak torque and the rated one, being considered representative of HDV engine operation. At each speed, 5 loads were under test, practically covering the whole load range. Each of the 13 modes was held from 4.5–6 min. Measured emissions during the 13 mode were HC, NO_x, CO and CO_2, and the results were obtained in g/bhp-h. Prior to the U.S. 13-mode cycle (period between 1970 and 1973), only smoke emissions from diesel-engined HDVs were regulated [19, 20]. Unlike the urban-oriented gasoline 9-mode test, the diesel 13-mode schedule was intended to simulate suburban and highway truck driving.

Table 4.8 Speed/load points of the U.S. 13-mode cycle

Mode	Engine speed	Engine load (%)	Weighting factor (%)
1	Idle	–	6.67
2	Intermediate	2	8.00
3		25	8.00
4		50	8.00
5		75	8.00
6		100	8.00
7	Idle	–	6.67
8	Rated	100	8.00
9		75	8.00
10		50	8.00
11		25	8.00
12		2	8.00
13	Idle	–	6.67

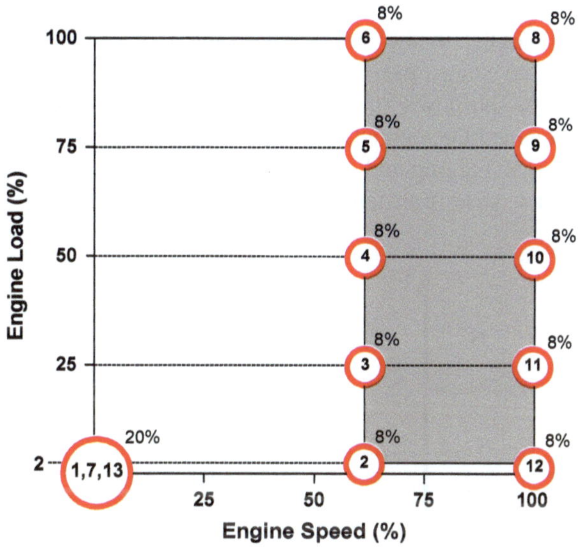

Fig. 4.18 Illustration of the speed/load points of the U.S. 13-mode cycle

The differences between the R49 (Fig. 4.3) and the U.S. 13-mode tests are located in the minimum load level employed (2 %, instead of 10 % for the R49), and the different weighting factors. As regards the latter, the European legislation emphasized slightly on idling and primarily on the 100 % load at both tested speeds, whereas in the United States, greater emphasis was placed on the points at the rated-power speed (weighting factors 8 % for the U.S. cycle compared to 2 % for the European). Overall, the weighting factors during the 13 mode were equally distributed between the intermediate and rated-speed modes (40 %), whereas in Europe, the weighting of the intermediate-speed modes was 57 % and of the rated 'only' 18 %. As regards the average weighted loading, for the U.S. 13 mode it was 40 % compared to 51 % for the European R49 cycle; obviously, exhaust gas temperatures during the latter test were higher.

An experimental 78-min, 23-mode steady-state schedule was also considered in the early 70s. The cycle was defined in terms of engine speed (intermediate and rated), and load %. Particularly for gasoline engines, the test speeds were 1200 and 2300 rpm [25]. The cycle practically contained all modes of the 13-mode diesel cycle. In any case, it was claimed that it did not offer any substantial improvement over the existing cycles, therefore it was never adopted [24].

In the United States, the legislation for heavy-duty engines soon moved from the steady-state 9 and 13-mode tests to a true transient cycle (Sect. 4.3.5), as was also the case with light-duty vehicles (from the synthetic California 7 mode, to the transient LA4 in 1972, as discussed in Sect. 2.2).

4.3.3 Federal Smoke Test

In parallel to the steady-state (and transient) engine-dynamometer cycles in the United States, the smoke test demonstrated in Fig. 4.19 has been valid for certification purposes of diesel-engined HDVs since 1974 (the emission limits have remained unchanged). The sequences to be followed are: idle, acceleration in three phases, lugging and engine unloading (40 CFR 86.884-7). The test aims to simulate worst case transient maneuvers for acceleration and engine lug-down [22].

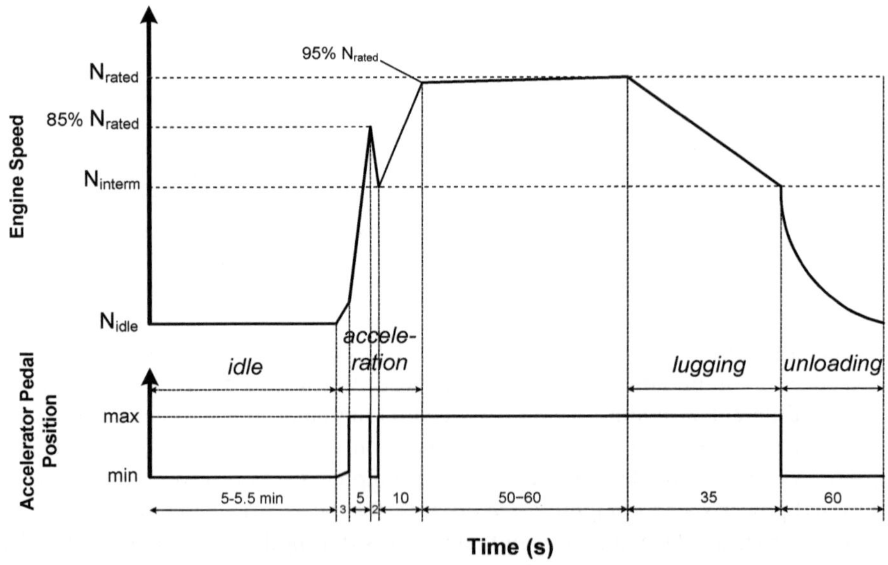

Fig. 4.19 Sequences during the U.S. federal smoke test (40 CFR 86.884-7)

In Fig. 4.19, the test engine speed and the accelerator pedal position are provided, showing also the average time duration for each mode. The cycle is repeated three times, with the engine warmed-up; from the measured values during all three runs, the average opacity for acceleration, lugging and peak are calculated.

4.3.4 Chassis-Dynamometer Heavy-Duty UDDS

The CAPE-21 program, jointly sponsored by the Coordinating Research Council (CRC) and the U.S. EPA, was a two-phase project to conduct a heavy-duty vehicle driving pattern and use survey in the 70s. The goal was to identify the composition, function and travel behavior of urban trucks, and characterize the function and use patterns of trucks operating over urban streets and freeways. The purpose of Phase I was the development of sampling plans for the collection of heavy duty vehicle operating data in New York City and the Los Angeles Basin. Phase II of the program consisted of instrumenting heavy-duty trucks and buses, selected in accordance with the sampling plans, to record operating data during performance of the normal vehicle function. The New York data sample was collected in 1973 and 1974 and the Los Angeles data sample in 1975. The two cities were selected because they represented extreme air-pollution cases, and were substantially different with respect to driving conditions and amount of traffic. In total, 50 trucks and 5 buses were instrumented in each area [26, 27].

During the data processing phase of the program, however, some vehicles were deleted from the sample because of instrumentation problems or lack of reliable calibration. The CAPE-21 database ultimately used for the cycles generation consisted of the operating data from 44 trucks and 4 buses in New York and 44 trucks and 3 buses in Los Angeles. In total, data from 95 vehicles for 290 truck-days and 21 bus-days (1365 h) of operation was gathered. The truck sample in each city contained gasoline-fueled and diesel-fueled engines in each of three vehicle configurations: 2-axle single unit (highest occurrence in the population), 3-axle single unit, and tractor-trailer; all buses, on the other hand, were equipped with diesel engines. All the concerned vehicles had a GVWR of at least 10,000 lbs (4530 kg), with the vast majority being naturally aspirated (approximately 80 %). The measured data was: vehicle speed, engine speed, road type (freeway, arterial, local and unknown), traffic density (light, medium, heavy and unknown), throttle position (open or closed), and a load factor indicative of the engine power (manifold vacuum for SI, and fuel pump rack position or rail pressure for diesel engines) [26–30].

After the database was constructed, another EPA subcontractor was awarded the 'Heavy-duty vehicle cycle development' program for the purpose of developing appropriate software to process the database and generate the candidate cycles representative of the data for trucks and buses (1974–1978). For the cycles generation, each transition from engine condition A–B was assigned a probability of occurrence. These probabilities reflected the frequency of occurrence observed in the CAPE-21 database. The cycle generation technique produced a continual progression of engine transitions, which represented the actual transitions and their frequency of occurrence observed in the 95 trucks and buses in New York and Los Angeles. Engine and chassis-cycle segments for trucks with nominal 5-min durations were computer-generated with a Monte Carlo model for each one of eight categories (NY and LA, freeway and non-freeway, SI and diesel). For buses, the generated cycles had nominal durations of 5, 10, 20, 30, and 60 min [31]. An important question with respect to the composite SI and diesel cycles concerned vehicle operation when cold. An analysis of cold/hot operation was conducted to determine if there was need for a unique cold operation segment. The results of the analysis indicated that the difference between cold and hot heavy-duty vehicle operation was not very significant. It was found, however, that the initial idle following a cold start was longer than normal. It was thus concluded that a longer idle at the start of the cycle was the unique requirement for the cold-start portion of an emission test. From the numerous cycle segments computer generated, a few passed the statistical criteria imposed, ultimately leading to three candidate cycle segments of 5-min duration for each of the 8 categories mentioned above. These were submitted to the EPA in order to choose the most appropriate for each category [31]. Since these cycles were derived from a sequence of computer-generated random numbers, it is questionable whether any of them actually corresponded to the actual time data from the test vehicles. It is also possible that some cycle segments might actually represent operating sequences that have never taken place in the real world [30].

The procedure for selecting a best of the three candidate cycles in each category involved (as discussed in Sect. 1.4.2) ranking the cycles according to three main criteria: (1) overall summary statistics of the cycle, (2) agreement of the distribution functions for power and engine speed with the input data as screened by a Kolmogorov-Smirnov test, and (3) visual rating of the density plots [28]. From the best cycle for each category, the final cycle was then formulated based on engineering judgment and predetermined goals regarding 'balance' (New York vs. Los Angeles, freeway vs. non-freeway) and length of the cycle. The resulting *chassis-dynamometer* cycle is referred to as the EPA UDDS (Urban Dynamometer Driving Schedule) or heavy-duty UDDS, to distinguish it from the light-duty vehicles FTP-72/LA4, or 'd test' (because of its heading in the CFR 86 Appendix I), and is illustrated in Fig. 4.20, with its speed/acceleration frequency distribution provided in Fig. 4.21. Notice that the NY freeway segment was not included in the final cycle; this segment represented lower than 10 % of the total input in the dataset, and, more importantly, was tainted by discrepancies in the input data.

The HD UDDS lasts 1060 s (17.67 min) for a traveled distance of approximately 9 km, starting with a 24-s idle phase. The cycle was designed by EPA with the intent to be a 'realistic engine exercise', and to produce a nearly 50–50 weighting of urban and highway driving. Its combination of idle, full throttle acceleration, rapid transitions, and low and high-speed cruising were meant to represent the full range of vehicle/engine operating modes [28]. Upon its introduction, however, it was heavily criticized by vehicle manufacturers. In particular, the model used to process the data was not deemed to result in an adequate

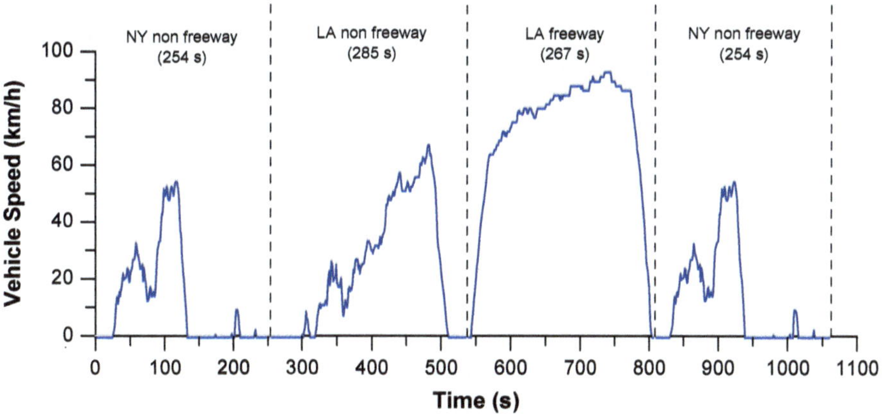

Fig. 4.20 Speed profile of the U.S. heavy-duty UDDS (also known as the EPA 'd test')

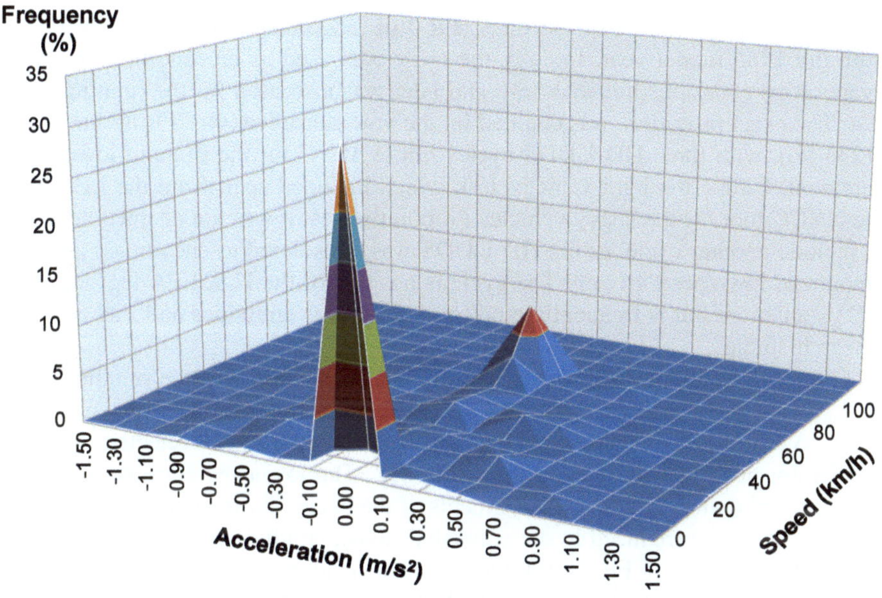

Fig. 4.21 3D speed/acceleration frequency distribution of the heavy-duty UDDS driving cycle

representation of actual engine and vehicle operation. Furthermore, the manufacturers contended that there are excessive speed and acceleration fluctuations throughout the cycle (cf. Fig. 4.60), as well as random up and down shifts in speed. Thus, driving the cycle would result in abnormal manipulations of the accelerator and brake pedals, possibly leading to not representative emissions [32].

Table 4.9 Comparison of the technical specifications between the segments of the HD UDDS (notice that the maximum speed and acceleration are higher than the light-duty vehicles FTP-72)

Segment	Duration (s)	Distance (m)	Maximum speed (km/h)	Average speed (km/h)	Driving speed (km/h)	Maximum accel. (m/s^2)	Idling time (%)	RPA (m/s^2)
UDDS	1060	8935	93.3	30.3	44.6	1.96	31.9	0.143
NY (NF)	254	859	54.7	12.2	24.3	1.96	50.0	0.285
LA (NF)	285	1854	67.6	23.4	32.4	1.77	27.7	0.186
LA (F)	267	5363	93.3	72.3	73.7	1.32	1.87	0.082

As evidenced in Fig. 4.21, most points are located around zero acceleration, with a noteworthy density also at higher than 60 km/h speeds (LAFY segment). Technical specifications for each part of the UDDS are provided in Table 4.9, with the NY urban segment exhibiting 50 % idling time (cf. the New York Bus cycle later in the text).

The HD UDDS cycle has been employed from the 80s for evaporative testing of SI-engined HDVs, namely gasoline-fueled, natural gas-fueled, liquefied petroleum gas-fueled and methanol-fueled ones (FR Vol. 88, January 12, 1983), beginning with the 1985 model year. Initially the cycle was used as preconditioning for the evaporative emissions hot-soak test, and later was also used for the running losses test. The exact procedure was detailed in, the now cancelled, 40 CFR 86 Subpart M (1234-96), with three HD UDDS cycles run at 35 °C for the running losses measurement. Unlike the light vehicles test, which included apart from the LD UDDS the NYCC too, for the heavy vehicles evaporative test it was not deemed necessary to include another cycle, as the HD UDDS was considered to encompass adequate amount of low-speed driving. The cycle is not part of the U.S. EPA legislation for HDVs anymore but is frequently employed in various emissions and fuel economy research studies.

Figures 4.22 and 4.23 illustrate data from a heavy-duty vehicle running on the HD UDDS, highlighting engine and vehicle performance during real-world operation. A clear inter-dependence between vehicle speed and selected gear and accelerator pedal results is evident in Fig. 4.22. The lower the engaged gear, as is the case during the early and last segments of the cycle, the smaller the total vehicle moment of inertia (reduced to the crankshaft axis), resulting in higher acceleration rates (emissions too in Fig. 4.23). On the other hand, the higher the engaged gear, the lower the acceleration, for example during the 550–800 s cycle segment, owing to now greater vehicle inertia but, at the same time, the higher the respective vehicle speed.

The development of the engine speed in Fig. 4.22 is closely inter-related to the vehicle acceleration. Engine torque (representative of the applied loading) as well as turbocharger speed, develop in a similar manner, following the engine speed demands throughout the urban part of the cycle.

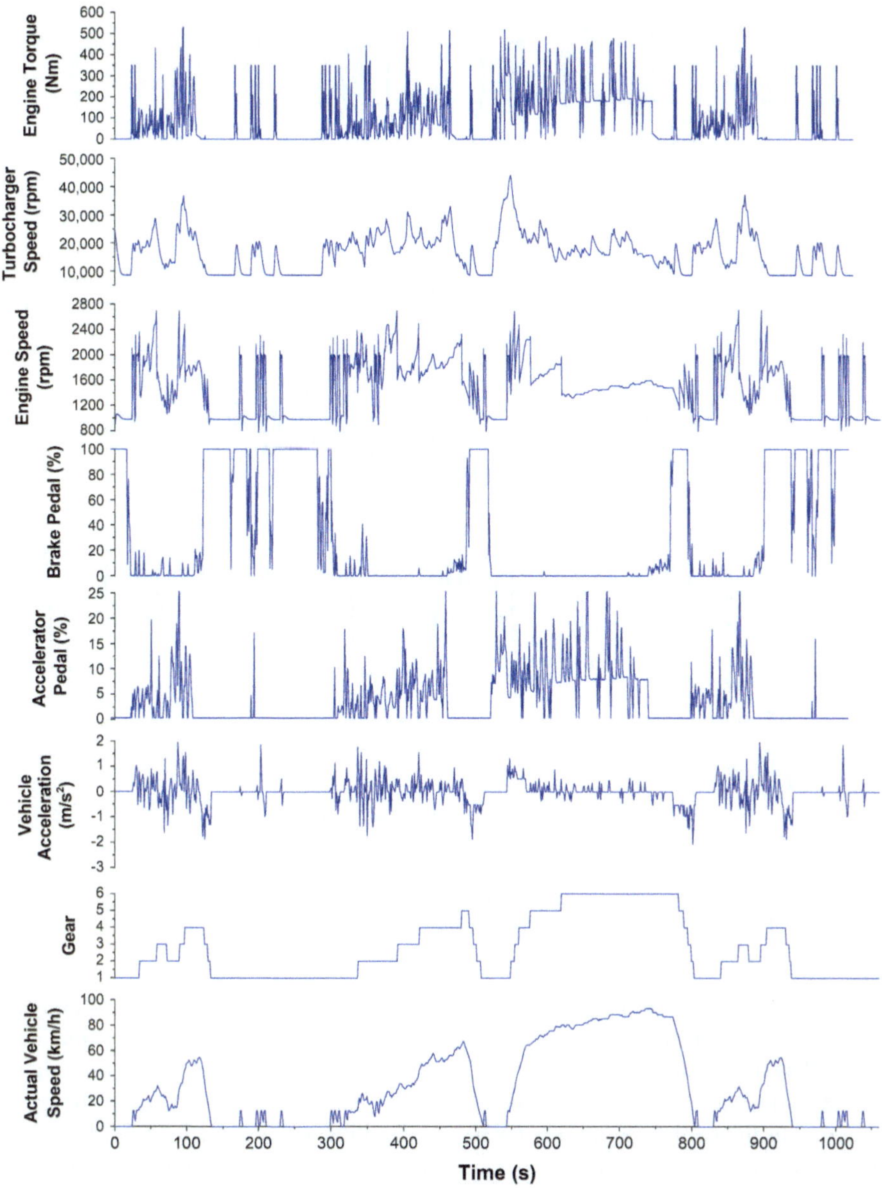

Fig. 4.22 Development of vehicle and engine parameters during the UDDS (simulation results from the GT-Power® software); reprinted from [33] with permission from ASCE)

The points where the most abrupt accelerations are experienced lead in Fig. 4.23 to serious fueling increase, as well as to overshoot in NO_x and soot (indicative of PM) emissions [33, 34]. Net soot production is mainly dependent on engine load. As the load increases, more fuel is injected into the cylinders, increasing the

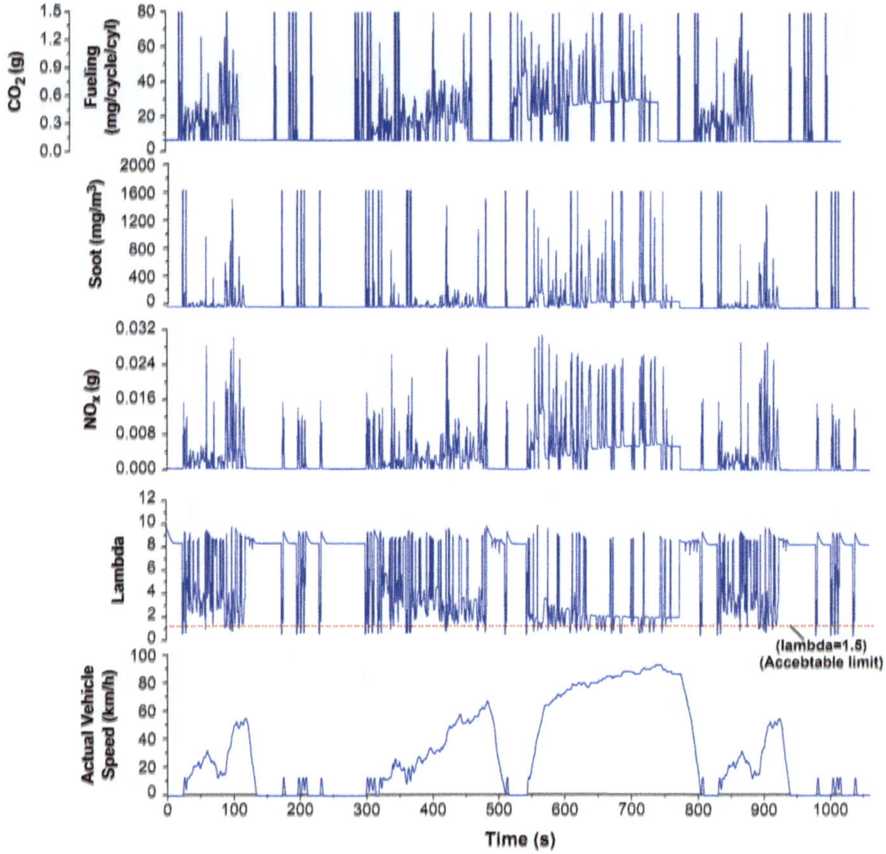

Fig. 4.23 Development of fueling and emissions during the UDDS (simulation results derived from the GT-Power® software); reprinted from [33] with permission from ASCE)

temperature in the fuel-rich zones. Moreover, the duration of diffusion combustion is increased favoring soot formation, whereas the remaining time after combustion as well as the availability of oxygen—both of which enhance the soot oxidation proces—decrease; thus, the production of soot is favored [35]. During the accelerations in the cycle, the above mechanism is remarkably enhanced by the locally very low values of air-fuel ratios experienced during turbocharger lag, as is evidenced in the lambda sub-diagram in Fig. 4.23. For NO_x emissions, it is the lag between increased fueling and the response of the air-supply system that is responsible for the overshoot in emissions. Since the main parameter affecting NO_x formation is the burned gas temperature, all the turbocharger lag cycles at the onset of each acceleration event, where lambda approached unity, provoked locally high gas temperatures that promoted increase in the NO_x emissions. The EGR starvation during the first cycles of a load increase or acceleration transient event enhances the above trend.

4.3.5 Transient FTP

Since 1984 (optionally) and 1985 (mandatory), the primary test cycle used for emission certification of heavy-duty on-road engines in the United States is the FTP (Federal Test Procedure) transient cycle. The engine FTP cycle essentially concerns HDVs of GVWR higher than 14,000 lbs. Class 2b MDPVs beginning with the LDV Tier 2 standard (2004), and Class 3 HDVs from LDV Tier 3 standard (2017), primarily SI engined, are chassis-tested. Contrary to the European ETC (and the worldwide WHTC), two versions of the engine FTP cycle are applicable, one for diesel engines, illustrated in Fig. 4.24, and one for Otto-cycle engines (Fig. 4.25), both in normalized speed/torque versus time form, but with different time schedule and duration.

As discussed in the previous section, the cycles were developed in the late 70s from data obtained from trucks and buses, mostly naturally aspirated ones, operating in the New York and the Los Angeles area under varying traffic conditions (FR Vol. 45, No. 14, Jan. 21, 1980). From the same database, the chassis dynamometer UDDS was also developed. EPA believed that the transient tests would be important in assuring compliance with the forthcoming NO_x and particulate emission limits. As regards the increase in the vehicle cost, it was estimated by EPA at \$6 per HD gasoline vehicle and \$97 for each HD diesel vehicle [36].

When the cycles were introduced in 1979/80, they were faced with strong criticism by the manufacturers. For example, since most of the diesel engines at the

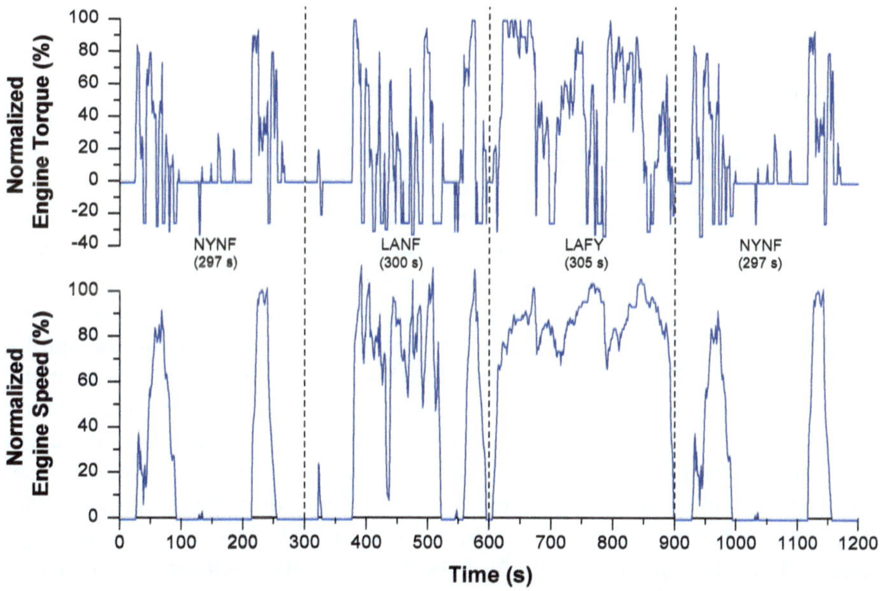

Fig. 4.24 Engine speed and torque profile of the U.S. FTP transient cycle for diesel engines (NY: New York, LA: Los Angeles, FY: freeway, NF: non freeway) (40 CFR 86, App. I)

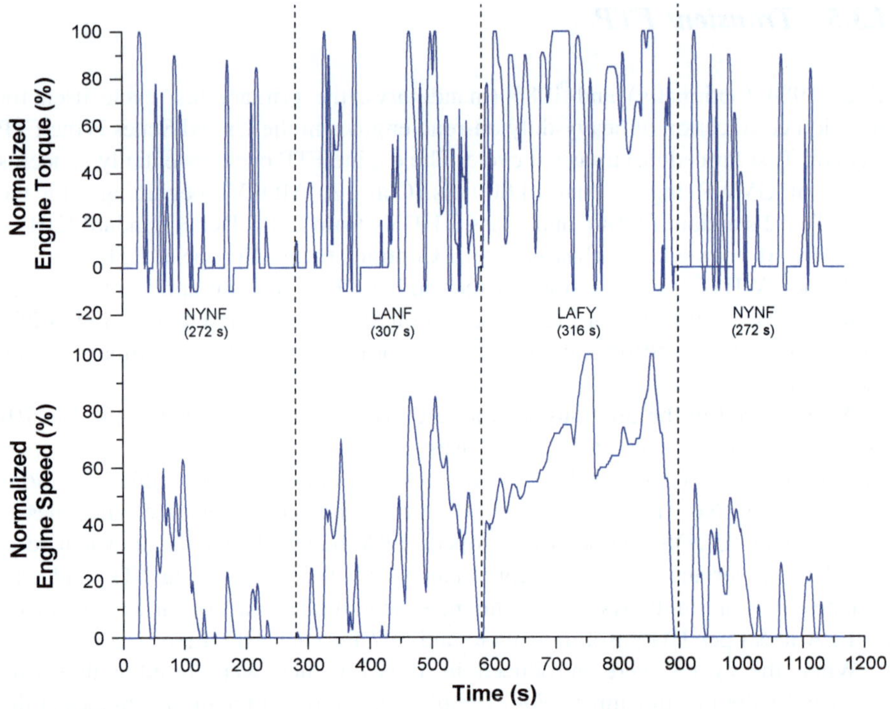

Fig. 4.25 Engine speed and torque profile of the U.S. FTP transient schedule for Otto-cycle engines (NY: New York, LA: Los Angeles, FY: freeway, NF: non freeway) (40 CFR 86, App. I)

time were naturally aspirated, the transition from the U.S. 13 mode steady-state cycle to a transient one was deemed (by the manufacturers) unnecessary. Another reason for concern was the required new test equipment, a fact that led to a delay in the implementation of the cycles [37]. After the rule was finalized in January 1980, the industry continued to press the Administration to make changes to the transient tests. In April 1982, MVMA proposed a revised version of the gasoline-fueled transient test. This version had no speed points with higher than 100 % normalized values, and was, in general, softer. More importantly, a strong correlation between the MVMA and EPA test cycle emission results was established. In the spirit of moving forward, EPA decided to allow the MVMA test to be used as an alternative, hence two variants of the Otto-cycle engines test schedule have been provided over the years in the federal regulations. More specifically, for 1984, either the steady-state 9 mode or the 'EPA' (i.e., the original) transient cycle could be used. For 1985/86, either the 'EPA' or the 'MVMA' transient cycles (higher emission limits for the harsher 'EPA' cycle). From 1987, the 'MVMA' has been the applicable cycle for HD SI engines (FR Vol. 48, Nov. 16, 1983) [19, 36, 38]. This cycle is referenced in Appendix I, § (f) (1) of the code of federal regulations, and is the one illustrated in Fig. 4.25. The original 'EPA' Otto cycle has been referenced for

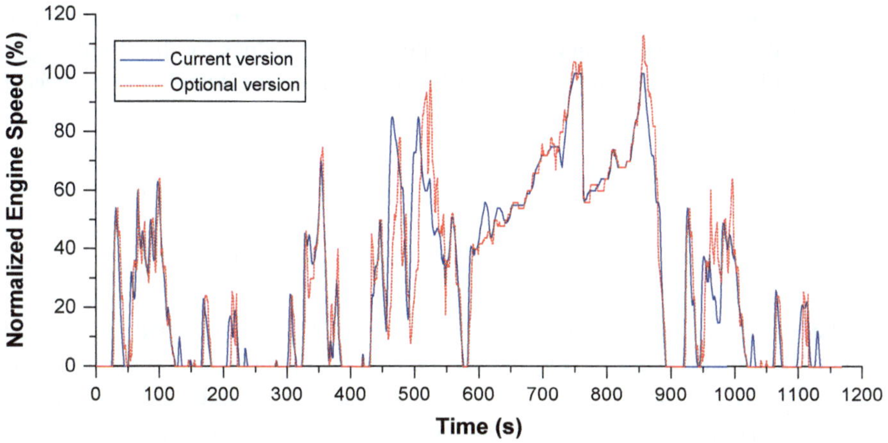

Fig. 4.26 Comparison of the engine speed profile between the two versions of the engine-dynamometer schedules for Otto-cycle HDEs, both of 1167 s duration. The optional (now cancelled) version is the cycle originally developed from the CAPE-21 database [29], and the current version is the subsequent cycle proposed by MVMA; total engine work is about 10% higher on the MVMA compared to the original EPA cycle

many years as an optional test in Appendix I, § (f) (3) of the CFR, and has been cancelled as of 2016 (FR Vol. 81, No. 206, October 25, 2016). Figure 4.26 compares the normalized speed profiles of the 'EPA' and 'MVMA' cycle versions.

Regarding the diesel engine cycle, Caterpillar, on behalf of the EMA, and based on the original EPA cycle (Fig. 4.24), proposed a 'softer' version, named real-time cycle (RTC); the latter, however, was never adopted [38].

Details about the FTP transient cycles are given in 40 CFR 86 Subparts A and N, with the corresponding emission measurement procedure described in 40 CFR 1065. Initially, only gaseous emissions were to be measured during the transient tests; PM limits for diesel engines were included from model year 1988 [1, 2, 19].

As is made obvious, and contrary to the European ETC, the emphasis during the FTP cycles is on urban driving with extended periods of idling (approximately 30 % of the total time) but also segments with elevated speeds, even higher than 100 % of the maximum. As was the case with the chassis UDDS, three distinct phases are identified in each of the FTP cycles diagrams, namely: New York Non-freeway (NYNF—simulating congested urban conditions, once at the beginning and a second time at the end), Los Angeles Non-freeway (LANF—simulating uncongested urban conditions), and Los Angeles Freeway (LAFY); the latter simulating extra-urban driving with higher vehicles/engine speeds.[4] Obviously, the greater part of the work is produced during the LAFY segment (approximately 55–60 %), compared to only 25 % during the two NYNF segments [39].

[4] It is noted that the chassis-dynamometer UDDS (Sect. 4.3.4) lasts 1060 s, whereas the engine-dynamometer FTP 1199 s for the diesel engine version and 1167 s for the Otto-cycle one.

For the purposes of the test, a denormalization applies in order to interpret the normalized speed/torque data into actual (rpm and Nm) values for the tested engine. Equation (4.5) describes the expression for determining the actual engine speed N (40 CFR 86.1333)

$$N = \frac{N_{norm} \times (N_{\max} - N_{idle})}{112} + N_{idle} \quad \textit{for diesel engines}$$
$$N = \frac{N_{norm} \times (N_{\max} - N_{idle})}{100} + N_{idle} \quad \textit{for SI engines} \tag{4.5}$$

The calculation of the maximum test speed N_{max} is provided in 40 CFR 1065.610 (denoted as f_{ntest}). Specifically, N_{max} is the engine speed at the maximum of the sum of the squares $\left(N^2_{normi} + P^2_{normi}\right)$, where: i an indexing variable that represents one recorded value of an engine map; N_{normi} an engine speed normalized by dividing it by N_{Pmax}, and P_{normi} an engine power normalized by dividing it by P_{max} (P_{max} being the maximum engine power and N_{Pmax} the speed corresponding to maximum power).

For engines with a high-speed governor, which are subject to duty cycles with normalized speeds greater than 100 % (e.g., the diesel FTP or the NRTC in Sect. 5.2), an alternate maximum speed is calculated as follows:

$$N_{\max,alt} = \frac{(N_{hi,idle} - N_{idle})}{N_{norm,\max}(>100\,\%)} + N_{idle} \tag{4.6}$$

This alternate $N_{max,alt}$ value is selected if lower than the N_{max} calculated from the sum of squares.

The latest EPA regulation (40 CFR 86.1333-2010) specifies that heavy-duty highway diesel engine manufacturers determine the maximum test speed as described in Part 1065 (with the sum of squares), and set this speed equal to the highest speed in the FTP diesel cycle, which is the 112 % speed. All other speeds are normalized relative to Eq. (4.5). For the denormalization of torque, Eq. (4.4) is applicable.

In Table 4.10, a comparison of some technical specifications between the two versions of the FTP cycle is presented. As evidenced, the average load of the FTP is approximately 24 % for the diesel-engined version and 32 % for the Otto-cycle one; in contrast, for the ETC it is 36.5 %. Taking into account that during the FTP diesel, 51 % of the time is spent at closed rack conditions (idling and motoring), it follows that for the remaining 49 % of the cycle, the engine operates at an average of 45–50 % load. Further, Fig. 4.27 illustrates the 3D (normalized) speed/torque distribution of the FTP. Obviously, the increased frequency at idling and motoring conditions dominates the distribution.

Figure 4.28 compares the steady-state U.S. 13-mode cycle with the transient FTP for a heavy-duty diesel engine. Apart from the obvious advantage of the FTP in terms of transient testing, it seems that low-speed/load points are also under investigation compared to the medium to high-speed oriented 13-mode cycle.

Table 4.10 Comparison of some technical specifications of the heavy-duty FTP cycles

	Otto	Diesel
Duration (s)	1167	1199
Average normalized speed (%)	30.7	41.5
Maximum normalized speed (%)	100	111.9
Average normalized torque (%)	32.4	24.1
Maximum normalized torque (%)	100	100
Idling period (%)	28.8	36.5
Motoring period (%)	12.2	14.8
Maximum load increase	m–73.3 %	0–78.5 %
Average load increase (%)	14.2	14.0
Load increases (min^{-1})	13.3	11.9
Time accelerating/decelerating/cruising (%)	26.5/30.4/43.0	26.4/26.7/46.8
Time w. load increasing/decreasing/constant (%)	21.6/39.9/38.4	20.4/34.9/44.6
Supplemental test (SET)	No	Yes
Not-to-Exceed standards	No	Yes
Smoke limits	No	Yes
Cycle generation	–	Divide by 1.12

m denotes motoring point

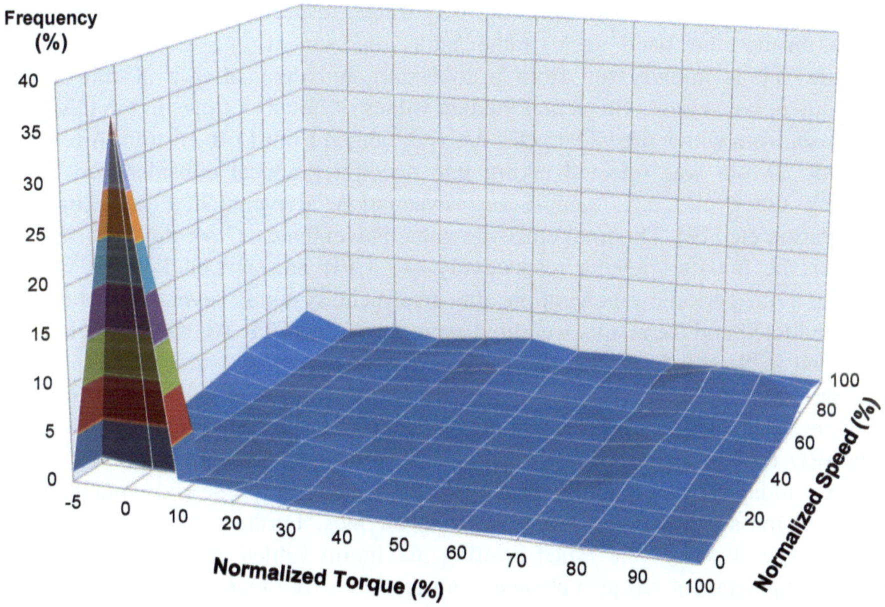

Fig. 4.27 3D normalized speed/torque frequency distribution of the FTP cycle for diesel engines (idle/motoring dominate the frequency spectrum)

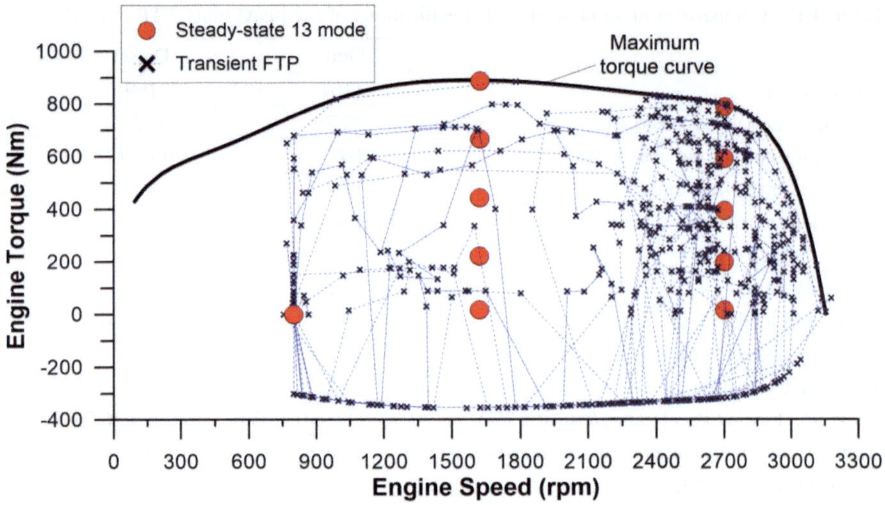

Fig. 4.28 Comparison of the tested engine speed/torque range between the steady-state U.S. 13 mode and the transient FTP for a heavy-duty diesel engine

The FTP manifests increased speed and load density at low and high engine speeds, characteristic of urban buses and highway trucks respectively; the medium-speed range, on the other hand, seems rather under-represented.

The FTP is initially run from a cold start and then, after a 20-min engine shut-down, hot started. The actual median values of hot soak times in the CAPE-21 database, from which the FTP cycles derived, ranged from 11 to 16 min. A hot soak time of 20 min was selected taking into account practical considerations when running the cycle, e.g., sample bag evacuation, dynamometer and computer preparation etc [29]. The final emission value is weighted 1/7 for the cold start and 6/7 for the hot-start results respectively (40 CFR 86.007-11). For the case of engines with periodic regeneration after-treatment systems, several hot-start tests are conducted and the results weighted accordingly (40 CFR 86.004-28); details are provided in Sect. 6.5.

The FTP cycles represent vehicles from the 70s. Most of the instrumented vehicles used for the cycles development were naturally aspirated, and even the relatively few turbocharged ones had significantly lower compressor boost than is typical today. Consequently, the involved engines possessed much lower power-to-mass ratios compared to modern engines. Further, some of the instrumented vehicles had gear ratios limiting maximum vehicle speed. Transient fuel and air management has also changed substantially from that time, since electronic control has become commonplace [9, 37]. All the above, probably render the cycles outdated for most of the capabilities of today's modern trucks and buses. Not surprisingly, the FTP cycles' failure to capture typical in-use driving conditions has been exploited to design engines that obtain good fuel consumption in the field at the expense of NO_x emissions, while meeting the NO_x limit during the certification procedure. This abuse (i.e., cycle beating) led to a significant tightening of U.S. legislation, in the form of supplemental testing and not-to-exceed emission limits (for diesel engines only), following the consent decree of 1998 with various heavy-duty engine manufacturers [20].

For the purposes of supplemental emission testing (SET) in the United States, the use of the European steady-state ESC test cycle (Fig. 4.6) was adopted in 2000 and modified in ramped form in 2005. The mid speeds and mid to high loads (e.g., highway cruise operation) encompassed in the cycle represent a significant portion of the speed/load regime HDV engines are designed to operate for maximum efficiency. Thus, for the 2007–2009 model year HD diesel engines, either the European steady-state ESC could be employed (40 CFR 86.1363-2007), or, alternatively, a 40-min ramped version, where all engine loads were tested sequentially for each speed A, B and C (40 CFR 86.1362-2007). From model year 2010, the 40-minute *ramped* ESC is the sole applicable cycle (40 CFR 86.1362), illustrated in Fig. 4.29 for a specific HD diesel engine. Ramped modal test cycles (RMC) are cycles, where gaseous and particulate emissions are measured and sampled continuously in a 'pseudo-transient' manner with the engine running fully warmed up. The time length of the modes is selected in such a way to be equivalent to the weighting factors of the corresponding discrete steady-state test. The change in engine speed and load from one mode to the next has to be linearly controlled within 20 ± 1 s, with the mode change time being part of the new mode (including

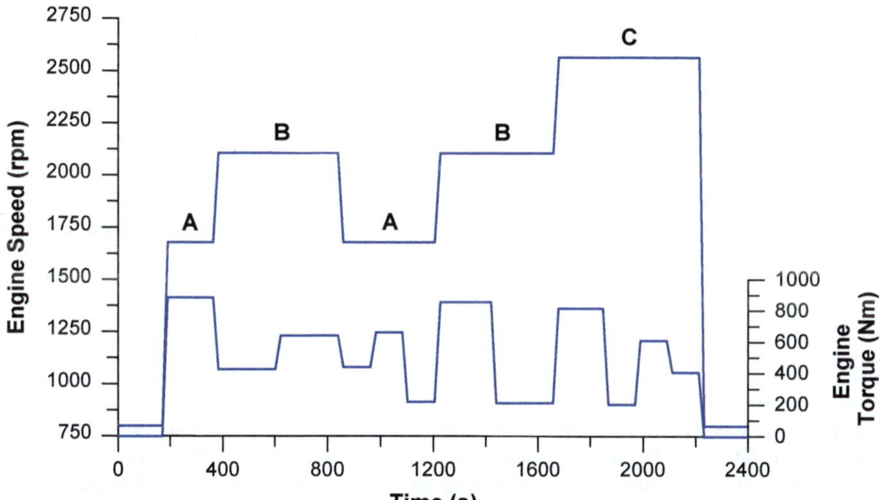

Fig. 4.29 Ramped ESC version applicable in the United States as a Supplemental Emission Test from 2010, for a specific heavy-duty engine (speeds A, B, and C are determined in Eq. (4.1) and Fig. 4.7); notice that idle is split into two equal modes, one at the beginning and the other at the end, helping facilitate certain technical aspects of emission sampling

the first mode). In order to validate the test run, a regression analysis between reference and actual speed, torque and power values are conducted after the test has been completed.

There are a number of advantages to running a steady-state test as a RMC. As mentioned earlier, the procedure for conducting a steady-state test allows emission sampling periods as short as the last minute of each mode. Discrete after-treatment regeneration events, e.g., NO_x regeneration for NO_x adsorption catalysts and forced PM regeneration for DPFs, typically cause short-duration sharp increases in NO_x, HC and PM emissions. Thus, it may be challenging to gather good, repeatable emissions from the steady-state procedures since a regeneration event may or may not be sampled in a given (steady-state) mode. Furthermore, the RMC reduces the number of sampling system starts and stops. This is significant at low emission standards when considering that a previous mode's emissions may be incorrectly included in the next mode due to an unavoidable dead volume in the sampling system. The longer sampling period of a RMC also increases the mass of the particulates sampled. This is significant because the nowadays applicable PM standards approach the minimum detection limits for many current PM microbalances [40].

The above-mentioned SET, in the form of the ramped ESC, is also employed for testing the compliance of combination tractor engines with the applicable fuel consumption (from MY 2016) and greenhouse gases (from MY 2014) standards; greenhouse gases meaning CO_2, N_2O, CH_4 etc.[5] For the rest of the HDEs, such as those installed in tank trucks, refuse trucks, buses, utility vehicles etc, for which a considerable part of their duty cycle involves urban (hence highly transient) driving, the hot-started FTP is the applicable test cycle for fuel consumption and GHGs testing; for the whole vehicle GHG emissions estimation, the GEM simulation tool is employed [1, 40, 41].

Along with the additionally implementing ESC steady-state or ramped test, a region of the engine map was stipulated representing engine speeds and loads expected to be encountered in normal vehicle operation and use by diesel heavy-duty engines. In this Not-to-Exceed (NTE) area, emissions may not exceed the respective limits multiplied by one of the NTE factors, which are dependent on the level of the emission limits [1, 2].

[5]NHTSA (National Highway Traffic Safety Administration) is the relevant authority for the fuel consumption rule and EPA for the greenhouse gases one. According to the Phase II rule of the heavy-duty GHG emission limits, the weighting factors of the SET cycle have changed to better reflect real-world engine operation, which has been shifted to lower engine speeds during highway cruising. This is accomplished by moving most of the C-speed weighting to the A speed. More specifically, 45% of the total weighting has been assigned to the A-speed modes, 38 % to the B speed, and only 5 % to the high-speed C modes; idle weighting has also been reduced from 15 to 12 %. These mode weightings apply to CO_2/fuel consumption measurements only and not criteria pollutants (FR Vol. 81, No. 206, October 25, 2016).

Table 4.11 Speed/load points of the AVL 8-mode cycle for diesel engines [42]

Mode	Engine speed (%)	Engine load (%)	Relative weighting factor (%)
1	Idle	–	35.00
2	11	25	6.34
3	21	63	2.91
4	32	84	3.34
5	100	18	8.40
6	95	40	10.45
7	95	69	10.21
8	89	95	7.34

Further to the above cycles, the steady-state AVL 8-mode, presented in Table 4.11, has been utilized in certain cases. Although this was a non-legislated, European developed cycle, it is presented here as its aim was to simulate the U.S. federal FTP transient test for diesel engines for gaseous pollutants only, so the exhaust emission results are closely correlated for the two test cycles for HC and NO_x [42].

Lastly, Fig. 4.30 demonstrates total particulate matter concentration during the U.S. FTP cycle for a heavy-duty diesel engine. A sharp increase in the particle concentration at the onset of each acceleration in the cycle is the most obvious observation.

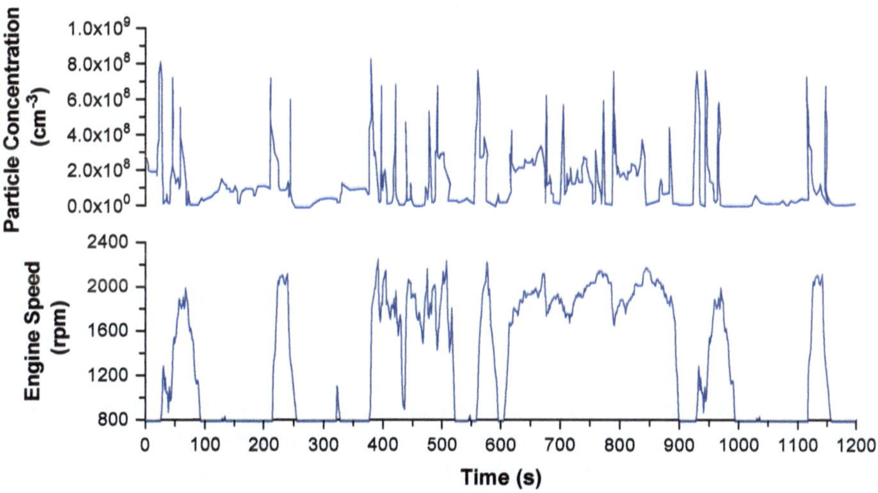

Fig. 4.30 Total particle number concentration during the heavy-duty FTP transient cycle measured via an Engine Exhaust Particle Sizer (adapted from Wang et al. [43])

4.3.6 Snap Acceleration Test

On April 3, 1997, EPA recommended (but did not mandate) the use of the free acceleration test described in SAE J1667 as the basis for diesel-engined HDV inspection (similar to the one valid in the EU). SAE refers to the test as a 'snap-acceleration procedure', but is also commonly called the 'snap-idle test', the 'J1667 test' or the 'free-acceleration test'. It consists of the following steps (the vehicle is conditioned, fully warmed-up and at idle speed) [44]:

- The operator moves the accelerator pedal to the fully open position as rapidly as possible;
- The operator holds the pedal in the fully open position until the time the engine reaches its maximum governed speed, plus an additional 1 to 4 s;
- The operator releases the accelerator pedal and allows the engine to return to the low idle speed;
- Once the engine reaches its low idle speed, the operator allows the engine to remain at idle for a minimum of 5 s, before initiating the next snap-acceleration sequence. The time period at low idle allows the engine's turbocharger (if present) to decelerate to its normal speed at engine idle, helping reduce the smoke variability between snap-acceleration cycles.

At least three preliminary test cycles are performed initially to allow the operator to familiarize with the proper accelerator pedal movement and also remove any loose soot which may have accumulated in the exhaust system during previous operation. Prior to this preconditioning test, the vehicle should be operated under load for at least 15 min to ensure the engine is warmed-up. Within 2 min of the execution of the preliminary cycles, three snap-acceleration test cycles are conducted, actuating the vehicle accelerator pedal in the manner and sequence described above, during which smoke is measured [44]. The test can be conducted either at the roadside or in a test facility, and is employed in many U.S. States, with the EPA having defined specific opacity cut-points [7].

4.3.7 Other U.S. Non-legislated Chassis-Dynamometer Cycles

Owing to EPA's reliance on engine testing for HDVs certification, there is no chassis-dynamometer test for exhaust emissions in the United States (as already mentioned, the HD UDDS was applicable for evaporative emissions only). Largely because of this, there are many specialized transient cycles available in the United States for heavy-duty trucks and/or buses conducted on the chassis dynamometer, such as for example the Manhattan Bus, the New York Composite, the New York Bus, the Central Business District, the Highway Line Haul, the City Suburban Heavy Vehicle, the Neighborhood Refuse Truck, the Utility Service Truck, etc. The most important of them are illustrated in Figs. 4.31, 4.32 and 4.33. Table 4.12 provides some of their technical specifications; more detailed data is available in the Appendix. These cycles have been employed in various research projects over the years relating to emissions and CO_2/fuel consumption.

From the CAPE-21 database, from which the UDDS cycle was developed, a first version of a bus cycle was constructed (duration 1191 s, average speed 14 km/h, traveled distance 4.6 km). This cycle, later termed the 'unfiltered' version, contained small, closely spaced rapid changes in vehicle speed, most probably not representative of real bus operation, therefore a second 'filtered' version (i.e., with softer accelerations) was developed. The NY (Bus) Composite cycle is a shortened version, 1029 instead of 1191 s, of the EPA unfiltered bus test cycle. The first part of the cycle simulates non-freeway driving with large amounts of idle time, while the second part simulates freeway driving of a highly variable transient nature as the cycle graphical illustration in Fig. 4.31 indicates [45]. RPA is 0.243 owing to many stops, frequent and, more importantly, steep accelerations, and short covered distance. The average speed is the same as for the 'unfiltered' version, namely 14 km/h.

From the same database, the New York truck cycle of 1016 s duration, maximum speed 55 km/h and approximately 3.5 km distance was also developed (Fig. 4.31); it consists of four similar parts [31].

The New York Bus cycle, Fig. 4.31, attempts to simulate some of the toughest bus driving conditions in the United States. The cycle consists of very rapid accelerations to speeds ranging from 15 to 50 km/h and then rapid decelerations to idle. The bus idles for very long periods (65 % of the total time) and accelerates again quickly [45]. Not surprisingly, as Table 4.12 confirms, this cycle has a very high value of relative positive acceleration (positive kinetic energy too), owing to steep accelerations and many stops. Average driving vehicle speed in only 5.9 km/h for this 10-min test (excluding idling). The New York Bus cycle is incorporated in the U.S. EPA MOVES simulation tool.

The Manhattan Bus cycle is based, as the name suggests, on driving patters of urban transit buses in Manhattan. It is characterized by operation under low speeds due to frequent stops and increased idling phases (35 % of the time) due to traffic congestion. The cycle is illustrated in Fig. 4.31, and exhibits many accelerations from no load. Cruising time is less than 2 %.

The Orange County Bus cycle is another test for heavy vehicles illustrated in Fig. 4.31. It was developed by the West Virginia University (WVU) based on real bus operating data from the Orange County Transportation Authority in California. It is an intermediate-speed cycle of more than 30-min duration, reflecting transit bus use in urban and suburban areas [7], with a high (0.233) RPA value owing to frequent and steep accelerations. Transient time is 80% of the total, with less than 1 % cruising. Each one of the 30 intermediate stops lasts 11 s.

The HHDDT schedule is a chassis dynamometer suite of cycles developed by the California Air Resources Board with the cooperation of WVU for heavy heavy-duty diesel trucks (GVW > 33,000 lbs/approx. 15,000 kg). Heavy-duty

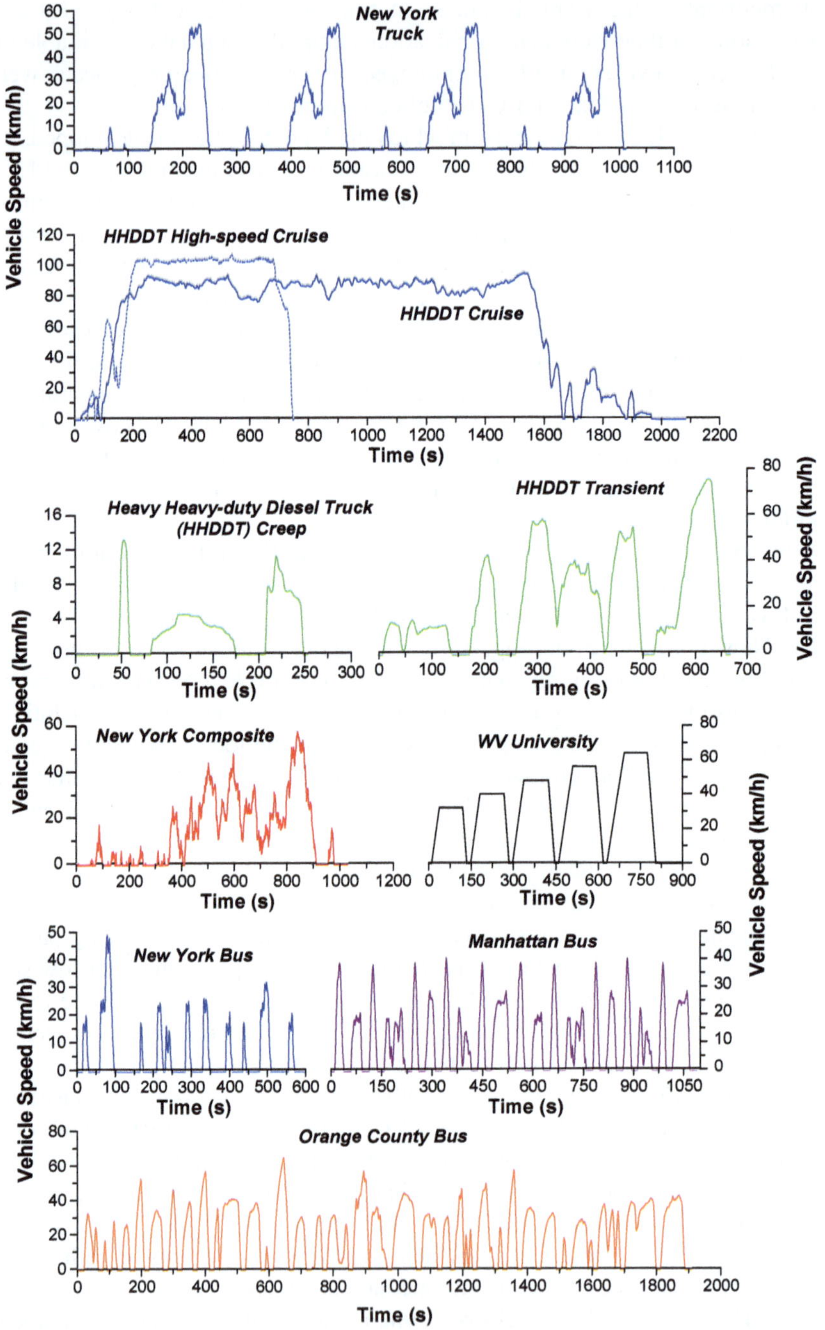

Fig. 4.31 Speed profile of various U.S. HDV chassis-dynamometer driving cycles

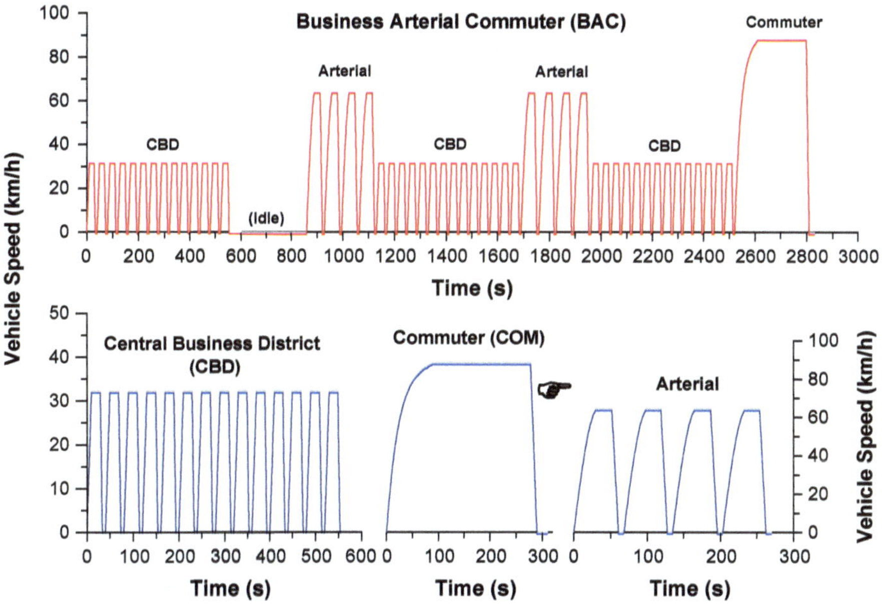

Fig. 4.32 Speed profile of the advanced design bus project driving cycles [50]

truck data was gathered between 1997 and 2000 from randomly selected vehicles. More than 1600 h of data from 84 trucks fitted with data loggers equipped with GPS, and traveling more than 60,000 miles was used for the development of the cycles [46]. The HHDDT test consists of five speed/time modes (known also as the CARB 5 modes): idle, 'creep', transient, cruise and high-speed cruise (Fig. 4.31):

- The creep mode represents very low speed truck operation (average speed 2.8 km/h);
- The transient mode mimics truck stop-and-go operation at a moderate speed (average velocity 25 km/h) with sharp accelerations and decelerations;
- The cruise mode represents freeway driving with almost constant speed at 96.5 km/h (60 mph) for 1400 s;
- In contrast to the cruise mode, which has lower speed and more speed variations (between 50 and 60 mph), the high-speed cruise represents fairly smooth higher-speed operation at 105 km/h (65 mph).

Apart from the idle segment, the other four parts last 1012 (=4 × 253), 668, 2083 and 760 s respectively [9]. The modal approach to the cycles provides flexibility in cycle weightings to accommodate a variety of truck applications. A downside is that the cycles were developed from truck activity in California only. The HHDDT creep cycle is one of the 49 driving schedules incorporated in the U.S.

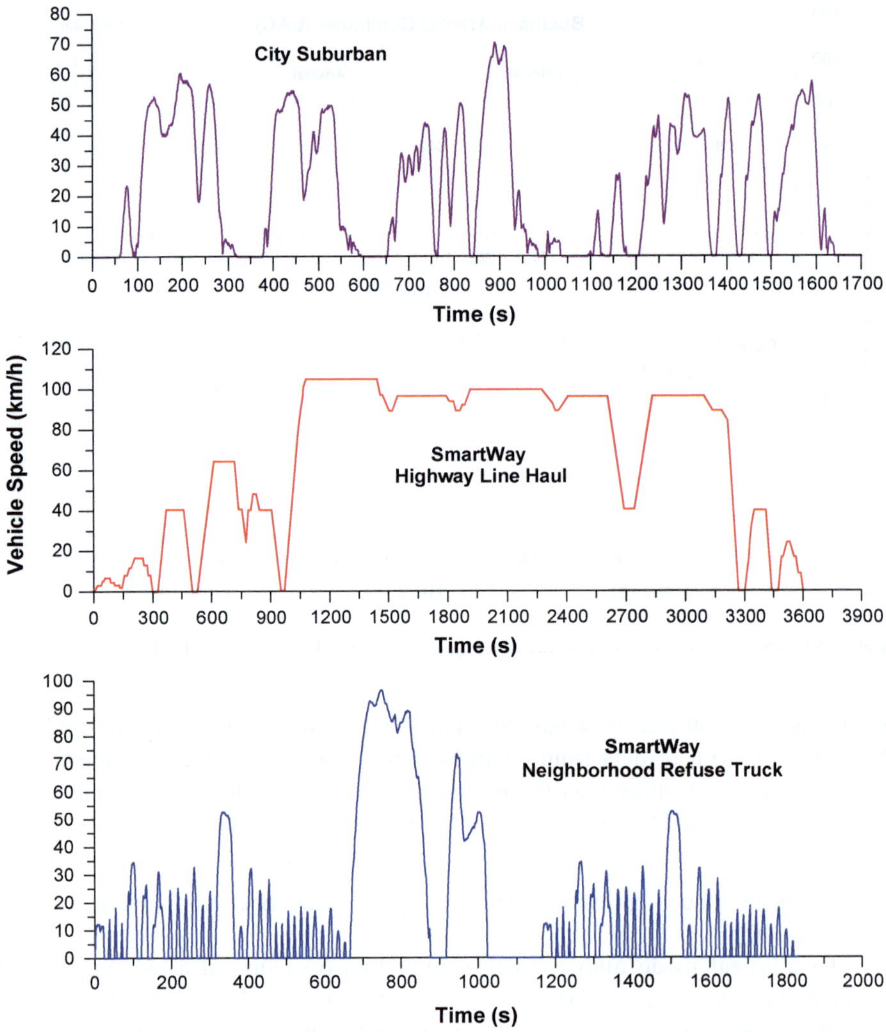

Fig. 4.33 Speed profile of various U.S. heavy-duty chassis-dynamometer cycles (Neighborhood Refuse Truck, Highway Line Haul and City Suburban) [50, 51]

EPA MOVES simulation tool. The HHDDT transient schedule, together with a 55 mph and a 65 mph steady-state cruise without grade were utilized in the U.S. EPA GEM (greenhouse gas emissions model) simulation tool[6] to estimate HDV

[6]Owing to difficulties in testing the whole HDV and great variety in engine/transmission/chassis combinations, the countries that apply HDV fuel consumption and/or GHG limits (at the time of writing, Japan, the United States, Canada and China, with the EU working on a proposal) rely on simulation tools [4]; separate engine-dynamometer tests for the HD engine alone are sometimes applied, as is the case in the U.S. discussed earlier.

Table 4.12 Summary of technical specifications of various U.S. (and European) HDV chassis-dynamometer cycles (see Fig. 4.59 for a graphical representation of all cycles' RPA values with respect to average driving speed); values for the NY truck cycle correspond to four runs

Cycle	Application	Duration (s)	Distance (m)	Maximum speed (km/h)	Average speed (km/h)	Idling time (%)	RPA (m/s^2)
City Suburban	Truck	1700	10,750	70.6	22.8	22.6	0.177
HHDDT Creep		253	200	13.3	2.8	41.1	0.079
HHDDT Transient		668	4588	76.4	24.7	15.0	0.154
HHDDT Cruise		2083	37,130	95.4	64.2	7.7	0.044
Neighbor. Refuse		1832	9150	96.6	18.0	44.8	0.235
NY Truck		1016	3440	54.7	12.2	51.9	–
UDDS		1060	8935	93.3	30.3	31.9	0.143
Arterial (BAC)	Bus	270	3155	64.4	42.1	10.4	0.210
CBD (BAC)		560	3292	32.2	21.2	15.0	0.187
Commuter (BAC)		310	6434	88.5	74.7	6.5	0.048
Manhattan Bus		1089	3324	40.7	11.0	34.3	0.308
NY Bus		600	989	49.6	5.9	65.3	0.459
NY Composite		1029	4032	57.9	14.1	31.2	0.243
Orange County		1909	10,526	65.4	19.9	19.6	0.233
Braunschweig		1740	10,873	58.2	22.5	23.7	0.235
Paris		1898	5678	47.9	10.8	30.5	0.276

fuel consumption and CO_2 emissions (Phase I of the GHG HDV regulation). The weighting factors for each one of the three schedules varied according to the application, e.g., for vocational vehicles they were 42/21/37 % for transient, 55-mph and 65-mph cruise respectively, while for sleeper cabs, the respective percentages were 5/9/86 % [47]. For Phase II of the GHG HDV regulation, effective model year 2018 and beyond, the HHDDT transient cycle remains the applicable test schedule regarding transient HDV operation, along with the 55-mph and 65-mph cruise incorporating now road grade effects, plus additional idling schedules and new weightings (FR Vol. 81, No. 206, October 25, 2016).

The WVU cycle, also demonstrated in Fig. 4.31, is another test sequence used in various research studies. It was developed in 1994 by the Vehicle Emissions Testing Laboratory at WVU for Class 8 trucks, which typically have unsynchronized transmissions. As is evidenced in Fig. 4.31, the cycle has five peaks, each consisting of an acceleration, cruise period, and deceleration, with speeds and acceleration requirements claimed to be met by virtually all vehicles in common service. It is termed the 'WVU 5 peak truck test', the five peak speeds being 20, 25, 30, 35, and 40 mph, respectively (cf. the European SORT cycles in Fig. 4.17). Cycle length is 8 km (5 miles) with total duration 900 s [48].

The Advanced Design Bus (ADB) 3-phase cycle was created by the U.S. Department of Transportation and the Urban Mass Transit Association as part of the specifications for assessing bus performance [21]. The three cycles, illustrated in Fig. 4.32, are:

- The Central Business District (CBD) cycle (SAE J1376); it consists of fourteen segments of equal length, wherein the bus accelerates from idle to 32 km/h (20 mph), holds at that speed for 20 s, and then decelerates back to idle (total length 3.2 km or 2 miles). The CBD phase simulates the boarding and exiting of passengers in the business district, where frequent stops must be made and heavy traffic is encountered [49]. Obviously, the cycle is biased at the 32 km/h speed; a derivative, unique feature is that the vehicle cruises for 47.5 % of the whole time.
- The Arterial (ART) phase; it simulates passenger activity in less congested areas where traffic is lighter, and higher vehicle speeds are attained [49]. This sub-cycle consists of four identical modes (idle, acceleration to 40 mph, 20-s constant speed, deceleration to idle, and 7-s idle) lasting 270 s for a traveled distance of 3.2 km (2 miles); maximum speed is 64 km/h (40 mph). An interesting feature of the cycle is that the vehicle spends 44 % of the time accelerating and only 13 % decelerating.
- The Commuter (COM) single-segment cycle; it models boarding of passengers in suburban areas and transportation to metropolitan areas [49]. The cycle consists of a single acceleration (maximum speed 88.5 km/h (55 mph)), speed held for 200 s (61.5 % of the time is spent cruising), deceleration and 20-s idle; time length is 310 s for 6.4 km (4 miles) traveled distance.

All three phases are combined in a single test sequence—Transit Coach Operating Duty Cycle or BAC (Business-Arterial-Commuter)—originally designed to measure fuel economy from heavy-duty vehicles. The BAC, illustrated in Fig. 4.32, is composed of a CBD phase—idle period of 5 min—Arterial phase—CBD phase—Arterial phase—CBD phase—Commuter phase, for a total time length of 2830 s (47.1 min), traveled distance of 22.6 km, 88.5 km/h maximum speed, and RPA 0.154 m/s^2.

The Highway Line Haul cycle is part of the U.S. EPA SmartWay Program,[7] which also includes other specialized cycles, such as the Regional Haul, Local Pick

[7]In 2004, EPA initiated the SmartWay Transport Partnership to accelerate the deployment of fuel-efficient clean technologies for heavy duty vehicles. Through SmartWay, EPA works in collaboration with industry and other stakeholders to provide incentives for adopting cleaner, more fuel efficient transportation technologies to benefit the environment. An important aspect of the SmartWay Transport Partnership is to determine, through testing and analysis, the environmental benefits of heavy truck technologies, and to provide this information to partners and to the general public [51].

Up and Delivery, Neighborhood Refuse Truck, Utility Service Truck, Transit Bus and Intermodal Drayage Truck cycles [51]. The Highway Line Haul, of which one version is illustrated in Fig. 4.33, simulates a combination tractor-trailer commercial freight delivery truck engaged in interstate (predominantly highway) operation, and is based on the California HHDDT cruise mode shown in Fig. 4.31. This type of vehicle operation is characterized by long trip lengths (it is not uncommon for the cruise control to be applied for hours), high average speeds, with steady-state like engine operation. The vehicles are powered by heavy-duty engines typically used in Class-8 trucks.

In the version of the cycle demonstrated in Fig. 4.33, the most significant modifications made to the HHDDT cruise mode were to add in some transient and idling operation, and to reduce deceleration and acceleration rates, considered more appropriate for a Class-8 combination tractor-trailer, especially when loaded to typical weights of 15,000–30,000 kg. For this version, average speed is 75.6 km/h (47 mph), maximum speed 104.5 km/h (65 mph) and the traveled distance is 67 km. Another version of the cycle—NESCAUM, developed by the Southwest Research Institute—is extended to 165 km (103 miles) traveled distance with 112 km/h (70 mph) maximum and 80 km/h (50 mph) average speed, incorporating also grade and altitude effects [52].

Another SmartWay cycle is the Neighborhood Refuse Truck (Fig. 4.33) developed by the NREL (National Renewable Energy Laboratory). It simulates a trash collection and compacting truck operating in urban-type driving. The vehicle operation is characterized by very frequent stops and starts, with limited 'stem' highway or suburban driving to and from the refuse drop-off site. During its operation, the engine is typically used to operate hydraulic and/or other lifting, compacting, and dumping functions. A significant amount of the fuel used by neighborhood refuse trucks is consumed by these non-motive functions. The cycle was developed using data collected by the Ohio State University from instrumented neighborhood refuse trucks in 5 different cities [51]. Average speed is 18 km/h, maximum speed 96 km/h, for a traveled distance of 9.15 km. Only 3 % of the time is spent cruising (indicative of the transient nature of the cycle), and 45 % idling. Overall, there are 60 stops during the cycle.

Lastly, the City Suburban is a HDV cycle used in various research projects in the U.S. The cycle, illustrated in Fig. 4.33, was developed by the WVU, and is highly transient. Approximately 23 % of the time is spent idling. Although maximum speed is 70 km/h, average vehicle speed is only 22.8 km/h.

In particular as regards the bus cycles discussed so far, as is obvious from Table 4.12 as well as the preceding speed traces, their overall profile incorporates low speeds and extended idling phases. These result in the buses operating highly transient (hence the increased RPA values in Table 4.12), and at relatively low average speeds and loads. Indicative here is the cycle acceleration profile, illustrated in Fig. 4.34 for some of them. Consequently, for diesel-engined vehicles, it is expected that the low exhaust gas temperatures may lead to poor NO_x abatement in

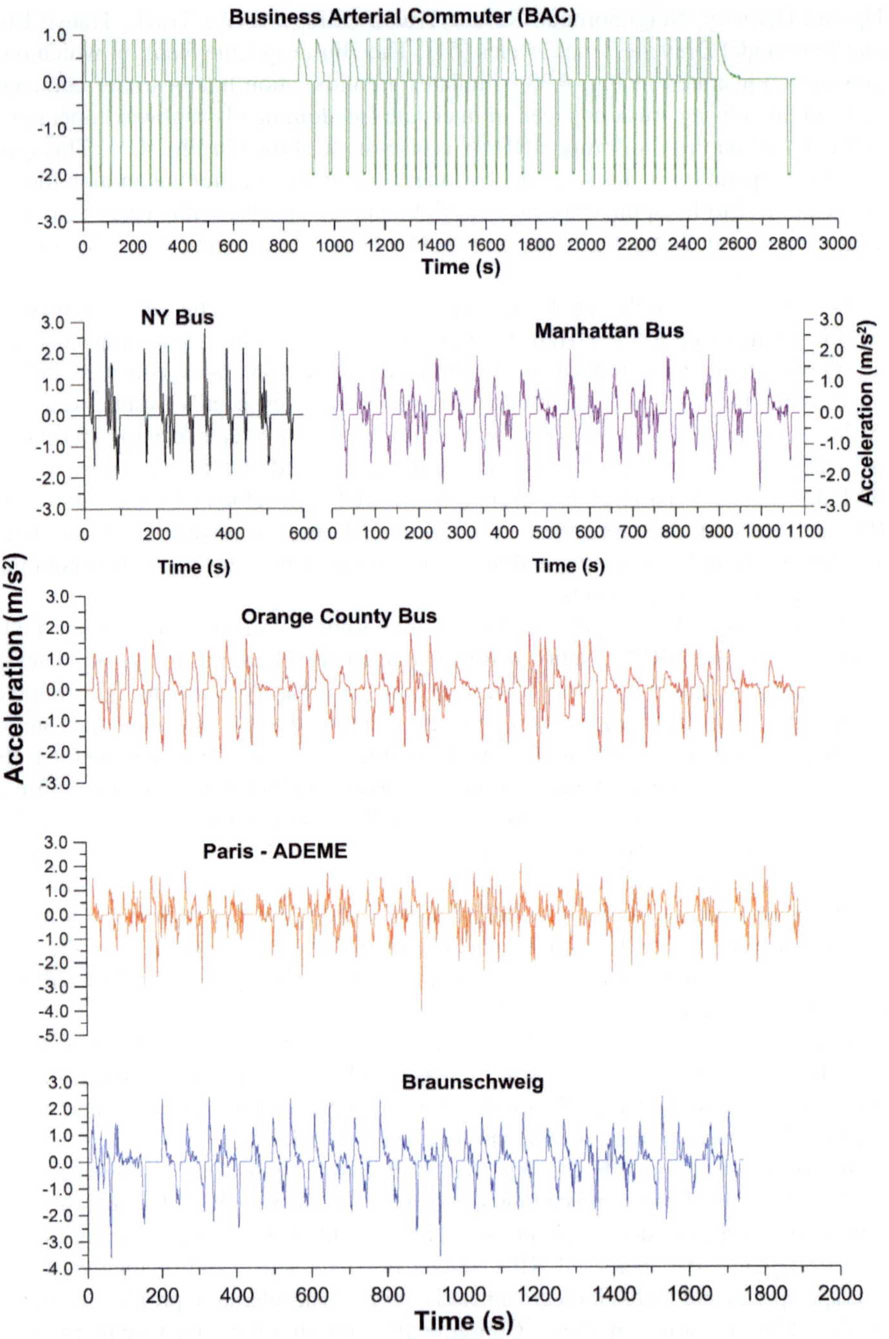

Fig. 4.34 Acceleration profile of various U.S. and European bus cycles

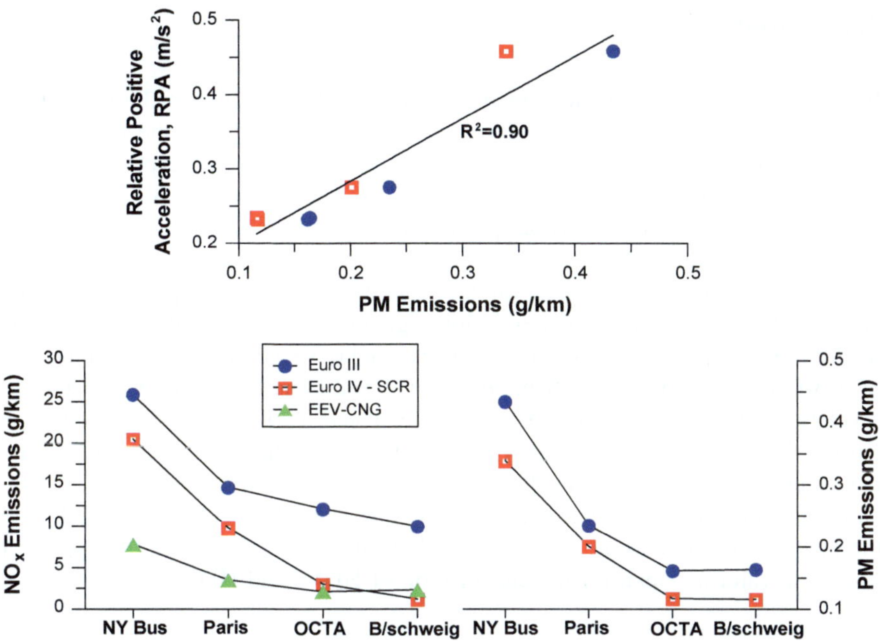

Fig. 4.35 PM and NO_x emissions from Euro 3, Euro 4 and EEV-CNG full-loaded (18,000 kg) buses tested on four driving cycles (data from [16]); notice the strong correlation between RPA and PM emissions in the upper sub-diagram

the SCR catalyst. A comprehensive comparative analysis of various technology urban buses running on many European and U.S. chassis-dynamometer cycles is available in [16]; a representative result from this study is demonstrated in Fig. 4.35.

At the moment, there are no international standards for HDV chassis dynamometer testing. Society of Automotive Engineers (SAE) has published a document titled 'Recommended Practice for Measuring Fuel Economy and Emissions of Hybrid-Electric and Conventional Heavy-Duty Vehicles' (SAE J2711), with the aim to provide an accurate, uniform and reproducible procedure for simulating use of heavy-duty hybrid-electric vehicles and conventional vehicles on dynamometers for the purpose of measuring emissions and fuel economy. Although the recommended practice can be applied using any driving cycle, the practice recommends three U.S. cycles: the Manhattan Bus, representing low-speed transit bus operation; the Orange County Transit cycle, representing intermediate-speed bus operation; and the Urban Dynamometer Driving Schedule (UDDS) representing high-speed operation for buses and tractor-trailers [53].

4.4 Japan

In Japan, heavy-duty vehicles are those with GVW greater than 3500 kg. Limit values for new truck and bus engines are set by the Ministry of the Environment under authority of Japan's Air Pollution Control Law. Two types of exhaust emission standards are established denoted as 'mean' and 'max'. The 'mean' standards are to be met as a type approval limit and as a production average. The 'max' standards are to be met generally as an individual limit in series production and as type approval limit if sales are less than 2000 per year and per vehicle model [1, 2]. Japan introduced its first emission regulation for heavy-duty diesel vehicles in 1972, for the purpose of controlling black smoke (soot). Testing was on the basis of a 3-mode test, and as of September 1974 additionally on the basis of a free acceleration test. CO, HC and NO_x were regulated as of 1988. Subsequent regulations for HDEs targeted mainly NO_x emissions until 1994, when particulate matter emissions were regulated for the first time [5]. During the 2000s, exhaust emission standards were strengthened four times, in 2003, 2005, 2009 and 2016; the latter being the current standard [1, 2]. Table 4.13 summarizes the applicable test cycle valid in Japan for the certification of HDVs and HDEs.

4.4.1 Steady-State 6 Mode

The first standards for gaseous emissions from heavy-duty engines in Japan became applicable on the basis of a 6-mode test in October 1988, when NO_x emissions were for the first time controlled [5]. There were two variants of the test modes and weighting factors, one for diesel engines and another for gasoline/LPG ones. The engine was tested over six different speed/load conditions that are detailed in Table 4.14 for diesel and Table 4.15 for gasoline/LPG engines, and illustrated in Fig. 4.36 for the case of diesel engines. Emissions were measured at each mode and averaged over the cycle using weighting factors also demonstrated in Tables 4.14 and 4.15 and Fig. 4.36. The final test result was expressed as volumetric

Table 4.13 Japanese test cycles and controlled pollutants for HDVs and their engines

Model year	Test cycle		Controlled pollutants
	SI engines	Diesel engines	
1988	Steady-state 6 mode (ppm)	Steady-state 6 mode (ppm)	CO, HC, NO_x
1993	Steady-state 13 mode (g/kWh)		CO, HC, NO_x, PM
1994		Steady-state 13 mode (g/kWh)	
2005	Chassis-dynamometer JE05 (HDV GVW > 3500 kg)		CO, NMHC, NO_x, PM
2016	Transient WHTC		

Table 4.14 Speed/load points of the Japanese 6-mode cycle for diesel engines (duration of each mode: 3 min) [21]; notice the increased idle percentage

Mode	Engine speed (% of nominal)	Engine load (%)	Weighting factor (%)
1	Idle	–	35.5
2	40	100	7.1
3	40	25	5.9
4	60	100	10.7
5	60	25	12.2
6	80	75	28.6

Table 4.15 Speed/load points of the Japanese 6-mode cycle for gasoline or LPG engines (duration of modes 1–5: 3 min) [21]

Mode	Mode description	Engine speed (rpm)	Inlet manifold vacuum (kPa)	Weighting factor (%)
1	Idle	Idle	–	12.5
2	Stabilized	2000	17	11.4
3	Stabilized	3000	17	27.7
4	Stabilized	3000	27	25.4
5	Stabilized	2000	56	13.9
6	Deceleration, throttle valve closed in 10 s	2000–1000	56	9.1

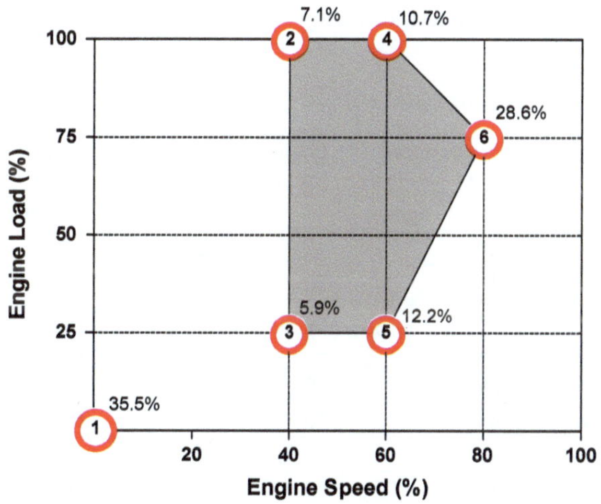

Fig. 4.36 Illustration of engine speed/load points of the Japanese 6-mode cycle for diesel engines (Table 4.14)

concentration in ppm. This test was used for vehicles over 2500 kg GVW or transporting more than 10 passengers. The 6-mode test was intended for the measurement of gaseous pollutants only [5, 21].

4.4.2 Steady-State 13 Mode

The 6-mode cycle was subsequently replaced by the 13 mode for gasoline/LPG engines in 1993, and diesel engines in 1994 (known also as D13) [1, 7, 54]; at the same time, simultaneous NO_x and PM control became effective. The two versions of the cycle are detailed in Table 4.16 and illustrated in Figs. 4.37 and 4.38, having small differences in the tested modes and significant differences in the weighting factors. As the name suggests, the 13-mode test included a sequence of 13 steady-state operating conditions (similar to the European R49 and ESC and the U.S. 13 mode); the averaged cycle emissions were expressed in g/kWh.

The test emphasized on driving conditions at low, and 40 and 60 % of rated-power speeds (cf. remarks on the chassis-dynamometer cycles for light-duty vehicles in Japan in Sect. 2.3). For example, for the modes at 80–95 % load, the weighting factors vary from 3.7–5.5 % for diesel engines and 2.8–3.4 % for gasoline/LPG ones. In contrast, for the European R49 (Fig. 4.3) and the U.S. 13-mode (Fig. 4.18) schedules, the weights at high power (75–100 % load) were considerably higher, namely 8 and 8–25 % respectively. Further, and in contrast to the European R49 and ESC, lower engine loads were under test. Compared to the 6-mode, the 13 mode has more measurements at high speeds (60 and 80 % of the rated), but with considerably lower weightings for the 80 % loading. As was the case with the 6 mode, the 13-mode steady-state test was applicable to engines installed on heavy vehicles with GVW greater than 2500 kg.

The 13-mode cycle was valid in Japan up to 2005, at which point it was replaced by the chassis-dynamometer JE05. At the same time, the HDV category for diesel engines changed, covering now vehicles with GVW greater than 3500 kg.

Table 4.16 Speed/load points of the Japanese 13-mode cycle

Mode	Engine speed (% of nominal)	Engine load (%)	Weighting factor (%)	Engine speed (% of nominal)	Engine load (%)	Weighting factor (%)
	Diesel engines			Gasoline or LPG engines		
1	Idle	–	20.5	Idle	–	15.7
2	40	20	3.7	40	40	3.6
3	40	40	2.7	40	60	3.9
4	Idle	–	20.5	Idle	–	15.7
5	60	20	2.9	60	20	8.8
6	60	40	6.4	60	40	11.7
7	80	40	4.1	80	40	5.8
8	80	60	3.2	80	60	2.8
9	60	60	7.7	60	60	6.6
10	60	80	5.5	60	80	3.4
11	60	95	4.9	60	95	2.8
12	80	80	3.7	40	20	9.6
13	60	5	14.2	Deceleration to idle		9.6

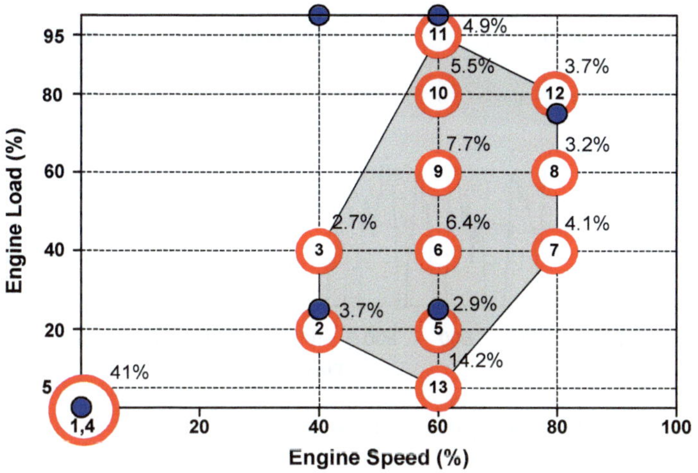

Fig. 4.37 Illustration of engine speed/load points of the Japanese 13-mode cycle for diesel engines (*solid-filled circles* designate the earlier 6-mode cycle points from Table 4.14)

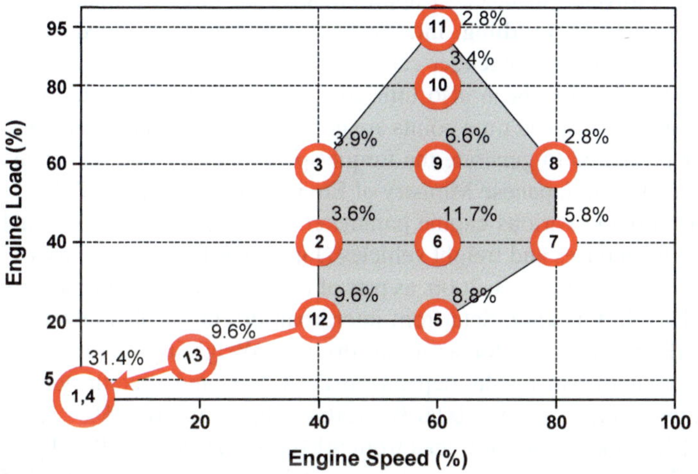

Fig. 4.38 Illustration of engine speed/load points of the Japanese 13-mode cycle for gasoline and LPG engines

4.4.3 Chassis-Dynamometer JE05

In 2005, the Japanese Ministry of Environment introduced a new test schedule for emission certification of heavy-duty vehicles with GVW higher than 3500 kg (both diesel and gas engined). The cycle, which was a chassis test, is graphically illustrated in Fig. 4.39. It is a highly transient driving schedule based on typical Tokyo

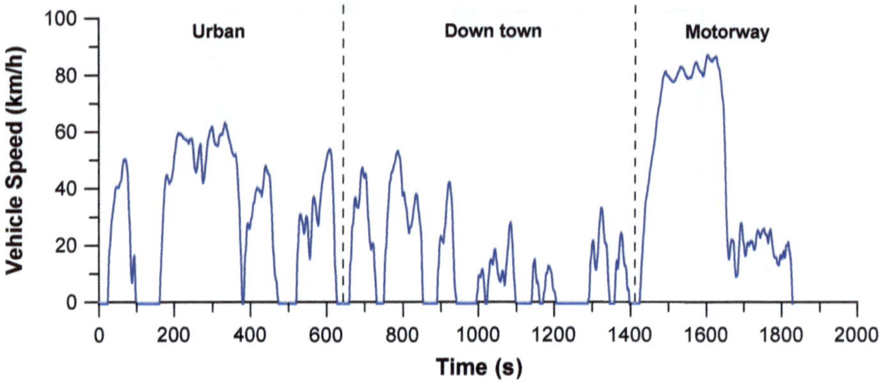

Fig. 4.39 Speed profile of the Japanese JE05 driving cycle

traffic/driving conditions. Three parts can be identified: urban, down town (congested urban with lower speeds and increased idling) and motorway. The JE05 replaced the steady-state engine-dynamometer 13-mode cycle. The cycle lasts 1829 s for a traveled distance of approximately 14 km with many stops (13) and 25 % of the time spent idling. Interestingly, the maximum speed of 87.6 km/h is higher than that of the passenger cars JC08.

The JE05 cycle has been also utilized for engine-dynamometer testing. In this case, engine torque/speed time points are generated based on the vehicle speed ones. Computer programs to generate the torque/speed data for both gasoline and diesel are provided by the Japanese Ministry of Environment based on GVW, frontal area, total gear ratio and various engine parameters. The cycle is used for fuel economy estimation from buses and freight vehicles (Japan was the first country to implement HDV efficiency standards in 2006, as part of the country's commitment to the Kyoto Protocol [4]). In this case, data from interurban driving is also taken into account, with the JE05 accounted for as urban driving. Weighting factors between urban (JE05) and interurban results depend on the GVW and specific vehicle use. For buses, they range from 100/0 % (urban bus) to 65/35% (coaches with GVW > 14,000 kg), and for trucks from 90/10 % (GVW < 20,000 kg) to 70/30 % (GVW > 20,000 kg); road grade effects are taken into account as well [55].

Figure 4.40 demonstrates the 3D speed/acceleration frequency distribution of the JE05 (cf. Fig. 2.46 for the JC08), with most of the operating points located at or around zero acceleration, and few occurrences of high speed and acceleration (RPA is 0.126). The profile illustrated in Fig. 4.40 is expected to have a negative effect on the after-treatment devices' operation for diesel-engined HDVs, owing to the relatively low loads, hence exhaust gas temperatures during the test.

Particle concentration and size/mass distribution for a heavy vehicle running on the JE05 is illustrated in Fig. 4.41. Heavy accelerations throughout the cycle are responsible for overall high particle number concentrations, practically for each examined diameter bin.

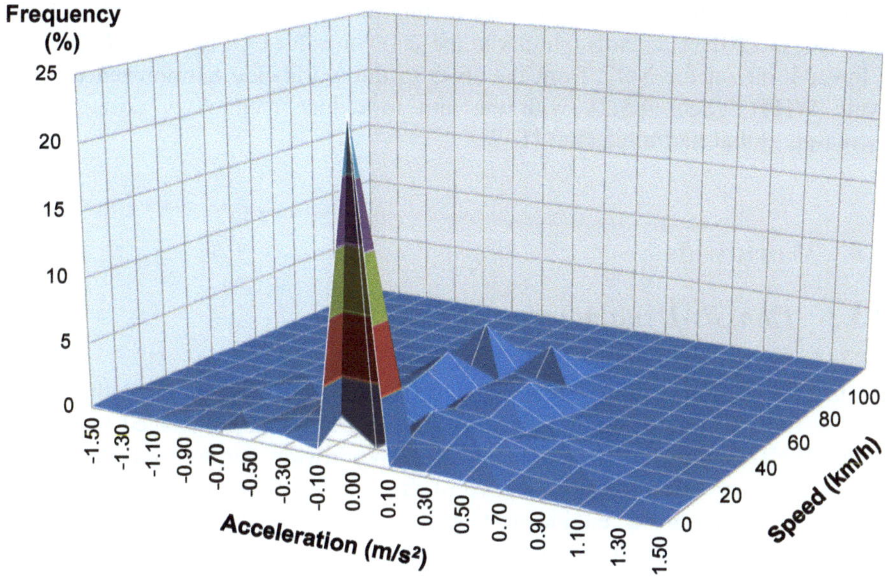

Fig. 4.40 3D speed/acceleration frequency distribution of the JE05 driving cycle

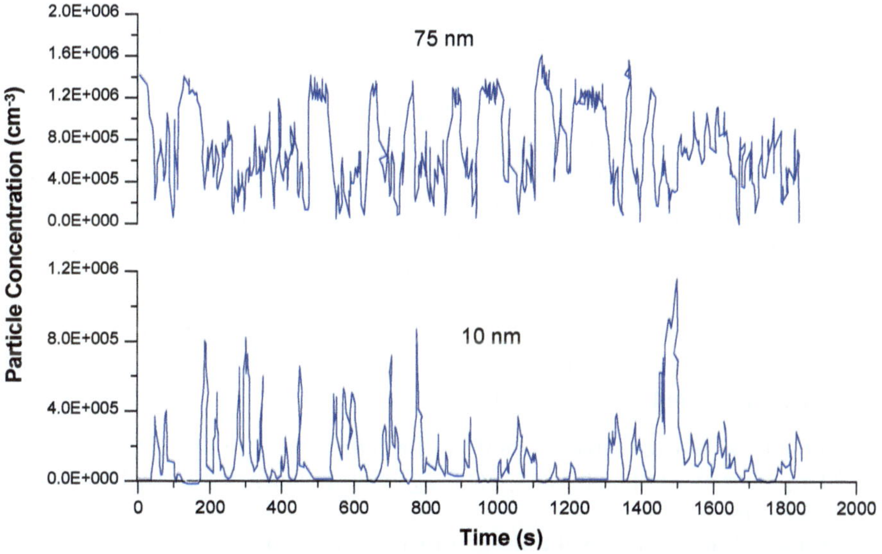

Fig. 4.41 Particle number concentration for two distinct aerodynamic diameters during the JE05 measured with a differential mobility analyzer (DMA) (experimental results adapted from Takeuchi et al. [56])

In July 2010, the Japanese Ministry of Environment changed the applicable test cycle for the 2016 standards (similar to the 2009 ones for CO, HC and PM but with a lower limit set for NO_x) from the JE05 to the worldwide harmonized transient cycle WHTC (Sect. 4.5.2), with the aim to reduce technology costs through increased global harmonization [1, 2].

4.5 Worldwide

4.5.1 Chassis-Dynamometer WTVC

At its 34th session in June 1997, the UNECE Group of Experts on Pollution and Energy (GRPE), under the guidance of Working Party 29, mandated the ad hoc group WHDC with the development of a 'Worldwide harmonized Heavy-duty Certification' procedure. The objective of the research program was to develop a worldwide harmonized engine test cycle for the emissions certification of heavy-duty vehicles/engines that would [57]:

- become a uniform global basis for engine exhaust emissions certification;
- be representative of worldwide, real-life, heavy-duty engine operation;
- give the highest potential for the control of real-life emissions;
- be applicable to future state-of-the-art technology; and
- match emissions in relative terms for accurate ranking of different engines/technologies.

The procedure followed is demonstrated in Fig. 4.42.

In order to proceed to the development of the worldwide harmonized test cycle, the UN research group initially conducted a collection and analysis of driving behavior data as well as a statistical investigation of heavy-duty vehicle usage in different regions of the world. In contrast to light-duty vehicles, HDVs are much more varied in applications, with the individual driving style playing a secondary role, as the vehicle power to mass ratio is considerably lower compared to LDVs. Specifically, TNO and TÜV, which were the partners responsible for the driving behavior database, received data from 65 different vehicles from Australia, Europe, Japan and the United States. This dataset comprised:

- 10 light trucks (max. mass <7500 kg) for a total mileage of 2850 km;
- 20 rigid trucks (max. mass $\geq$7500 kg) and 1 coach for a total mileage of 13,430 km;
- 23 trailer trucks for a total mileage of 61,370 km; and
- 11 public transport buses for a total mileage of 2500 km [58].

The data consisted of driving from various road categories (urban, rural, motorway (see also Fig. 4.43) for the exact composition), vehicle classes, vehicle loads and power to mass ratios; road gradient data as well as traffic load were initially considered but not included.

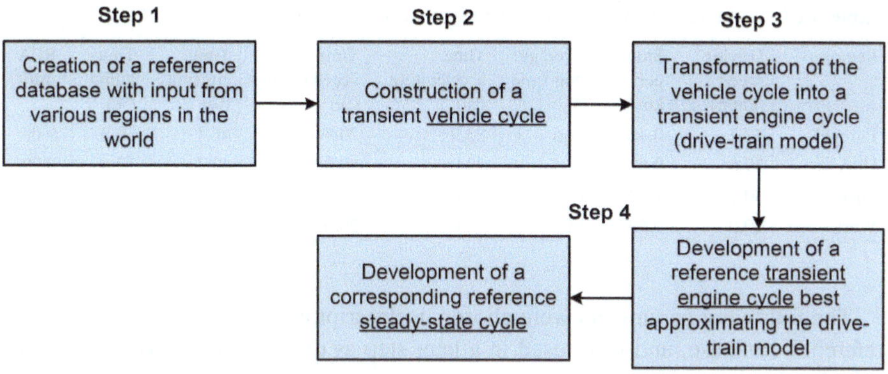

Fig. 4.42 Four-step approach during the development of the worldwide HDV cycles from the UNECE WHDC group [58]

Fig. 4.43 Composition of road categories of the regional vehicle cycles [58]

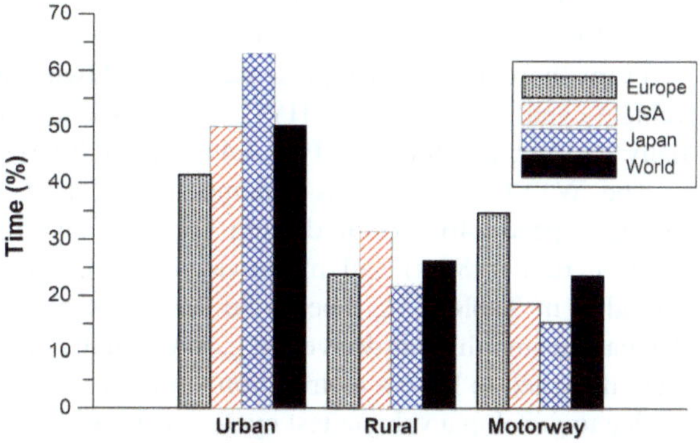

Interestingly, the European database contained a high fraction of motorway traffic (Fig. 4.43), and the average normalized engine power and vehicle speed were relatively higher compared to those of the other regions. In contrast, the RPA was significantly lower. Unlike Europe, the Japanese database contained a considerable fraction of urban traffic, with significantly higher RPA values (cf. comments regarding the JE05). The parameters of the U.S. database, on the other hand, were closest to the worldwide database. Technical characteristics of the regional databases compared to the worldwide are provided in Table 4.17.

The next task was to determine weighting factors for each combination of region, vehicle class, power to mass ratio sub-class and road category. This was determined on the basis of the total operating time of heavy-duty vehicles in real life, established from statistical information on worldwide heavy-duty vehicle use. External help from traffic consultancies, transport associations and the heavy-duty vehicle industry was given at this stage.

Table 4.17 Technical specifications of the regional databases [58]

Region	Driving speed (km/h)	Stops per km	Average stop time (s)	Time accelerating (%)	Time decelerating (%)	Time cruising (%)	Time idling (%)	RPA (m/s^2)
Europe	47.2	0.40	17.6	23.9	20.8	46.1	9.3	0.08
USA	39.6	0.52	21.4	24.0	20.3	43.3	12.4	0.09
Japan	30.2	1.12	18.0	24.0	20.6	38.6	16.9	0.12
Worldwide	40.0	0.59	18.8	24.0	20.6	43.1	12.3	0.09

The following parameters were chosen as descriptors of the characteristics of the reference database, and were used in a later step as quality criteria for the transient vehicle cycle: average vehicle speed, average stop time, number of stops per km, average sequence time, power P and dP/dt distribution, speed/acceleration distribution, average power during the time the engine delivers power to drive-shaft, RPA, propulsion factor (% of time the engine delivers power to drive-shaft during operation), and relative energy demand (energy demand of engine, related to distance driven and rated engine power) [58].

From the database, a representative worldwide transient chassis-dynamometer vehicle cycle—WTVC or WHVC (worldwide harmonized vehicle cycle) was initially derived, illustrated in Fig. 4.44, although never enforced into legislation.

The WTVC is a chassis-dynamometer test of 1800 s duration and 40 km/h average speed (46.1 km/h driving speed), consisting of three parts, i.e., urban (900 s), rural (480 s) and motorway (420 s). Specifics about each segment are provided in Table 4.18. The cycle has the same statistical characteristics as the database from which it derived, e.g., the lengths of the micro-trips and of the stops were designed to have a distribution similar to those of the reference database. The choice to develop a vehicle test cycle was based on the fact that a vehicle duty cycle is much more stable over time than an engine one; the latter changes significantly

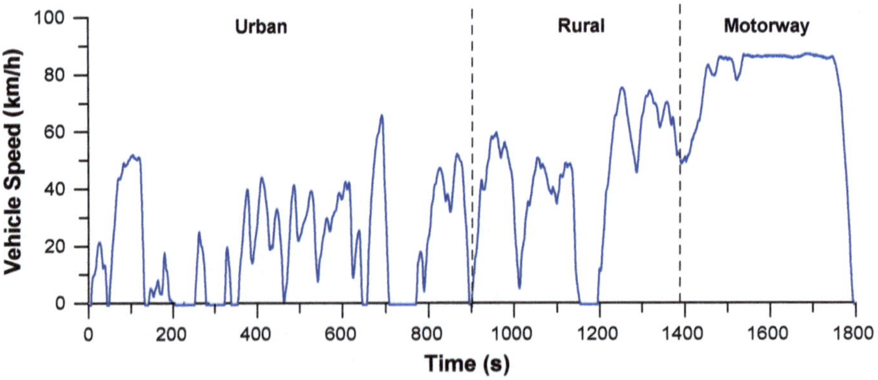

Fig. 4.44 Speed profile of the worldwide WTVC driving cycle [58]

Table 4.18 Technical specifications for each part of the WTVC and the whole cycle

(sub) Cycle	Duration (s)	Distance (m)	Maximum speed (km/h)	Average speed (km/h)	Driving speed (km/h)	Maximum accel. (m/s^2)	Idling time (%)	RPA (m/s^2)
Urban	900	5320	66.2	21.3	27.0	1.59	20.9	0.181
Rural	480	5810	75.9	43.6	47.7	1.17	8.5	0.134
Motorway	420	8940	87.8	76.6	77.4	0.41	1.0	0.070
WTVC	1800	20,072	87.8	40.1	46.1	1.59	12.9	0.107

with engine and drive-train technology, whereas the former only with significant changes in traffic conditions [58].

The weighting factors applied to each of the three segments can be varied between 0–100 % to represent different classes of vehicle operation.

Figure 4.45 presents the 3D speed/acceleration frequency distribution of the WTVC, which exhibits much broader and uniform profile compared with the rather 'flat' one of the JE05 and UDDS (no specific bin dominates the frequencies).

A comparison between the worldwide heavy-duty WTVC and its light-duty counterpart WLTC (Class 3-2) is on display in Fig. 4.46. As expected, the light-duty WLTC exhibits speeds higher than the WTVC. Owing to its distinct motorway part, the WTVC has high frequency in the 85–95 km/h speed range. The accelerations for the light-duty test are also steeper (even steeper for the motorcycle WMTC).

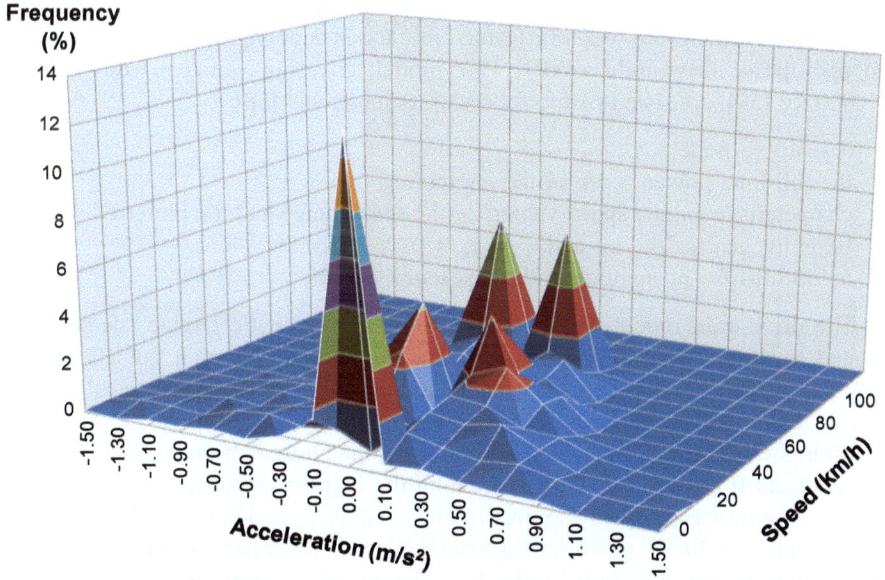

Fig. 4.45 3D speed/acceleration frequency distribution of the WTVC driving cycle

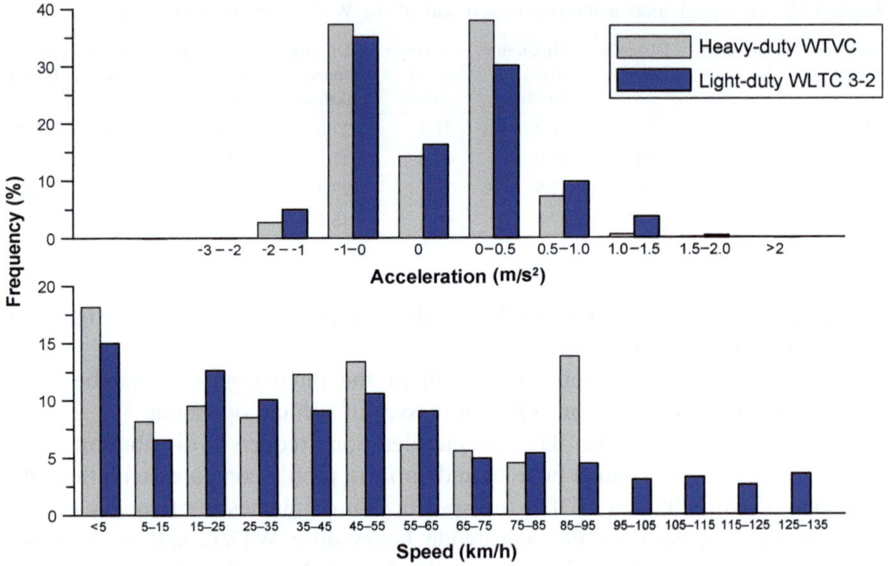

Fig. 4.46 Comparison of the speed and acceleration frequencies between the heavy-duty WTVC and the light-duty WLTC (Class 3-2)

4.5.2 Transient WHTC

In order for the WTVC cycle to be performed on an engine rather than a chassis dynamometer, the cycle was transformed into a transient engine test cycle applying an appropriate drive-train model. This model, covering vehicles from 4600 to 40,000 kg [59], consisted of the following components [58]:

- a gearbox model;
- the full-load power curve of the engine;
- characteristic engine speed values for the speed range and the preferred speed as basis for gear selection; and
- algorithms for plausibility and consistency checks.

The drive-train model was based on three characteristic engine speed values related to the full-load power curve of the engine, and as such not affected by changes in engine technology. These are the N_{lo}, N_{pref}, and N_{hi} speeds as illustrated in Fig. 4.47.

From the engine mapping curve of Fig. 4.47, the integral of the maximum torque is calculated from N_{idle} to N_{95h}; N_{95h} being the highest speed (after the rated one) where the power is 95 % of the maximum. Then, N_{pref} is defined as the speed that corresponds to 51 % of the whole integral, as illustrated in Fig. 4.47 (right sub-diagram). This approach for the N_{pref} definition was preferred with the aim to

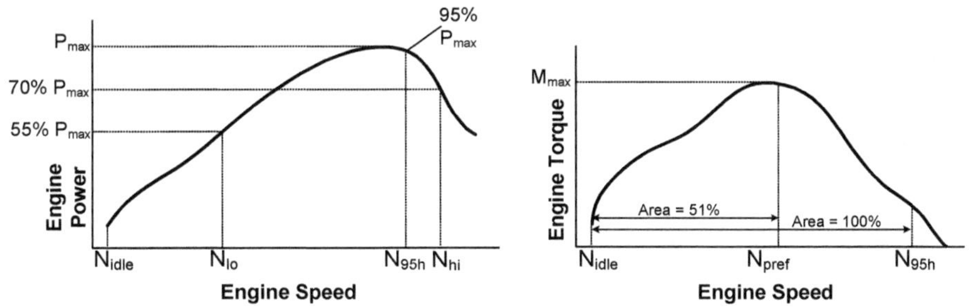

Fig. 4.47 Typical engine speed/power and speed/torque mapping for the definition of the WHTC and WHSC test speeds

limit possible torque-curve design tampering (a strong possibility with electronically controlled engines).

The definition of the upper end of the engine speed range is mainly oriented on cycle bypass prevention. Following the results of the in-use data, the upper limit of the engine speed range could be described by the rated engine speed. For modern electronically controlled engines, however, the rated speed might not be clearly defined because the maximum power can be provided for a range of engine speeds. This leaves the door open for cycle bypass measures. To avoid this, N_{hi} was defined as the highest engine speed where the power at full load is 70 % of rated power [58].

An eight-speed gearbox was chosen for the drive-train model representing a good compromise over the whole range of engines, with the gear ratios derived from the vehicle sample of the in use-data. Gears 3–8 were used for normal operation, whereas gears 1 and 2 only for uphill driving with extremely high gradients. Finally, an overall transmission ratio was defined, assuming that it is linked to the target speed on motorways so that the engine runs at the speed where the fuel consumption is minimized. Normally, this is the lower end of the speed range, where the engine torque is maximum. Therefore, the overall transmission ratio was defined as the quotient of the minimum engine speed for maximum torque (at full load) and 87 km/h, the target speed on motorway. As a result, the highest gear was designed as an overdrive [58].

The developed drive-train model was tested extensively, with the results claimed to be good; however, as it is difficult to implement a computer-based calculation module into regulation, the substitution of the drive-train model with a reference cycle in terms of normalized engine speed and load was then investigated. For this purpose, a reference engine was selected, the test cycle of which, as computed applying the drive-train model, could be used to define the reference test cycle, i.e., an engine whose speed distribution represented the average of all engines. This engine test cycle, provided in terms of normalized speed/torque vs. time points, is schematically represented in Fig. 4.48, and is known as the Worldwide Harmonized Transient engine Cycle—WHTC, detailed in UNECE Global Technical Regulation No. 4 [57].

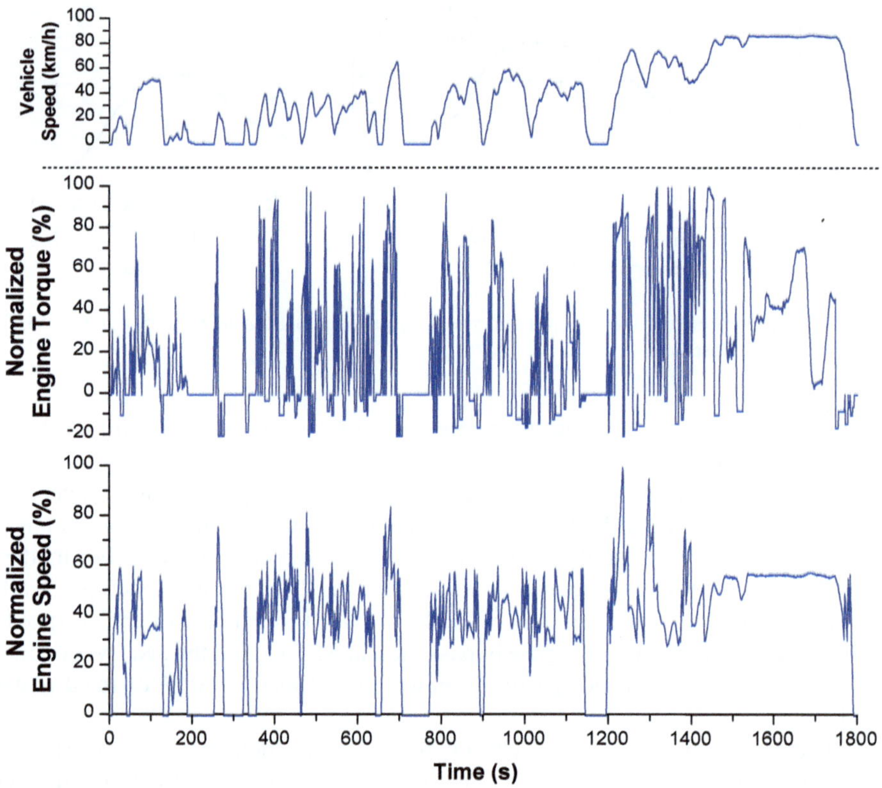

Fig. 4.48 Engine speed/torque profile of the worldwide WHTC transient cycle (the upper sub-diagram illustrates the chassis-dynamometer WTVC from which the WHTC was derived) [57]

For the cycle to be executed on the dynamometer, a denormalization procedure is first performed to obtain the actual engine speeds and torques at each operating point. Equation (4.7) describes this procedure

$$N = N_{norm} \times \ (0.45 \times N_{lo} + 0.45 \times N_{pref} + 0.1 \times N_{hi} - N_{idle}) \times 2.0327 + N_{idle} \qquad (4.7)$$

where:

- N is the actual engine speed (rpm),
- N_{norm} is the normalized speed (%),
- N_{lo} is the lowest engine speed, where the engine produces 55 % of the maximum power (cf. the 50 % value of the European ESC),
- N_{pref} is the engine speed where the maximum torque integral is 51 % of the whole integral (Fig. 4.47),
- N_{hi} is the highest engine speed where the engine produces 70 % of the maximum power (same as for the European ESC).

For torque denormalization, Eq. (4.4) is valid.

Detailed technical characteristics of the WHTC will be presented in Sect. 4.7.2 in comparison to the other two legislated heavy-duty engine cycles (FTP and ETC). The 3D normalized speed/load distribution of the WHTC is illustrated in Fig. 4.49.

For the WHTC test in the laboratory, the procedure consists of a cold-start test (temperatures of the engine's lubricant, coolant, and after-treatment systems all between 20 and 30 °C) following either natural or forced cool-down of the engine, a hot-soak period, and a hot-start test. The final emission value E_i is then determined from the weighted average of the two runs [57]

$$E_i(g/kWh) = \frac{wf_{cold} \cdot M_{i,cold} + wf_{hot} \cdot M_{i,hot}}{wf_{cold} \cdot W_{act,cold} + wf_{hot} \cdot W_{act,hot}} \qquad (4.8)$$

where wf is the weighting factor assigned to each run, M_i the mass of pollutant i (gaseous or PM) in grams per test (see Sect. 6.5 for the exact calculation procedure), and W_{act} the actual work produced by the engine.

Selection of the hot-soak period (5, 10 or 20 min) and the weighting factors for the cold and hot-start tests are to be decided by each country. For an exhaust after-treatment based on a periodic regeneration process, the emissions are to be measured on at least three WHTC tests, one during and two outside a regeneration event on a stabilized after-treatment system, and the results weighted (see Sect. 6.5).

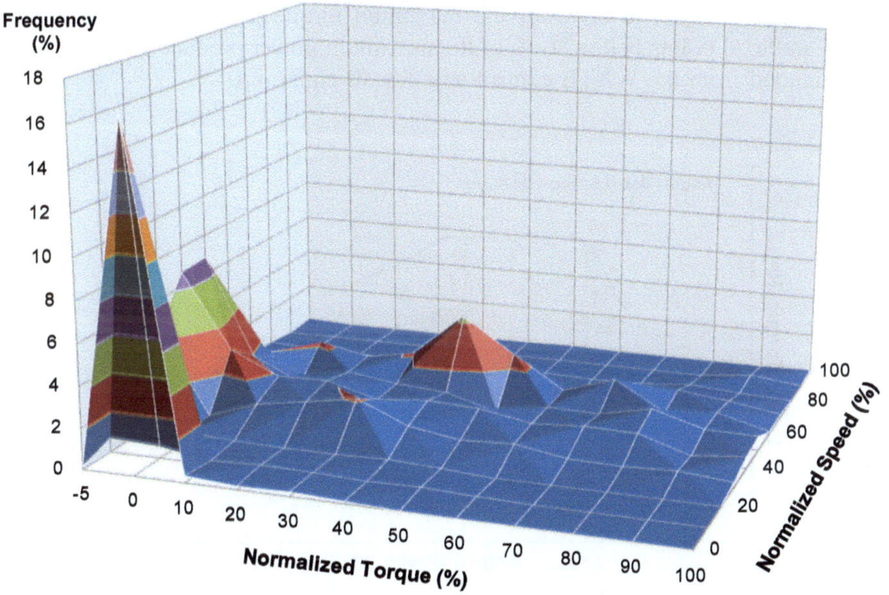

Fig. 4.49 3D normalized speed/torque frequency distribution of the WHTC

In Japan, the WHTC has been implemented from 2016; regulated emissions are CO, NO_x, NMHC and PM. Weighting factors for the cold and hot-start test are 14 and 86 % respectively [60, 61]; off-cycle measures have been adopted too [61].

In Europe, the WHTC has replaced the ETC from the Euro VI emission standard (2013 for new type approvals and 2014 for all registrations). The weighting factors needed in Eq. (4.8) are 14 % for the cold start and 86 % for the hot-start one, with a hot-soak period of 10 min. Apart from the gaseous (NO_x, CO and HC) and particulate mass emissions, particle number limits for diesel engines have been legislated too for the WHTC and the WHSC test, as well as an ammonia NH_3 10-ppm concentration limit for both diesel and gas engines (applicable to engines equipped with SCR systems); PN limit for gas engines will be defined at a later stage [1, 2].

The Euro VI regulation has also legislated new testing requirements for off-cycle (OCE) emissions (following UNECE GTR No. 10) in a stationary *ramped* engine test cycle with 15 random modes, as well as in-service conformity on-road tests with PEMS (Portable Emission Measurement Systems).

OCE measurements are performed during the type approval testing and follow the NTE limit approach. A control area is first defined on the engine map; there are two definitions, one for engines with rated speed lower than 3000 rpm (Fig. 4.50), and another for engines with rated speed greater than 3000 rpm. The WNTE control area includes all operating speeds between the 30th percentile cumulative speed distribution over the WHTC test cycle, including idle (designated N_{30} in Fig. 4.50), and the highest speed where 70 % of the maximum power occurs (N_{hi}). As regards the loading, the WNTE control area includes all engine load points with values greater than or equal to 30 % of the maximum torque produced by the engine. Speed and load points below 30 % of the maximum power produced by the engine are excluded from the WNTE control area for all emissions.

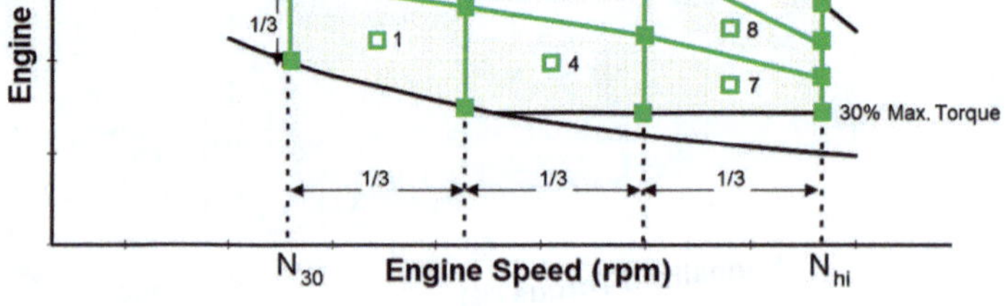

Fig. 4.50 WNTE test cycle grids for engines with rated speed less than 3000 rpm (for engines with rated speed higher than 3000 rpm, the engine-speed range is divided into four regions, instead of three, hence there are overall 12 grids) [61]

The control area is divided into grids; testing involves random selection of three grid cells and emission measurement at 5 points per cell sequentially (cf. the three extra NO_x measurement points during the ESC certification procedure in Fig. 4.6). The engine is initially warmed-up at mode 9 of the steady-state WHSC (55 % speed–50% load) for a minimum period of 10 min, and then the test commences, with 2 min operation at each random test point (transitions between the points are to be accomplished linearly within 20 s following the ramped modal cycle logic). Thus, total duration of the test is 30 min. The average specific mass of gaseous pollutants and PM should not exceed the corresponding WNTE limit values [62].

In-Service Conformity Testing, on the other hand, involves field measurements using PEMS. The testing is conducted over a mix of urban (0–50 km/h), rural (50–75 km/h) and motorway (>75 km/h) conditions, with the exact percentages of these conditions depending on the vehicle category as detailed in Regulation 582/2011/EU. First in-use test should be conducted at the time of type approval testing [60]. The minimum duration of each in-use test is based on the amount of power produced by the engine; each test must continue until the engine produces at least five times as much work as it would produce when operating over the WHTC. The in-use test data are analyzed using a 'moving average window' method (i.e., not over the entire test). Last amendment is to be found in Regulation 2016/1718/EU.

4.5.3 Steady-State WHSC

Apart from the transient WHTC, the UNECE WHDC group also developed a steady-state cycle, the world harmonized stationary cycle WHSC. This was too included in UNECE Regulation No. 49/06 and Global Technical Regulation (GTR) No. 4 [57]. The steady-state WHSC is detailed in Table 4.19 and illustrated in Fig. 4.51.

The development of the stationary WHSC was based on the following requirements [58]:

- 12 engine speed/torque combinations (modes): one idle point and up to four torque levels at five different engine speeds;
- Engine speed/load points based on the speed/load distribution of the reference transient engine cycle;
- Similar engine load distribution compared to the reference transient cycle;
- Test speeds expressed in normalized form dependent on engine characteristics;
- Denormalization of reference speeds in an analogous to the transient engine cycle way;
- Weighting factors for each mode of the same order of magnitude;
- Same set of weighting factors for different engines;
- Possibility to apply 'Control Area', in which emissions should be 'controlled'.

Table 4.19 Speed/load points of the WHSC cycle for HD engines (mode 0 is not run, but is only accounted for mathematically by a weighting factor of 24 % and zero emissions/power) [57]

Mode	Engine speed (%)	Engine load (%)	Weighting factor (%—for reference)	Mode length (s) (incl. 20 s ramp)
0	Motoring	–	24.0	–
1	0	0	8.5	210
2	55	100	2.0	50
3	55	25	10.0	250
4	55	70	3.0	75
5	35	100	2.0	50
6	25	25	8.0	200
7	45	70	3.0	75
8	45	25	6.0	150
9	55	50	5.0	125
10	75	100	2.0	50
11	35	50	8.0	200
12	35	25	10.0	250
13	0	0	8.5	210
Total			100	1895

Fig. 4.51 Illustration of engine speed/load points of the WHSC cycle

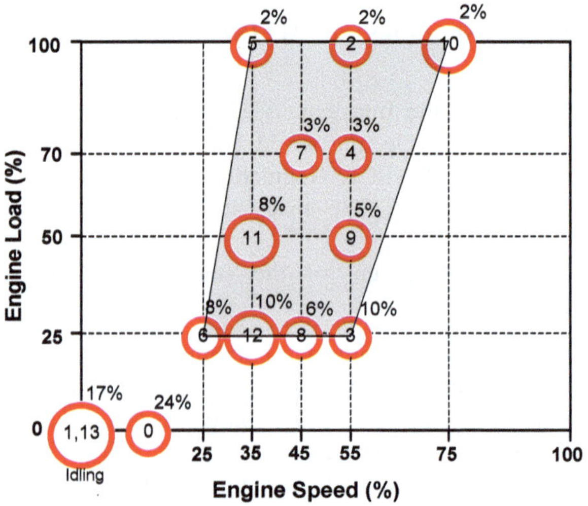

As is evident from Table 4.19, there are 13 different engine speed/load combinations under test, including one under motoring conditions, each with a specific duration and weighting factor. The idling phase is separated into two modes, one at the beginning and the second (mode 13) at the end, with a total weighting factor 17 %; this is similar to the transient WHTC's idling segment; the motoring mode's weighting factor (24 %) is similar to the transient WHTC's motoring segment as well.

The engine speed/torque pattern for an individual engine under test has to be derived through denormalization of the reference speed/torque pattern of the reference cycle. For the denormalization, Eq. (4.7) for speed and Eq. (4.4) for torque are applicable, as is also the case with the WHTC (version 1 of the cycle had different denormalization equations that led to lower than desired engine speeds, hence version 2 (final version) was developed) [58].

Figure 4.52 compares the transient WHTC to the stationary WHSC for a heavy-duty engine in terms of engine speed in rpm and torque in Nm. The previously mentioned fact that the WHSC was developed based on the joint frequency distribution of engine speeds/loads of the transient WHTC is evident in Fig. 4.52.

Comparing the WHSC with the ESC (Fig. 4.6), since the former replaced the latter from the Euro V to Euro VI standard, the following remarks can be made:

- Unlike the three speeds in the ESC, the tested speeds in the WHSC are five (excluding idle);
- The speeds are determined in a different way (different equation and different speed N_{lo}), resulting also in different power produced at each mode;
- Not all loads are tested at each engine speed during the WHSC;
- Testing at 70 % load has replaced the 75 % load mode. The other loads, 25, 50 and 100 % are applicable in both tests;
- The weighting factors are different;
- The ESC (as did the R49 before it) focused on medium to high speeds and loads; on the contrary, the WHSC includes testing points at lower engine speeds and loads that are more typical during urban driving (the same applies to the transient WHTC compared to the ETC as will be discussed later in Sect. 4.7.2)—even lower engine-speed points are tested during the Japanese 13 mode. Table 4.20 compares some technical (load) specifications of the legislated

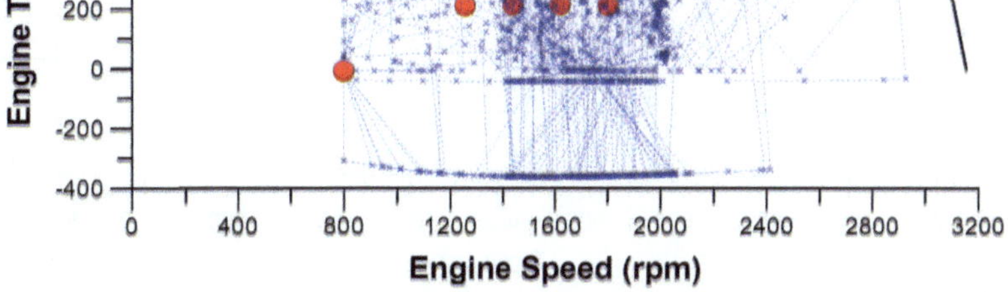

Fig. 4.52 Comparison between the transient WHTC and the stationary WHSC cycles for a heavy-duty diesel engine

Table 4.20 Technical specifications of stationary cycles for diesel engines

Cycle	Number of modes (excl. idling)	Idling (%)	Average (weighted) load (%)	Maximum load (%)
European/UNECE R49	10	25	51.0	100
European/UNECE ESC	12	15	55.0	100
U.S. 13 mode	10	20	40.3	100
Japanese 6 mode	5	35.5	43.8	100
Japanese 13 mode	11	41	25.9	95
Worldwide WHSC	11	17	25.2	100

stationary cycles employed in the past. The high average loading of the ESC compared to the WHSC is, perhaps, the most important finding from this table;

- Contrary to the 'truly' steady-state ESC, the WHSC is actually a ramped cycle (Fig. 4.53 serves as a graphical representation of the cycle's ramped form).

During the WHSC sampling process, the engine is operated for the prescribed time in each mode defined in Table 4.19, whereby engine speed and load are changed linearly within 20 s (in the order listed in Table 4.19), Fig. 4.53.

The weighting factors in Table 4.19 are given for reference only since the sampling time for each mode is different. The WHSC is run with the engine fully warmed up. Specifically, the engine is conditioned before the test in mode 9 (55 %

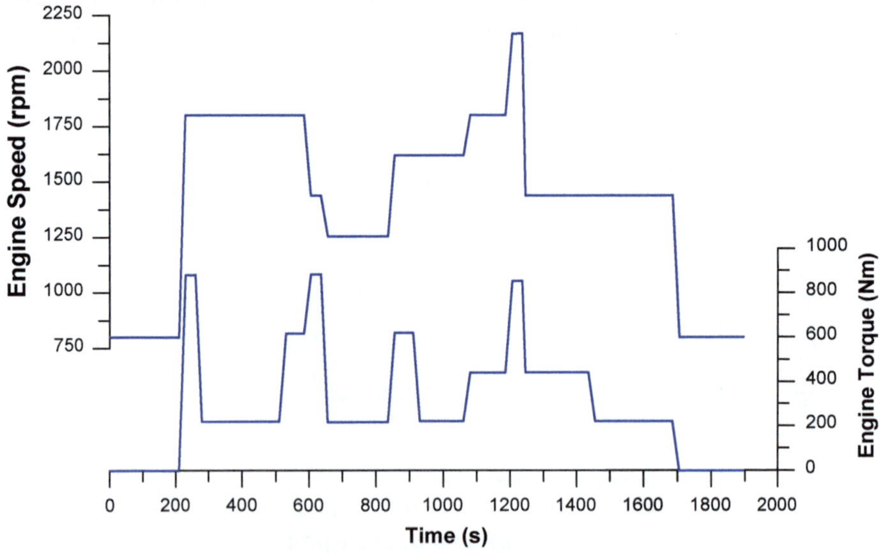

Fig. 4.53 Speed and torque sequences during the ramped WHSC for a heavy-duty diesel engine; emissions are measured continuously, including the 20-s speed/load changes

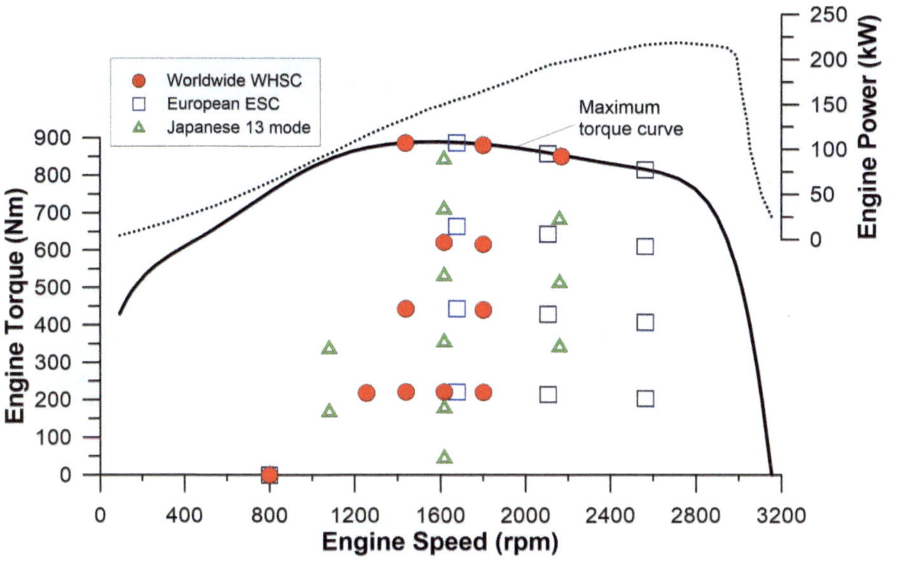

Fig. 4.54 Torque mapping of a heavy-duty turbocharged diesel engine demonstrating the points tested during the WHSC compared to the ESC and to the Japanese 13 mode (cf. Fig. 4.8 for the same engine comparison between the R49 and the ESC); the Japanese 13 mode is the cycle with the lowest engine speed under test (excluding idle) and the ESC the one with the maximum for the present engine

engine speed–50 % load); 5 min after completion of the preconditioning, the engine is started and the test sequence is performed [57].

A comparison of the engine control area between the ESC and the WHSC is on display in Fig. 4.54 for the same engine used in Fig. 4.8 for a similar comparison between the ESC and the R49. Clearly, the WHSC compared to the ESC focuses on lower speed ranges, having maintained testing at high loads.

The WHSC (replacing the ESC) has been enforced in the European legislation together with the transient WHTC (replacing the ETC) from 2013 with the emission standard Euro VI, after emission correlation factors with respect to their predecessors were defined (Regulations 582/2011/EU and 133/2014/EU).

4.6 Other Countries

In other areas of the world, the following (European, U.S. or Japanese) steady-state or transient cycles have been valid for testing the compliance of heavy-duty engines and vehicles with the corresponding emission standards over the years [1–3]; see also Fig. 4.55 for a graphical representation:

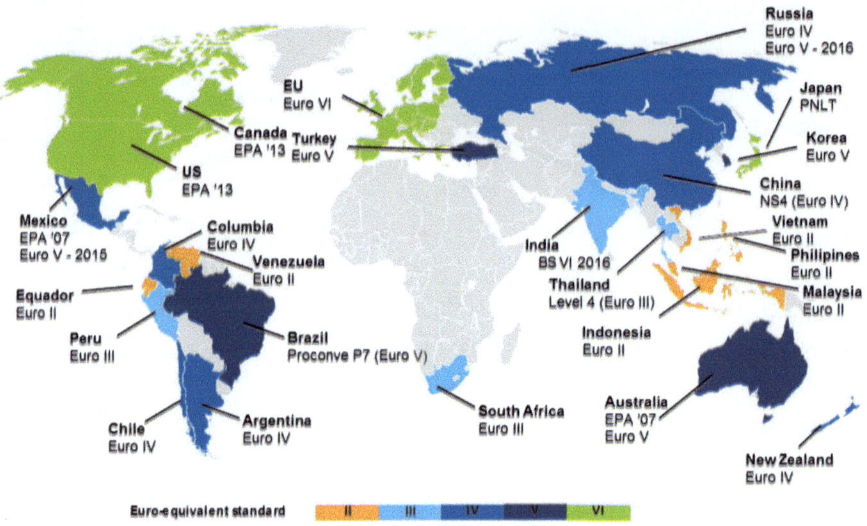

Fig. 4.55 Overview of world emission legislation as of 2016 (copyright by IAV); countries that follow the Euro II standard employ the R49 cycle, those at Euro III, IV or V standards employ the ESC/ELR + ETC, whereas those having adopted the Euro VI standard employ the WHSC + WHTC test procedures

- Argentina follows EU standards and test cycles.
- Australian Design Rules ADR 80 specifies three test cycles for examining heavy-duty engines, the ESC, the ETC and the load response ELR, effective 1/1/2003 for all vehicles, replacing the ECE R49 [63]. The JE05 cycle has been accepted in Australia together with U.S. standards as alternatives. Further to the above, four HDV chassis cycles have been developed for inventory purposes, each designed for a specific diesel-engine vehicle category (see Sect. 2.4 regarding details on how the composite urban emission driving cycles-CUEDC were constructed). Figure 4.56 illustrates the speed/time trace of these cycles, and Table 4.21 provides some important technical specifications [64].
- Canada follows U.S. regulations and test procedures.
- Brazil follows the European regulations beginning with the R49 cycle for MY 1993 trucks and buses, and moving to ESC/ELR+ETC from MY 2006.
- Chinese regulations for HDVs have followed closely the European ones over the years (R49, ESC/ELR, ETC, WHSC/WHTC). China regulates HDVs fuel consumption employing a softer with regards to accelerations version of the WTVC, depicted in Fig. 4.57 (known as C-WTVC). Local cycles for durability purposes of engines and whole vehicles are employed too.
- Indian regulations are also based on the European standards beginning with the R49 cycle in 1992 and later (from 2010) adopting the ESC and ETC; the worldwide WHSC and WHTC are proposed to be valid from 2020 together with the Bharat Stage VI standard.

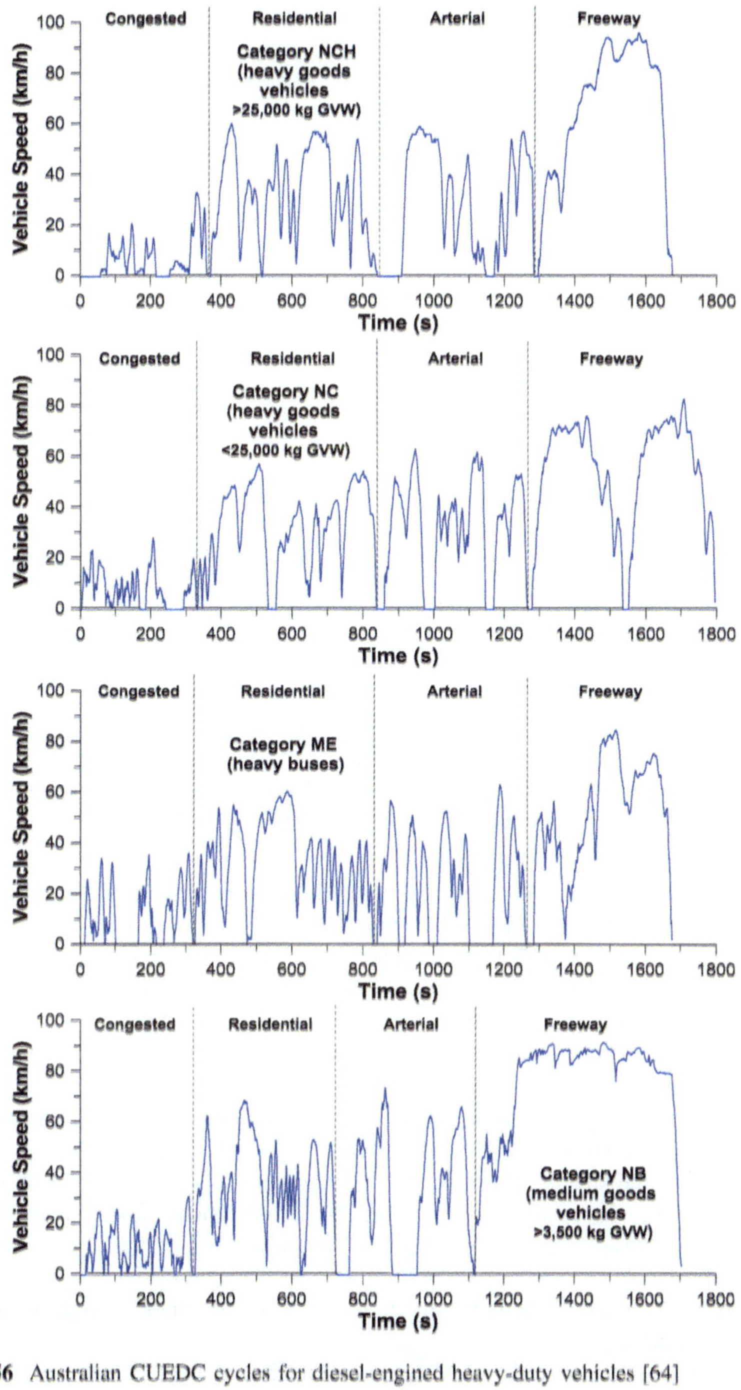

Fig. 4.56 Australian CUEDC cycles for diesel-engined heavy-duty vehicles [64]

Table 4.21 Technical specifications of the Australian CUEDC cycles for HDVs

Category	Duration (s)	Distance (m)	Maximum speed (km/h)	Average speed (km/h)	Driving speed (km/h)	Maximum accel. (m/s^2)	Idling time (%)	RPA (m/s^2)
NB	1705	20,590	91.6	43.5	47.2	4.73	7.9	0.161
ME	1677	14,387	85.0	30.9	36.1	3.80	14.4	0.230
NC	1796	17,035	82.9	34.1	38.1	3.59	10.4	0.156
NCH	1675	15,524	96.1	33.4	37.9	1.67	11.9	0.143

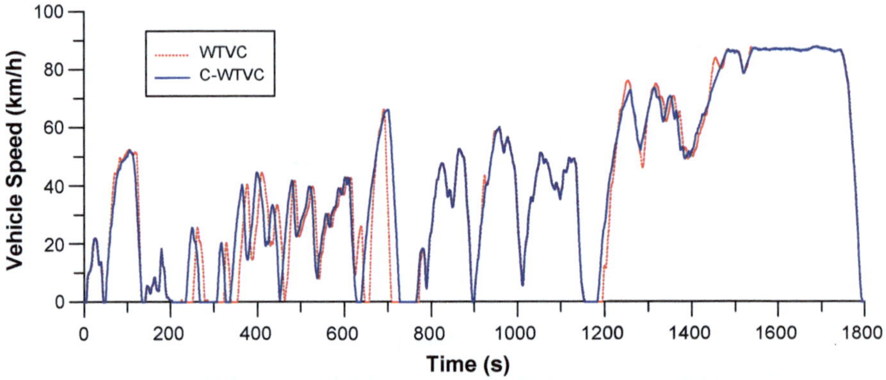

Fig. 4.57 Speed profile of the Chinese version of the WTVC heavy-duty chassis-dynamometer cycle used for fuel consumption purposes only; average positive acceleration is 0.238 m/s^2 for the C-WTVC (0.279 for the WTVC), and maximum acceleration 0.92 m/s^2 instead of 1.59 for the WTVC (detailed technical specifications are provided in the Appendix)

- In Mexico, the U.S. FTP transient cycle is employed for vehicles with GVW greater than 3857 kg (8500 lbs) that are diesel or gasoline powered; the European cycles ESC/ETC are also valid. FTP-75 chassis-certification is available for SI-engined vehicles with GVWR lower than 14,000 lb (6350 kg). Euro VI emission standards will have to be met from 2020.
- Russia follows the European standards, with Euro IV in effect from January 1, 2013 and Euro V from January 1, 2016.
- South Africa's legislation for HDVs follows the European regulation (Euro V planned for 2020), with U.S. and Japanese standards as alternatives.
- From 2003, the steady-state ESC and the transient ETC have been employed in S. Korea too, with a move to WHSC/WHTC from January 2013 as was the case in the EU; previously (1993–2002) the Japanese 6-mode and 13-mode cycles were applicable.
- Switzerland follows the EU standards as does another European non-EU country, Norway.
- Turkey's legislation is also closely related to that of the EU's.

4.7 Comparative Results

4.7.1 Chassis-Dynamometer Cycles

In the following pages, a comparison between the four most important heavy-duty chassis-dynamometer cycles discussed earlier in the chapter, the European FIGE, the U.S. UDDS, the Japanese JE05 and the Worldwide WTVC, will be presented. Table 4.22 provides some technical specifications for these cycles in comparison.

Apart from the UDDS, all the other cycles last approximately 30 min. Maximum speed varies slightly between them (from 87.6 to 93.3 km/h), with the average and driving speed, however, differentiating by a lot. In particular as regards the FIGE, noteworthy attributes are its minimal idling phase (3.4 %), highest average and driving speeds, hence its longest distance compared to the other three, indicative of long-distance traveling (highway) trucks. On the other hand, the UDDS and JE05 have 32 and 24 % idling phase respectively, focusing more on urban driving. It seems that the WTVC falls somewhere between the other three cycles (perhaps because it was derived from data from all three regions from where the other cycles originate).

Speed/acceleration frequency distributions for all cycles are presented in Fig. 4.58. Apart from the rather unique pattern of the FIGE, all the other cycles exhibit increased density at lower than 30 km/h speeds and −1.5 to 1.5 m/s^2 accelerations. Both the FIGE and the WTVC manifest overall higher speed values owing to their more pronounced motorway segment (particularly FIGE). It is not surprising then that the more urban-oriented UDDS and JE05, which are also characterized by shorter traveled distance, have also higher RPA values compared to the FIGE and WTVC (RPA values for many heavy-vehicle cycles are presented in Fig. 4.59 with respect to each cycle's average driving speed). The harsh urban schedule of the UDDS, combined with increased idling phases, is confirmed in Fig. 4.60 that illustrates the respective accelerations during the first 600 s of each cycle.

Table 4.22 Comparison of the technical specifications between the FIGE, UDDS, JE05 and WTVC

Cycle	Duration (s)	Distance (m)	Maximum speed (km/h)	Average speed (km/h)	Driving speed (km/h)	Maximum accel. (m/s^2)	Idling time (%)	RPA (m/s^2)
FIGE	1800	29,482	91.1	59.0	61.0	3.83	3.4	0.070
UDDS	1060	8935	93.3	30.3	44.6	1.96	31.9	0.143
JE05	1829	13,892	87.6	27.3	36.2	1.59	24.4	0.126
WTVC	1800	20,072	87.8	40.1	46.1	1.59	12.9	0.107

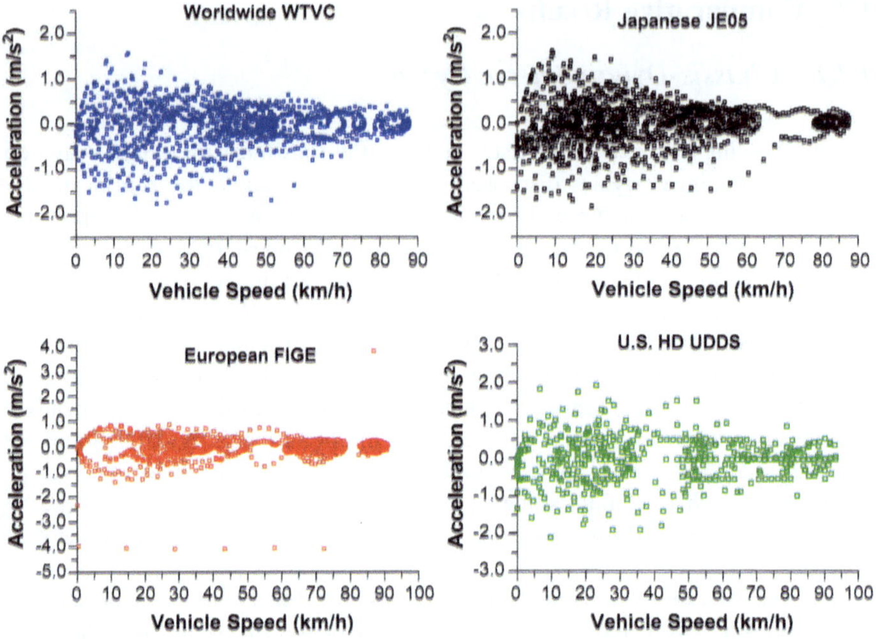

Fig. 4.58 Comparison of the speed/acceleration frequency distributions between the FIGE, UDDS, JE05 and WTVC

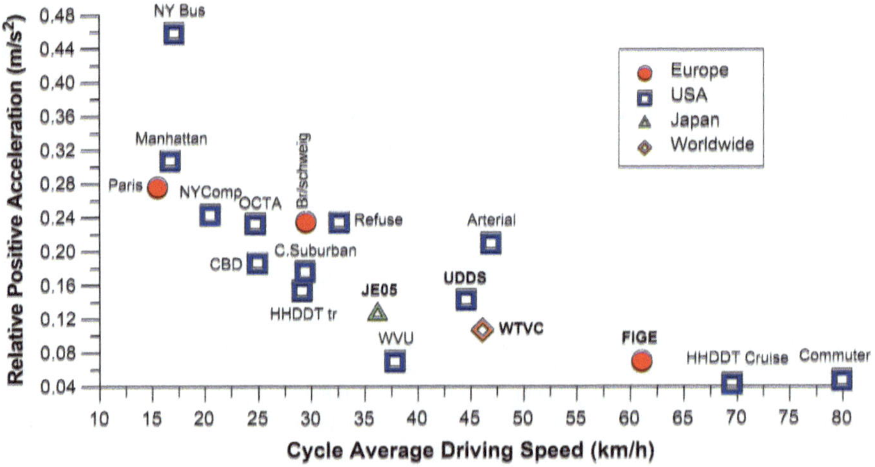

Fig. 4.59 Relative positive acceleration for various HDV chassis-dynamometer cycles (strictly urban (bus) cycles such as the Braunschweig, Paris, NY and Orange County, which exhibit frequent and sharp accelerations, are the ones with the highest RPA values, in contrast to cycles such as FIGE or WTVC, which also include extra-urban segments; the extreme case here is the exclusively highway-driving Commuter and HHDDT cruise cycles, with RPA values of 0.04)

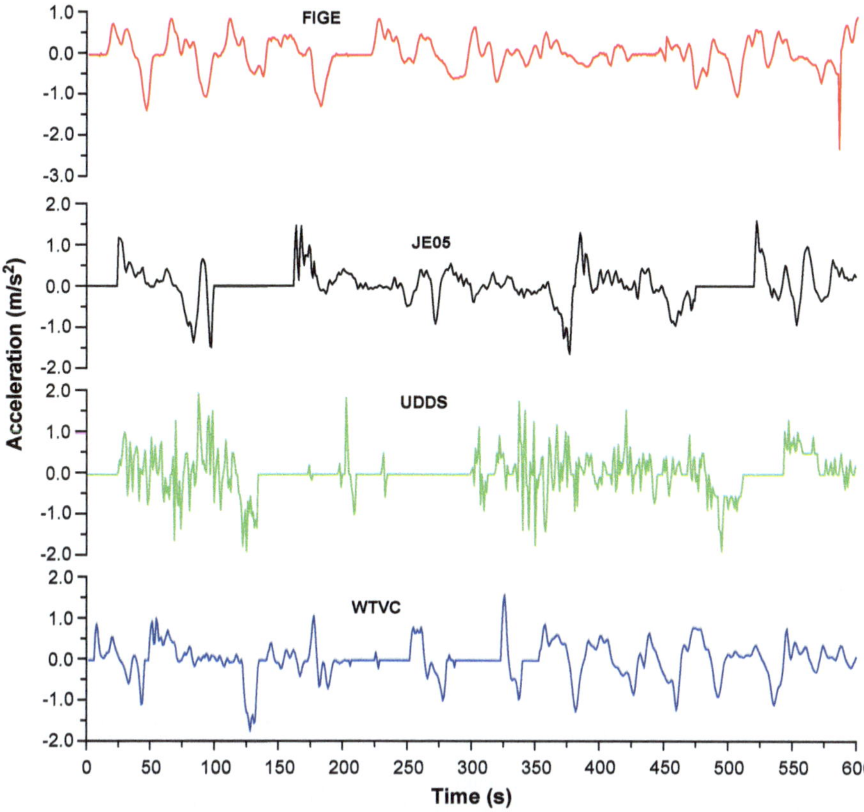

Fig. 4.60 Comparison of the acceleration profile during the first 600 s between the FIGE, UDDS, JE05 and WTVC

Lastly, Fig. 4.61 demonstrates the actual engine speed/torque distribution experienced by a turbocharged, diesel-engined, heavy-duty vehicle running on all four cycles. In general, the distribution seems quite similar between the FIGE and the WTVC (the latter exhibiting a few higher-torque points, more points at idling and, in general, more 'uniform' pattern). The UDDS manifests less 'dense' pattern owing to its shorter duration (fewer operating points), but has also operating points at higher engine torques resulting from more abrupt accelerations. A similar figure was provided in Chap. 2 comparing the engine speed/power distribution of a passenger car running on four legislated light-duty cycles (Fig. 2.65). From that figure it became obvious that considerable differences between the passenger-car cycles exist, something which cannot be said, however, for the heavy-duty cycles on display in Fig. 4.61.

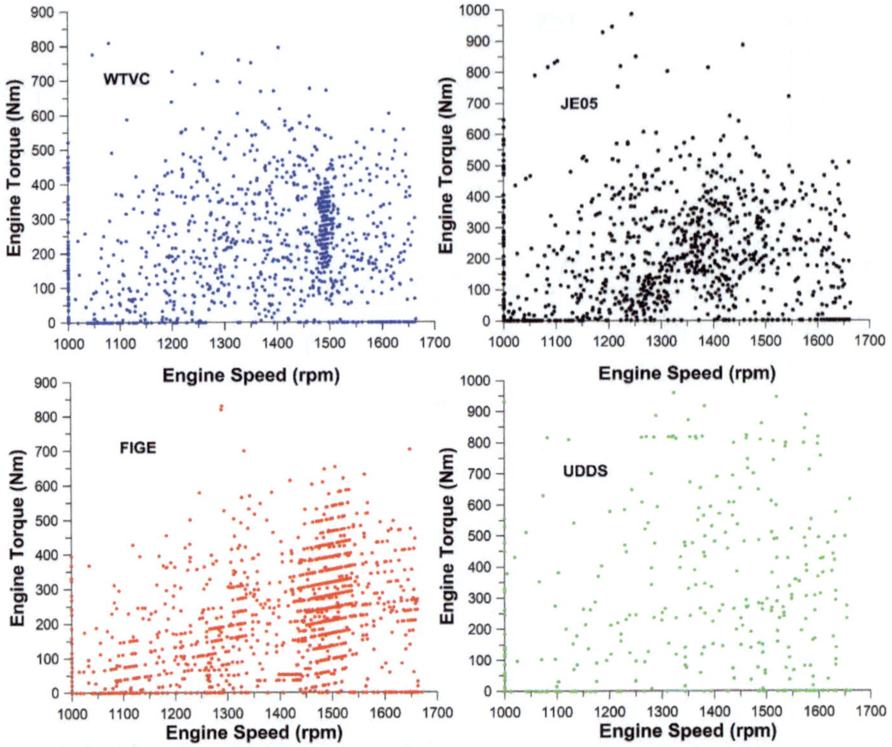

Fig. 4.61 Comparison of the actual engine speed/torque distribution for a heavy-duty turbocharged diesel-engined vehicle between the FIGE, WVTC, JE05 and UDDS

4.7.2 Engine-Dynamometer Cycles

In the following paragraphs, a comparison between the engine tests ETC, FTP and WHTC will be performed based on their technical specifications as well as results from actual heavy-duty engines operation. Table 4.23 presents some of the cycles' technical characteristics; these are further illustrated in the two-page Fig. 4.62. Data from the non-road NRTC cycle (Sect. 5.2) is also provided for the sake of completeness.

Based on the data presented in Table 4.23 and Fig. 4.62, the following comments can be made regarding the technical attributes of the engine-dynamometer on-road cycles:

- Both the ETC and the WHTC last for 30 min in contrast to the 20-min duration of the FTP.
- The FTP has by far the longest idling period (37 % of the total time), with the ETC idling for less than 7 % of the time, and the WHTC being somewhere in the middle.

Table 4.23 Comparison of some technical specifications between the ETC, FTP and WHTC

	ETC	FTP	WHTC
Duration (s)	1800	1199	1800
Average normalized speed (%)	50.9	41.5	37.1
Maximum normalized speed (%)	90.1	111.9	100
Average normalized torque (%)	36.7	24.1	23.9
Maximum normalized torque (%)	100	100	100
Idling period (%)	6.6	36.5	16.2
Motoring period (%)	18.0	14.8	22.3
Maximum load increase	0–98%	0–78.53 %	m–84.7 %
Average load increase (%)	17.6	14.0	13.6
Load increases (min^{-1})	18.9	11.9	16.3
Time accel./decel./cruising (%)	43.8/43.4/12.7	26.4/26.7/46.8	42.4/37.6/20.0
Time w. load incr./decreas./const. (%)	35.9/54.3/9.7	20.4/34.9/44.6	30.6/52.3/17.1

m denotes motoring point

- The FTP has the greatest number of stops (19) and stops per minute.
- For all three cycles, maximum torque is 100 %.
- The European ETC proves to be the harshest cycle, exhibiting highest values for both average normalized engine speed and torque, and, most importantly, highest frequency of load increases.
- Similarly, it is during the ETC that the engine spends most of the time accelerating and with increasing load, i.e., the ETC is the most transient from the three.

An engine cycle with steep and frequent transients, such as the European ETC, is also expected to produce high *engine-out* PM and NO_x emissions (in g). The lower sub-diagram of Fig. 4.63 highlights this for a heavy-duty turbocharged diesel engine running on all three transient cycles. Unsurprisingly, the ETC turns out to be the most polluting of the three cycles in terms of *absolute* engine emissions, particularly soot and particularly during its first segment, i.e., urban driving. On the other hand, during the U.S. FTP cycle, the current engine produces 40 % less NO, 50 % less soot, and has 24 % less fuel consumption compared to the ETC. For the worldwide WHTC, the reduction in NO emissions is 29 %, in soot 38.5 % and in fuel consumption 31 % compared to the European ETC. The fact that the average engine speed during the WHTC is lower than during the FTP is responsible for its lower fuel consumption compared to its U.S. counterpart.

Things differentiate—sometimes considerably—when the normalized emissions are evaluated, i.e., when the absolute emission (or fueling) values are related to the total work produced by the engine during the cycle, and the emissions and fueling are calculated in g/kWh (upper sub-diagram of Fig. 4.63). There are three additional characteristics of a transient cycle that influence strongly the normalized emissions (apart from the actual rate of emission production during each engine

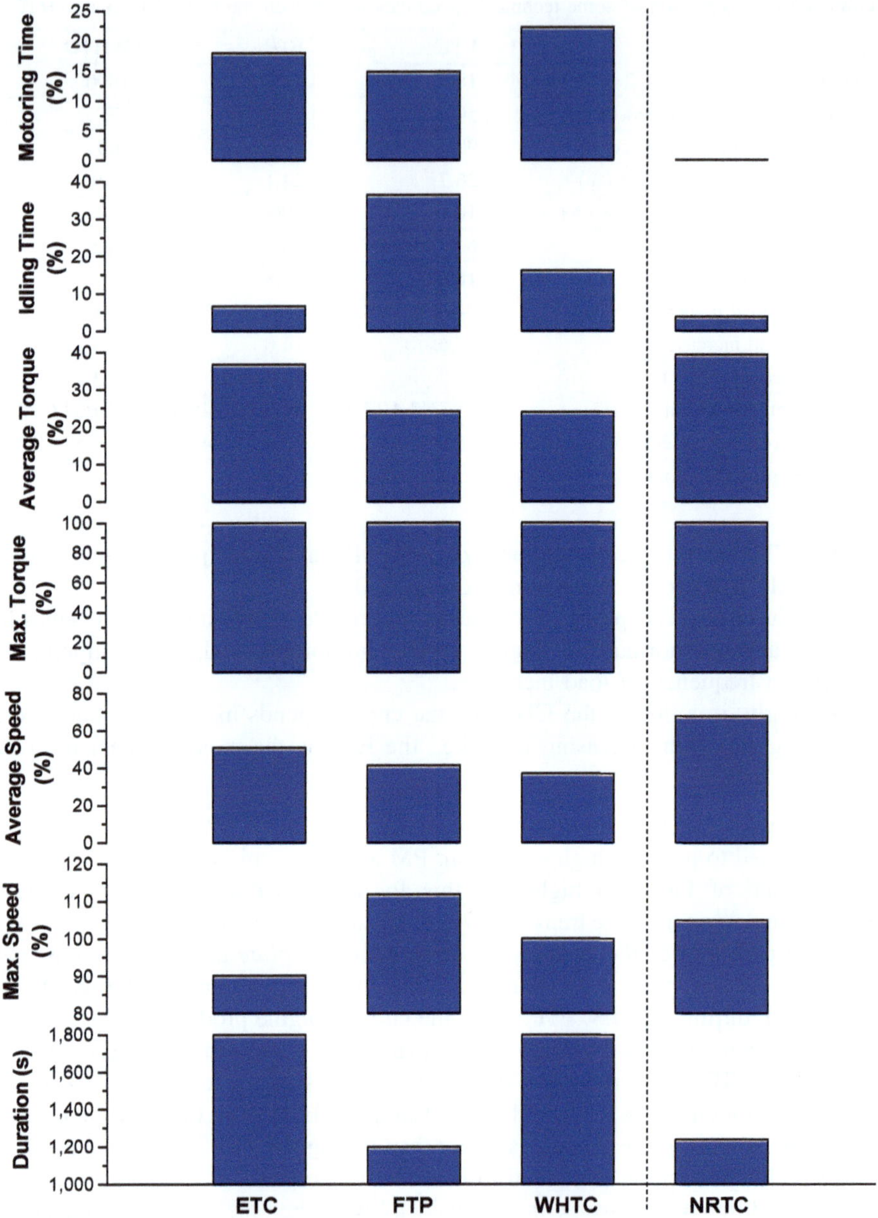

Fig. 4.62 Comparison of some important technical specifications between the three heavy-duty (ETC, FTP, WHTC) and the non-road NRTC (Sect. 5.2) engine cycles

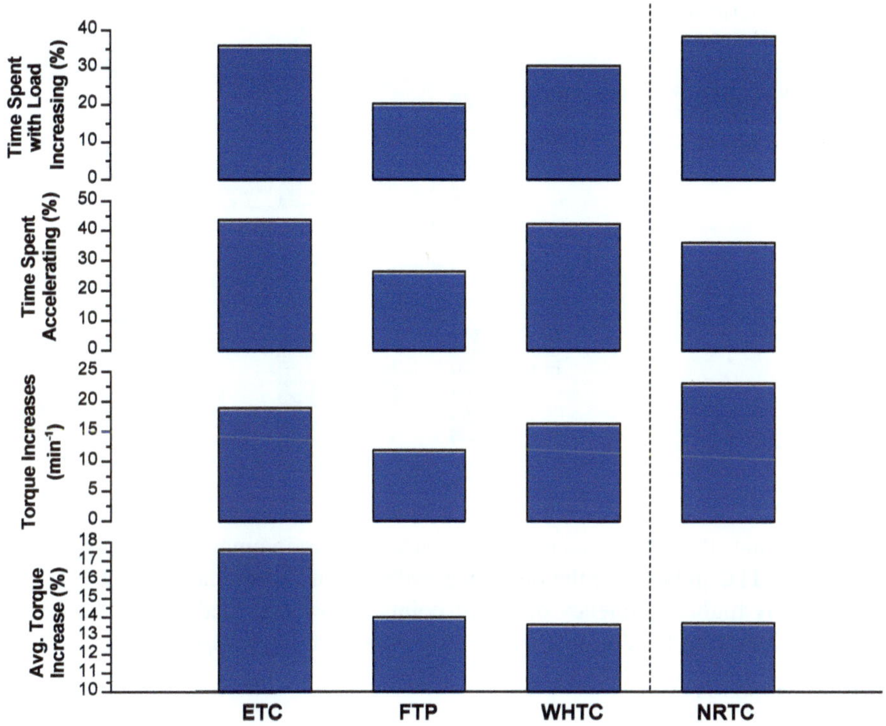

Fig. 4.62 (continued)

cycle or 'second'), namely the instantaneous power produced by the engine, as well as the duration of the idling and the motoring segments. The FTP and the WHTC have produced lower emissions (in g/min) than the ETC but are also characterized by longer idling and motoring periods and lower average engine loading. Eventually, the total work produced by the engine is much smaller compared to the ETC. As a result, the cumulative *normalized* emissions during the FTP and the WHTC are much more cohesive compared to the ETC.

All the emission values provided in Fig. 4.63 are hot-started engine-out emissions. When cold starting as well as temperature-dependent after-treatment devices are involved, however, the vehicle-out emission profile may change (sometimes dramatically) as will be discussed in more detail later in the section.

Figure 4.64 illustrates the normalized speed/torque frequency distribution of the three engine-dynamometer cycles. Following the arguments presented earlier, the increased urban segment of the FTP is confirmed in Fig. 4.64 as regards both speed and torque. Interestingly, it is the FTP that reaches the highest engine speeds too, although its average normalized speed is lower than that from the ETC. For the FTP, there is an apparent sparse distribution of speeds in the range between 10 and 50 % (recall that the cycle consists of congested urban and freeway segments). On

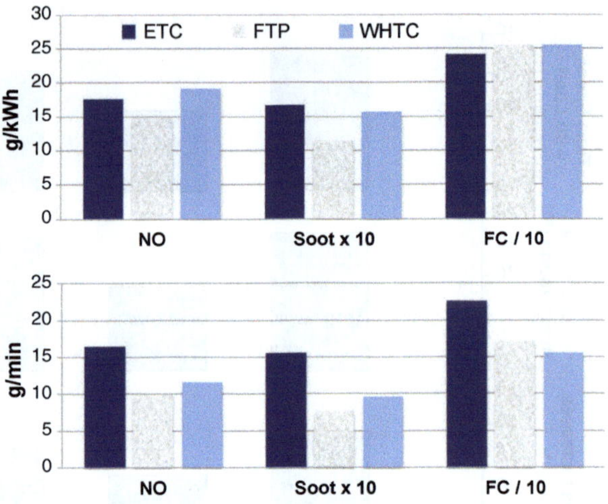

Fig. 4.63 Comparison of engine-out NO and soot emissions and fuel consumption (FC) between the ETC, FTP and WHTC for a heavy-duty turbocharged diesel engine (hot-started results)

the other hand, the ETC reaches higher loads and predominantly medium speeds, with the WHTC in between the other two cycles. More specifically, the WHTC has considerably higher frequency of tested points at low and medium engine speeds and loads, covering thus more appropriately urban driving compared with the ETC.

WHTC versus ETC

Since the WHTC has already replaced the ETC as the certification cycle for heavy-duty engines in the EU and other countries, beginning with the Euro VI emission standard, the following paragraphs concentrate on a WHTC versus ETC comparison.

Firstly, the denormalization procedure, as regards the engine speed is different between the two cycles. The ‘low’ speed (N_{lo}) corresponds to a different engine speed, and an overall different conversion formula is employed (torque denormalization remains the same).

The WHTC is run twice, once with the engine cold-started and the second with the engine hot started. Further to this, the hot started test is not run immediately after completion of its cold-start counterpart, but after a 10-minute soak, during which the thermal condition of the after-treatment devices alters. On the other hand, the ETC was run with the engine fully warmed up. This has influenced exhaust emissions (and motivated manufacturers to thermal management techniques for temperature-dependent after-treatment devices), even though the weighting factor of the cold-started WHTC run is small compared to the hot-started one (see also Fig. 4.69 later in the text) [65].

Different denormalization procedures and original vehicle speed trace for the two cycles result in a considerably different overall engine speed/torque tested area as illustrated in Fig. 4.65.

An instructive analysis has been conducted by TNO (with support from other partners from the EU and the U.S.) on five six-cylinder engines, ranging from

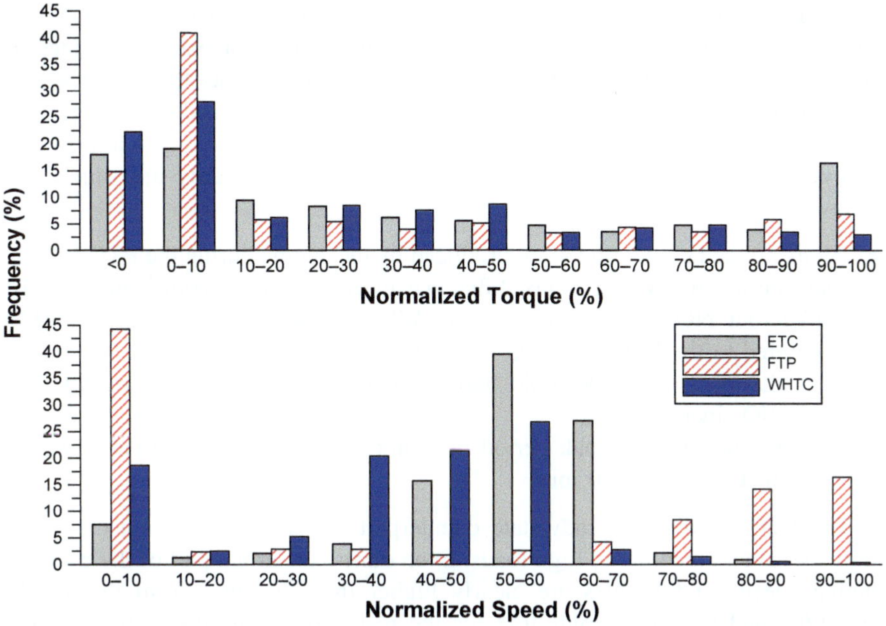

Fig. 4.64 Normalized speed and torque frequency distribution for the three heavy-duty transient cycles (ETC, FTP, WHTC)

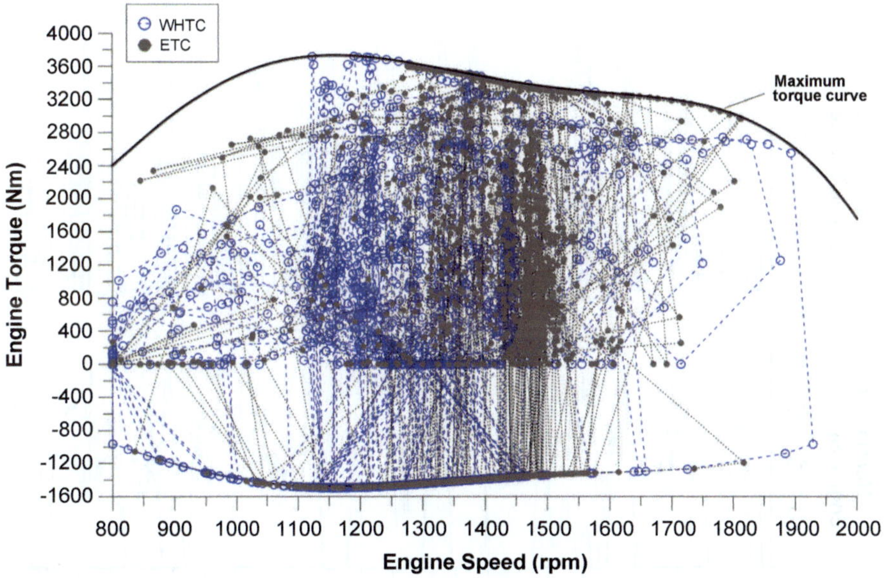

Fig. 4.65 Comparison of the actual speed/torque operating range between the WHTC and the ETC for a heavy-duty turbocharged diesel engine (ETC and WHTC denormalized using the same equations)

Euro III to Euro VI technology, regarding WHTC/WHSC versus ETC/ESC effects on emissions. By choice, only NO_x emissions were under investigation, since this pollutant is usually close to the emission limit because of its trade-off with fuel consumption and PM (in the sense that an engine calibration leading to higher NO_x level but lower fuel consumption and engine-out PM might be a typical choice for off-cycle operating points). The results derived from the study are summarized below [66]:

- As mentioned earlier, the work produced by the engine during the ETC is higher than during the WHTC. This difference in engine cycle work leads to a 3 % vehicle-out NO_x increase during the WHTC, and 1.6 % during the WHSC compared to the ETC and ESC respectively;
- The additional test under cold-started conditions of the WHTC results in 8–13 % NO_x increase; and
- The intermediate hot soak period does not contribute to additional NO_x, provided that it is up to 10 min.

The ETC (as did its steady-state counterpart the ESC) focuses on medium to high speeds and medium to high loads. Even during the urban part of the cycle, the engine speeds and torques are clearly higher than the ones from the WHTC, Fig. 4.66, and the same holds true for the respective transients demonstrated in Fig. 4.67. This has a direct effect on exhaust emissions from temperature-dependent after-treatment devices, such as SCRs. Since during the ETC the engine operates at

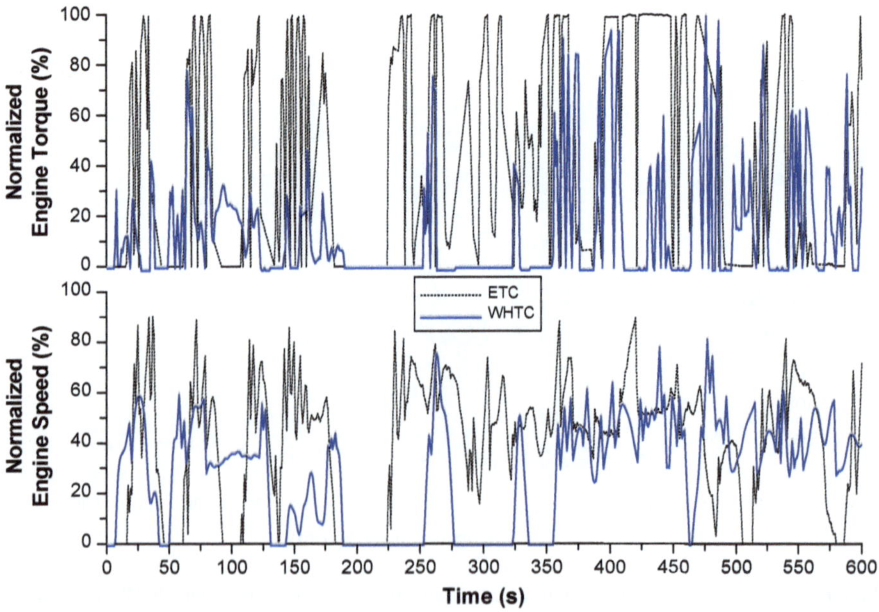

Fig. 4.66 Comparison between the WHTC and the ETC during the first 600 s (urban segment)

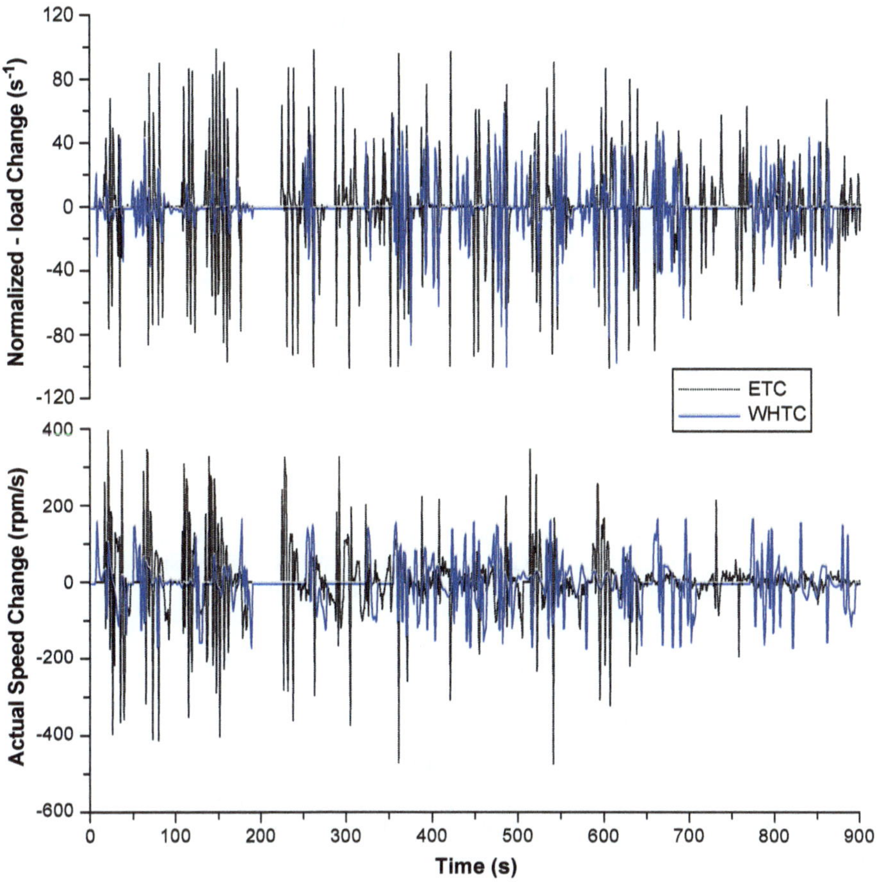

Fig. 4.67 Comparison of the actual speed change and the normalized load change between the WHTC and the ETC for a heavy-duty diesel engine during the first 900 s

high temperatures, both the conversion of urea into ammonia, as well as the ammonia-based reduction reactions inside the SCR are accomplished at a much higher rate compared to actual in-use engine operation. Not surprisingly, measurements conducted on the road (with a considerable portion including driving at low loads/speeds as well as idling) showed much higher amount of emitted NO_x during actual truck or bus operation compared to the ETC certification results [67]. For example, Fig. 4.68 shows how Euro IV and V (ETC-certified) trucks usually emit higher NO_x compared to the Euro VI (WHTC-certified) ones [68].

Contrary to the ETC, the WHTC incorporates testing points that are more typical during urban driving as Figs. 4.64 and 4.65 indicate. The high volume of operating

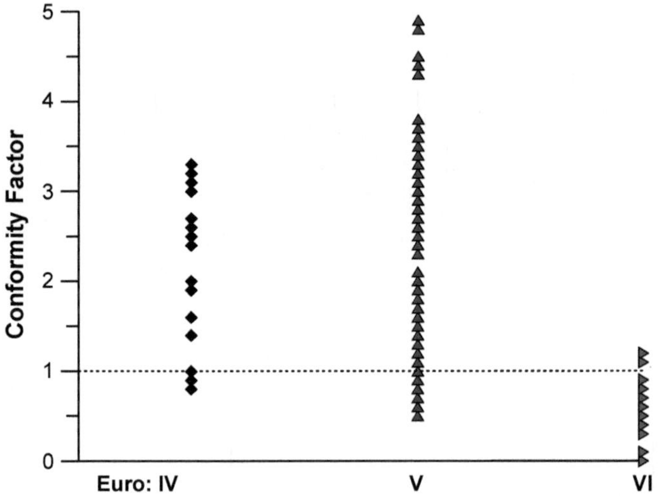

Fig. 4.68 NO_x emissions conformity factors for a range of HDVs tested on various chassis-dynamometer cycles (e.g., FIGE, Braunschweig, WTVC) with respect to Euro level; horizontal line indicates conformity factor of one (data from [68])

points at relatively low engine speeds and loads during the WHTC results in lower NO_x to N_2 conversion rates inside the SCR owing to the relatively lower gas temperatures involved.[8] The cold-started run of the WHTC further increases the amount of emitted NO_x during the certification procedure compared to the ETC [69, 70]. For example in [70], and comparing the SCR behavior between the ETC and the WHTC, it was found that whereas during the hot-started ETC, urea injection started immediately, during the cold-started WHTC it took 9 min for the exhaust gas recirculation system to start operating (in order to avoid engine wear and condensation in the EGR cooler) and approximately 13 min to start the urea injection (Fig. 4.69). In view of the above (i.e., different test cycle and inclusion of cold-started run), manufacturers have been enforced to modify the engine calibration in order to meet the NO_x certification targets from their heavy-duty engines. Furthermore, since the WHTC more accurately represents in-use vehicle operation, it is expected that real-world emissions will be much closer to type approval ones as Fig. 4.68 suggests. Unfortunately, there is a high population of Euro IV and V certified trucks and buses still in circulation, as Fig. 4.55 eloquently shows, which, calibrated on the high load/speed ETC cycle, are exhibiting higher actual NO_x emissions on the road particularly in heavy congested areas in China and India [67].

[8]It is reminded that load control in a diesel engine is qualitative, meaning that the increase in load is accomplished through injection of greater fuel quantity; this in turn reduces the (global and always lean) air-fuel equivalence ratio, hence gas temperatures are increased.

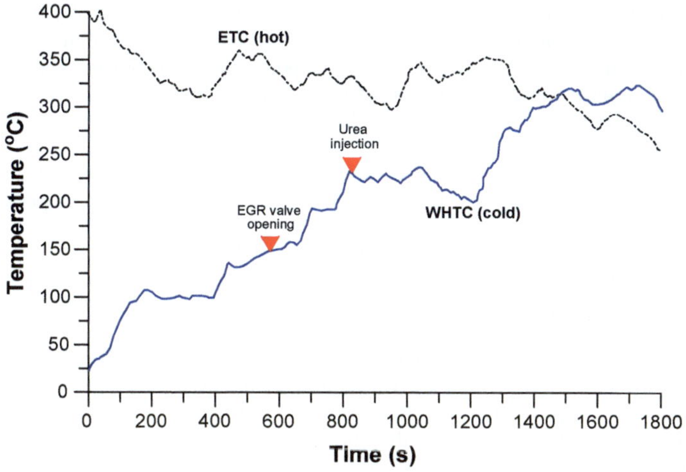

Fig. 4.69 Comparison of the SCR inlet temperature from a US2007, 6-cylinder, 7.5 L turbocharged diesel engine running on the WHTC and the ETC cycles (AECC Euro VI test program presentation given at AECC technical seminar on 25 October 2007 [70])

References

1. TransportPolicy.net. http://www.transportpolicy.net/index.php?title=Category:Heavy-duty_Vehicles. Accessed 4 Sept 2016.
2. Worldwide emission standards. Heavy duty and off-highway vehicles. Delphi, 2015–2016. http://delphi.com/docs/default-source/worldwide-emissions-standards/delphi-worldwide-emissions-standards-heavy-duty-off-highway-15-16.pdf. Accessed 11 Aug 2016.
3. Exhaust emission legislation–Diesel and gas engines. VDMA, April 2010.
4. Sharpe B. Testing methods for heavy-duty vehicle fuel efficiency: Trends from regulatory programs around the world and implications for India. ICCT Working Paper 2015-3; 2015.
5. Berg W. Legislation for the reduction of exhaust gas emissions. In: Gruden D, editor. The handbook of environmental chemistry. Traffic and environment, Vol. 3. Berlin: Springer; 2003. p. 175–253.
6. Council Directive 88/77/EEC of 3 December 1987 on the approximation of the laws of the Member States relating to the measures to be taken against the emission of gaseous pollutants from diesel engines for use in vehicles.
7. Aarnink TJ, Flöysand SÅ, Gadd P, Le Breton D, Mann N, Marotta A et al. Motor vehicle emission regulations and fuel specifications—part 2 detailed information and historic review (1970–1996). Prepared for the CONCAWE automotive emissions management group; 1997.
8. Steven H, Bachurski J, Enz W, Kirstein KG, Schneider D, Skrzipczyk E. Influence of the transient operating mode of commercial vehicles and its consideration in emission measurement according to ECE R 49. FIGE Report No. 104 05 316, FIGE Institute, Aachen, Germany; 1994.
9. Zhen F, Clark NN, Bedick CR, Gautam M, Wayne WS, Thompson GJ, et al. Development of a heavy heavy-duty diesel engine schedule for representative measurement of emissions. J Air Waste Manage Assoc. 2012;59:950–9.

10. Directive 1999/96/EC of the European parliament and of the Council of 13 December 1999 on the approximation of the laws of the Member States relating to the measures to be taken against the emission of gaseous and particulate pollutants from compression ignition engines for use in vehicles, and the emission of gaseous pollutants from positive ignition engines fuelled with natural gas or liquefied petroleum gas for use in vehicles and amending Council Directive 88/77/EEC.
11. Reducing NO_x emissions on the road. European Conference of Ministers of Transport. Report prepared by the ECMT group on transport and the environment in co-operation with the OECD environment policy committee's working group on transport. 2006. http://www.itf-oecd.org/sites/default/files/docs/06nox.pdf. Accessed 3 July 2016.
12. Giakoumis EG, Alafouzos AI. Study of diesel engine performance and emissions during a transient cycle applying an engine mapping-based methodology. Appl Energy. 2010;87:1358–65.
13. Dreyer W. Stochastischer Fahrzyklus für Stadt-Linienbusse. Sonderforschungsbereich 97, Fahrzeuge und Antriebe, Technische Universität Braunschweig, Bericht Nr. 34, Februar 1975.
14. Begleitendes Prüfprogramm im Rahmen der 'Effizienz- und Kostenanalyse für den Linienbetrieb von Hybridbussen'. Abschlussbericht im Auftrag des Bundesministeriums für Umwelt, Naturschutz und Reaktorsicherheit, 2012.
15. Coroller P, Plassat G. Comparative study on exhaust emissions from diesel and CNG-powered urban buses. In: Proceedings of the diesel engine emissions reduction conference (DEER); 2003.
16. Nylund N-O, Erkkilä K, Clark N, Rideout G. Evaluation of duty cycles for heavy-duty urban vehicles. Final report of IEA AMF Annex XXIX, Espoo; 2007.
17. Mezghani M. UITP's standardised on-road test (SORT) cycles for measuring fuel consumption. 4th Business Forum in the MENA region, Cairo, Egypt, 4–6 Feb 2007.
18. UITP project 'SORT'—Standardised on-road test cycles. International Association of Public Transport; 2014.
19. Supplement A to Compilation of air pollutant emission factors, Vol. II: Mobile sources, Jan 1991.
20. Morriss AP, Yandle B, Dorchak A. Regulating by litigation: the EPA's regulation of heavy-duty diesel engines. Admin L Rev 403; 2004.
21. Degobert P. Automobiles and pollution. Paris: Editions Technip; 1995.
22. U.S. Environmental Protection Agency. An examination of interim emission control strategies for heavy duty vehicles (a regulatory support document), Mar 1976.
23. France CJ. Substitution of percent load for manifold vacuum in the heavy duty gasoline test procedure. EPA report, Sept 1974.
24. Springer KJ. Baseline characterization and emissions control technology assessment of HD gasoline engines. Southwest Research Institute, Final report, EPA Contract EHS 70-110, Nov 1972.
25. Characterization and control of emissions from heavy duty diesel and gasoline fueled engines. Bureau of Mines, Final report, EPA IAG-0129(D), Dec 1972.
26. Wilbur Smith and Associates. Truck driving pattern and use survey Phase II. Final Report, Part I, EPA-460/3-77-009, June 1977.
27. Higdon L. Truck driving pattern and use survey Phase II. Final Report, Part II, Los Angeles. EPA HDV 78-03, May 1978.
28. Wysor T, France C. Selection of transient cycles for heavy-duty engines. Interim Report, U.S. EPA AA-HDV 77-01, Nov 1977.
29. France CJ. Transient cycle arrangement for heavy-duty engine and chassis emission testing. U.S. EPA HDV 78-04, 1978.
30. Brown JE, Harris DB, King FG Jr. Heavy-duty truck test cycles: Combining driveability with realistic engine exercise. Heavy Vehicle Syst. 2000;7:299–316.
31. Smith M. Heavy-duty vehicle cycle development. U.S. EPA, 460/3-78-08, July 1978.

32. Motor Vehicle Manufacturers Association. A review of the heavy-duty transient certification test cycle. Cambridge, MA: A.D. Little; 1981.
33. Giakoumis EG, Alysandratou A. Performance and emissions of a heavy-duty truck during the UDDS driving cycle—simulation analysis. ASCE, J Energy Eng. Article No. E4015011; 2016.
34. Rakopoulos CD, Giakoumis EG. Diesel engine transient operation. London: Springer; 2009.
35. Heywood JB. Internal combustion engine fundamentals. New York: McGraw-Hill; 1988.
36. Proposed gaseous emission regulations for 1983 and later model year heavy-duty engines: Regulatory analysis. U.S. EPA, Dec 1978.
37. Summary and analysis of comments to the NPRM: '1983 and later heavy-duty engines proposed gaseous emission regulations'. U.S. EPA, December 1979.
38. Personal communication with Mr. Chester France, September 2016.
39. Ullman TL, Hare CT. Emission characterization of a spark-ignited, heavy-duty, direct-injected methanol engine. EPA 460/3-82-003, Nov 1982.
40. U.S. Environmental Protection Agency. Final regulatory analysis: control of emissions from nonroad diesel engines. Office of Transportation and Air Quality, EPA 420-R-04-007, May 2004.
41. Environmental Protection Agency, 40 CFR Parts 85, 86, 600, et al. and Department of Transportation, National Highway Traffic Safety Administration, 49 CFR Parts 523, 534, and 535. Greenhouse gas emissions standards and fuel efficiency standards for medium- and heavy-duty engines and vehicles; Final rule. FR Vol. 76, No. 179, 15 Sept 2011.
42. Cartellieri WP, Ospelt WM, Landfahrer K. Erfüllung der Dieselabgasgrenzwerte von Nutzfahrzeug-Dieselmotoren der 90er Jahre. MTZ. 1989;50:440–51.
43. Wang J, Storey J, Domingo N, Huff S, Thomas J, West B. Studies of diesel engine particle emissions during transient operations using an engine exhaust particle sizer. Aerosol Sci Technol. 2006;40:1002–15.
44. J1667 SAE surface vehicle recommended practice. Snap acceleration smoke test procedure for heavy-duty diesel powered vehicles. Society of Automotive Engineers, Feb 1996.
45. Kitchen M, Damico W. Development of conversion factors for heavy-duty bus engines g/bhp-hr to g/mile. Technical report EPA-AA-EVRB-92-01, July 1992.
46. Gautam M, Clark NN, Riddle W, Nine R, Wayne WS, Maldonado H et al. Development and initial use of a heavy-duty diesel truck test schedule for emissions characterization. SAE Paper No. 2002-01-1753; 2002.
47. EPA GEM software. https://www3.epa.gov/otaq/climate/gem.htm. Accessed 29 Aug 2016.
48. Clark NN, Messer JT, McKain DL, Wang WW, Bata RM, Gautam M et al. Use of the West Virginia University truck test cycle to evaluate emissions from class 8 trucks. SAE Paper No. 951016; 1995.
49. Booz-Allen and Hamilton Inc. Energy use in ground transportation. Final report, U.S. Dept of Transportation DOT-TSC-UMTA-83-12, Urban Mass Transportation Authority UMTA-MA-06-0153-83-2, June 1983.
50. Dieselnet. https://www.dieselnet.com//standards/cycles. Accessed 11 Aug 2016.
51. U.S. Environmental Protection Agency. SmartWay fuel efficiency test protocol for medium and heavy duty vehicles. Working Draft, EPA420-P-07-003, Nov 2007.
52. Capps G, Franzese O, Knee B, Lascurain MB, Otaduy P. Class-8 heavy truck duty cycle project final report. Oak Ridge National Laboratory, ORNL/TM-2008/122, Dec 2008.
53. SAE J2711. Recommended practice for measuring fuel economy and emissions of hybrid-electric and conventional heavy-duty vehicles. Truck and Bus Powertrain Steering Committee, Sept 2002.
54. Japanese Industrial Safety and Health Association. Technical standard for 13-mode exhaust emission test procedure for diesel powered motor vehicles. JISHA 899; 1983.

55. Tachibana H. Heavy vehicle fuel efficiency. Energy Efficiency Standards Subcommittee of the Advisory Committee for Natural Resources and Energy. Final Report; 2006.
56. Takeuchi K, Yabumoto J, Okada Y, Kawai T, Montajir RM, Goto Y. A new dual-type DMA for measuring nanoparticles emitted from combustion engines. J Nanopart Res. 2005;7:287–93.
57. Global Technical Regulation No. 4. Test procedure for compression-ignition (C.I.) engines and positive-ignition (P.I.) engines fuelled with natural gas (NG) or liquefied petroleum gas (LPG) with regard to the emission of pollutants. Established in the Global Registry on 15 Nov 2006.
58. TRANS/WP29/GRPE/2001/2. Development of a worldwide harmonised heavy-duty engine emissions test cycle, Final report, Informal document No. 2, GRPE 42nd session, 28 May–1 June 2001, ECE-GRPE WHDC Working Group, Convenor: Dr. Cornelis Havenith, Author: Heinz Steven, Apr 2001.
59. Nakajima M. Harmonization activities of regulation and international standardization for vehicle exhaust emissions (Heavy Duty Vehicles). JSAE seminar on vehicle environment and safety issues, Jan 2005.
60. Engeljehringer K. Emission: heavy-duty and off-road—emission test systems. AVL-List GmbH; 2010.
61. Japan Automobile Manufacturers Association. 2011 report on environmental protection efforts; 2011.
62. Global Technical Regulation No. 10. Off-cycle emissions (OCE). Established in the Global Registry on 24 June 2009.
63. Effects of fuel composition and engine load on emissions from heavy duty engines. Department of the environment, Water, Heritage and the Arts, Orbital, Australia, Final report, July 2010.
64. Brown S, Bryett C, Mowle M. In-service emissions performance—drive cycles, Vol. 1. NSW Environment Protection Authority, Mar 1999.
65. Merkisz J, Pielecha J, Radzimirski S. New trends in emission control in the European Union. Springer tracts on transportation and traffic, Vol. 4. Cham: Springer; 2014.
66. Verbeek RP, Ligterink NE, Dekker HJ. Correlation factors between European and world harmonised test cycles for heavy-duty engines. TNO Report, MON-RPT-033-DTS-2008-03854, Dec 2008.
67. Lowell D, Kamakate F. Urban off-cycle NO_x emissions from Euro IV/V trucks and buses: Problems and solutions for Europe and developing countries. ICCT White Paper No. 18, Mar 2012.
68. Muncrief R. Comparison of real-world off-cycle NO_x emissions control in Euro IV, V, and VI. ICCT Briefing, Mar 2015.
69. Myung C-L, Ko A, Kim J, Choi K, Kwon S, Park S. Specific engine performance and gaseous emissions characteristics of European test cycle and worldwide harmonized driving cycle for a heavy-duty diesel engine. J Mech Sci Technol. 2013;27:3893–902.
70. Rickert G. Overview of emissions on world-harmonised and non-road cycles from the AECC Euro VI programme. AECC technical seminar on heavy-duty engine emissions, Brussels, 25 Oct 2007.

Chapter 5
Non-Road Engines

Abstract This chapter reviews the cycles used for the certification of non-road engines/vehicles in the world, which are all of the engine-dynamometer type. A distinction is made with respect to whether the cycles are steady-state or transient. For the former case, standard ISO 8178 describes most of the relevant procedures; for the latter, the non-road transient cycle, NRTC, is the applicable test. Historical review, 2D and 3D frequency diagrams and comparative results are presented for all non-road steady-state and transient cycles developed over the years, combined with results from actual engines operation.

In contrast to engines employed for on-road applications, engines for non-road, or off-road, use are produced in a much wider range of configurations and power outputs, and are employed in a variety of applications, ranging from small hand-held SI to large construction or marine (inland waterway vessels) diesel engines; Fig. 5.1 illustrates some examples regarding construction and agricultural machinery. Non-road vehicles, in effect wheeled industrial equipment, form a very unique vehicle category, manifesting very low velocities (much lower than their HDV counterparts). Their exact emission profile is determined by the individual operational/duty cycle, and not vehicular performance.

Emission limits for non-road engines were set at a later date compared to other vehicle types. In the United States, EPA was given authority under the 1990 Clean Air Act Amendments to regulate emissions from non-road engines. The first federal standards for new non-road diesel engines (defined as 'all internal combustion engines except motor vehicle (highway) engines, stationary engines, engines used solely for competition, or engines used in aircraft') were adopted in 1994 (Tier 1), phased in from 1996 to 2000 [1]; Tier 4 is the current standard, adopted in 2004 and phased-in between 2008 and 2015. The regulated pollutants are CO, NMHC, NO_x and PM. The relevant citation is 40 CFR 1039 for non-road CI engines, 1048 for large (>19 kW) and 1054 for small non-road SI engines, 1051 for recreational engines and vehicles, 1033 for locomotives, and 1042/43/45 for marine engines.

In the EU, the first directive dealing with emissions from non-road (diesel) engines was published in 1997 (97/68/EC). In this directive, non-road mobile

E.G. Giakoumis, *Driving and Engine Cycles*,
DOI 10.1007/978-3-319-49034-2_5

Fig. 5.1 Various non-road construction and agricultural machinery

machinery was defined as 'any mobile machine, transportable industrial equipment or vehicle with or without bodywork, not intended for the use of passenger or goods-transport on the road, in which an I.C. engine has been installed' [2]. Stage III and IV limits, effective at the time of writing, are harmonized with U.S. Tier 3 and 4 ones, and apply to engines with power between 19 and 560 kW [1]. In July 2016, the European Parliament and the Council of the EU adopted the final draft for the Stage V regulation, to be phased in from 2018/19 [3]; the standard incorporates particle number limits apart from the already regulated mass emissions of CO, HC, NO_x and PM, and also defines various engine categories, e.g., NRE: engines for mobile non-road machinery; NRSh: SI engines below 19 kW for use in hand-held machinery etc. (EU Regulation 2016/1628).

In Japan, regulated non-road engines are those with power between 19 and 560 kW, having license plates (special motor vehicles) or not (non-road motor vehicles). The current standards, although similar in stringency to the U.S. Tier 3 (2006–2008) and the EU Stage III/A (2005–2007), are not harmonized with U.S. and EU regulations. The standards do not require the use of exhaust after-treatment devices, such as diesel particulate filters [1].

The test cycles used for the certification of non-road equipment, all of the engine-dynamometer type, were initially stationary (Sect. 5.1). In the early 2000s, the transient NRTC cycle was developed, discussed in Sect. 5.2.

5.1 Steady-State Cycles

Steady-state cycles applicable to non-road engines are defined in the ISO 8178-4 standard, known as ISO 8178 cycles. These have been employed in Europe, the United States, Japan, China, India, Canada, South Korea and other countries [4].

According to the standard, the ISO 8178 cycles are used for the testing of reciprocating internal combustion engines intended for 'mobile, transportable and stationary use, excluding those for motor vehicles primarily designed for road use'. Owing to significant disparity in application and power range of non-road equipment, different versions of the 8178 exist, which are defined with a letter at the end, e.g., 8178 C is the cycle for off-road vehicle engines, 8178 E for marine engines etc. Furthermore, a number after the letter is used for additional classification, e.g., 8178 C1 is the cycle for diesel-engined off-road vehicles, 8178 D2 the cycle for generating sets with intermittent load etc.

Similarly to all stationary cycles, the ISO 8178 ones are formed from a combination of speed/load operating conditions/modes (load meaning torque or, in some cases, power), each with its own weighting factor. The engine under study undergoes the test by passing from each mode in a pre-defined order, allowing for adequate time to stabilize at each one. Emissions are measured at each mode, and the final amount of emitted pollutants is calculated based on the specific weighting factor of each point in g/kWh (see Eq. 6.30). Predictably, the tested points depend on the specific application and so do the corresponding weighting factors. For example, vehicular or locomotive engines are tested at idling, something that does not apply to engines intended to be used in power plants. Tables 5.1 and 5.2 detail the speed/load points for the various non-road engine applications covered by the ISO 8178 standard, namely

- off-road vehicles, such as excavators and crawlers (categories C1, C2);
- constant-speed engines (D1, D2);
- locomotives (F) (diesel engines for railcars can be tested on the 8178 C1);
- utility, lawn and garden (G1, G2 and G3);
- marine engines (E1, E2, E3, E4 and E5); and
- snowmobiles (H).

Most of the ISO 8178 cycles follow the same principles as the UNECE R49 13-mode steady-state HDV test cycle, known as ISO 8178 A (notice, for example, the similarities between the 8178 C1 and the ECE R49 in Table 4.2).

The engine types that belong to each category are itemized in Table 5.3. Further, Figs. 5.2 (non-marine) and 5.3 (marine engines and snowmobiles) illustrate schematically the test modes for each application/category; Fig. 5.4 provides an alternative illustration of the cycle modes, focusing on the power at each tested point for the non-propeller-law cycles.

Apart from the test modes of cycles E3, E4, E5 and H, which are calculated based on a propeller-law curve, the test modes of the other cycles can be combined into a universal cycle (ISO 8178 B) without weighting factors. From emissions data for each of the modes of this universal cycle, emission values for each of the other individual cycles may be calculated using the appropriate weighting factors for the respective application. By doing so, duplication of test runs can be avoided if the same engine is to be certified for different applications [4].

Table 5.1 Speed/torque points and weighting factors (%) of the ISO 8178 cycles for non-road engines (excl. marine engines with propeller law and snowmobiles); all test cycles are performed in ascending order of the mode number of the cycle in question [4]

MODE	1	2	3	4	5	6	7	8	9	10	11
Torque (%)	100	75	50	25	10	100	75	50	25	10	0
Speed	Rated Speed					Intermediate Speed					Low Idle
Off-road Vehicles											
C1	15	15	15	–	10	10	10	10	–	–	15
C2	–	–	–	6	–	2	5	32	30	10	15
Constant Speed											
D1	30	50	20	–	–	–	–	–	–	–	–
D2	5	25	30	30	10	–	–	–	–	–	–
Marine Applications (excl. propeller law)											
E1	8	11	–	–	–	–	19	32	–	–	30
E2	20	50	15	15	–	–	–	–	–	–	–
Locomotives											
F	15	–	–	–	–	–	–	25	–	–	60
Utility, Lawn and Garden											
G1	–	–	–	–	–	9	20	29	30	7	5
G2	9	20	29	30	7	–	–	–	–	–	5
G3	85	–	–	–	–	–	–	–	–	–	15

– engine torque is expressed as % of the maximum possible torque at a given engine speed (Fig. 1.27)
– rated speed is the speed at which the manufacturer specifies the rated engine power
– intermediate speed is the peak-torque speed, in general between 60 and 75 % of the rated speed
The most commonly tested point is the 100 % load at rated speed, and the least tested modes are the 10 and 25 % load at intermediate speed

Before sampling, a preconditioning is performed, which involves warming up of the engine under load conditions higher than 80 % in order to stabilize the engine parameters according to the recommendations of the manufacturer. Except for the category G cycles, the minimum test mode length is 10 min, and the gaseous exhaust emission concentration values are measured and recorded for at least 3 min anywhere in the mode if the engine is stabilized and meets the speed and torque requirements of the respective mode; only the last 60 s of the 3-min period are used for emissions calculation. For spark ignition engines in the category G cycles, when only gaseous emissions are measured, each mode lasts 3 min; the gaseous exhaust emission concentration values are measured and recorded for the last 2 min of the respective test mode [4].

For the case a regulator requires that emissions from the engine be controlled in areas not covered by the test cycle, the standard also defines the appropriate engine

Table 5.2 Speed/power or torque modes, and weighting factors (%) of the ISO 8178 cycles for marine engines following propeller law (category E), and snowmobiles (category H) [4]

E3	1	2	3	4	
Power	100	75	50	25	
Speed	100	91	80	63	
Weighting factor	20	50	15	15	
E4	1	2	3	4	5
Torque	100	71.6	46.5	25.3	0
Speed	100	80	60	40	idle
Weighting factor	6	14	15	25	40
E5	1	2	3	4	5
Power	100	75	50	25	0
Speed	100	91	80	63	idle
Weighting factor	8	13	17	32	30
H	1	2	3	4	5
Torque	100	51	33	19	0
Speed	100	85	75	65	idle
Weighting factor	12	27	25	31	5

The E3, E4 and E5 cycles are based on the earlier ICOMIA 5-mode cycle [5]

control area according to the application, namely: for the cases of (a) multiple-speed engines (C, E1, F), (b) constant-speed engines or, in general, engines operating very close to their design operating speed (D, E2, G), and (c) engines operated on a propeller curve (E3, E5, H).

Further to the above, a special duty cycle applicable in the United States for discrete mode testing of engines used in transport refrigeration units (TRUs) is defined in [6]. TRUs are units fitted to highway trucks used to transport refrigerated goods. These engines operate at two speeds, the low one to maintain the temperature when the compartment is sealed, and the high one when a higher cooling rate is needed (e.g., refrigeration compartment opened). Due to their practically steady-state nature of operation, the U.S. EPA decided to introduce in 2004 a two-speed cycle (instead of testing them on the ISO 8178 C1). As is demonstrated in Table 5.4, this cycle consists of four discrete modes, and is expected to be included in a future ISO 8178 update [7].

A special case of non-road machinery is locomotive engines. These diesel engines are typically connected in a series arrangement to an electrical generator or alternator that converts mechanical energy to electricity; the latter is used to power traction motors which turn the wheels. The effect of this arrangement is that the diesel engine operation is not constrained by the locomotive speed. Thus, the engine can be operated at a desired power output and speed irrespective of vehicle

Table 5.3 Engines that fit into the ISO 8178 categories (the list in not exhaustive) [4]

Class	Application	Specifications	Typical examples
C1	Off-road vehicles	diesel powered off-road industrial equipment	industrial drilling rigs, compressors; construction equipment incl. wheel loaders, bulldozers, crawler tractors and loaders, truck-type loaders, dumpers, hydraulic excavators; agricultural equipment, rotary tillers; forestry equipment; self-propelled agricultural vehicles; material handling equipment; fork-lift trucks; road maintenance equipment (graders, road rollers, asphalt finishers); snow plough equipment; snow tractors; airport supporting equipment; aerial lifts; mobile cranes
C2		SI powered off-road industrial equipment >20 kW	fork-lift trucks; airport supporting equipment; material handling equipment; road maintenance equipment; agricultural equipment
D1	Constant speed	power plants	
D2		generating sets with intermittent load	gas compressors; irrigation pumps; generating sets with intermittent load including generating sets on board of ships and trains (not for propulsion), refrigerating units, welding sets; turf care, chippers, snow removal equipment, sweepers
E1	Marine applications (excl. propeller law)	diesel engines for craft less than 24 m in length except those of tug boats and push boat	
E2		heavy-duty constant-speed engines for ship propulsion without length limitation	
E3	Marine applications (propeller law)	heavy-duty marine engines for ship propulsion without length limitation	
E4		spark ignition engines for craft less than 24 m in length except for tug boats and push boats	
E5		diesel engines for craft less than 24 m in length when operated on a propeller law except for tug boats and push boats	
F	Locomotives	locomotives; railcars; shunting locomotives	
G1	Utility, lawn and garden	non-hand-held intermediate-speed applications	pedestrian-controlled rotary or cylinder lawn mowers; front or rear engine riding lawn mowers; rotary tillers; edge trimmers; lawn sweepers; waste disposers; sprayers; snow removal equipment; golf carts
G2		non-hand-held rated-speed applications	portable generators, pumps, welders, air compressors; lawn and garden equipment at engine-rated speed
G3		hand-held rated-speed applications	edge trimmers; string trimmers; blowers; vacuum equipment; chain saws; portable saw mills
H	Snowmobiles		

speed. The range of possible combinations of locomotive speed and engine power vary from the engine at idle speed with no propulsion power and the locomotive at maximum speed to the engine at rated power and speed and the locomotive speed approaching zero. Following the lack of a direct, mechanical connection between CI engine and wheels, the engine practically operates at steady-state conditions, which are defined in a number of discrete modes (power settings) or notches [8].

Prior to the ISO 8178 F locomotive cycle, a dedicated European railroad cycle existed, consisting of four modes: (1) nominal speed/100 % power, (2) intermediate speed 1/50 % power, (3) intermediate speed 2/25 % power, and (4) idle speed/5 % power, with weighting factors 20, 10, 10 and 60 % respectively [5]. The cycle had been developed by the Union Internationale des Chemins de Fer (UIC).

For testing locomotives in the United States, use is made of two steady-state cycles, detailed in Table 5.5; these are more elaborate compared to the ISO 8178 F cycle described in Table 5.1. The U.S. cycles further distinguish between line-haul and switch locomotive operation. The term line-haul refers to the movement of freight between cities; switching, on the other hand, refers to the process of assembling and disassembling trains in a relatively small area (known as switch-yard) [8]. Notice the idle operation weighting factor of the switch test cycle, of the order of 60 %, being similar to the one defined for the ISO 8178 F locomotive cycle.

Ramped Modal Cycles

UNECE Global Technical Regulation (GTR) No. 11 [9] as well as U.S. regulations [6] incorporate, apart from the steady-state tests, the ramped modal versions of these cycles. Ramped modal test cycles (RMC) are cycles where gaseous and PM emissions are measured and sampled continuously in a 'pseudo-transient' manner, with the engine operating fully warmed-up. In order to avoid extreme changes in the after-treatment temperature during the ramped modal cycle test, the mode sequence may not be the same as the corresponding steady-state discrete mode cycle.

The time length of the modes is selected in such a way to be equivalent to the weighting factors of the corresponding discrete steady-state test. The change in engine speed and load from one mode to the next one has to be linearly controlled within 20 ± 1 s; this mode change time is part of the new mode (including the first mode). As already mentioned in Chap. 4, a RMC test cycle is better suited to report engine emissions from PM traps and NO_x adsorbers more reliably and consistently than the respective steady-state cycle [6]. Table 5.6 for variable-speed engines and Table 5.7 for constant speed engines detail two examples of ramped modal cycles, graphically illustrated in Fig. 5.5. Regulations in the United States define RMCs for

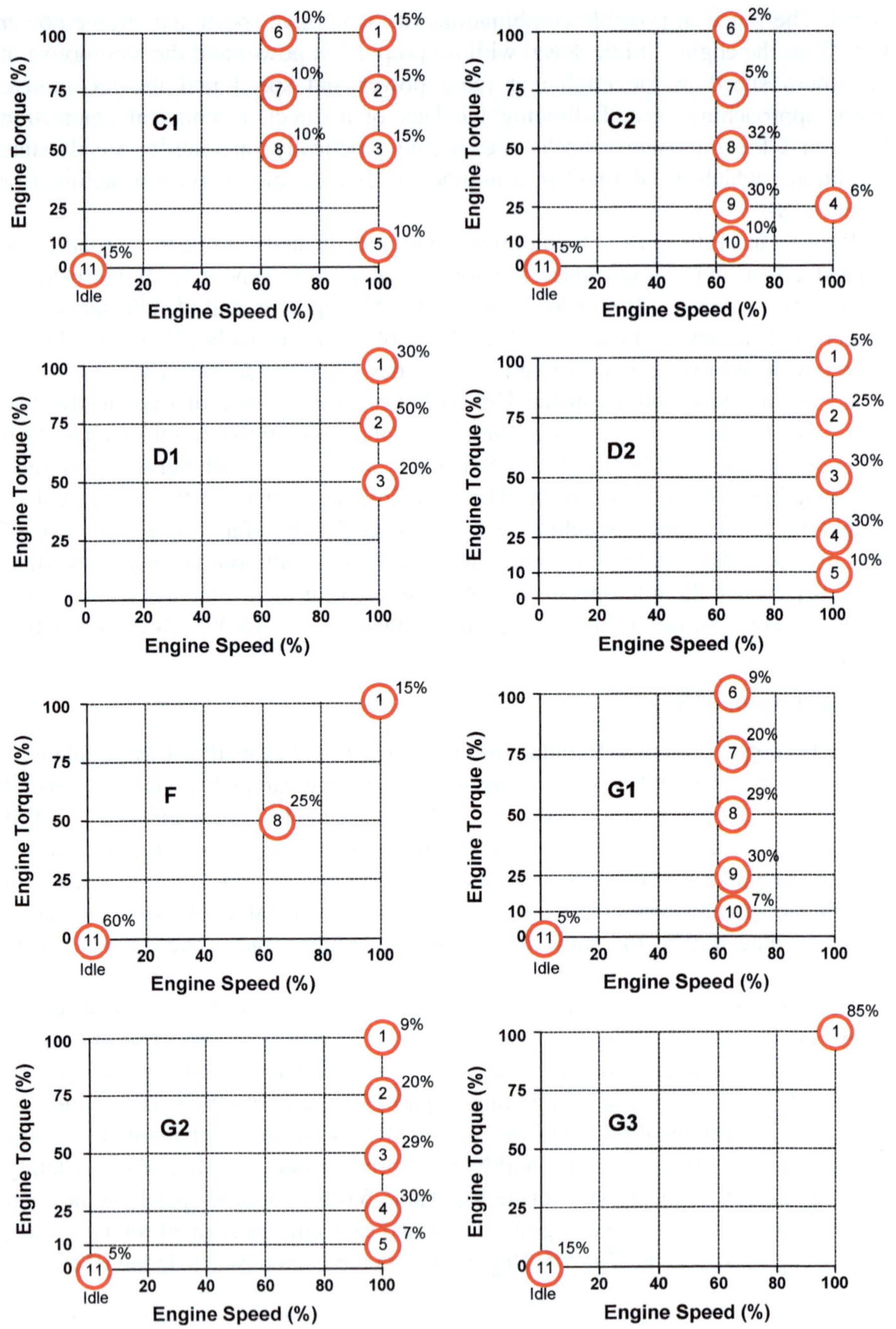

Fig. 5.2 Engine speed/torque points of categories *C* (off-road vehicles), *D* (constant speed), *F* (locomotives) and *G* (utility, lawn and garden) of the ISO 8178 cycles; percentages denote weighting factors for each mode

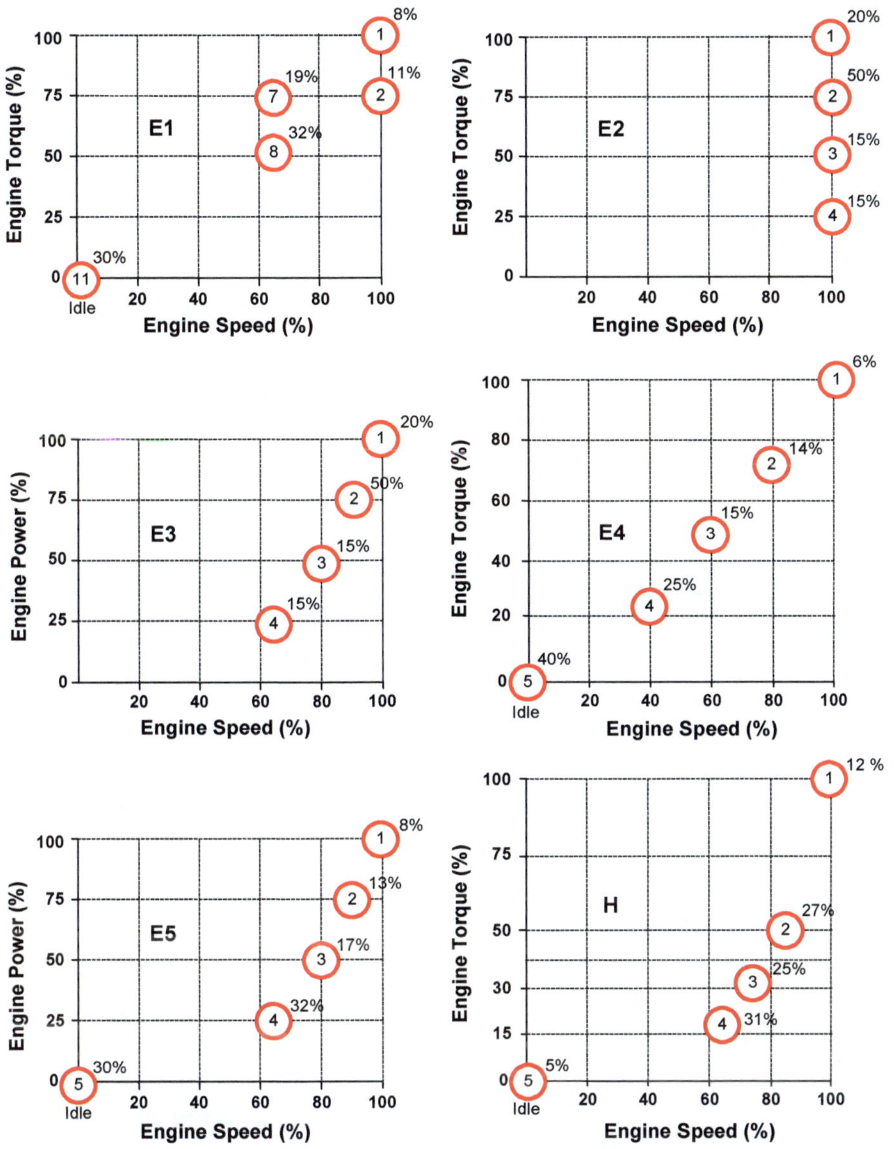

Fig. 5.3 Engine speed/power or torque points of category *E* (marine engines) and *H* (snowmobiles) of the ISO 8178 cycles; percentages denote weighting factors for each mode

the ISO 8178 G2 (variable speed engines below 19 kW), and the TRU operation (40 CFR 1039.645), as well as for locomotive engines (40 CFR 1033.520). It is expected that RMCs will be part of a future ISO 8178 amendment [7].

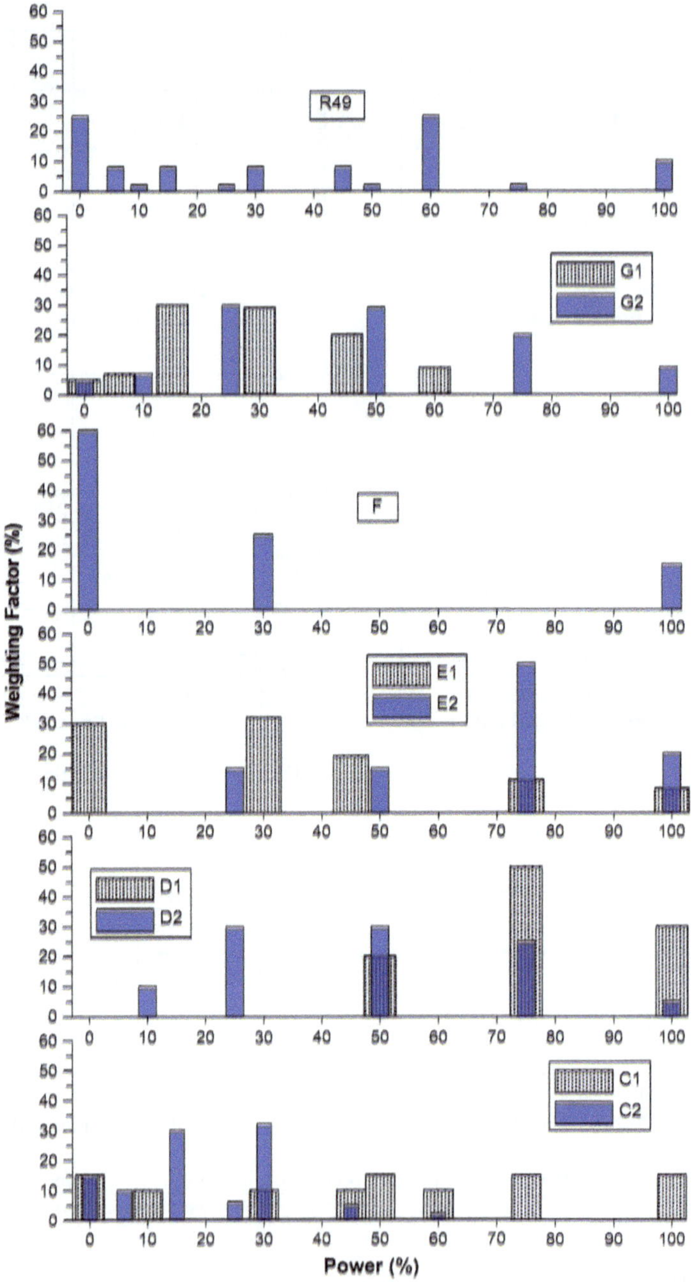

Fig. 5.4 Comparative illustration of the non-road ISO 8178 and HDV R49 modes' power and weighting factors (assumption: intermediate speed is 60 % of rated)

Table 5.4 U.S. EPA cycle for transport refrigeration units (TRUs) [6]

Mode	Engine speed	Engine torque (%)	Weighting factor (%)
1	Maximum	75	25
2	Maximum	50	25
3	Intermediate	75	25
4	Intermediate	50	25

Table 5.5 U.S. diesel-electric locomotive test cycles (40 CFR 1033.515 and 530); all-electric locomotives are not subject to these tests

Test Mode	Throttle setting	Duration (min)	Weighting factors (%)	
			Line Haul	Switch
Warm-up	Lowest idle	10–15	–	–
A	Low idle	5–10	19.0	29.9
B	Normal idle*	5–10	19.0	29.9
C	Dynamic braking	5–10	12.5	0
1	Notch 1	5–10	6.5	12.4
2	Notch 2	5–10	6.5	12.3
3	Notch 3	5–10	5.2	5.8
4	Notch 4	5–10	4.4	3.6
5	Notch 5	5–10	3.8	3.6
6	Notch 6	5–10	3.9	1.5
7	Notch 7	5–10	3.0	0.2
8	Notch 8	10–15	16.2	0.8

*for applications with only one idle speed, the weighting factor is doubled

For the line-haul cycle: if the locomotive is equipped with two idle settings and no dynamic brake, the normal-idle weighting factor is 31.5 %; if the locomotive is equipped with only one idle setting and no dynamic brake, the idle weighting factor is 50.5 %

5.2 Transient Cycles

The previously discussed steady-state duty cycles provide reasonably good coverage of the range of activity seen by non-road diesel applications. Particularly the mid-to-high speeds and loads represented by steady-state testing requirements, are the speeds and loads at which many of these engines are designed to operate for extended periods for maximum efficiency and durability (e.g., excavator and agricultural tractor). However, stationary cycles are incomplete, in the sense that the range of non-road activity is much broader and much more varied than what can be captured by a set of steady-state points (see Fig. 5.10 later in this chapter) [10]. It is further acknowledged that the contribution of the highly transient turbocharged diesel engine operation of non-road equipment to the total exhaust emissions is significant. In light of these facts, and as was the case with the heavy-duty engines, research was initiated on a transient cycle for non-road engines. Obviously, no

Table 5.6 Thirty-minute ramped modal cycle based on the ISO 8178 C1 test for variable speed engines (notice that the order of modes is different from the discrete-mode test in Table 5.1, and the idle mode is split, hence referred to as 9-mode RMC) [6, 9]

RMC mode	Time (s)	Engine speed	Engine torque (%)
1a—Steady state	126	Warm idle	0
1b—Transition	20	Linear transition	Linear transition
2a—Steady state	159	Intermediate	100
2b—Transition	20	Intermediate	Linear transition
3a—Steady state	160	Intermediate	50
3b—Transition	20	Intermediate	Linear transition
4a—Steady state	162	Intermediate	75
4b—Transition	20	Linear transition	Linear transition
5a—Steady state	246	Rated	100
5b—Transition	20	Rated	Linear transition
6a—Steady state	164	Rated	10
6b—Transition	20	Rated	Linear transition
7a—Steady state	248	Rated	75
7b—Transition	20	Rated	Linear transition
8a—Steady state	247	Rated	50
8b—Transition	20	Linear transition	Linear transition
9—Steady state	128	Warm idle	0

Table 5.7 Twenty-minute ramped modal cycle based on the ISO 8178 D2 test for constant speed engines (notice that the order of modes is different from the discrete-mode test in Table 5.1) [6, 9]

RMC Mode	Time (s)	Engine speed	Engine torque (%)
1a—Steady state	53	Engine governed	100
1b—Transition	20	Engine governed	Linear transition
2a—Steady state	101	Engine governed	10
2b—Transition	20	Engine governed	Linear transition
3a—Steady state	277	Engine governed	75
3b—Transition	20	Engine governed	Linear transition
4a—Steady state	339	Engine governed	25
4b—Transition	20	Engine governed	Linear transition
5—Steady state	350	Engine governed	50

single transient cycle could capture the full spectrum of non-road diesel operation from all relevant equipment applications. Nonetheless, it is possible to capture typical operation of non-road engines, and extrapolate the applicability to other similarly-operating equipment for certification and modeling purposes. This is crucial in order to drive the development of engine design strategies to focus on emission control on a wider set of activity [10, 11].

During the second half of the 90s, two research endeavors were conducted in the United States by EPA and its subcontractors focused on studying and characterizing

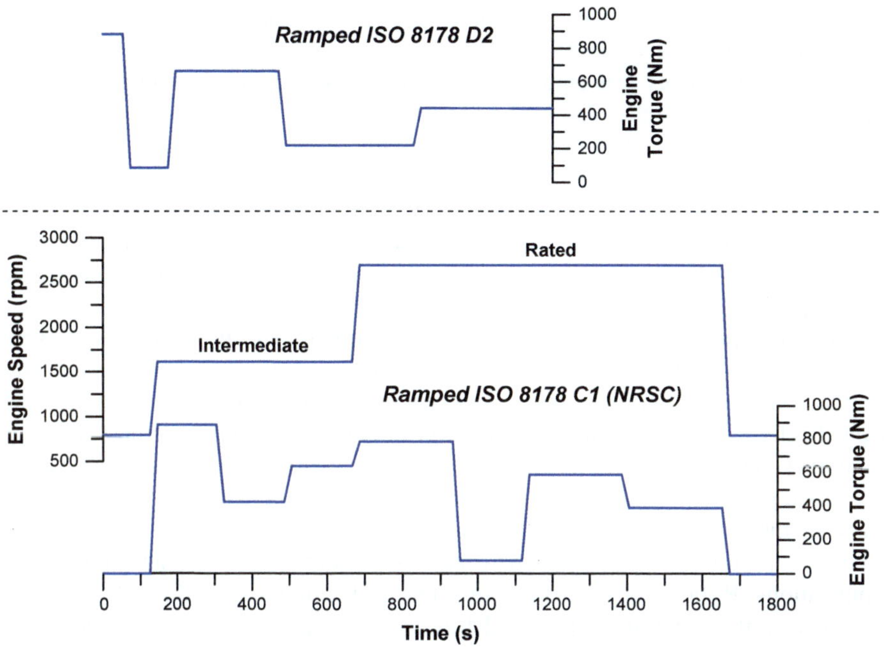

Fig. 5.5 Illustration of the ramped ISO 8178 C1 from Table 5.6 and ISO 8178 D2 from Table 5.7 for a turbocharged diesel engine

the duty cycles of typical diesel engines used in various non-road mobile applications [12, 13].

In [12], three pieces of non-road equipment were instrumented to obtain data on how the engine operated during work activities, commonly performed by each application, as these were defined by their manufacturers. More specifically, a 85 kW crawler tractor, a 77 kW backhoe loader and a 190 kW agricultural tractor were selected. The manufacturers weighted the activities for each application according to the fraction of time they estimated the equipment spent performing each of the various activities. For example, a crawler tractor was projected to spend 34, 47, 11 and 8 % of its time performing activities categorized as pit activity, road-bed preparation, clearing, and idling respectively. Resulting speed and torque data from the three instrumented vehicles over a period from 2 to 4 days were separated by activity for each corresponding piece of machinery, and characteristics of engine operation were established. For each activity, a micro-trip was chosen and validated to be representative of the larger body of data for that activity. The duration of the representative micro-trip was proportional to the weighting given the activity for the piece of equipment, relative to the total length of the test cycle being developed for that piece of equipment. EPA established that the backhoe cycle was to be 8 min, the crawler 12, and the agricultural tractor 10 min. Validated micro-trips were then assembled to build a cycle representative of the corresponding engine application after normalizing the speed and torque data. With this approach, representative cycles for the crawler tractor, the backhoe/loader, and the agricultural tractor were developed [12].

A different strategy was employed for a second set of non-road duty cycles; this approach was termed the 'day-in-the-life' strategy. Instead of asking the equipment operators to perform typical duties characteristic of each application activity, data was collected during work that would have otherwise been done during those days. Four duty cycles were developed based on this approach, namely the arc welder cycle, the rubber-tire loader cycle, the excavator cycle (assembled from three equal time segments of operating data from two different excavators), and the skid-steer loader [10, 13].

The previously mentioned seven non-road component application duty cycles were selected on the basis of frequency of finding such engines in the non-road population, and summing those with engines/equipment doing related work. Specifically, agricultural tractors were seen to have operations generally similar to combines and off-highway trucks, in addition to tractors. Arc welders represented the broad group of constant-speed applications. The backhoe-loader group included most of the lawn/garden/commercial turf tractors, commercial lifts and sweepers. The crawler/dozer application matched with other dozer, grader and scraper applications. Rubber-tire loaders were found to be similar to industrial and rough terrain forklifts, aircraft support and forestry equipment. Skid-steer loaders were seen, at the time, as a unique application/category. Finally, excavators and cranes were grouped together as similar applications. In any case, the body of data for neither approach was meant to be all encompassing to the extent that no other activity would be expected from that piece of equipment over its lifetime [10].

Representative cycles from each equipment are illustrated in Figs. 5.6 as regards speed profile, 5.7 torque profile and 5.8 speed/torque distribution. Further, Tables 5.8 and 5.9 demonstrate various technical specifications of these cycles.

The following remarks can be made with reference to Tables 5.8 and 5.9 and Fig. 5.8:

- Most of the cycles have 20-min duration; the shortest duration is noticed for the backhoe loader (515 s) and the agricultural tractor (629 s).
- The arc welder has noticeably the highest average speed (91 %), indicative of its stationary-like duty cycle, followed by the excavator (86 %) and the agricultural tractor (79 %). The backhoe loader and the wheel-loaders, on the other hand, exhibit the lowest average engine speed (33–37 %). For all cycles, the maximum normalized engine speed is close to 100 %, and for some of them even higher.
- As regards the achieved torque, the backhoe loader and the arc welder are the extreme low ones (approx. 25 %), with the agricultural tractor having the highest average torque of the order of 79 % and the excavator 70 %. It can be concluded, therefore, based also on the speed/torque distribution in Fig. 5.8 that, generally speaking, the arc welder operates at very high engine speeds and low loading, whereas the agricultural tractor and the excavator at both high speeds and loads.

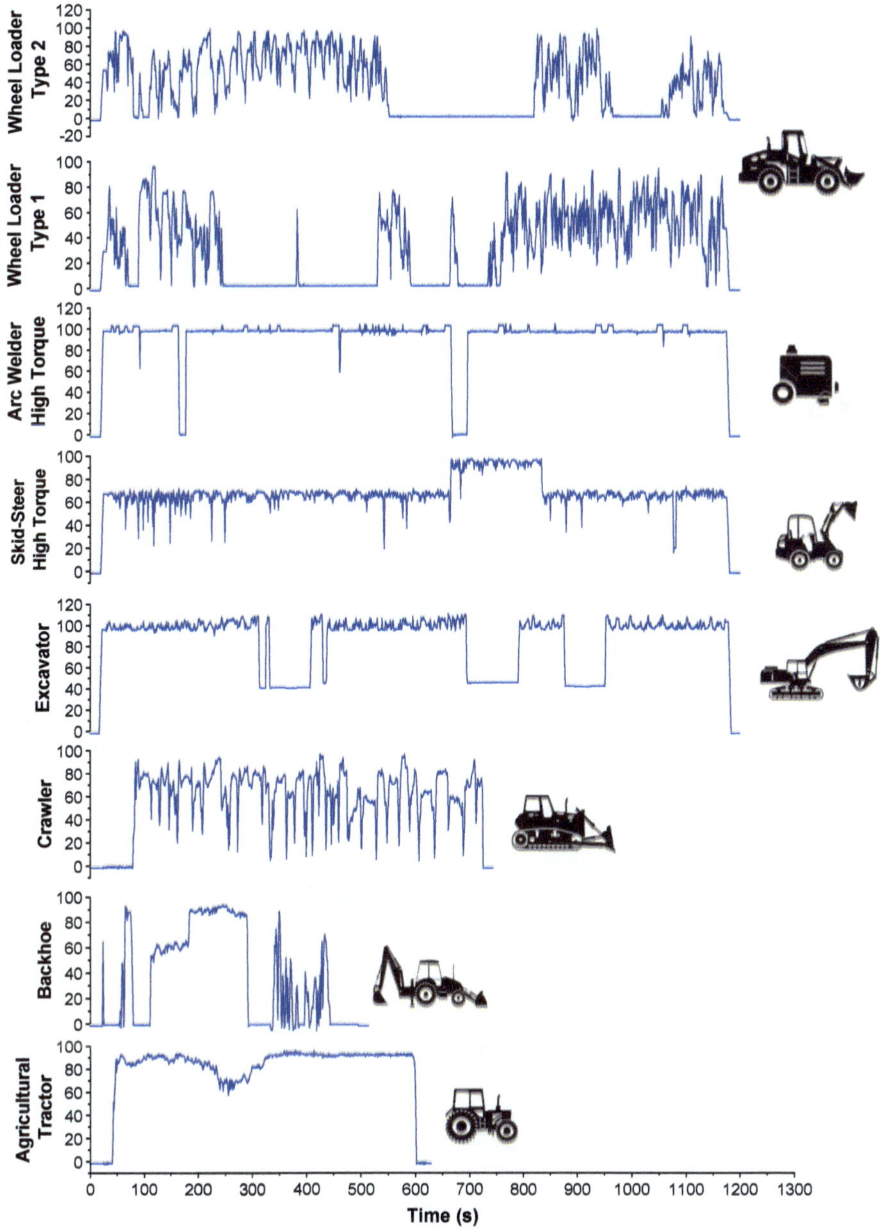

Fig. 5.6 Normalized engine speed profile of various U.S. non-road test cycles [14]

- The idling period is generally short for all cycles (at least compared to their on-road counterparts in Chap. 4), and so is the motoring phase, with the exception of the wheel loaders. This is justified based on the different duty cycle of non-road equipment compared to on-road vehicles.

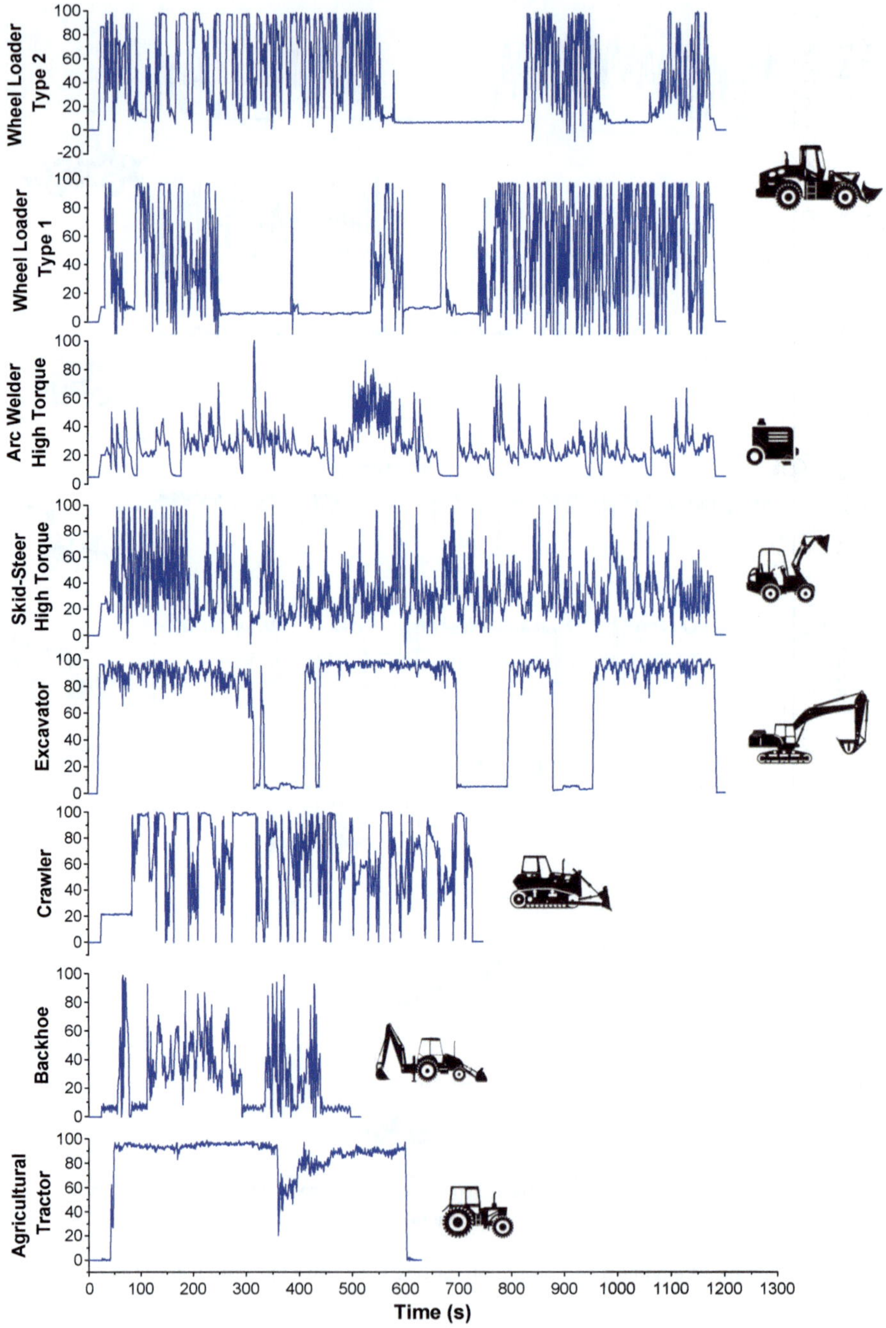

Fig. 5.7 Normalized engine torque profile of various U.S. non-road test cycles [14]

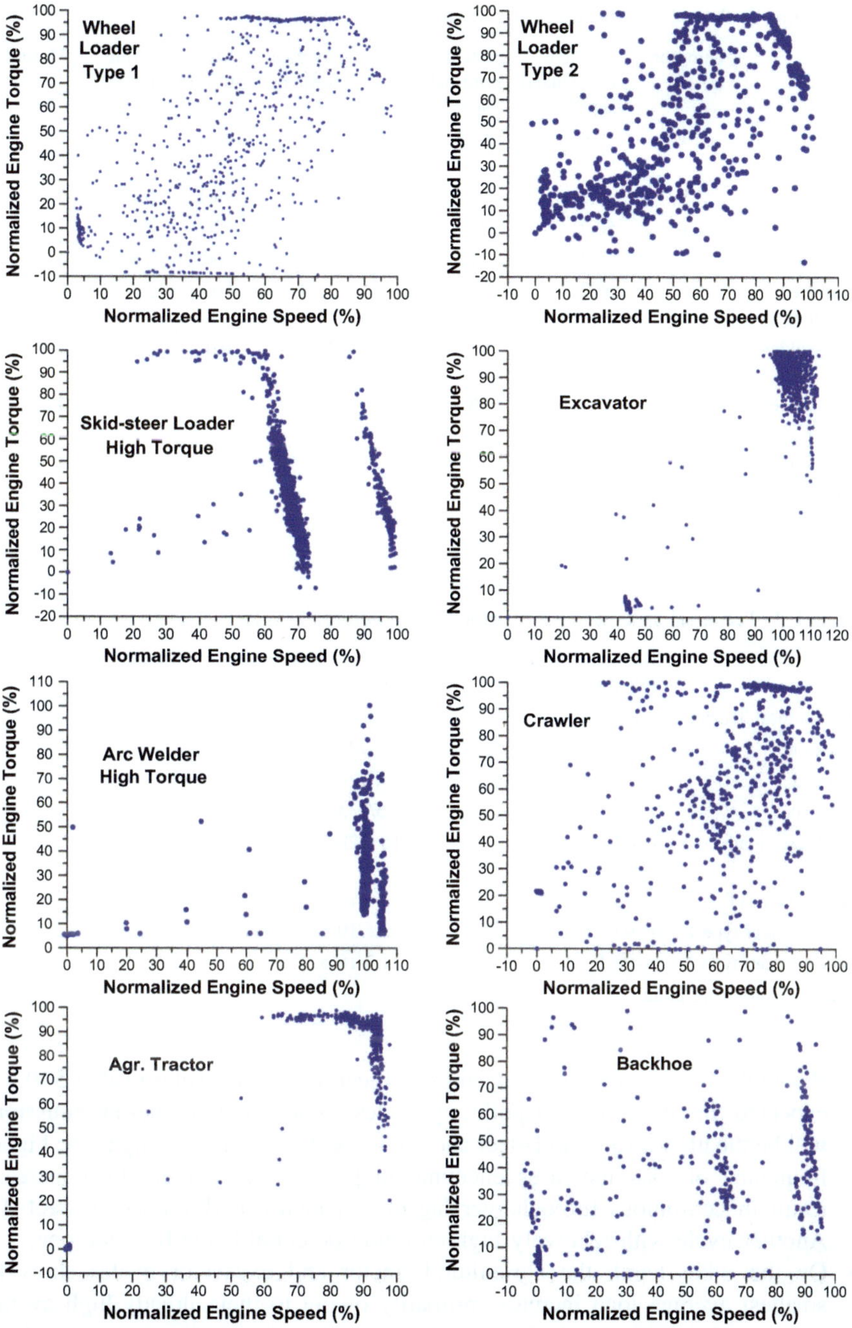

Fig. 5.8 Normalized speed/torque distribution of the U.S. non-road cycles from Figs. 5.6 and 5.7

Table 5.8 Technical specifications of various U.S. non-road cycles

Cycle	Duration (s)	Average normalized speed (%)	Maximum normalized speed (%)	Average torque (%)	Max. torque (%)	Idling period (%)	Motoring period (%)
Agricultural Tractor	629	78.9	97.9	78.6	98.9	5.2	1.1
Backhoe	515	35.7	95.5	25.2	99.1	8.2	0
Crawler	744	58.2	98.6	60.3	100	5.6	0
Excavator	1199	86.2	112.7	69.8	100	2.8	0
Skid-steer loader HT	1199	68.2	99.4	32.8	100	3.3	0.3
Arc welder HT	1199	90.9	96	25.7	101	0	0
Wh. loader type 1	1199	33.1	98	34.7	97.7	3.3	4.6
Wh. loader type 2	1199	37.6	100.4	39.1	99.1	3.3	1.3

HT high torque

Table 5.9 Technical specifications of various U.S. non-road cycles

Cycle	Maximum load (%) Increase	Average load increase (%)/Load increases (min^{-1})
Agricultural Tractor	1.36–56.9	3.7/24.6
Backhoe	0–92.8	12.2/25.4
Crawler	0–100	19.7/20.7
Excavator	0–19.4	5.5/23.0
Skid-steer loader HT	m–42.3	16.6/28.2
Arc welder HT	5.6–53	8.5/17.3
Wheel loader type 1	m–97.2	23.6/19.5
Wheel loader type 2	m–98.1	21.3/18.6

m denotes motoring point

- If we take into account the transient specifications of the cycles (Table 5.9), it is expected that the most PM polluting cycles (concerning engine-out emissions) will be the high-torque skid-steer and the wheel loader ones owing to the highest frequency of load increases encountered [15]; these frequent load increases result in continuous turbocharger lag discrepancies as discussed in Sect. 1.2. Another cycle with relatively high emission potential is the backhoe one.
- On the other hand, the agricultural tractor and excavator cycles show the smallest average load increase, primarily owing to their already high average speed and torque level. During these cycles, the engine operates almost

continuously at high speeds and loads (average values of more than 70 %), inevitably leading to rather low speed/load changes. As a result, it is expected that these vehicles will produce relatively low PM owing to the small magnitude of the turbocharger lag phases that are primarily responsible for elevated particulates in turbocharged diesel engines.
- The wheel-loader and the crawler cycles have the broadest speed/torque distribution; on the other hand, the distribution of the excavator, agricultural tractor, arc welder and skid-steer loader is considerably narrower as Fig. 5.8 indicates.

Having all seven application cycles in hand, including different versions for some of them (arc welder, skid-steer loader and rubber-tire loader, in the form of high-speed or high-torque variants), in total 16 cycle versions [14], work began on the creation of a composite non-road cycle. Initially, assigning different weighting factors for each equipment in the composite cycle was being considered, based on population or inventory grounds. However, it was ultimately decided that the weighting of contributions from each equipment type to the composite cycle be equal. An important decision was the selection of the appropriate data segments from each component to be incorporated in the composite cycle. This task was primarily accomplished based on uniqueness of operation (avoidance of replicate data in the composite cycle), and level of transient operation. To this aim, load or speed changes lower than 2 % (essentially considered stationary operation), were not included in the final cycle. In order to link the various extracted cycle segments, transition periods of 3-s duration were added when needed to ensure smooth and realistic transitions within the cycle. Such transition periods were considered necessary when the change in the magnitude of the speed or torque value was greater than 20 percentage points (normalized data) [10].

The next step was the harmonization, in order for the cycle to be adopted initially in both the United States and Europe, and perhaps universally. Apart from technical staff from the United States and the EU, representatives from Japan's government periodically participated in this process as well. Throughout the development of the cycle, manufacturers associations from Europe (EUROMOT), the United States (EMA) and Japan (LEMA) provided detailed technical input. Firstly, data-collection effort confirmed that, as expected, non-road diesel activity in Europe was comparable to the results obtained in the United States. It was also important to ensure that concerns about global suitability were addressed. The context used for this assessment in Europe was the existing ETC (Fig. 4.11). While this cycle was developed for on-road diesel applications, it was considered as an adequate basis for which European industry and government staff could assess EPA's proposed non-road cycle. After modifications suggested by the European Commission (e.g., not allowing the instantaneous transient speed and torque changes to be greater than those experienced within the European ETC), the resulting composite non-road cycle, known as the non-road transient cycle, or NRTC, was finalized in 2002 [10]; the cycle, illustrated in Fig. 5.9, is detailed in:

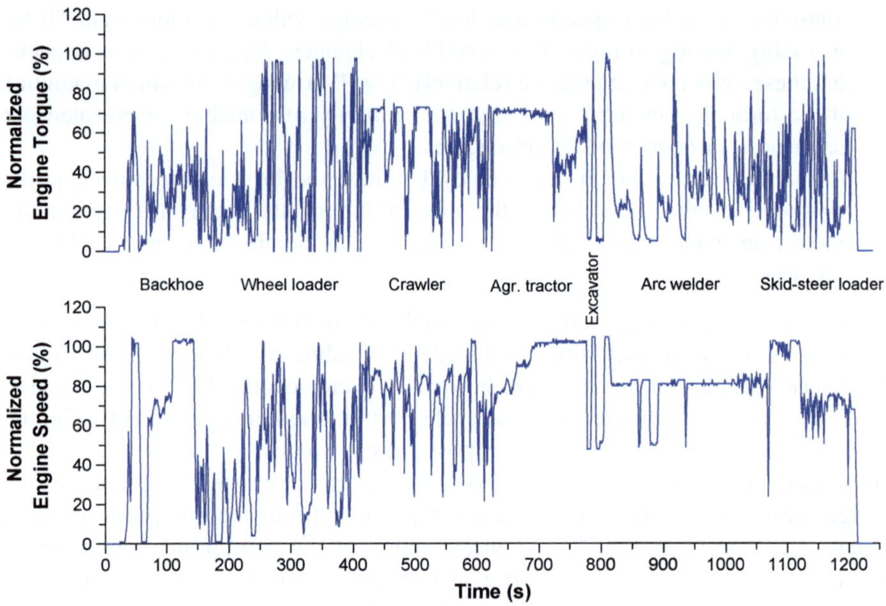

Fig. 5.9 Engine speed and torque profile of the non-road NRTC composite transient cycle [9]

- European Directive 2004/26/EC, published on April 21, 2004, amending the initial non-road engines Directive 97/68/EC [16]. The last amendment is Regulation 2016/1628/EU;
- Federal Register Vol. 69, No. 124, June 29, 2004, 'Control of emissions of air pollution from non-road diesel engines and fuel; Final Rule' [6], and 40 CFR 1039, with the emission procedure detailed in 40 CFR 1065; and
- UNECE Global Technical Regulation No 11 (2010) regarding 'Engine emissions from agricultural and forestry tractors and from non-road mobile machinery' [9], and the implementation of its content in Regulation 96/03.

The NRTC is an engine dynamometer transient schedule of 1238 s total duration. The cycle has two idling periods, 23 s at the beginning and 25 s at the end (reduced from the initial U.S. version's 28 and 34 s), for a total idling phase of 3.9 %. Average normalized engine speed is 68 % and torque 39 % (changed from the original 63 % and 47 % respectively of the U.S. version).

As mentioned earlier, the NRTC is based closely on the operating conditions of diesel engines installed in various non-road applications, as presented in Figs. 5.6 and 5.7, with duty cycles from backhoe, wheel loader, crawler, agricultural tractor, excavator, arc welder and skid-steer loader having been used to form the cycle. Close inspection of the various segments of the NRTC in Fig. 5.9 indicates the parts that have been used from the individual on-road equipment cycles from Figs. 5.6 and 5.7. As evidenced in Fig. 5.9, the cycle is highly transient; hence, testing over the NRTC shifted new engine design and calibration to the dynamic operation compared with the steady-state, which was the case with the NRSC/ISO 8178 C1. In fact, Fig. 5.10 demonstrates in an unambiguous way the significant dissimilarities encountered when the engine is certified using the transient cycle compared to the steady-state one or even its ramped version.

The NRTC, being a composite cycle formed from duty cycles of common non-road machinery, can be considered typical for non-road mobile engines as it includes specific segments indicative of the duty cycle from many of them. However, the data comes exclusively from the United States (European data confirmed the American), and most of the instrumented engines were American as well. Quite surprisingly, rather few instrumented vehicles were used (practically only one for each non-road activity), unlike the vast amount of data used in the development of the WHTC or the WLTC (obviously, the limited amount of engines used for the development of the NRTC can be attributed to the rather low disparity in the operating schedule for each specific non-road application). In any case, no composite cycle could realistically be representative of each and every engine application it intends to cover. For example, and as observed in Figs. 5.6, 5.7 and 5.8, the typical speed and torque profile of the agricultural tractor (actually, more stationary than transient) is fundamentally different from that of the wheel loader.

In order for the NRTC to be reproduced in the laboratory, a denormalization procedure needs to be performed first. Similar to the heavy-duty cycles discussed in Chap. 4, for the specific engine under transient testing, one option for speed denormalization is using the following equation [9]

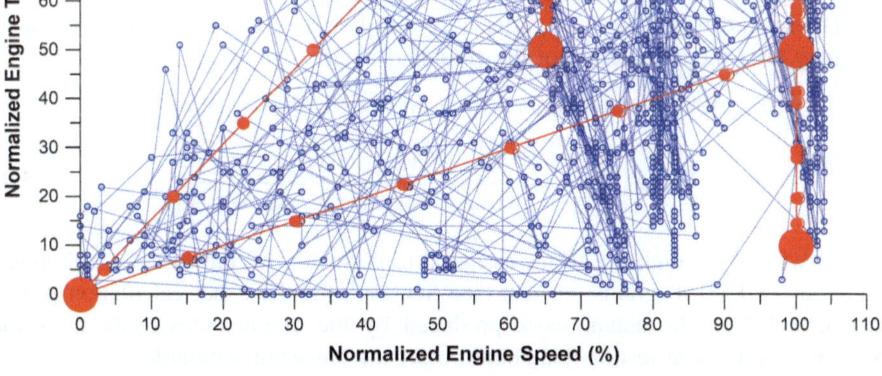

Fig. 5.10 Comparison of the normalized engine speed/torque distribution between the NRTC, the steady-state NRSC (ISO 8178 C1) and the ramped NRSC

$$N = \frac{N_{norm}(N_{ref} - N_{idle})}{100} + N_{idle} \tag{5.1}$$

with N the actual speed (rpm), N_{norm} the normalized one, and the reference speed N_{ref} corresponding to the 100 % speed values specified in the engine dynamometer schedule, defined as follows:

$$N_{ref} = N_{lo} + 95\,\%(N_{hi} - N_{lo}) \tag{5.2}$$

with N_{hi} the highest engine speed above the rated one, where 70 % of the declared (by the manufacturer) maximum power occurs, and N_{lo} the lowest engine speed below the rated one, where 50 % of the declared maximum power occurs (see Fig. 4.7). U.S. regulations for speed denormalization are defined in 40 CFR 1065.610.

Similarly, engine torque is denormalized using the following equation (M is the actual torque (Nm) and M_{norm} the normalized one)

$$M = \frac{M_{norm} \times M_{max}}{100} \tag{5.3}$$

with the maximum torque value M_{max} at each speed found from the respective engine mapping curve.

The NRTC is run twice, the first time with the engine cold started and the second with the engine hot started. Taking into account cold starting effects is another important field where the transient testing differentiates from its steady-state predecessor. With the application of emission-control devices having optimal temperature operating windows, the cold-start test has become a design challenge for non-road engine manufacturers, just as it has been a design challenge for light-duty and heavy-duty vehicle manufacturers. In particular, the NRTC is run:

- The first time (cold start), after the engine has soaked to room temperature and the engine coolant and oil temperatures, after-treatment systems and all auxiliary engine control devices are stabilized between 20 and 30 °C.
- The second time (hot start), after a 20-min hot soak that commences immediately after the completion of the cold-start cycle.

The final emission is determined from the weighted average of the two runs [9]

$$E_{wi}\,(g/kWh) = \frac{wf_{cold} \cdot M_{i,cold} + wf_{hot} \cdot M_{i,hot}}{wf_{cold} \cdot W_{act,cold} + wf_{hot} \cdot W_{act,hot}} \tag{5.4}$$

where wf is the weighting factor assigned to each run, M_i the mass of pollutant i (gaseous or PM) in grams per test (see Sect. 6.5 for the exact calculation procedure), and W_{act} the actual work produced by the engine. Obviously, it is the weighted composite results that should meet the emission standards.

The weighting factors (hot–cold) are 90–10 % in the EU [16] and 95–5 % in the United States [6]. The weighting of the cold-starting emissions on the composite cycle results might seem rather small. Nonetheless, and since the typical daily use of non-road engines extends over many hours, the effect of cold starting on the total exhaust emissions is indeed limited (the equivalent of a light-duty vehicle covering a very long distance every day). More specifically, the legislated percentage weight for the cold-start run of the cycle was based on a typical scenario of engine operation involving an overnight soak, 20-min warm-up and a total of seven hours of operation over the course of a workday [10].

In the U.S., the NRTC was introduced together with the EPA Tier 4 emission standards in 2004. Its implementation started in the model year that the trap-based Tier 4 PM standards and/or adsorber-based Tier 4 NO_x standards were first applied, i.e., with MY 2011 for 130–560 kW engines, MY 2012 for 56–130 kW and MY 2013 for engines below 56 kW. Standards have still to be met on the 8-mode ISO 8178 C1; engines above 560 kW are not tested on the transient test [10].

A separate test for constant-speed engines, based upon the operation of a 30 kW, naturally aspirated, diesel-engined arc welder, was also proposed at some point. This cycle was intended for pumps, electrical power generator sets, pavement saws and cement mixers. For various reasons, however, the CSVL (constant-speed, variable-load) duty cycle, illustrated in Fig. 5.11, was never adopted [6].

In the European Union, the NRTC is employed for the measurement of gaseous and particulate matter emissions for Stages III/B (2011) and IV (2014), in parallel

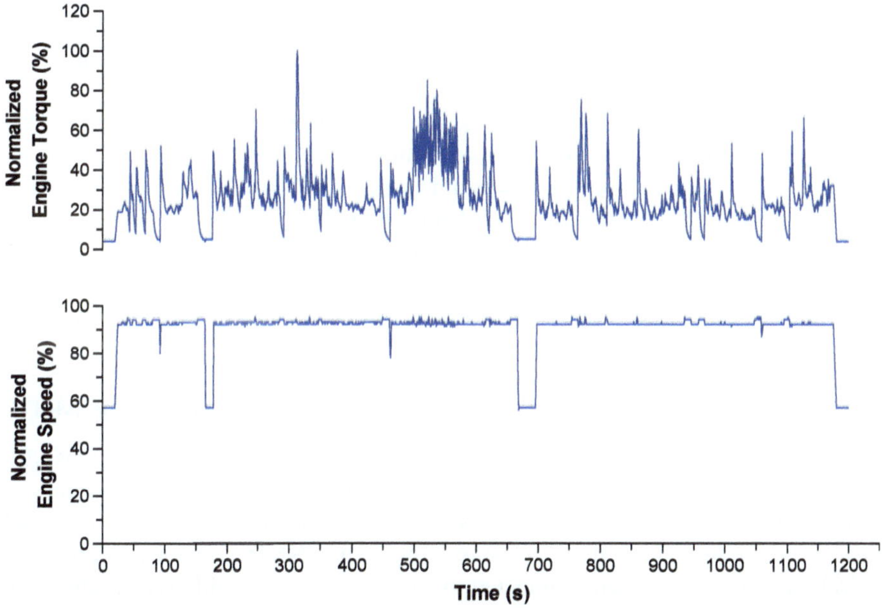

Fig. 5.11 U.S. EPA proposed constant-speed, variable-load duty cycle (never legislated) [17]; duration 1199 s, average speed 90.9 %, average torque 25.7 % (notice the similarities with the arc-welder speed and load traces in Figs. 5.6 and 5.7)

with the steady-state ISO 8178 C1/NRSC. Obviously, non-road constant-speed engines (i.e., those engines which may not operate at a speed outside a single fixed reference one set by the engine's governor) are excluded [1]. The non-road transient cycle NRTC is also employed in Japan, South Korea, China and Canada [1, 18, 19].

Special provisions are applicable for engines with continuous or periodic regeneration after-treatment systems according to UNECE GTR No. 11 (and Regulation 96/03). For an exhaust after-treatment system, based on a continuous regeneration process, the emissions are to be measured on an after-treatment system that has been stabilized so as to result in repeatable emissions behavior. For the case of exhaust systems regenerated on a periodic basis, the emissions are to be measured on several NRTC (or RMC) hot start tests, at least one during and two outside a regeneration event on a stabilized after-treatment system (Sect. 6.5) [9].

Lastly, and acknowledging that many non-road engines may operate in-use in a way different from the NRTC, specific not-to-exceed (NTE) standards are needed to assure that emissions are indeed controlled during the full range of possible in-use operating conditions. Such not-to-exceed standards have been legislated by the EPA and are also part of the UNECE GTR No. 10 [20].

Comparison with Other Engine-Dynamometer Transient Cycles

Table 5.10 summarizes and compares some important technical specifications for the NRTC and the three heavy-duty engine-dynamometer cycles (ETC, FTP, WHTC) discussed in Chap. 4.

The following remarks can be made with reference to Table 5.10:

- The NRTC exhibits both higher average speed and torque compared to all the legislated heavy-duty engine cycles.
- Owing to the different duty cycle of non-road machinery compared to on-road engines, the NRTC has (much) lower idling phase, and no motoring segments.
- The NRTC is also the most dynamic one (23.1 load increases per minute compared to 18.9 for the ETC, 16.3 for the WHTC and 'only' 12 for the FTP, which also has the longest idling period).
- However, the ETC seems the most aggressive, with the highest average load increase of the order of 17.6 % (14 % for the FTP, 13.7 % for the NRTC and 13.6 % for the WHTC).
- Comparing the NRTC with the component cycles from Tables 5.8 and 5.9, it seems that the skid-steer loader's specifications resemble those of the NRTC.

Following the arguments made with reference to Table 5.10, Figs. 5.12 (regarding normalized values), and 5.13 (for an actual diesel engine) compare the speed/torque distribution between the NRTC and the worldwide WHTC. It is made clear from these figures that, not surprisingly, the NRTC operates primarily at overall higher engine speeds compared to the (on-road, hence differently structured) WHTC; during the latter, predominantly medium and low engine speeds are encountered. In these speed ranges (i.e., during urban driving), the WHTC exhibits higher engine loads than its non-road counterpart.

Table 5.10 Comparison of some technical specifications between the non-road NRTC and the heavy-duty ETC, FTP and WHTC

Cycle	Duration (s)	Average normalized speed (%)	Max. normalized speed (%)	Average torque (%)	Max. torque (%)	Idle period (%)	Motoring period (%)
NRTC	1238	67.7	105	39.3	100	3.8	0
ETC	1800	50.9	90.1	36.7	100	6.6	18.0
FTP	1200	41.5	111.9	24.1	100	36.5	14.8
WHTC	1800	37.1	100	23.9	100	16.2	22.3

Cycle	Maximum load increase (%)	Average load increase (%)/Load increases (min^{-1})
NRTC	0–44	13.7/23.1
ETC	0–98	17.6/18.9
FTP	0–78.5	14.0/11.9
WHTC	m–84.7	13.6/16.3

m denotes motoring point

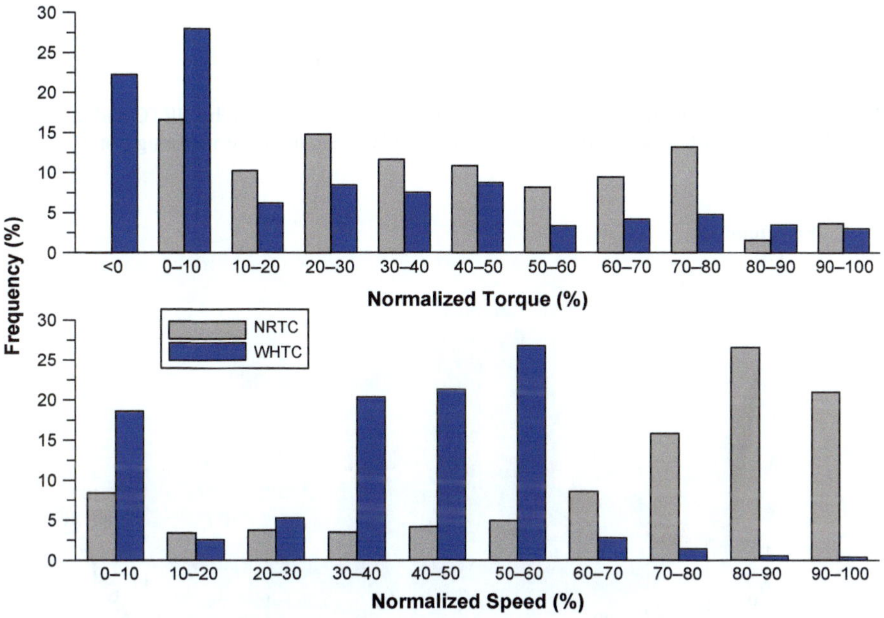

Fig. 5.12 Comparison of the speed and torque frequencies between the NRTC and the WHTC

Unlike the WHTC, the NRTC has relatively few operating points at low or no-load conditions and none at motoring. Figure 5.14, demonstrating the 3D normalized speed/torque frequency distribution of the NRTC, supports the previous arguments.

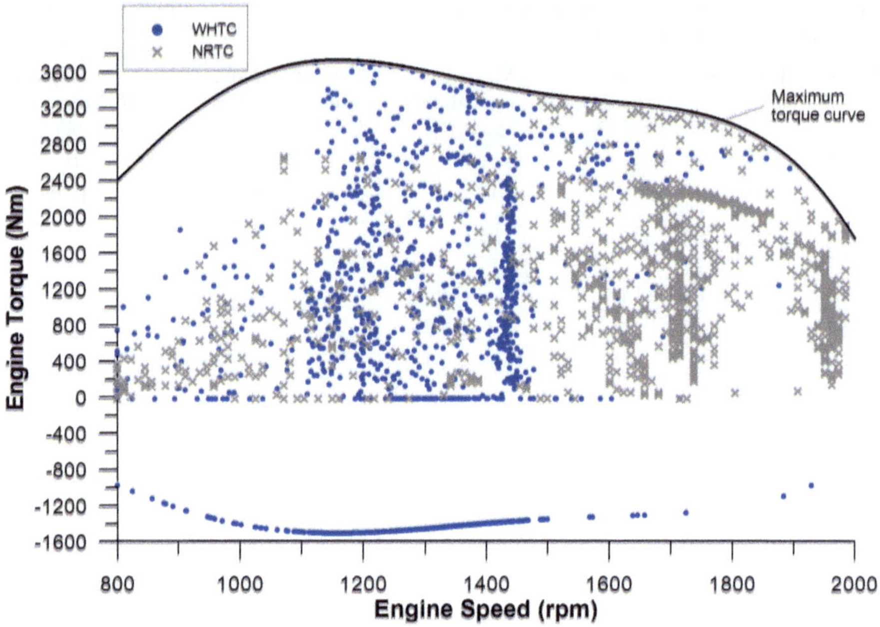

Fig. 5.13 Comparison of the engine speed/torque distribution between the NRTC and the WHTC for a heavy-duty turbocharged diesel engine (the NRTC does not have motoring points)

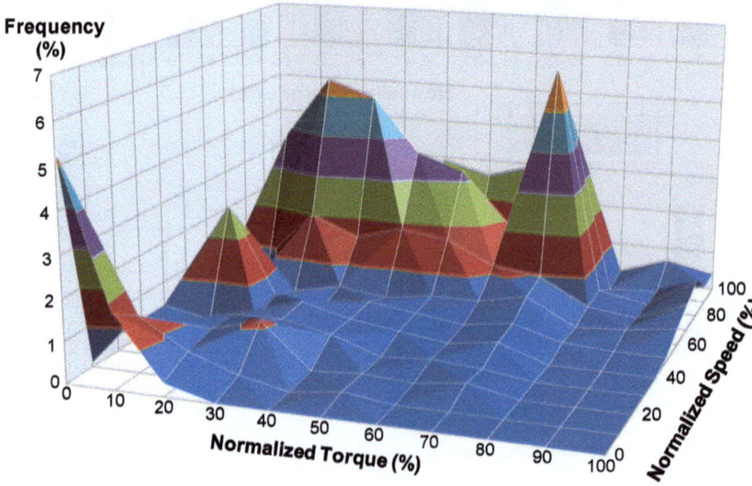

Fig. 5.14 Normalized speed/torque frequency distribution of the NRTC

Whereas Figs. 5.13 and 5.14 are indicative of the specific engine operating range during the test, Fig. 5.15 focuses more on the transient aspect by comparing the NRTC with the WHTC in terms of speed and load changes second-by-second

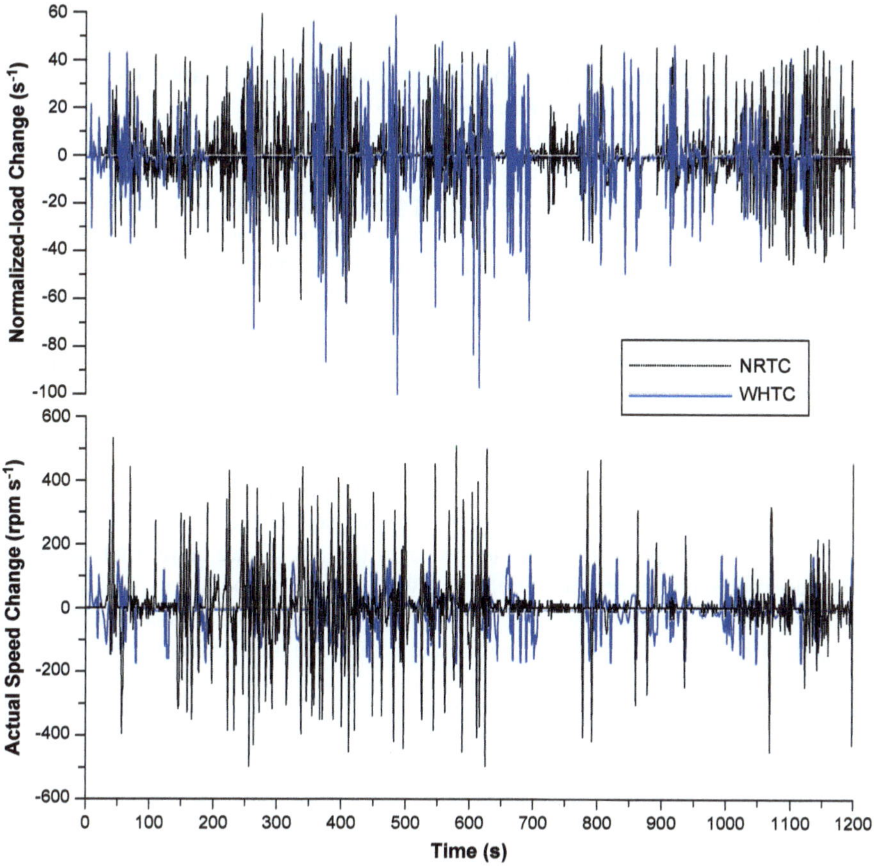

Fig. 5.15 Comparison of the actual speed change and the normalized load change between the NRTC and the WHTC for a heavy-duty diesel engine during the first 1200 s

(cf. the data in Table 5.10). Since, following Eq. (5.1), the acceleration during the cycle is dependent on the specific tested engine's speed range, the lower sub-diagram of Fig. 5.15 illustrates the speed changes (rpm/s) for the engine used for this exercise. It is obvious that the NRTC is much more transient and filled with steeper accelerations compared to the WHTC, resulting from the much more different duty cycle of non-road engines compared to on-road ones.

As regards the loading during the cycles, the upper sub-diagram of Fig. 5.15 attempts to shed some light, demonstrating the normalized torque changes per second (owing to many load increases commencing from zero or even negative load, load changes are not provided in percentage form). The NRTC exhibits again more frequent load changes (increases) compared to the WHTC, at least in part, owing to the latter's significant idling phase. However, the WHTC seems to have some load increases that are more abrupt than those of the NRTC during its urban part.

Typical engine-out NO and soot (indicative of PM) emission development during the NRTC is demonstrated in Fig. 5.16. Practically at every speed and load increase throughout the cycle, an emission overshoot is noticed; the lower the initial engine (and turbocharger) operating point and the steeper the acceleration and/or load increase, the higher the observed peak. Notice the absence of emission overshoots during the time period from approximately 630 to 720 s. This is the agricultural tractor contribution to the NRTC, characterized by high speed/load, hence very small speed/load changes.

While the results in Fig. 5.16 correspond to engine-out emissions from an old-technology engine, for newer engines with PM traps and deNO$_x$ catalysts, it is expected that the higher level of engine loading throughout the NRTC, as well as its overall higher speeds and absence of idling phases, will lead to fast warm-up of the engine and after-treatment control; this is representatively illustrated in Fig. 5.17 for the SCR of a turbocharged diesel engine compared to the same catalyst's operation during the WHTC [21]. On this subject, analysis conducted by the EPA indicated that the exhaust temperatures experienced over the NRTC are better matched to the NO$_x$ adsorber catalyst temperature window than the temperatures that would be expected over the highway FTP test cycle (Sect. 4.3.5). This suggests (when coupled with the fact that PM filters function with equal effectiveness at essentially

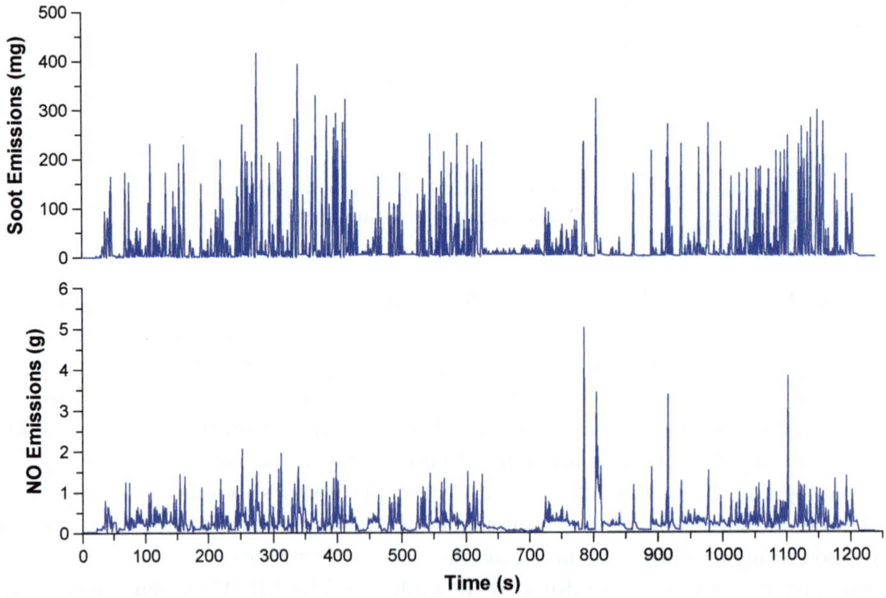

Fig. 5.16 Engine-out NO and soot development during the NRTC for a non-road turbocharged diesel engine with no in-cylinder antipollution measures

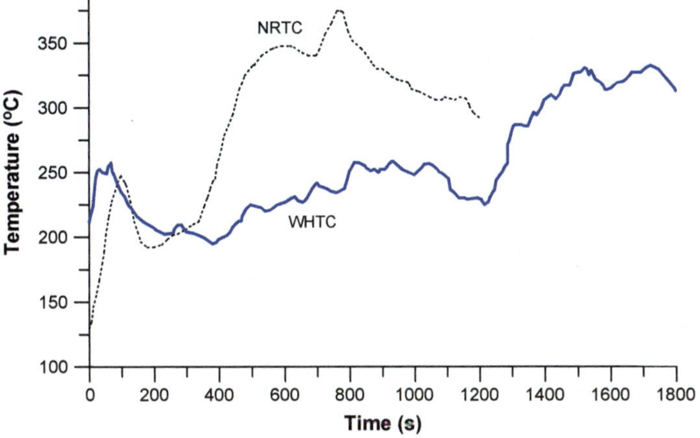

Fig. 5.17 Comparison of the SCR inlet temperature from a US2007, 6-cylinder, 7.5 L turbocharged diesel engine running on the WHTC and the NRTC cycles (hot starting) (AECC Euro VI test program presentation given at AECC technical seminar on 25 October 2007 [21])

all conditions) that compliance based on testing with the U.S. non-road Tier 4 standards on the NRTC will be somewhat easier, using similar technology, than complying with the highway 2007 emission standards on the highway transient FTP cycle [6].

References

1. TransportPolicy.net website. http://www.transportpolicy.net/index.php?title=Category:Nonroad. Accessed 22 Aug 2016.
2. Directive 97/68/EC of the European Parliament and of the Council of 16 December 1997 on the approximation of the laws of the Member States relating to measures against the emission of gaseous and particulate pollutants from internal combustion engines to be installed in non-road mobile machinery.
3. Proposal for a regulation of the European Parliament and of the Council on requirements relating to emission limits and type-approval for internal combustion engines for non-road mobile machinery. European Commission, COM(2014) 581 final, 25 Sept 2014.
4. ISO 8178-4:2007. Reciprocating internal combustion engines—Exhaust emission measurement, Part 4: Steady-state test cycles for different engine applications, Dec 2007.
5. Degobert P. Automobiles and pollution. Paris: Editions Technip; 1995.
6. U.S. Environmental Protection Agency. 40 CFR Parts 9, 69, et al. Control of emissions of air pollution from nonroad diesel engines and fuel; Final rule, Federal Register Vol. 69, No. 124, 29 June 2004.
7. EUROMOT response on the stakeholder consultation on the revision of Directive 97/68/EC on emissions from non-road mobile machinery engines. European Association of Internal Combustion Engine Manufacturers, Apr 2013.

8. U.S. Environmental Protection Agency. Locomotive emission standards. Regulatory support document, EPA-420-R-98-101, Apr 1998.
9. UNECE Global Technical Regulation No. 11. Engine emissions from agricultural and forestry tractors and from non-road mobile machinery. Established in the Global Registry on 12 Nov 2009.
10. U.S. Environmental Protection Agency. Final regulatory analysis: Control of emissions from nonroad diesel engines. U.S. EPA, Office of Transportation and Air Quality, EPA420-R-04-007, May 2004.
11. Lindgren M. Engine exhaust gas emissions from non-road mobile machinery—effects of transient load conditions. PhD Thesis, Dept. of Biometry and Engineering, Swedish University of Agricultural Science, Uppsala, 2004.
12. Ullman TL, Webb CC, Jackson Jr CC, Doorlag MH. Nonroad engine activity analysis and transient cycle generation. SAE Paper No. 1999-01-2800, 1999.
13. Starr ME, Buckingham JP, Jackson Jr CC. Development of transient test cycles for selected nonroad diesel engines. ASME Paper No. 99-ICE-171, 1999.
14. U.S. Environmental Protection Agency. Nonregulatory nonroad duty cycles. https://www3.epa.gov/otaq/regs/nonroad/nrcycles.htm. Accessed 20 June 2016.
15. Rakopoulos CD, Giakoumis EG. Diesel engine transient operation. London: Springer; 2009.
16. Directive 2004/26/EC of the European Parliament and of the Council of 21 April 2004 amending Directive 97/68/EC on the approximation of the laws of the Member States relating to measures against the emission of gaseous and particulate pollutants from internal combustion engines to be installed in non-road mobile machinery.
17. Memorandum from Kent Helmer to Cleophas Jackson, Speed and load operating schedule for the constant speed variable load (CSVL) transient test cycle, e-Docket OAR-2003-0012-0993, 2003.
18. Worldwide emission standards—heavy duty and off-highway vehicles. Delphi, 2015–2016. http://delphi.com/docs/default-source/worldwide-emissions-standards/delphi-worldwide-emissions-standards-heavy-duty-off-highway-15-16.pdf. Accessed 7 Aug 2016.
19. Exhaust emission legislation—diesel and gas engines. VDMA, Apr 2010.
20. UNECE Global Technical Regulation No. 10. Off-cycle emissions (OCE). Established in the Global Registry on 24 June 2009.
21. Rickert G. Overview of emissions on world-harmonised and non-road cycles from the AECC Euro VI programme. AECC Technical seminar on heavy-duty engine emissions, Brussels, 25 Oct 2007.

Chapter 6
Driving Cycles Test Procedure

Abstract This chapter summarizes the most important features of the driving/engine cycles test procedure in the laboratory. A brief discussion of the type approval process is first conducted, followed by dynamometer fundamentals, analysis of the road-load equation and description of the coast-down procedure. The main part of the chapter is devoted to the exhaust emissions measurement using the constant volume sampling system, providing description of the facility as well as the basic equations for steady-state and transient testing during both vehicle and engine certification. Emphasis is placed on the procedures/equations valid in the European Union through UNECE regulations, which are also applicable in many other areas in the world; certain differentiations valid in the United States are discussed.

6.1 Introduction

In this chapter, the fundamentals of the laboratory measurement procedure during driving/engine cycles will be presented. The term test procedure describes the exact methodology of the vehicle testing to be followed during the type approval process with respect to

- ambient conditions,
- determination of vehicle mass and road load,
- preconditioning (e.g., whether the test cycle is run with the engine hot or cold started),
- measurement equipment,
- analyzers calibration and accuracy,
- reference fuels,
- tolerances,
- possible gear-shift schedule,
- exhaust emissions measurement and calculations.

E.G. Giakoumis, *Driving and Engine Cycles*,
DOI 10.1007/978-3-319-49034-2_6

Table 6.1 Regulations in the EU and the U.S. relating to vehicles and engines laboratory testing for certification purposes

	EU	U.S.A.
Motorcycles	GTR No. 2	40 CFR 86 Subpart F
Light-duty vehicles	GTR No. 15	40 CFR 86 Subpart B (through model year 2021) 40 CFR 1066 (optional through model year 2021, mandatory from model year 2022) 40 CFR 600 (fuel consumption)
Heavy-duty and non-road engines	GTR No. 4 (HDVs)	40 CFR 1065 (including field testing and PEMS)
	GTR No. 11 (non-road engines)	
	ISO 8178	

Japan has already adopted GTRs No. 2, 4 and 11, and will adopt GTR No. 15 as well

Such test procedures are described in detail in the corresponding regulatory documents. Table 6.1 lists the current regulations valid in the EU and the United States regarding laboratory testing for vehicles and engines, which are also followed by many other countries in the world. It should be noted that in the EU, the regulations define only the emission limits; the test cycle, test procedure and calculation formulas are not included in the regulations anymore but only in the corresponding UNECE GTRs.

During the early years of vehicle certification on the chassis dynamometer, a preconditioning phase was usually applied, which secured that the engine (at that time, exclusively spark ignition) was already fully warmed up when the sampling began. The controlled pollutants were initially CO and HC only, and it was their concentration (ppm) that was originally measured. Sampling took place throughout all or certain runs of the repetitive/modal cycle, with specific weighting factors assigned to each run or certain parts of the cycle.

In the next decades, the testing was extended to diesel-engined vehicles, heavy-duty engines and later to non-road engines and motorcycles too, whereas the authorities in many countries gradually changed the testing procedure with respect to two important aspects. Firstly, the constant volume sampling (CVS) method was adopted (discussed in Sect. 6.5), with the results expressed now in g/test, or g/km (or g/mile) for the pollutants concerned. The latter are usually unburned hydrocarbons, or non-methane hydrocarbons, carbon monoxide, nitrogen oxides, particulate matter (in Europe particle number too), and also fuel consumption and carbon dioxide. Second, sampling begins at the moment the vehicle is cold started. An engine start is considered to be 'cold' if it is preceded by a relatively long uninterrupted soak, such as those that occur after an overnight soak, typically at a temperature of 20–30 °C. Through cold-starting test procedures, it is acknowledged that a significant amount of pollutants is emitted during the early minutes of the engine's operation before the after-treatment devices reach their operating temperature and begin to function efficiently (cf. Figs. 1.11 and 1.12). Regardless, since it is fairly unnatural to enter a motorway with the engine cold, highway drive cycles (if separately run from the urban part) are always executed with the engine fully warmed up. When a test procedure incorporates both urban and motorway driving phases, it is not uncommon for different weighting factors to be assigned to each part as discussed in Chaps. 2 and 3. Consequently, the total amount of emissions (or fuel consumption and CO_2 for that matter) is calculated using the weighted averages from each part.

The legislation in some regions (for example, EU and the United States) further requires a separate cycle run at a very low temperature (−7 °C). This requirement is met in specially-equipped, temperature-controlled laboratory facilities. Another case of such climatic test cell is when the vehicle is subject to a high temperature and solar loading test (e.g., the U.S. SC03 cycle).

In Europe, each country has a dedicated type approval authority,[1] which in turn appoints specific technical services/laboratories to carry out homologation tests. The framework directive relating to type approval of motor vehicles is 2007/46/EC (which is in the process of being updated). The emissions component of the type approval test takes place at the premises of the manufacturer under the auspices of the technical service of the corresponding EU Member State, or at the certified laboratories. Approvals obtained in any member are recognized by other Member States. Not all models of a manufacturer are required to undergo the homologation process; only new vehicle types do. Vehicles are categorized as being of the same type if they have the same inertia mass and the same engine and vehicle characteristics. All further details of a vehicle define different vehicle variants and versions. If the prescribed procedures have been followed, required tolerances are met and limits are not exceeded, type approval will be granted by the national type approval authority. Within strictly specified conditions, type approval for one vehicle model can also be extended to similar models of the same manufacturer. Initially, a 'representative of the type to be approved' vehicle is selected for testing. Alternatively, in agreement with the type approval authority, the manufacturer may also select a vehicle that is not representative of its type but 'combines a number of the most unfavorable features with respect to the required level of performance.' As a result, for exhaust pollutant emissions, often only one vehicle of a type is tested, the one with the (expected) highest exhaust pollutant emissions. The obvious advantage being that type-approval is granted for several vehicle models at the same time [1, 2]. During the homologation procedure, up to three Type I (tailpipe emission) tests may be conducted, depending on the emission results. For carbon dioxide emissions, on the other hand, since fleet-average emissions are of concern, each vehicle model is tested individually. In this regard, it is not uncommon for the exhaust emissions and the CO_2 testing to be carried out separately.

---

[1]The approval authorities of the EU Member States are listed in the following link http://ec.europa.eu/DocsRoom/documents/15493/attachments/1/translations/en/renditions/native.

In the United States, the procedure is quite different, with vehicle manufacturers conducting their own certification tests, whereas the relevant authorities (EPA) carry out only confirmatory checks—most of them randomly selected—to validate the results reported by the manufacturers. Prior to the 2015 Volkswagen scandal discussed in Sect. 1.3.1, another notable violation was resolved on November 3, 2014. Automakers Hyundai and Kia were fined $100 million civil penalty to resolve alleged Clean Air Act violations based on their sale of more than 1 million vehicles that collectively emitted 4.75 million metric tons of greenhouse gases in excess of what the automakers certified to the EPA [3].

For engine testing, again not all engines are tested. Instead, for each engine family, a parent engine is chosen, the emissions of which are considered representative of the specific group of engines (ISO 16185). For example for CI engines, the one with the highest fuel delivery per stroke at the maximum torque speed is selected. The term engine family refers to a manufacturer's grouping of engines (approved by the certification authorities), which have similar design specifications, e.g., combustion cycle, aspiration, cylinder arrangement, cooling medium, fuel type, ECU strategy, after-treatment systems. Hence, they can be considered to exhibit similar exhaust emission behavior.

Returning to chassis-dynamometer testing, during the certification process, the vehicle is tied (it remains stationary) but at least two of its wheels are spinning on the dynamometer rolls to simulate driving at different speeds. By so doing, the whole drive-train is under test. The hood is open and a cooling fan is positioned so that cooling air is appropriately directed to the radiator and to the vehicle. The tested car is driven by a trained driver (or even an automated driving system ('robot') in some cases), who follows closely the defined cycle schedule within certain tolerances defined in the legislation. During the whole test, the driver receives real-time information regarding the demanded and achieved vehicle speed, as well as the difference between them. A key difference between engine testing and chassis testing can be located exactly in the extent of driver input control. Whereas a chassis-dynamometer facility relies exclusively on an experienced driver to operate the vehicle throughout the test cycle, an engine dynamometer facility controls all inputs, including the engine accelerator pedal.

6.2 Dynamometers

Dynamometers are devices intended to measure power from a connected engine by absorbing and converting it into another form, e.g., heat or electricity. More specifically, the torque produced by the engine is resisted, and measured, by the connected dynamometer. Calculation of power is then conducted using the rotational speed signal. Obviously, the accuracy of speed and torque measured by the dynamometer is of paramount importance for all the derivative experimental results [4].

Dynamometers can be hydraulic, eddy current (the electrical power generated is dissipated as heat) or electric motor/generator AC or DC (the engine power absorbed is converted into electrical energy). In the early days of vehicle certification, dynamometers with constant load curve and relatively small diameter rollers were used. For example in the United States, the Clayton dynamometer was mostly employed in the 70s, having 220 mm twin rolls. One noteworthy limitation, as discussed in Sect. 2.2, was the tire slip observed at accelerations/decelerations higher than 1.5 m/s^2 (3.3 mph/s). Nowadays, however, dynamometers have larger rolls providing more realistic traction conditions, possess an adjustable load curve, and enable inertia simulation in multiple 'steps'. Chassis dynamometers can have one or two rolls per driven wheel, Fig. 6.1, while different dynamometer configurations can be applied for the case of FWD (four-wheel drive) vehicles, where power is transmitted to all wheels, hence both axles. A special case is mileage accumulation dynamometers comprising a single large roll of at least 48 in (122 cm) diameter, directly coupled to a DC or AC motor. Typical technical attributes of a chassis dynamometer are: base inertia, minimum and maximum simulated inertia, rated power, torque range, number and diameter of rolls, maximum speed etc. [4].

As mentioned, a potential problem with dynamometers is tire slip, which may be experienced at high accelerations. To deal with this, an option is to place additional weight in or on the vehicle. Another option is to apply a hub dynamometer, i.e., the tire is entirely eliminated during the test and the dynamometer is attached to the hubs of the wheels. This, however, results in the test not simulating the real vehicle operation on the road, hence hub dynamometers are not foreseen in the legislation; hub dynamometers have been employed in various (non-regulatory) tests of heavy-duty vehicles. Another potential issue is vehicle tie-down, which, if not conducted properly, may lead to incorrect test results (hub dynamometers do not require tying of the vehicle). Once acceleration and braking are involved, the vehicle motion on the dynamometer is quite different than on the road. On the road, a vehicle is subjected to braking forces on all wheels, which provoke a couple about the centre of gravity; the latter results in a transfer of load from the rear to the front wheels causing pitching of the car. On a chassis dynamometer, on the other hand,

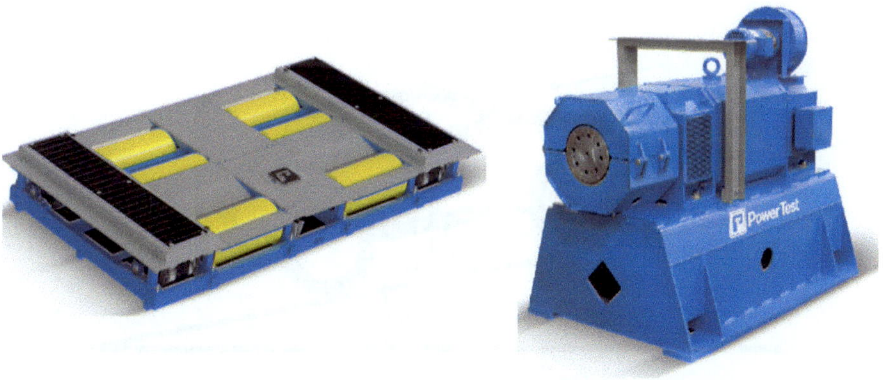

Fig. 6.1 AC Power Test chassis dynamometer (*left*) and engine dynamometer (*right*)

the braking force is applied only to the driven wheels. The forward deceleration force acting through the centre of gravity is absent, and replaced by tension forces in the front-end vehicle restraints [4].

Special requirements arise in the case of engine testing. First of all, since the engine, and not the whole vehicle, is under test, the dynamometer measures the power from the engine crankshaft applying a torque-meter for the engine torque and a shaft encoder for the engine speed. Furthermore, as discussed in Sect. 1.3.2, there are motoring segments during an engine-dynamometer cycle. Thus, the dynamometer should be able to not only absorb the power from the engine but also motor it. This can be achieved with (AC or DC) electric motor/generator dynamometers, Fig. 6.1 [5].

6.3 Road-Load Equation

The dynamometer rolls, on which the car rests, have the task to simulate the resistances the vehicle experiences when being driven on the road, Fig. 6.2. In general, these forces are the rolling resistance force F_r, the grade-dependent force F_{gr}, the aerodynamic force F_a, as well as the inertia force F_{in}.

Total traction force F_{tr} is given by [6]

$$F_{tr} = F_r + F_{gr} + F_a + F_{in} \tag{6.1}$$

The various terms in Eq. (6.1) are defined as follows:

- The rolling resistance force F_r is expressed by

$$F_r = c_{fr} W \cos\theta = c_{fr} m_V g \cos\theta \tag{6.2}$$

The term c_{fr} is the tire rolling resistance coefficient. It is influenced by tire deformation (biggest contribution), tire penetration and slippage. W is the vehicle

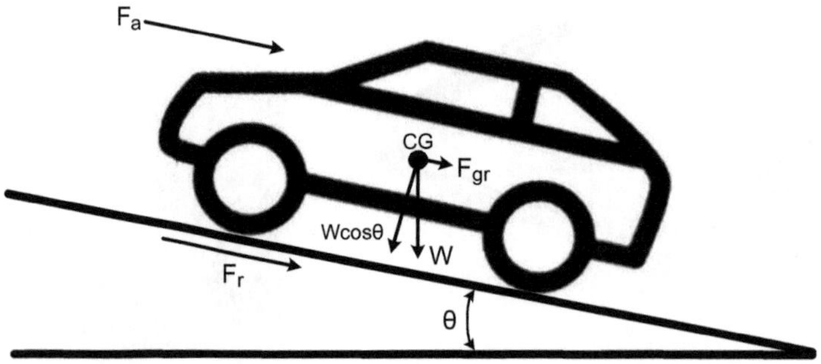

Fig. 6.2 Simplified diagram showing the forces affecting vehicle motion (*CG* center of gravity)

weight, m_V the vehicle loaded mass and g the gravitational acceleration (the latter depending on the geographic latitude). Angle θ accounts for the road grade, which may vary with time. Grade is defined as the 'rise' over the 'run' (i.e., vertical distance over the horizontal one), being the tangent of the grade angle θ. It is often expressed as a percentage after multiplying by 100; grade may reach up to 4 % in highways and up to 10–12 % in primary and secondary roads [6]. For straight roads, $\cos\theta = 1$. From Eq. (6.2), the rolling resistance force appears to be constant. However, this is not entirely true, as the friction coefficient c_{fr} is speed dependent. In general, c_{fr} increases with increasing loading, speed and decreasing tire pressure.

- The resistance due to road grade is given by

$$F_{gr} = W \sin\theta = m_V g \sin\theta \tag{6.3}$$

- Assuming negligible ambient wind, the aerodynamic force F_a is given by

$$F_a = \frac{1}{2}\,\rho_a\, c_d\, A_f\, V^2 \tag{6.4}$$

  with A_f the vehicle frontal area, ρ_a the air density, and c_d the vehicle aerodynamic resistance (drag) coefficient; V is the vehicle speed. The aerodynamic drag is composed of turbulent air flow and air friction around the vehicle, as well as flow resistance in the radiator and front grille. Parameters that affect drag coefficient c_d are side mirrors, open windows and sunroof, flow through radiator etc.
- When the vehicle accelerates, the engine has in addition to overcome an inertia term related to the longitudinal motion of the vehicle and rotating motion of various parts such as wheels and axles

$$F_{in} = \left(m_V + \frac{\Theta_{wheel}}{r_{wheel}^2}\right)\frac{dV}{dt} = \left(1 + \frac{\Theta_{wheel}}{m_V r_{wheel}^2}\right) m_V a \tag{6.5}$$

with a the vehicle acceleration = dV/dt. Ideally, in Eq. (6.5), the rotating inertia terms for the engine, flywheel, clutch and propeller shaft should be incorporated (depending on the gear and final-drive ratio), although their contribution is much smaller compared to the wheel term. The second term inside the parenthesis in Eq. (6.5) contributes approximately 5 % [7].

Total force F_{tr} is then given by

$$F_{tr} = \left(\overbrace{c_{fr}(V) m_V g \cos\theta + m_V g \sin\theta + \frac{1}{2}\,\rho_a\, c_d\, A_f V^2}^{\text{road-load force}}\right) + \overbrace{\left(1 + \frac{\Theta_{wheel}}{m_V r_{wheel}^2}\right) m_V \frac{dV}{dt}}^{\text{acceleration}} \tag{6.6}$$

More advanced analyses of the vehicle motion on the road incorporate additional terms in Eq. (6.6), accounting, for example, for wheel spin or cornering [6]. For certification purposes, Eq. (6.6) is transformed into

$$F_{tr} = b_0 + b_1 V + b_2 V^2 + m_V \frac{dV}{dt} \tag{6.7}$$

Constants b_0, b_1 and b_2 depend on the rolling, grade (if applicable) and aerodynamic resistance, commonly referred to as road-load coefficients. The contribution of the b_1V term is smaller than the b_0 one (although not negligible at high speeds), hence sometimes it is neglected. It is the above described forces in Eq. (6.6) that the dynamometer has to simulate, however with the exception of the grade component. Legislative test cycles assume straight roads with zero gradient, thus no account is taken of the road-dependent resistance force from Eq. (6.3). Consequently, the b_0 term in Eq. (6.7) incorporates rolling resistance effects only. Notice that due to the speed depending nature of the c_{fr} coefficient, it is not feasible to directly correlate coefficients b_i with the corresponding vehicle parameters; instead, an experimental procedure is required to determine the road-load coefficients. This is detailed in the next section.

6.4 Coast-Down Procedure

Estimation of the b_0, b_1 and b_2 coefficients of the road-load Eq. (6.7) is usually accomplished through the coast-down procedure. This is conducted on a flat and sufficiently long and dry test track with average wind speed less than 3 m/s as follows: The vehicle, run-in for at least 3000 km, clean and with the windows and any covers of the air-condition system closed, is loaded to its reference mass according to the requirements of the specific test, and is driven up to a certain speed, and then the 'neutral' gear is selected. The maximum speed is usually above 100 km/h but this is something that depends on the certification cycle; for example in Europe, it is 130 km/h, and in the United States 75 mph. After the neutral gear is selected and the engine is decoupled from the drive-train, the vehicle begins to slow down due to friction and aerodynamic drag. During the coast-down procedure, Eq. (6.1) is transformed into

$$F_r + F_a = -m_V \frac{dV}{dt}$$

For a number of reference speeds between (e.g., in Europe) 20 and 120 km/h, and for 20 km/h increment (overall 6 reference speeds), the coast-down time is measured over an interval from +5 to –5 km/h of each reference speed, i.e., the time is measured that is needed for the vehicle to decelerate from 125 to 115 km/h (corresponding reference speed 120 km/h), then from 105 to 95, 85 to 75 etc. up to

25 to 15 km/h. The test is performed again with the vehicle moving in the opposite direction (to eliminate even the slightest wind or grade effects), and the average time between the two runs for each reference speed is calculated. The tests, in both directions, are repeated several times up to the point where a certain level of statistical accuracy is achieved. Using the coast-down time and the mass of the vehicle, the corresponding absorbed power and the sum of the resistance forces are then computed for each reference speed. Table 6.2 provides results from a typical cost-down test of a passenger car.

The next step is to derive the b_0, b_1 and b_2 terms in Eq. (6.7) applying a least squares method. For the coast-down procedure described in Table 6.2, Fig. 6.3 illustrates the development of the corresponding rolling, aerodynamic drag and total resistance with respect to vehicle speed.

Table 6.2 Results from a coast-down test according to European regulations for a passenger car

Vehicle Reference Speed (Speed Interval) (km/h)	Time Interval (s)	Absorbed Power = F × V (kW)	Deceleration Force F (N)
120 (125–115)	7.5	20.99	629.6
100 (105–95)	9.5	13.81	497.1
80 (85–75)	12.1	8.67	390.3
60 (65–55)	15.5	5.08	304.7
40 (45–35)	19.5	2.69	242.2
20 (25–15)	23.5	1.12	200.9

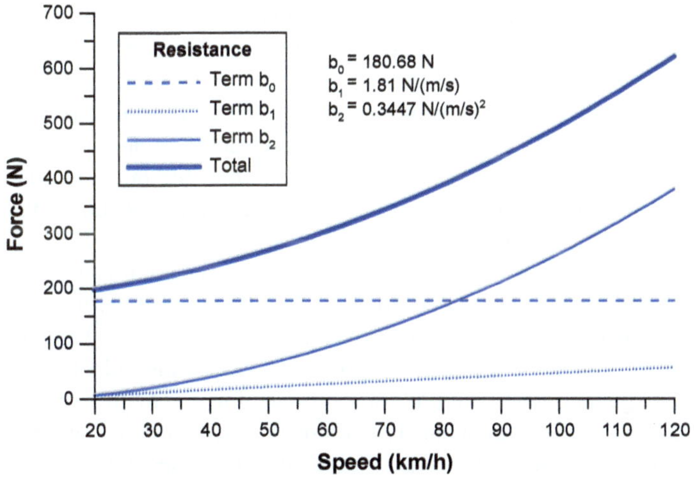

Fig. 6.3 Rolling, aerodynamic and total resistance with respect to vehicle speed from the coast-down results of Table 6.2

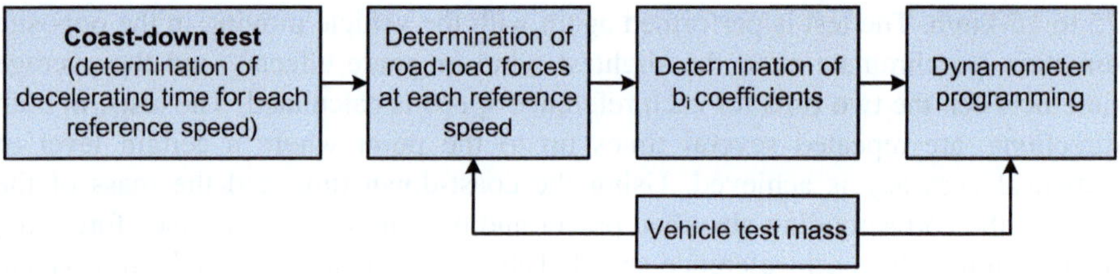

Fig. 6.4 Process for determination of road-load forces and coefficients

Coefficients b_0, b_1 and b_2 are corrected with respect to the ambient conditions (20 °C, 1 bar), and subsequently simulated in the laboratory by appropriately adjusting the resistance of the chassis-dynamometer rollers. Figure 6.4 illustrates the entire road-load determination process. For adjustable load curve dynamometers, coefficients b_i can be adjusted independently from each other [1, 8]. The legislation also provides the applicable equations for the not preferred case a fixed-curve dynamometer is employed. Oddly, in the EU, the results of the road-load test are kept secret, being only available in the type approval documents, and no other authority, independent or not, conducts retesting for confirmatory purposes (although this may change in the future), even though the coefficients are of paramount importance for the subsequent emissions test [9, 10]. On the contrary in the United States, the coast-down results are made public. A recent study conducted by ICCT on a number of SI and diesel passenger cars identified significantly underestimated road-load coefficients in the EU, which in turn resulted in clearly higher real-world fuel consumption and CO_2 emissions, of the order of 7.2 % on average. Interestingly, the gap was much smaller (1.8 % with regard to CO_2 emissions) when the U.S. road-load coefficients were investigated, which indicates that the procedure in the United States is much more realistic [9]. There are certain flexibilities/loopholes in the legislation (particularly in Europe) that allows manufacturers to 'manipulate' the coast-down test and define non-realistic coefficients. For example, road inclination of the test track, depth of tire tread, inertia of rotating parts such as wheels in Eq. (6.5) not taken into account, and no humidity effects considered in the air density calculation [9].

Further to the resistance forces, the inertia of the dynamometer rolls is adjusted based on the vehicle's mass in order to also take into account the inertia component ($m_V dV/dt$) from Eq. (6.7). Prior to the WLTP procedure, each vehicle in Europe was categorized for the test into a discrete inertia class instead of using its actual mass for the dynamometer inertia; for example, for a reference weight between 1250 and 1470 kg, the equivalent dynamometer inertia was set at 1360 kg. This, however, will change when the WLTP procedure becomes effective from 9/2017, with the dynamometer inertia reflecting the actual vehicle test mass [11].

In the United States, 40 CFR 1066 describes the relevant coast-down procedure for speeds between 115 and 15 km/h, which is based on SAE J1263 [12] and SAE J2263 [13] regarding road-load coefficients evaluation, and SAE J2264 regarding adjustment of an electric chassis-dynamometer to provide accurate simulation of the resistances [14]. The U.S. regulation also details the coast-down procedure to be followed for the case of HDVs having GVWR above 14,000 lbs (for use in the U.S. EPA 'GEM' greenhouse gas emissions model simulation tool). These vehicles are not chassis but engine certified. Term b_1, or B in the U.S. regulation, is assumed to be zero for the GEM calculations.

Vis-à-vis the cold-temperature test at −7 °C required in some countries, U.S. and European regulations state the following: either a coast-down test is conducted at a nominal ambient temperature of −7 °C, or alternatively, the coefficients of the coast-down test at normal conditions are multiplied by 1.1 to approximate a 10 % decrease in coast-down time for the test vehicle.

For motorcycles, the procedure is, in general, the same. For example, EU Regulation 134/2014/EU [15] and UNECE GTR No. 2 [16] describe the relevant coast-down details in terms of motorcycle preparation, rider and riding position etc. Since it can be dangerous to have the transmission shifted to neutral during motorcycle coasting, the test may be performed with the clutch disengaged.

As an alternative to the coast-down test, the torque meter method can be used. In this procedure, the running resistance is determined by measuring wheel torque on the driven wheels at various constant speeds with time periods of at least five seconds. The newly developed WLTP additionally foresees road coast-down with on-board anemometry and aerodynamic drag measurement in a wind tunnel as alternatives to the classic coast-down procedure [11].

After the determination of the vehicle resistances and the appropriate programming of the dynamometer, the exhaust emission test is carried out according to the specific requirements, e.g., with the engine hot or cold-started, with the air-conditioning system on or off, at a specific ambient temperature, loading etc. Part of the exhaust of the vehicle is collected and analyzed to calculate emission levels and fuel consumption, with a test protocol containing the details of the cycle and the test procedure as is discussed in the next paragraph. Before the emissions measurement, however, the legislation requires that the vehicle 'soaks' at a specific temperature (−7 °C for the cold test), or range of temperatures (20–30 °C for the Type I test). This is accomplished in a climatically controlled space, where the temperature is controlled, and the vehicle is not subject to direct sunlight.

6.5 Emissions Measurement

The exhaust gas sampling system must be able to measure the actual quantities of pollutants emitted in the exhaust gases. In general, either raw exhaust gas sampling or dilution sampling can be employed. Sampling may be *continuous*, i.e., the pollutant's concentration is measured continuously from the diluted or raw exhaust gas, or *batch*, in which case the continuously extracted sample is stored in a bag for later analysis.

The system nowadays universally employed during the certification of vehicles is the constant volume sampling (CVS) system. It was utilized for the first time in the United States in the early 70s [17] and from 1983 in Europe [18]. CVS requires that the whole vehicle exhaust is continuously diluted with filtered ambient air under controlled conditions. Dilution is performed to prevent water condensation in the sampling/measuring system and to also simulate actual mixing conditions of exhaust gas with air in the atmosphere. In the CVS, two conditions must be satisfied: the total volume of the mixture of exhaust gases and dilution air must be measured, and a continuously proportional sample of the volume must be collected (sampled) for analysis. The quantities of gaseous pollutants emitted are determined from the sample concentrations and the totalized flow over the test period. For determination of the PM level, on the other hand, suitable filters are used to collect particulates, again from a proportional part flow throughout the test. Following the continuously increasing stringency in the emission limits, (vehicle-out) emissions have reached extremely low levels. Thus, it has become imperative to also measure the dilution air, in order to correct the actual pollutants concentration. It should be emphasized that since the volume of the diluted exhaust gas is kept constant, the diluent flow and dilution ratio vary throughout the test inversely with engine exhaust flow [19].

6.5.1 Chassis Testing Employing Constant Volume Sampling

Figures 6.5 and 6.6 illustrate two typical CVS systems for chassis testing of passenger cars and motorcycles. Further, Fig. 6.7 demonstrates in more detail a

Fig. 6.5 Chassis dynamometer facility for testing passenger cars (*Source* TÜV Hessen)

Fig. 6.6 Chassis dynamometer facility for testing motorcycles (*Source* TÜV Hessen)

schematic diagram of a CVS equipped with a positive displacement pump (PDP-CVS). The whole set-up consists of [20–22]:

- filter for the dilution of air, if necessary preheated. It consists of an optional activated charcoal filter (inlet side) and a high efficiency particulate air (HEPA) filter (outlet side). It is recommended that an additional coarse particle filter is situated before the HEPA filter and after the charcoal filter, if used. The purpose of the charcoal filter is to reduce and stabilize the hydrocarbon concentrations of ambient emissions in the dilution air. The amount of dilution air must be sufficient to cool the exhaust gas/air mixture at all engine operating conditions at a temperature of no more than 325 K (52 °C);
- mixing chamber in which the exhaust gas and air are mixed homogeneously;
- dilution tunnel for the case of diesel-engined vehicles; it is usually a polished and heated stainless steel tube of electrically conductive material. It ensures that no chemical reactions will take place that may alter the pollutants concentration in the exhaust gas. The tunnel's diameter (at least 200 mm) depends on the exhaust gas flow, being small enough to cause turbulent flow (Re > 4000). The tunnel is earthed;
- positive displacement pump to transport a constant-volume flow of exhaust gas + dilution air, e.g., Roots blower;
- heat exchanger to maintain the exhaust mixture temperature within the design operating values;
- probes and sampling lines;
- gas analyzers for the pollutants concerned;
- pumps;

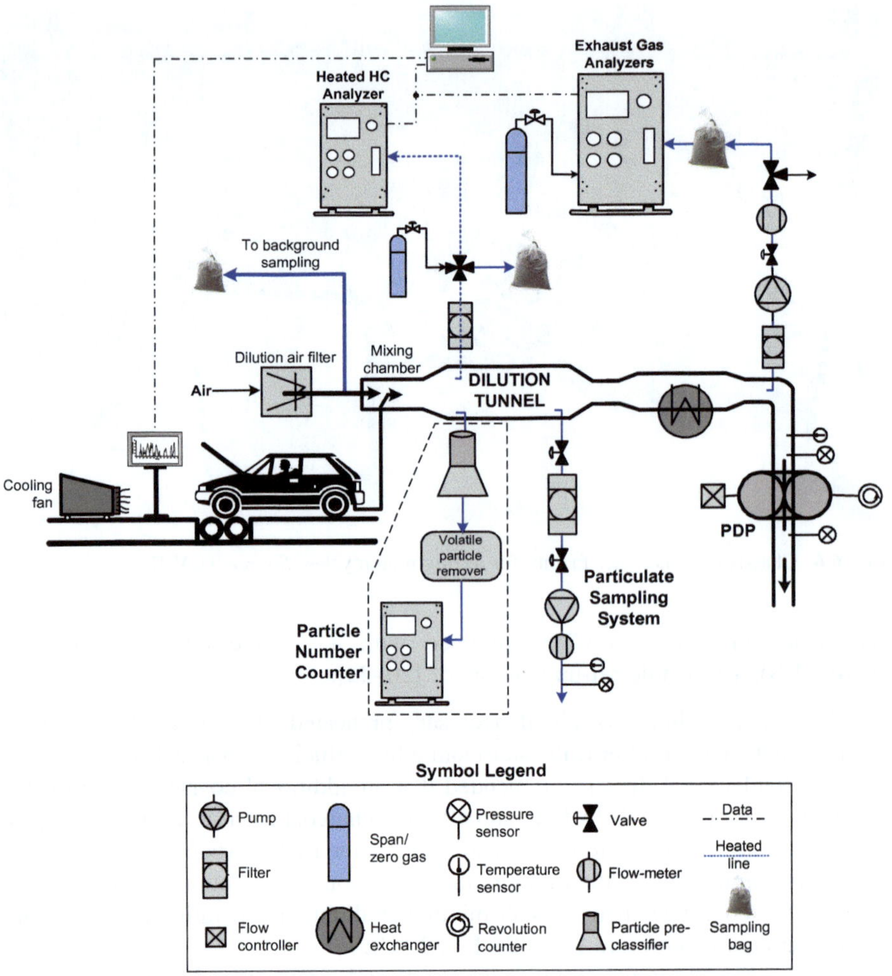

Fig. 6.7 Typical constant volume sampling configuration with positive displacement pump (PDP) for emissions measurement of vehicles in the EU; including pre-test and post-test activities, one test over the 1180-s NEDC typically lasts 1 h

- filters;
- flow controllers to ensure constant uniform flow;
- flow meters for adjusting and monitoring the constant flow of gas samples during the test;
- pressure/temperature/humidity sensors;
- bags for the collection of the diluted exhaust gas samples (made of fluorinated poly-hydrocarbons, or laminated poly-ethylene or polyamide films);
- digital counter to register the number of PDP revolutions.

Instead of the PDP, a critical flow venturi (CFV) can be utilized, which is based on the principles of flow mechanics at critical flow. In this case, the variable exhaust gas + dilution air mixture is maintained at sonic velocity; the latter being proportional to the square root of the gas temperature.

As mentioned, the CVS process is based on dilution of the total raw exhaust gas with ambient air, by keeping the flow constant through the PDP or CFV. Samples for continuous gas phase measurements are drawn from a point where the exhaust gas and dilution air mixture is homogeneous. A small fraction of the CVS flow (e.g., 0.1 %) is sampled into Teflon bags. Gas samples for the bags are drawn from downstream of the heat exchanger or, in the case of gases that may suffer losses in the heat exchanger (e.g., HC and NO_x), heated bag sampling from upstream of the heat exchanger is the preferred practice. Immediately after the test (e.g., within 20 or 30 min) the concentrations of the gases in the sample inside the bags are measured by means of the gas emission analyzers. Their operating principle ranges from electric or optical to gravimetric. Specifically [23]:

- for CO and CO_2 emissions, the non-dispersive infrared (NDIR) absorption type analyzer is used;
- for HCs, the flame ionization (FID) detector, appropriately heated to maintain all surfaces exposed to emissions at 191 °C. For separate calculation of NMHC, either a gas chromatograph FID, or a non-methane cutter installed upstream of the main FID can be utilized;
- for NO_x, the chemiluminescent, or alternatively the non-dispersive ultra-violet resonance absorption method (NDUV) is applied;
- for particulates, a gravimetric approach is followed;
- a particle number counter (PNC) is required by the European legislation for measuring the particle number concentration in the exhaust gas. The PNC efficiency is required to be 50 % at 23 nm and at least 90 % at 41 nm electrical mobility diameter.

For PM measurement, it is of paramount importance that there is no nucleation, coagulation or condensation since these lead to particles changes. In order to deal with these issues, on the one hand the temperature of exhaust gas should be maintained at specific levels (not more than 52 °C), and also the residence time in the dilution tunnel minimized. A sample of the total flow from the dilution tunnel is extracted and passed either directly ('single' dilution, Fig. 6.7) or through secondary dilution ('double' dilution) to the particulate sampling system, where the particles are collected on Teflon coated, fiber-glass filters. To ensure reproducible measurement of volatile components absorbed on the filter, it is required that the filter equilibrates at a defined temperature and humidity level. Particle emission is determined from the total particle mass collected onto the filters during the cycle (difference in the weight of the filter after and before the test) [24].

The CVS system is regularly calibrated, with the exact procedure defined in the relevant regulatory documents.

The following paragraphs provide the principal equations used to calculate the mass of emitted pollutants in Europe through UNECE regulations (hence are followed by many other countries in the world). Slight differentiations applicable in the United States are highlighted too.

The mass M_{gi} of each gaseous pollutant i emitted by the vehicle during the test is determined by obtaining the product of the volumetric concentration and the volume of the gas in question, taking into account the pollutant density under reference conditions (0 °C, 1 atm). The following equation applies [11, 16]

$$M_{gi} = V_{deg}\left(C_i \times 10^{-6}\right) K_h \rho_i \tag{6.8}$$

where:

- V_{deg} the volume of the diluted exhaust gas (L/test) (=volume of exhaust gas V_{eg} plus volume of ambient dilution air V_{da}). In the case of CVS with PDP, the volume of the diluted exhaust gas is the product of volume of gas V_{PDP} delivered by the PDP per revolution by the number of pump revolutions N_{PDP} (per test). The derived exhaust gas volume is then corrected to the standard reference conditions

$$V_{deg} = \left(\overbrace{V_{PDP}\,N_{PDP}}^{\text{exhaust gas volume}=V_{eg}+V_{da}}\right)\overbrace{\frac{273.15\,\text{K}}{101.325\,\text{kPa}} \times \frac{p_{PDP}}{\overline{T}_{PDP}}}^{\text{correction to reference conditions}} \tag{6.9}$$

with p_{PDP} the absolute pressure at PDP inlet (kPa) and $\overline{T}_{PDP}$ the average temperature of the diluted exhaust gas entering the PDP.

- C_i the measured emission concentration of pollutant i in the diluted exhaust gas (ppm) corrected by the amount of the pollutant i in the dilution air. If concentration C_i is measured %, then in Eq. (6.8) coefficient 10^{-6} is replaced by 10^{-2}. If the concentration of pollutant i is measured continuously (e.g., HC for CI engines), then its average concentration, before correction, during time period from t_1 to t_2 (e.g., from t = 0 to t_{cycle}) is calculated as follows

$$C_{i-deg} = \frac{\int_{t_1}^{t_2} C_{i-deg}(t)\,dt}{t_2 - t_1} \tag{6.10}$$

For the correction of the measured concentration, the following equation applies

$$C_i = C_{i-deg} - C_{i-da} \times \left(1 - \frac{1}{DF}\right) \tag{6.11}$$

where $C_{i\text{-}deg}$ concentration of pollutant i in the sample of diluted exhaust gas, $C_{i\text{-}da}$ concentration of pollutant i in the dilution air, and DF dilution factor (not to be confused with the CVS dilution ratio). DF depends on the measured concentrations of CO, HC and CO_2 in the diluted exhaust gas and the test fuel (general chemical type $C_1H_mO_z$)

$$DF = \frac{\left(\frac{100}{1+\frac{m}{2}+3.76\left(1+\frac{m}{4}-\frac{z}{2}\right)}\right)}{C_{CO_2} + (C_{HC} + C_{CO}) \times 10^{-4}} = \frac{F_s}{C_{CO_2} + (C_{HC} + C_{CO}) \times 10^{-4}} \qquad (6.12)$$

Parameter (stoichiometric factor) F_s corresponds to CO_2 concentration in the exhaust gas assuming complete combustion at stoichiometric conditions. For example in the EU, for B5 diesel fuel (m = 1.86, z = 0.005), F_s = 13.5; for natural gas (m = 4), F_s = 9.5 etc. For the U.S. 3-bag FTP-75 (Sect. 2.2.2), 40 CFR 1066 further defines an overall time-weighted dilution factor DF_w derived from the DFs from each phase of the test (cold-started, stabilization and hot-started), and each phase's time length. Owing to different chemical types for the reference fuels in the U.S. (see Table 6.4 later in the text), coefficient F_s assumes slightly different values compared to the ones valid in the EU.

- K_h humidity correction factor (for NO_x only) based on the absolute humidity H (grams of water per kg of dry air; 10.71 being the reference condition),

$$K_h = \frac{1}{1 - 0.0329 \times (H - 10.71)} \qquad (6.13)$$

Absolute humidity H can be calculated from the relative air humidity (%) and saturation vapor pressure at ambient temperature.

- ρ_i density of pollutant i (g/L) at standard pressure and temperature. The following values are used at 0 °C and 1 atm (the values for HCs depend on the test fuel):
  - for CO, ρ = 1.250 g/L,
  - for HCs, the assumption is made that the average composition of HCs in the exhaust is the same as in the fuel. HCs are expressed as $C_1H_mO_z$, with m the hydrogen to carbon atom ratio, and z the oxygen to carbon one. European regulations define the following:
    - for gasoline (equivalent hydrocarbon $C_1H_{1.85}$), ρ = 0.619 g/L;
    - for gasoline with 5 % ethanol, E5 ($C_1H_{1.89}O_{0.016}$), ρ = 0.631 g/L;
    - for ethanol E85 ($C_1H_{2.74}O_{0.385}$), ρ = 0.932 g/L;
    - for diesel ($C_1H_{1.86}$), ρ = 0.619 g/L;
    - for diesel with 5 % biodiesel, B5 ($C_1H_{1.86}O_{0.005}$), ρ = 0.622 g/L;
    - for diesel with 7 % biodiesel, B7 ($C_1H_{1.86}O_{0.007}$), ρ = 0.623 g/L;
    - for LPG ($C_1H_{2.525}$), ρ = 0.649 g/L;
    - for NG/bio-methane (CH_4), ρ = 0.714 g/L;
    - for NMHC, ρ is fuel dependent

(owing to slightly different fuel compositions defined in 40 CFR 1065 (see Tables 6.3 and 6.4), density values may differ in the U.S.);
- for NO_x (NO_2 equivalent) and NO_2, $\rho = 2.050$ g/L;
- for N_2O (nitrous oxide), $\rho = 1.964$ g/L;
- for CO_2, $\rho = 1.964$ g/L.

Specifically as regards NMHC, these are calculated from THC reduced by the amount of methane for the case a gas chromatography heated FID is utilized (Fig. 6.8)

$$C_{NMHC} = C_{THC} - (RF_{CH_4} \times C_{CH_4}) \tag{6.14}$$

where C_{NMHC} corrected concentration of NMHC in the diluted exhaust gas; C_{THC} concentration of THC in the diluted exhaust gas and corrected by the amount of THC in the dilution air; C_{CH_4} concentration of CH_4 in the diluted exhaust gas and corrected by the amount of CH_4 in the dilution air; RF_{CH_4} the FID response factor to methane.

Table 6.3 H/C and O/C atom ratio and carbon weight fraction of various reference fuels according to UNECE GTR No. 15 [11]

Fuel	H/C	O/C	Carbon weight fraction (CWF)
Petrol E0	1.85	0	0.866
Petrol E5	1.89	0.016	0.848
Ethanol E85	2.74	0.385	0.574
LPG	2.525	0	0.825
NG/biomethane	4	0	0.749
Diesel B0	1.86	0	0.866
Diesel B5	1.86	0.005	0.861
Diesel B7	1.86	0.007	0.859

Table 6.4 H/C and O/C atom ratio and carbon weight fraction of various reference fuels according to U.S. regulation 40 CFR 1065.655 (notice the difference in the natural gas composition between U.S. and European regulations)

Fuel	H/C	O/C	Carbon weight fraction (CWF)
Gasoline E0	1.85	0	0.866
Gasoline E10	1.92	0.03	0.833
Gasoline E15	1.95	0.05	0.817
Ethanol E85	2.73	0.38	0.576
Ethanol E100	3	0.5	0.521
Methanol M100	4	1	0.375
LPG	2.64	0	0.819
Natural gas	3.78	0.016	0.747
Diesel # 1	1.93	0	0.861
Diesel # 2	1.80	0	0.869

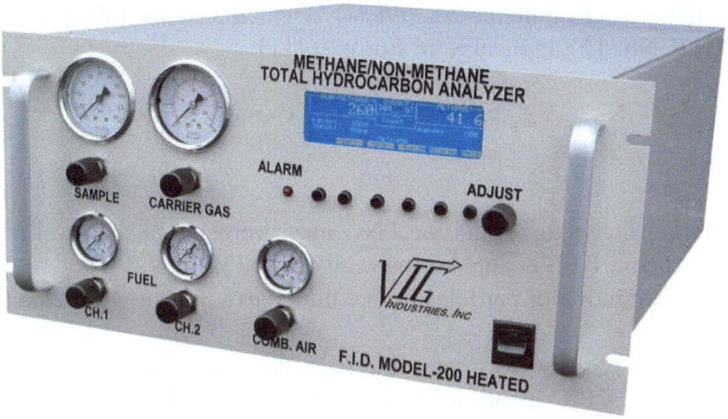

Fig. 6.8 Dual channel FID for HC measurement; one channel is utilized to measure total hydrocarbons and the second, coupled with a gas chromatography column, for the separation of the methane and non-methane components (*Source* VIG Industries)

Alternatively, a non-methane cutter (NMC) can be used. The NMC is a heated catalyst that oxidizes non-methane HCs. Ideally, the conversion for methane is 0 %, and for the other hydrocarbons (represented by ethane) 100 %, but the exact efficiencies are calculated for the certification procedure as described in the corresponding regulatory documents. Due to the FID measurement methodology utilized in Europe, the measured amount of THC practically includes organic compounds too, such as aldehydes and esters. Certain issues with the CVS testing procedure in the EU are highlighted and discussed in [21, 25].

In the United States, NMOG is one of the criteria pollutants (as opposed to NMHC or THC in the EU or Japan). Calculation of the amount of emitted non-methane organic gases is accomplished by further measuring alcohols and carbonyls as described in 40 CFR 1065.805 and 1065.845. Ethanol, formaldehyde and acetaldehyde emissions may be included during the WLTC test in Europe at a later stage, possibly based on the U.S. regulation [22].

For the particulate matter mass M_{PM} in g/test (primarily comprising carbon, condensed HCs, and sulfates with associated water)

$$M_{PM} = \frac{V_{deg}}{V_{ep}} M_{filter} \tag{6.15}$$

where:

- V_{ep} the volume of exhaust gas flowing through the particulate filter under standard conditions,
- M_{filter} mass of particulates collected on the filter after buoyancy correction.

Lastly, the particle number N_P (#/test) is calculated as follows according to UNECE GTR No. 15 (applicable in Europe) [11]:

$$N_P = V_{deg}\, k\left(C_{N-deg} f_{r-deg} - C_{N-da} f_{r-da}\right) \times 10^3 \tag{6.16}$$

with:

- k calibration factor to correct the PNC measurements to the level of the reference instrument (if it is applied internally, k = 1);
- $C_{N\text{-}deg}$ concentration of particles in the diluted exhaust gas corrected to standard conditions;
- $f_{r\text{-}deg}$ average particle concentration reduction factor of the volatile particle remover (the VPR incorporates devices for sample dilution and particle evaporation);
- $C_{N\text{-}da}$ concentration of particles in the dilution air corrected to standard conditions;
- $f_{r\text{-}da}$ average particle concentration reduction factor of the volatile particle remover for the background measurement.

Particle number concentration is highly susceptible to measurement variability, with (sometimes) considerable differentiations observed with changes in the dilution ratio, temperature and residence time of the exhaust gas at diluted conditions.

Distance-related Emissions

The mass emission for each gaseous pollutant i from Eq. (6.8) and for PM from Eq. (6.15), as well as the total particle number from Eq. (6.16), after appropriately accounting for any regeneration effects, are reduced to the actual distance S covered during the cycle (km or miles)

$$\begin{aligned} E_{gi}(g/km) &= \frac{M_{gi}\,(g/test)}{S} \\ E_{PM}(g/km) &= \frac{M_{PM}\,(g/test)}{S} \\ E_N(\#/km) &= \frac{N_P\,(\#/test)}{S} \end{aligned} \tag{6.17}$$

Distance S is obtained by multiplying the total number of revolutions shown on the dynamometer revolution counter by the circumference of the roller.

Special provisions are made for vehicles equipped with periodic regeneration systems. During regeneration of a DPF or NO_x adsorber, brief overshoot of emissions is experienced (the applicable limits may be exceeded in this case). For such vehicles, several test cycles are run so as for the regeneration system to be activated, and the final mass emission for each pollutant is calculated from the weighted average of these tests. In effect, the excess emissions generated during regeneration are spread out across all emission tests between successive events, providing a representative average emission level from the vehicle.

Fuel Consumption

The corresponding fuel consumption FC (L/100 km) in the EU regarding passenger cars, light-duty trucks and motorcycles is indirectly calculated from the measured emissions (g/km) of HC, CO and CO_2, with the final value depending on the test fuel chemical type and density ρ_f at 15 °C [11, 16, 26]. Specifically

$$FC = \left(\frac{1}{10 \times CWF \times \rho_f}\right) \times [(CWF \cdot E_{HC}) + (0.429 \cdot E_{CO}) + (0.273 \cdot E_{CO_2})] \quad (6.18)$$

with CWF the fuel carbon weight fraction. Table 6.3 provides typical values of CWF for various fuels taken from UNECE GTR No. 15.

In the United States, 40 CFR 600.113 provides the fuel consumption equations in mpg (miles per gallon) for all types of fuels, being similar in logic to the European ones, i.e.,

$$\begin{aligned} mpg_{gasoline} &= \frac{5174 \times 10^4 \times CWF \times SG}{[(CWF \cdot E_{HC}) + (0.429 \cdot E_{CO}) + (0.273 \cdot E_{CO_2})] \times [(0.6 \times SG \times NHV) + 5471]} \\ mpg_{diesel} &= \frac{2778}{(0.866 \cdot E_{HC}) + (0.429 \cdot E_{CO}) + (0.273 \cdot E_{CO_2})} \end{aligned} \quad (6.19)$$

with SG the fuel specific gravity, NHV the net heating value (Btu/lb), and all concentrations in g/mile. For the EPA 5-city overall fuel economy value (Sect. 2.2.4), 40 CFR 600.114 provides the applicable set of equations. It is noted that the U.S. regulation defines slightly different chemical types for the reference fuels, hence different CWF and coefficient F_s in the dilution factor Eq. (6.12) are derived; these are demonstrated in Table 6.4.

Other SI engine-related tests conducted for the emissions component of the type approval are the evaporative test (for a detailed historical overview of evaporative testing in Europe and the United States, see, for example, [27]), measurement of crankcase gases, and CO measurement at idling conditions. Details about these tests are provided in the relevant regulatory documents (Table 6.1). At the time of writing, UNECE GTR No. 15 (as of May 2014) only details the procedure for Type I emissions test and for fuel consumption/CO_2 measurement. Type II (CO concentration at idle), III (emissions of crankcase gases), IV (evaporative emissions), V (durability of antipollution devices) and VI (low ambient temperature) will be incorporated at a subsequent revision. These tests are detailed in previous regulations [8, 26]. Furthermore, Regulation 2016/427/EU describes the RDE (real driving emission) test procedure to be conducted in Europe from September 2017. This covers measurement and calculation of (initially) NO_x and PN during real-world driving at varying traffic conditions employing portable emission measurement systems (PEMS). Measurement is made from the raw exhaust gas, with NO_x emissions not corrected for ambient temperature and humidity. The distance-specific emissions are calculated based on the moving averaging window or the power binning method [28].

6.5.2 Engine-Dynamometer Testing

Constant Volume Sampling

For the case of heavy-duty engine testing, Figs. 6.9 and 6.10, either on a 'true' transient or a ramped modal cycle employing the CVS system, UNECE and European regulations define a similar, in general, approach, where the total engine exhaust gas is diluted with ambient air (or synthetic air or nitrogen) [29, 30]. Figure 6.11 demonstrates the whole test sequence during the engine emissions certification procedure (in this case, for the heavy-duty WHTC and non-road NRTC cycles).

Fig. 6.9 Engine-dynamometer test bed for emissions measurement (*Source* TÜV Hessen)

Fig. 6.10 Engine-dynamometer test bed for emissions measurement (*Source* AVL)

Fig. 6.11 Test sequence for engine emissions certification (e.g., WHTC or NRTC); for the equations regarding generation of the reference cycle, see Chaps. 4 for heavy-duty engines and 5 for non-road ones

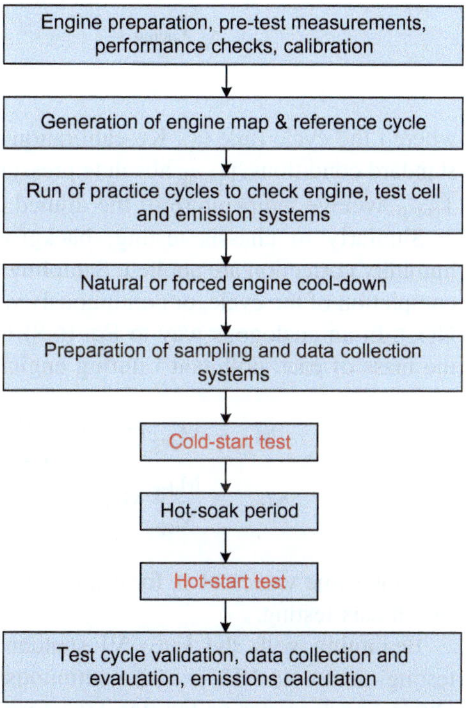

To keep the diluted exhaust temperature constant, a positive displacement pump can be used or a critical flow venturi, combined with a heat exchanger. Alternatively, a subsonic venturi or an ultrasonic flow meter can be employed.

If the emission of pollutant i is measured on a dry basis ($C_{i\text{-}d}$), a conversion to wet basis is applied

$$C_{i-w} = k_{dw} \times C_{i-d} \tag{6.20}$$

with k_{dw} the dry-to-wet correction factor; different equations apply for the correction of the raw exhaust gas, diluted exhaust gas and dilution air dry concentrations.

In the EU, one difference between chassis and engine CVS testing is that for the latter, the diluted exhaust gas is expressed on a mass basis (in kg/test) with the assumption of 1.293 kg/m^3 density (independent of test fuel since the dilution air assumes the greatest volume in the diluted exhaust gas). Thus, Eq. (6.9) is transformed into

$$\begin{aligned} M_{deg} &= 1.293 \cdot V_{deg} = 1.293 \cdot (V_{PDP} N_{PDP}) \frac{273.15\,\text{K}}{101.325\,\text{kPa}} \times \frac{p_{pump}}{\overline{T}_{pump}} \quad \text{for PDP systems} \\ M_{deg} &= 1.293 \cdot t \cdot K_V \frac{p_{pump}}{\sqrt{\overline{T}_{pump}}} \quad \text{for CFV systems} \end{aligned} \tag{6.21}$$

where t the cycle time (s), K_V calibration coefficient of the critical flow venturi for standard conditions, p_{pump} absolute pressure at pump (PDP or CFV) inlet (kPa), and $\overline{T}_{pump}$ average temperature of the diluted exhaust gas at pump inlet (K).

Similarly to chassis testing, background corrected concentrations and NO_x humidity correction are applied. Sampling can be made either in bags analyzed after completion of the cycle, or continuously with integration over the cycle (for HC and NO_x). In an analogous way to Eq. (6.8) for gaseous pollutants and (6.15) for PM, the mass of each pollutant i during engine CVS testing (g/test) is given by

$$\begin{aligned} M_{gi} &= M_{deg} C_{i-w} K_h \frac{\rho_i}{1.293} \quad \text{for gaseous pollutants} \\ M_{PM} &= \frac{M_{deg}}{M_{ep}} M_{filter} \quad \text{for PM} \end{aligned} \tag{6.22}$$

with the same values of ρ_i for each pollutant density as the ones mentioned earlier for chassis testing.

Beginning with the Euro VI standard in Europe, during heavy-duty engine testing, ammonia NH_3 is also continuously measured (in ppm) using either Laser Diode Spectrometer or Fourier Transform Infrared Analyzer [31]. This applies to engines equipped with SCR after-treatment system for NO_x abatement, and the measurement can be conducted in the undiluted exhaust gas [32].

Partial Flow Sampling

An alternative, cost effective to the CVS full flow dilution, option is to measure gaseous concentrations from raw exhaust gas, utilizing a partial flow dilution sampling system (PFSS) for PM, as demonstrated in Fig. 6.12. Here, gaseous emissions are measured continuously from the raw exhaust gas for all pollutants, whereas for PM (which is difficult to measure instantaneously), a small fraction of the engine exhaust gas is sampled, while dilution air (or synthetic air or nitrogen) is transiently added so that the diluted exhaust flow rate is constant through the filter [33]. The advantage when applying raw exhaust gas measurement is that pollutants concentration is preserved; on the other hand, the system deals with hot and humid samples, which may lead to losses, measurement interferences and even instrumentation damage. The usual choice is to use heated transfer lines to avoid water condensation ('wet' method). Raw sampling works well with FID, but may not be suitable for CO and CO_2. In such a case, dry measurements are preferred, subsequently corrected to wet as described earlier [19]. As regards the diluted exhaust gas, its temperature is maintained between 42 and 52 °C upstream of the filter holders. Since the flow of this sample must be proportional to the engine exhaust gas flow rate, the latter value must be known and has to be measured, e.g., by continuous measurement of air flow and fuel flow (the regulation provides various alternatives). Instantaneous measurements of exhaust emissions and mass flow rate should be performed at minimum 2 Hz sampling ratio.

Gaseous emission M_{gi} for pollutant i (of density ρ_i) is calculated from the integration of the instantaneous concentration in the exhaust gas (ppm) and the instantaneous exhaust mass flow

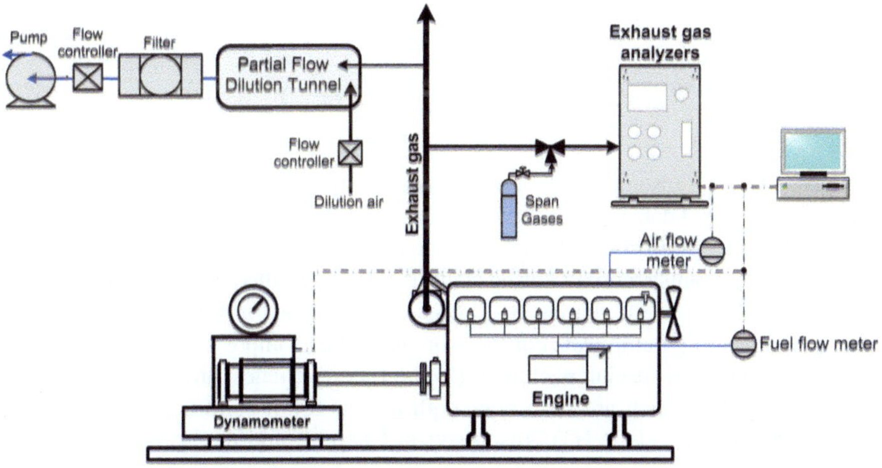

Fig. 6.12 Schematic arrangement of a partial flow sampling system for engine testing

$$M_{gi} = K_h \frac{\rho_i}{\rho_{eg}} \frac{1}{f} \sum_{j=1}^{n} \left[M_{eg}(j) \cdot C_{i-w}(j) \right] \tag{6.23}$$

with M_{eg} the instantaneous exhaust gas mass flow rate (kg/s) on a wet basis, ρ_{eg} the density of the exhaust gas (depending on the test fuel), f the data sampling rate (Hz) and n the number of measurements; K_h applies only to NO_x. An alternative procedure based on exact equations is also provided in Global Technical Regulation No. 4, which is more appropriate for the case the test fuel deviates from the reference ones; this procedure requires elementary analysis of the fuel composition [29].

In a similar to Eq. (6.23) manner, particulate mass emission calculation is based on the cycle integrated total exhaust mass flow, the integrated sample flow value through the PFSS and the filter mass loading.

Specific Emissions

Upon completion of the measurements, the mass of each gaseous pollutant M_i and the mass of particulates M_{PM} is reduced to the actual work W_{act} (kWh) produced by the engine

$$\begin{aligned} E_i\ (g/kWh) &= \frac{M_i(g/test)}{W_{act}} \\ E_{PM}\ (g/kWh) &= \frac{M_{PM}(g/test)}{W_{act}} \end{aligned} \tag{6.24}$$

The actual work is calculated using each pair of engine feedback speed and torque values recorded during the test, with at least 1 Hz sampling rate (the actual work should be within certain limits of the reference work, e.g., for the WHTC and the NRTC, between 85–105 %) [29, 30]

$$W_{act} = \sum_{j}^{n} P_j \Delta t_j = \frac{1}{f} \frac{1}{3600 \times 10^3} \frac{\pi}{30} \sum_{j}^{n} (N_j T_j) \tag{6.25}$$

with N_j, T_j, P_j, instantaneous engine speed (rpm), torque (Nm) and power (kW) respectively, f data sampling rate (Hz), and n the number of measurements during the test.[2] For the calculation of the actual work, all negative (motoring) torque values are set to zero; any points recorded during starting are also omitted. It should be noted that prior to certification, auxiliaries only necessary for the operation on the vehicle (e.g., power steering pump, A/C compressor, air-brakes compressor etc.) and/or linked with the operation of special machinery (for non-road vehicles) are removed.

For the cycles which are run under both cold and hot conditions (i.e., the U.S. FTP and the worldwide WHTC discussed in Chap. 4, and the non-road NRTC in

[2]It is reminded that the engine power (W) equals engine torque (Nm) times angular velocity (s^{-1}), $P = T\,\omega = T\,\pi\, n/30$.

Sect. 5.2), the final emission value of pollutant i is determined from the weighted average of the two runs

$$E_i\ (g/kWh) = \frac{wf_{cold} \cdot M_{i,cold} + wf_{hot} \cdot M_{i,hot}}{wf_{cold} \cdot W_{act,cold} + wf_{hot} \cdot W_{act,hot}} \tag{6.26}$$

where wf is the weighting factor assigned to each run. For the WHTC test in Europe and Japan, wf is 86 % for the hot start and 14 % for the cold start [29]; for the non-road NRTC, 90 and 10 % respectively [30] (95 and 5 % in the U.S. [34]); for the U.S. FTP, wf is 1/7 for the cold test and 6/7 for the hot one (40 CFR 86.007-11).

Special provisions apply for the case of engines with *periodic* regeneration systems, detailed in UNECE GTR No. 4 and 40 CFR 1065. For engines equipped with such after-treatment devices, average emissions between regeneration phases are determined as follows

$$E_{i,hot}\ (g/kWh) = F_r\overline{E}_{i,r} + (1 - F_r)\overline{E}_i \tag{6.27}$$

with

- F_r the frequency of the regeneration events expressed as fraction of tests during which regeneration occurs;
- $\overline{E}_{i,r}$ the average emission of pollutant i (g/kWh) during the regeneration cycle(s), and
- $\overline{E}_i$ the average emission of pollutant i (g/kWh) for the cycles outside regeneration.

An upward adjustment factor, UAF, can then be defined

$$UAF = E_{i,hot} - \overline{E}_i = F_r \cdot \left(\overline{E}_{i,r} - \overline{E}_i\right) \tag{6.28}$$

For example, if regeneration occurs once every 10 cycles, producing 1.6 g/kWh instead of 0.5 g/kWh during non-regeneration cycles, then the final amount of emitted pollutant i should be increased by

$$UAF = \frac{1}{10} \times (1.6 - 0.5) = 0.11\ g/kWh$$

Alternatively, a multiplicative regeneration factor $K_{r,i}$ can be defined for each pollutant i

$$K_{r,i} = \frac{E_{i,hot}}{\overline{E}_i} \tag{6.29}$$

For the previous example, $K_{r,i}$ = 0.61/0.5 = 1.22. The above-mentioned adjustment factors are applied to the weighted test result of Eq. (6.26). Table 6.5 presents the emission test results protocol to be filled during the WHTC test procedure.

Table 6.5 Emission test results protocol to be filled for the WHTC test [29]

	CO	THC	NO_x	PM	NH_3	PN
Deterioration Factor (DF)						
Emissions	mg/kWh				ppm	#
Cold start						
Hot start w/o regeneration						
Hot start w. regeneration						
Regeneration factor						
Weighted test result						
Final test result with DF						
CO_2 mass emission:................g/kWh						
Fuel consumption:................g/kWh						

In the United States, the relevant engine-dynamometer test procedure is described in 40 CFR 1065, being different to the older 40 CFR 86 regulation, Subpart N. One key change is that all calculations are based on mol (the same molar approach is also foreseen in UNECE GTR No. 11 concerning non-road engine emission calculations); mass-based calculations according to ISO 8178 are also approved. Using molar masses eliminates the confusion caused by applying different reference conditions for standard temperature and pressure. Further, the test system performance specifications according to 40 CFR 1065 are related to emission standards rather than to device ranges; continuous analyzers are employed without analog gains switch; low-quench chemiluminescent analyzers are required [32]. The amount of pollutants and specific emissions are calculated in the same manner as described above (40 CFR 1065.650).

Steady-State Engine-Dynamometer Testing

The procedure described above for transient engine testing is slightly different when *steady-state* engine testing is considered. As mentioned in Sect. 1.3.2, during 'true' stationary cycles, the engine under study undergoes the test by passing from each mode in a predefined order, allowing for adequate time to stabilize at each one. Measurements are undertaken during a certain time period, e.g., 30 s. Equation (6.30) describes the total amount of emitted pollutants measured from the raw exhaust gas (or, alternatively, with a CVS system) [29, 30, 35]

$$\begin{aligned} E_{gi} &= \frac{\sum_j \dot{M}_{gi-j} \times wf_j}{\sum_j P_j \times wf_j} \quad \text{(gaseous emissions)} \\ E_{PM} &= \frac{\dot{M}_{PMtotal}}{\sum_j P_j \times wf_j} \quad \text{(particulates)} \end{aligned} \tag{6.30}$$

with:

- E_{gi} the weighted average of gaseous emission i (g/kWh or g/HPh),
- $\dot{M}_{gi-j}$ the average gaseous emission of pollutant i at mode j (g/h); the average concentration of gaseous pollutants during each mode is determined from the average chart readings during a specified in the legislation time period

$$M_{gi} = K_h \frac{\rho_i}{\rho_{eg}} M_{eg} \cdot C_{i-w} \cdot 3600 \tag{6.31}$$

(M_{eg} is the exhaust gas flow rate on a wet basis (kg/s), and $C_{i\text{-}w}$ the emission concentration in the raw exhaust gas on a wet basis (kg/s))

- wf_j the weighting factor of mode j,
- P_j the net engine power (kW or HP) at mode j,
- E_{PM} the weighted average of particulate matter (g/kWh or g/HPh),
- $\dot{M}_{PMtotal}$ the total amount of particulates deposited on the filter (g/h).

It should be noted that many steady-state cycles are run as ramped modal tests (i.e., test cycles with a sequence of steady-state engine modes with defined speed/torque criteria at each mode and defined ramps between these modes). This means that the emission measurement is continuous, hence Eqs. (6.20–6.29) applicable to transient tests are valid. Typical examples of ramped modal testing

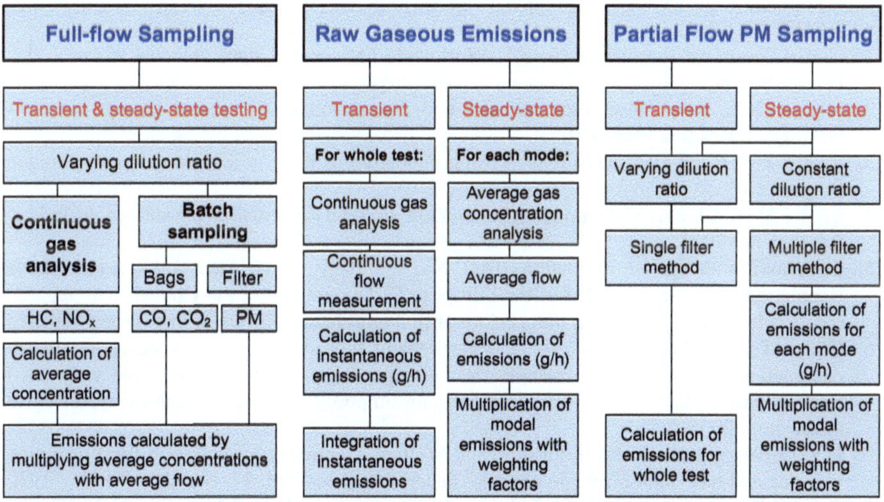

Fig. 6.13 Operations for the calculations of engine emissions during steady-state, ramped and transient engine-dynamometer cycles, utilizing either full-flow sampling (CVS) or PFSS (for PM) or raw gas sampling, according to GTR No. 11 [30]

are: in the United States, the heavy-duty engines ESC (as supplemental to the FTP transient test), in the EU the WHSC, as well as various non-road cycles (ISO 8178 C1 and D2 etc.) [36].

Lastly, Fig. 6.13 summarizes the exhaust emissions procedures during steady-state and transient testing for CVS, raw gas sampling, and PFSS for PM.

References

1. Mock P, German J. The future of vehicle emissions testing and compliance—How to align regulatory requirements, customer expectations, and environmental performance in the European Union. ICCT White Paper, Nov 2015.
2. Heinfellner H, Konrad C, Lichtblau G, Schodl B, Stranner G, Winter R. Legal obligations relating to emission measurements in the EU automotive sector. Directorate General for Internal Policies, IP/A/EMIS/2016-02, June 2016.
3. U.S. Environmental Protection Agency website. https://www3.epa.gov/. Accessed 6 Aug 2016.
4. Martyr AJ, Plint MA. Engine testing: the design, building, modification and use of powertrain test facilities. 4th ed. Oxford: Butterworh-Heineman; 2012.
5. Killedar JS. Dynamometer: theory and application to engine testing. Xlibris; 2012.
6. Gillespie TD. Fundamentals of vehicle dynamics. Warrendale PA: SAE International; 1992.
7. Wong JY. Theory of ground vehicles. 3rd ed. New York: Wiley; 2001.
8. UNECE Regulation No. 83. Uniform provisions concerning the approval of vehicles with regard to the emission of pollutants according to engine fuel requirements. Date of entry into force: 16 Oct 1995.
9. Kühlwein J. The impact of official versus real-world road loads on CO_2 emissions and fuel consumption of European passenger cars. ICCT White Paper, May 2016.
10. Kadijk G, Ligterink N. Road load determination of passenger cars. TNO Report 2012 R10237, Oct 2012.
11. Global Technical Regulation No. 15. Worldwide harmonized light vehicles test procedure. Established in the Global Registry on 12 Mar 2014.
12. SAE J1263. Road load measurement and dynamometer simulation using coastdown techniques. Revised Mar 2010.
13. SAE J2263. Road load measurement using onboard anemometry and coastdown techniques. Revised Dec 2008.
14. SAE J2264. Chassis dynamometer simulation of road load using coastdown techniques. Revised Jan 2014.
15. Commission delegated regulation (EU) No 134/2014 of 16 Dec 2013 supplementing Regulation (EU) No 168/2013 of the European Parliament and of the Council with regard to environmental and propulsion unit performance requirements and amending Annex V thereof.
16. Global Technical Regulation No. 2. Measurement procedure for two-wheeled motorcycles equipped with a positive or compression ignition engine with regard to the emission of gaseous pollutants, CO_2 emissions and fuel consumption. Established in the Global Registry on 22 June 2005.
17. U.S. Department of Health, Education and Welfare. Control of air pollution from new motor vehicles and new motor vehicle engines. Federal Register Vol. 35, No. 219, 10 Nov 1970.
18. Council Directive 83/351/EEC of 16 June 1983 amending Council Directive 70/220/EEC on the approximation of the laws of the Member States relating to measures to be taken against air pollution by gases from positive-ignition engines of motor vehicles.
19. Maricq MM. Exhaust emissions. In: Crolla D, Foster DE, Kobayashi T, Vaughan N, editors. Encyclopedia of automotive engineering. Wiley; 2014.

20. Council Directive 91/441/EEC of 26 June 1991 amending Directive 70/220/EEC on the approximation of the laws of the Member States relating to measures to be taken against air pollution by emissions from motor vehicles.
21. Merkisz J, Pielecha J, Radzimirski S. New trends in emission control in the European Union. Springer tracts on transportation and traffic, Vol. 4. Cham: Springer; 2014.
22. Engeljehringer K. Passenger car emission regulations. AVL-List GmbH, AVL-Italy Tech-day, Bologna, 21 Oct 2014.
23. Ladommatos N, Zhao H. Engine combustion instrumentation and diagnostics. Warrendale PA: SAE International; 2001.
24. Giechaskiel B, Maricq M, Ntziachristos L, Dardiotis C, Wang X, Axman H, et al. Review of motor vehicle particulate emissions sampling and measurement: from smoke and filter mass to particle number. J Aerosol Sci. 2014;67:48–86.
25. Majerczyk A, Radzimirski S. Analysis of measurement errors of CVS method used for tests of pollutant emissions from light-duty vehicles. J Kones Powertrain Transp. 2013;20:253–9.
26. Commission Regulation (EC) No 692/2008 of 18 July 2008 implementing and amending Regulation (EC) No 715/2007 of the European Parliament and of the Council on type-approval of motor vehicles with respect to emissions from light passenger and commercial vehicles (Euro 5 and Euro 6) and on access to vehicle repair and maintenance information.
27. Berg W. Legislation for the reduction of exhaust gas emissions. In: Gruden D, editor. The handbook of environmental chemistry, Vol. 3, Traffic and environment. Berlin: Springer; 2003. p. 175–253.
28. Commission Regulation (EU) 2016/427 of 10 March 2016 amending Regulation (EC) No 692/2008 as regards emissions from light passenger and commercial vehicles (Euro 6).
29. Global Technical Regulation No. 4. Test procedure for compression-ignition (C.I.) engines and positive-ignition (P.I.) engines fuelled with natural gas (NG) or liquefied petroleum gas (LPG) with regard to the emission of pollutants. Established in the Global Registry on 15 Nov 2006.
30. Global Technical Regulation No. 11. Engine emissions from agricultural and forestry tractors and from non-road mobile machinery. Established in the Global Registry on 12 Nov 2009.
31. Commission Regulation (EU) No 582/2011 of 25 May 2011 implementing and amending Regulation (EC) No 595/2009 of the European Parliament and of the Council with respect to emissions from heavy duty vehicles (Euro VI) and amending Annexes I and III to Directive 2007/46/EC of the European Parliament and of the Council.
32. Engeljehringer K. Emission: heavy-duty and off-road—emission test systems. AVL-List GmbH, 2010.
33. ISO 16183:2002. Heavy duty engines—measurement of gaseous emissions from raw exhaust gas and of particulate emissions using partial flow dilution systems under transient test conditions, Dec 2002.
34. U.S. Environmental Protection Agency. 40 CFR Parts 9, 69, et al. Control of emissions of air pollution from nonroad diesel engines and fuel; Final rule, Federal Register Vol. 69, No. 124, 29 June 2004.
35. Directive 2005/55/EC of the European Parliament and of the Council of 28 September 2005 on the approximation of the laws of the Member States relating to the measures to be taken against the emission of gaseous and particulate pollutants from compression-ignition engines for use in vehicles, and the emission of gaseous pollutants from positive-ignition engines fuelled with natural gas or liquefied petroleum gas for use in vehicles.
36. ISO 8178-4:2007. Reciprocating internal combustion engines—exhaust emission measurement—Part 4: steady-state test cycles for different engine applications, Dec 2007.

Appendix A—Technical Specifications of Chassis-Dynamometer Cycles

Abstract The following pages detail some of the most important technical specifications, together with the corresponding vehicle speed and vehicle acceleration diagrams, of the most influential chassis dynamometer driving cycles employed worldwide. Data for passenger cars/light-duty trucks cycles as well as heavy-duty vehicles and motorcycles ones is provided.

In the following pages, the most important chassis-dynamometer cycles will be summarized as regards speed profile and main technical specifications. Chassis-dynamometer cycles for passenger cars, light-duty trucks and motorcycles will be presented first (Sects. A.1–A.32). Afterwards, Sects. A.33–A.52 will focus on the heavy-duty test schedules. Lastly, the corresponding speed/acceleration distributions will be provided as well. Emphasis is placed on the worldwide light-duty WLTC and motorcycle WMTC, with all classes documented separately.

Since, apart from the speed profile, the vehicle acceleration throughout the cycle serves in highlighting the cycle's harshness, and is thus an important aspect of its overall structure being highly associated with the exhaust emissions, it was decided, for the purpose of completeness, to include the vehicle acceleration profile as well.

The order of presentation is: first European, then U.S., then Japanese, afterwards Indian, Australian and Chinese (if applicable), and finally worldwide cycles; also, cycles of passenger cars/light-duty vehicles first, then motorcycle ones, and lastly heavy-vehicle chassis-dynamometer test schedules.

Many of the cycles to be presented in the next pages are now obsolete (e.g., J10, heavy-duty UDDS) or have never been legislated (e.g., the ARTEMIS-project cycles or the WTVC). However, a decision has been made to reference them here for two reasons: First, they have played an important role over the past years (some even continue to play) in the research activity in many areas; one prominent example is the heavy amount of research carried out on the ARTEMIS-project cycles regarding the effects of biofuels, especially biodiesel, on diesel engine exhaust emissions. Second, some of these, obsolete now, cycles have been an integral part for the legislation of other, now universally applied, test schedules. One notable example is the worldwide WTVC for heavy-duty vehicles, which, although never legislated,

E.G. Giakoumis, *Driving and Engine Cycles*,
DOI 10.1007/978-3-319-49034-2

was the basis for the WHTC and WHSC (engine-dynamometer) cycles for heavy-duty engines.

The technical specifications provided in the next pages as well as the method employed for their calculation are described below:

- distance covered throughout the cycle S_{cycle} in km.
- cycle duration t_{cycle} in seconds and minutes.
- maximum vehicle speed V_{max} in km/h.
- average vehicle speed V_{avg} throughout the cycle (including idling).
- average driving speed V_{drive} (excluding idling).
- standard deviation σ of vehicle speed.
- average acceleration (positive values of acceleration included only) in m/s^2.
- maximum acceleration.
- average deceleration.
- minimum deceleration.
- standard deviation σ of (positive) acceleration.

It should be noted that the instantaneous accelerations are calculated based on the speed values between the current and the previous time-point. Some researchers, on the other hand, use the next and the previous time-point speed values to calculate the acceleration for the current second, i.e., the central difference method, and this yields slightly different results, which, in turn, influence all acceleration-related properties, e.g., RPA.

- total driving time (excluding idle periods) t_{drive}.
- percentage drive time spent with vehicle speed lower than 30 km/h (includes idle too), between 30 and 60, between 60 and 100, and higher than 100 km/h; sum of all these percentages is 100 %.
- total cruising time t_{cruise}, i.e., total time spent with constant speed; definition of cruising time can be dubious. For example, successive time points within a cycle with less than (e.g.) 0.3 km/h difference in vehicle speed could be considered as cruising. However, if the vehicle speed drops constantly from an initial value at 0.3 km/h increments, then this certainly cannot be considered cruising. In order to avoid such discrepancies, the strict definition of cruising has been chosen, that is speed remaining constant between two successive seconds.
- total time spent accelerating t_{accel} (speed increasing between two consecutive cycle points).
- percentage time spent accelerating between 0 and 1 m/s^2 and at higher than 1 m/s^2; sum of these two percentages is the total percentage time accelerating.
- total time spent decelerating t_{decel} (speed decreasing between two consecutive cycle points).
- percentage time spent decelerating at lower than −1 m/s^2 and between −1 and 0 m/s^2; sum of these two percentages is the total percentage time decelerating.
- total idle time t_{idle}, when both vehicle speed and acceleration are zero. Owing to a lot of cycles having many speed-points with lower than 1 km/h velocity, the

percentage of the total time spent during the cycle with speed lower than 1 km/h is also indicated in parenthesis, next to the 'true' idling time. A notable example here is the NYCC. During this cycle, the vehicle spends 32 % of the time at zero velocity and zero acceleration (which is the strict definition of idling). If, however, the lower than 1 km/h (or 1 mph) vehicle speeds are accounted for, the idling time increases to 40 % (therefore the time for acceleration and deceleration are equally reduced).

- all the above time periods (t_{drive}, t_{accel}, t_{decel}, t_{cruise} and t_{idle}) are provided both in seconds and as a percentage of the total cycle duration (e.g., % accelerating time = 100 * t_{accel}/t_{cycle} etc).
- number of accelerations throughout the cycle N_{accel} (a speed increase is counted here as an acceleration if it lasts at least 2 s). An important remark here is that a continuous speed increase, with up-gear shift changes, counts as more than one acceleration depending on the number of the intermediate gear changes. For example, the second speed increase in the European ECE-15 (from t = 49–61 s) is counted as two accelerations, since the driver changes the engaged gear. Likewise, the segment in the JC08 cycle from t = 1046–1087 s is a continuous speed increase part, with more than 70 km/h difference between final and initial vehicle speed; it is counted here as three accelerations, as one would expect at least two intermediate gear changes.
- number of accelerations per km = N_{accel}/S_{cycle}, and per min = N_{accel}/t_{cycle} (being strongly influenced by the preceding definition of acceleration).
- number of vehicle stops (including the initial and final idling period, if applicable) N_{stop}; it is noted that a stop is counted here if it lasts at least 4 s, hence no stops included, for example, in the U.S. HFET cycle.
- maximum (intermediate) stop duration, that is excluding the initial or final idle phase (hence, this value for all highway cycles is zero).
- number of stops per km = N_{stop}/S_{cycle}.
- average stop duration = t_{idle}/N_{stop} in seconds.
- relative positive acceleration $\mathrm{RPA} = \frac{1}{S_{cycle}} \int_{t=1}^{t_{cycle}} V(t) \cdot a(t) \cdot dt$.
- positive kinetic energy $\mathrm{PKE} = \frac{1}{S_{cycle}} \sum_{t=1}^{t_{cycle}} \left[V^2(t) - V^2(t-1)\right]$.

According to the above, total cycle duration consists of driving and idling time, that is $t_{drive} + t_{idle} = t_{cycle}$. Driving time t_{drive}, on the other hand, consists of the periods when the vehicle accelerates t_{accel}, cruises t_{cruise} and decelerates t_{decel}, that is $t_{drive} = t_{cruise} + t_{accel} + t_{decel}$.

Lastly, the cycles' data in all the figures and tables have been calculated based on the vehicle speed profile in km/h, even for the U.S. ones for which the speed points are provided by the EPA in mph; 1 mile = 1.609 km is the conversion ratio used for the calculations (EPA suggests that the mph speeds be multiplied with 1.6 and rounded up to the next decimal digit km/h value, an approach followed by many researchers from the U.S; for example, 12.55 mph * 1.609 = 20.19 km/h, whereas 12.55 * 1.6 = 20.08 km/h, rounded up to 20.10 km/h).

Passenger Cars and Light-duty Trucks

A.1 New European Driving Cycle for Passenger Cars and Light-Duty Trucks—NEDC

From year 2000, the cycle is run with the engine cold started (MVEG-B). Previously, there was a 40-s warm-up idling period. Cycle with constant accelerations, prolonged cruise and simple structure. In this form, valid from 1992 to 2017 (Fig. A.1 and Table A.1).

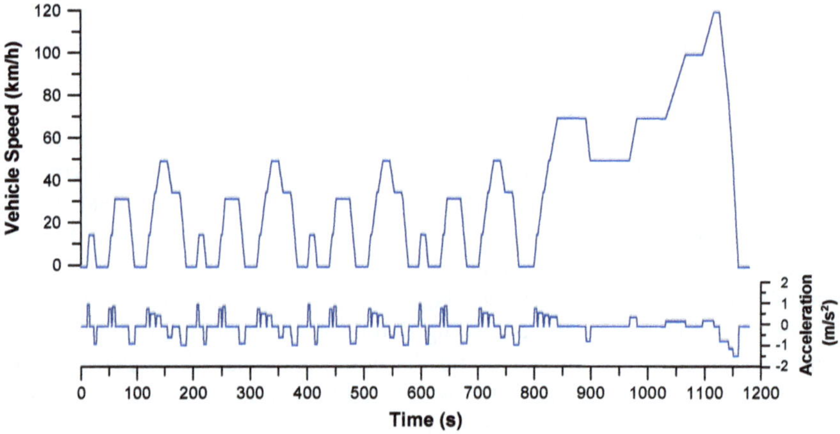

Fig. A.1 Vehicle speed and acceleration versus time of the European NEDC

Table A.1 Technical specifications of the European NEDC

Distance (m)/Duration (s or min)	11,000	1180 (19.67)
Maximum/Average vehicle speed (km/h)	120.00	33.56
Average driving speed (km/h)/Speed σ (km/h)	44.00	30.96
Average/Maximum acceleration (m/s^2)	0.594	1.042
Average/Minimum deceleration (m/s^2)	−0.789	−1.389
Driving time (s)/(%)	900	76.27
Driving time (%) V $\leq$ 30 km/h/30 < V $\leq$ 60 km/h	46.44	33.81
Driving time (%) 60 < V $\leq$ 100 km/h/V > 100 km/h	16.61	3.14
Cruising time (s)/(%)	467	39.58
Time spent accelerating (s)/(%)	247	20.93
Acceleration time (%) 0 < a $\leq$ 1.0 m/s^2/a > 1.0 m/s^2	19.58	1.35
Time spent decelerating (s)/(%)	186	15.76
Decel. time (%) a < −1.0 m/s^2/−1.0 $\leq$ a < 0 m/s^2	1.53	14.24
Idling time (s)/(%)	280	23.73 (24.83)
No. of accelerations/Positive acceleration σ (m/s^2)	31	0.252
Accelerations per km/per min	2.82	1.58
Number of stops/Max. interm. stop duration (s)	14	27
Stops per km/Average stop duration (s)	1.27	20.00
RPA (m/s^2)/PKE (m/s^2)	0.116	2.890

A.2 New European Driving Cycle for Passenger Cars and Light-Duty Trucks—Urban ECE-15 Section

Representative of heavy traffic in a European city. One simple schedule, comprising three micro-trips with long cruising time, repeated four times. Originally adopted in 1970. The same cycle, but with six repetitions instead of four, was the motorcycles R40 test cycle (Fig. A.2 and Table A.2).

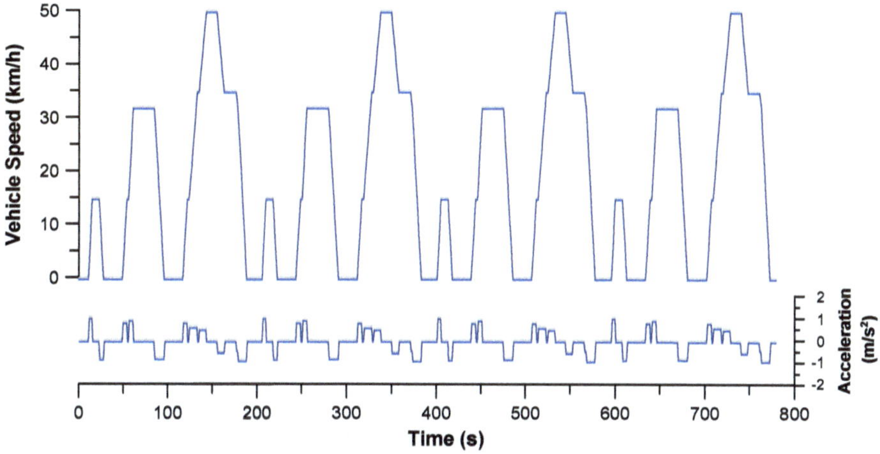

Fig. A.2 Vehicle speed and acceleration versus time of the urban ECE-15 segment of the NEDC

Table A.2 Technical specifications of the urban ECE-15 segment of the NEDC

Distance (m)/Duration (s or min)	4045.6	780 (13)
Maximum/Average vehicle speed (km/h)	50.00	18.67
Average driving speed (km/h)/Speed σ (km/h)	26.97	17.04
Average/Maximum acceleration (m/s^2)	0.748	1.042
Average/Minimum deceleration (m/s^2)	−0.748	−0.889
Driving time (s)/(%)	540	69.23
Driving time (%) V ≤ 30 km/h/30 < V ≤ 60 km/h	62.56	37.44
Driving time (%) 60 < V ≤ 100 km/h/V > 100 km/h	0	0
Cruising time (s)/(%)	252	32.31
Time spent accelerating (s)/(%)	144	18.46
Acceleration time (%) 0 < a ≤ 1.0 m/s^2/a > 1.0 m/s^2	16.41	2.05
Time spent decelerating (s)/(%)	144	18.46
Decel. time (%) a < −1.0 m/s^2/−1.0 ≤ a < 0 m/s^2	0	18.46
Idling time (s)/(%)	240	30.77 (32.31)
No. of accelerations/Positive acceleration σ (m/s^2)	24	0.181
Accelerations per km/per min	5.93	1.85
Number of stops/Max. interm. stop duration (s)	13	21
Stops per km/Average stop duration (s)	3.21	18.46
RPA (m/s^2)/PKE (m/s^2)	0.154	3.707

A.3 New European Driving Cycle for Passenger Cars and Light-Duty Trucks—Extra Urban EUDC Section

Introduced in 1992 with emission standard Euro 1. Accounts for motorway driving (high speeds, no intermediate idling); simply structured, modal cycle (Fig. A.3 and Table A.3).

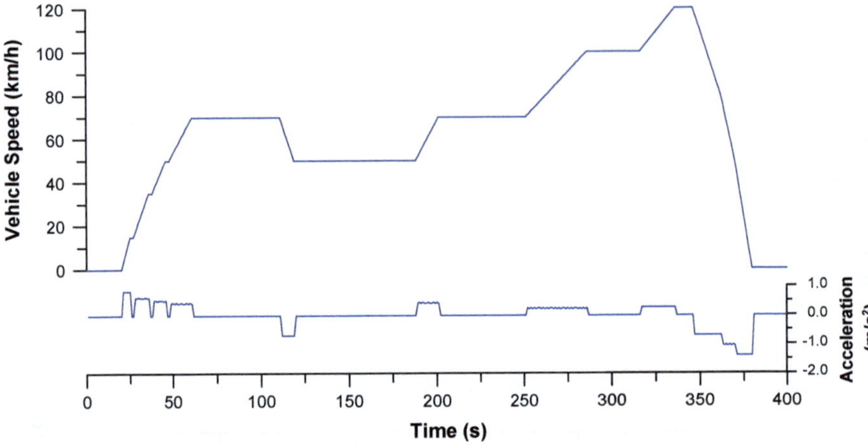

Fig. A.3 Vehicle speed and acceleration versus time of the EUDC segment of the NEDC

Table A.3 Technical specifications of the EUDC segment of the NEDC

Distance (m)/Duration (s or min)	6954.9	400 (6.67)
Maximum/Average vehicle speed (km/h)	120.00	62.59
Average driving speed (km/h)/Speed σ (km/h)	69.55	31.40
Average/Maximum acceleration (m/s^2)	0.378	0.833
Average/Minimum deceleration (m/s^2)	−0.926	−1.389
Driving time (s)/(%)	360	90.00
Driving time (%) V $\leq$ 30 km/h/30 < V $\leq$ 60 km/h	15.00	26.75
Driving time (%) 60 < V $\leq$ 100 km/h/V > 100 km/h	49.00	9.25
Cruising time (s)/(%)	215	53.75
Time spent accelerating (s)/(%)	103	25.75
Acceleration time (%) 0 < a $\leq$ 1.0 m/s^2/a > 1.0 m/s^2	25.75	0
Time spent decelerating (s)/(%)	42	10.50
Decel. time (%) a < −1.0 m/s^2/−1.0 $\leq$ a < 0 m/s^2	4.50	6.00
Idling time (s)/(%)	40	10.00 (10.25)
No. of accelerations/Positive acceleration σ (m/s^2)	7	0.161
Accelerations per km/per min	1.01	1.05
Number of stops/Max. interm. stop duration (s)	2	–
Stops per km/Average stop duration (s)	0.29	20.00
RPA (m/s^2)/PKE (m/s^2)	0.094	2.416

A.4 European ARTEMIS Urban

Part of the ARTEMIS-project cycles, this schedule is a much more realistic (than the legislated NEDC) simulation of the actual driving conditions in a European city (Fig. A.4 and Table A.4).

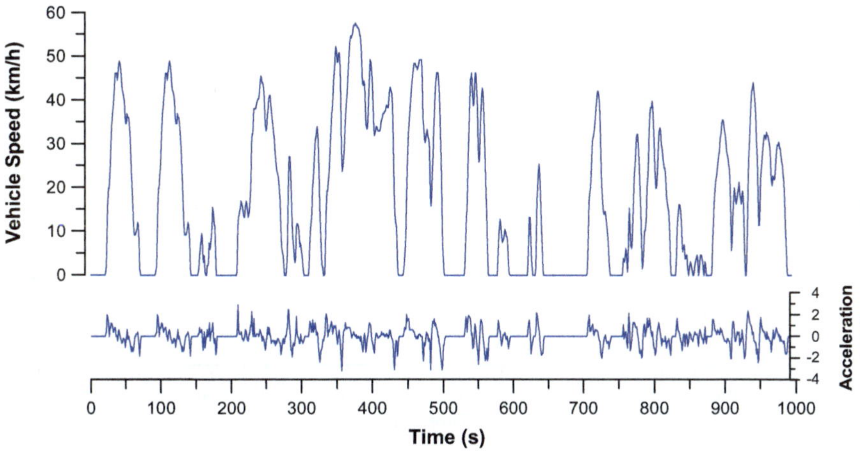

Fig. A.4 Vehicle speed and acceleration versus time of the ARTEMIS Urban

Table A.4 Technical specifications of the ARTEMIS Urban

Distance (m)/Duration (s or min)	4869.8	993 (16.55)
Maximum/Average vehicle speed (km/h)	57.70	17.65
Average driving speed (km/h)/Speed σ (km/h)	23.92	17.01
Average/Maximum acceleration (m/s^2)	0.732	2.861
Average/Minimum deceleration (m/s^2)	−0.782	−3.139
Driving time (s)/(%)	733	73.82
Driving time (%) V < 30 km/h/30 < V < 60 km/h	70.59	29.41
Driving time (%) 60 < V < 100 km/h/V > 100 km/h	0	0
Cruising time (s)/(%)	67	6.75
Time spent accelerating (s)/(%)	344	34.64
Acceleration time (%) $0 < a \leq 1.0$ m/s^2/a > 1.0 m/s^2	25.68	8.96
Time spent decelerating (s)/(%)	322	32.42
Decel. time (%) $a < -1.0$ m/s^2/$-1.0 \leq a < 0$ m/s^2	9.77	22.66
Idling time (s)/(%)	260	26.18 (30.41)
No. of accelerations/Positive acceleration σ (m/s^2)	46	0.536
Accelerations per km/per min	9.45	2.78
Number of stops/Max. interm. stop duration (s)	14	61
Stops per km/Average stop duration (s)	2.87	18.57
RPA (m/s^2)/PKE (m/s^2)	0.342	8.117

A.5 European ARTEMIS Rural Road

Part of the ARTEMIS-project cycles; simulates driving in rural roads (Fig. A.5 and Table A.5).

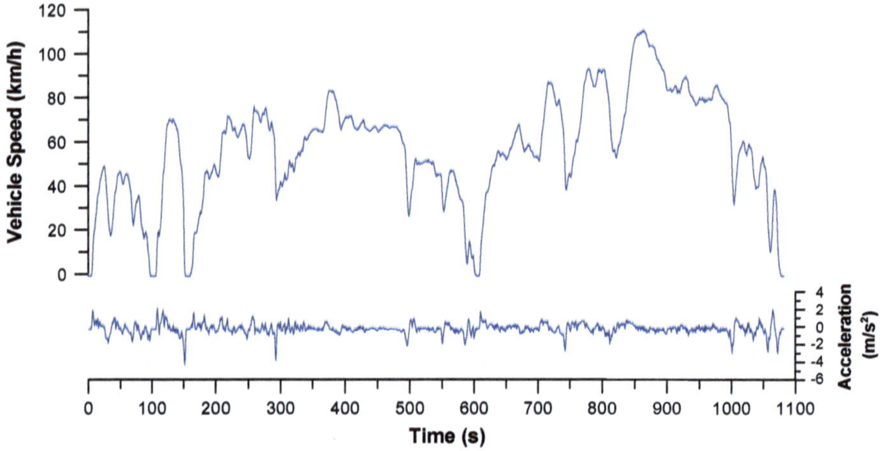

Fig. A.5 Vehicle speed and acceleration versus time of the ARTEMIS rural road

Table A.5 Technical specifications of the ARTEMIS rural road

Distance (m)/Duration (s or min)	17,272.5	1082 (18.03)
Maximum/Average vehicle speed (km/h)	111.50	57.46
Average driving speed (km/h)/Speed σ (km/h)	59.05	24.55
Average/Maximum acceleration (m/s^2)	0.494	2.361
Average/Minimum deceleration (m/s^2)	−0.516	−4.083
Driving time (s)/(%)	1053	97.32
Driving time (%) V $\leq$ 30 km/h/30 < V $\leq$ 60 km/h	13.31	37.43
Driving time (%) 60 < V $\leq$ 100 km/h/V > 100 km/h	45.84	3.42
Cruising time (s)/(%)	174	16.08
Time spent accelerating (s)/(%)	449	41.50
Acceleration time (%) 0 < a $\leq$ 1.0 m/s^2/a > 1.0 m/s^2	37.71	3.79
Time spent decelerating (s)/(%)	430	39.74
Decel. time (%) a < −1.0 m/s^2/−1.0 $\leq$ a < 0 m/s^2	5.08	34.66
Idling time (s)/(%)	29	2.68 (3.51)
No. of accelerations/Positive acceleration σ (m/s^2)	43	0.399
Accelerations per km/per min	2.49	2.38
Number of stops/Max. interm. stop duration (s)	5	9
Stops per km/Average stop duration (s)	0.29	5.80
RPA (m/s^2)/PKE (m/s^2)	0.182	4.571

A.6 European ARTEMIS Motorway 150

Part of the ARTEMIS-project cycles; simulates highway driving. There is also a version with maximum speed 130 km/h (Fig. A.6 and Table A.6).

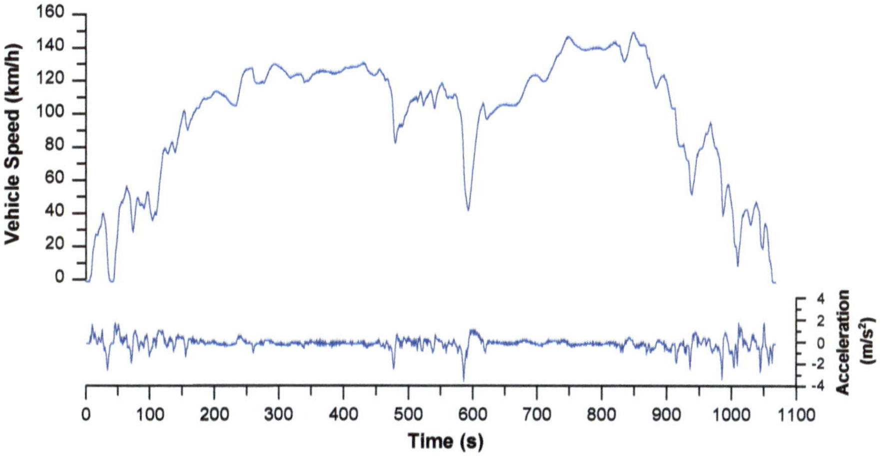

Fig. A.6 Vehicle speed and acceleration versus time of the ARTEMIS Motorway 150

Table A.6 Technical specifications of the ARTEMIS Motorway 150

Distance (m)/Duration (s or min)	29,545.0	1068 (17.8)
Maximum/Average vehicle speed (km/h)	150.40	99.59
Average driving speed (km/h)/Speed σ (km/h)	100.91	37.58
Average/Maximum acceleration (m/s^2)	0.426	1.917
Average/Minimum deceleration (m/s^2)	−0.509	−3.361
Driving time (s)/(%)	1054	98.69
Driving time (%) $V \leq 30$ km/h/$30 < V \leq 60$ km/h	6.37	13.95
Driving time (%) $60 < V \leq 100$ km/h/$V > 100$ km/h	14.23	65.45
Cruising time (s)/(%)	268	25.09
Time spent accelerating (s)/(%)	428	40.07
Acceleration time (%) $0 < a \leq 1.0$ m/s^2/$a > 1.0$ m/s^2	36.99	3.09
Time spent decelerating (s)/(%)	358	33.52
Decel. time (%) $a < -1.0$ m/s^2/$-1.0 \leq a < 0$ m/s^2	3.93	29.59
Idling time (s)/(%)	14	1.31 (1.78)
No. of accelerations/Positive acceleration σ (m/s^2)	39	0.345
Accelerations per km/per min	1.32	2.19
Number of stops/Max. interm. stop duration (s)	3	4
Stops per km/Average stop duration (s)	0.10	4.67
RPA (m/s^2)/PKE (m/s^2)	0.134	3.423

A.7 European ARTEMIS Urban-Road-Motorway 150

Part of the ARTEMIS-project cycles. This is the entire cycle with urban, rural and motorway parts. Very long distance (approx. 52 km) and duration (more than 52 min), and very high maximum vehicle speed (Fig. A.7 and Table A.7).

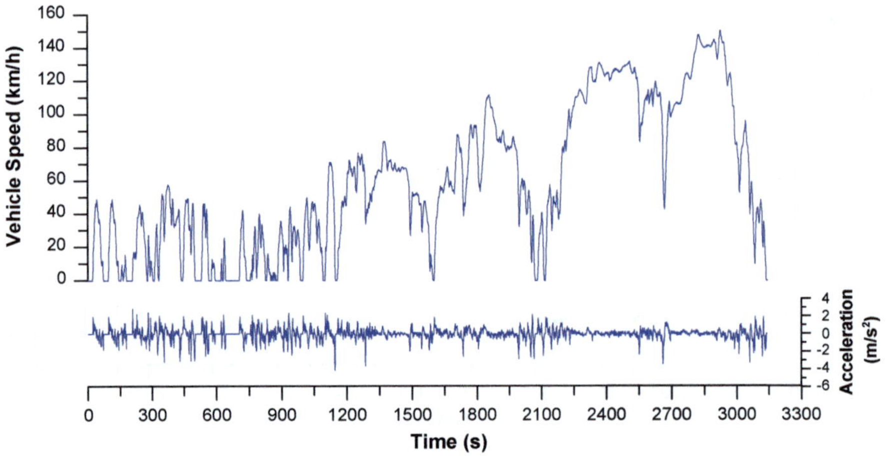

Fig. A.7 Vehicle speed and acceleration versus time of the ARTEMIS Urban-Rural-Motorway 150

Table A.7 Technical specifications of the ARTEMIS Urban-Rural-Motorway 150

Distance (m)/Duration (s or min)	51,687.4	3143 (52.38)
Maximum/Average vehicle speed (km/h)	150.40	59.20
Average driving speed (km/h)/Speed σ (km/h)	65.52	43.35
Average/Maximum acceleration (m/s^2)	0.537	2.861
Average/Minimum deceleration (m/s^2)	−0.591	−4.083
Driving time (s)/(%)	2840	90.36
Driving time (%) $V \leq 30$ km/h/$30 < V \leq 60$ km/h	29.05	26.92
Driving time (%) $60 < V \leq 100$ km/h/$V > 100$ km/h	20.62	23.42
Cruising time (s)/(%)	509	16.19
Time spent accelerating (s)/(%)	1221	38.84
Acceleration time (%) $0 < a \leq 1.0$ m/s^2/$a > 1.0$ m/s^2	33.66	5.19
Time spent decelerating (s)/(%)	1110	35.32
Decel. time (%) $a < -1.0$ m/s^2/$-1.0 \leq a < 0$ m/s^2	6.17	29.14
Idling time (s)/(%)	303	9.64 (11.42)
No. of accelerations/Positive acceleration σ (m/s^2)	128	0.444
Accelerations per km/per min	2.48	2.44
Number of stops/Max. interm. stop duration (s)	20	61
Stops per km/Average stop duration (s)	0.39	15.15
RPA (m/s^2)/PKE (m/s^2)	0.170	4.249

A.8 U.S. FTP-72 Cycle for Passenger Cars, Light-Duty Trucks, and Motorcycles with Engine Capacity > 170 cm³

Also known as UDDS or LA4; highly transient urban/suburban cycle. Valid from model year 1972 in the United States for passenger cars. Notice that the highest engine speed occurs relatively early in the cycle. For motorcycles with engine capacity lower than 170 cm^3, see Sect. A.26 (Fig. A.8 and Table A.8).

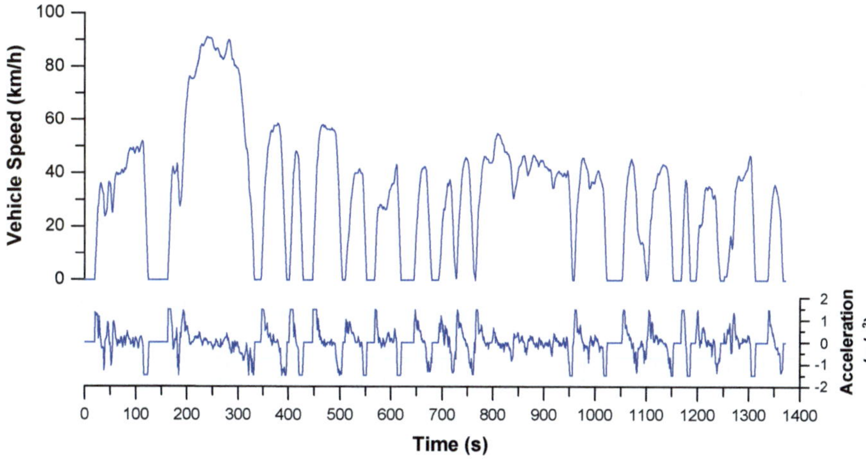

Fig. A.8 Vehicle speed and acceleration versus time of the U.S. FTP-72

Table A.8 Technical specifications of the U.S. FTP-72

Distance (m)/Duration (s or min)	11,990.2	1372 (22.87)
Maximum/Average vehicle speed (km/h)	91.25	31.46
Average driving speed (km/h)/Speed σ (km/h)	38.27	23.66
Average/Maximum acceleration (m/s^2)	0.505	1.475
Average/Minimum deceleration (m/s^2)	−0.578	−1.475
Driving time (s)/(%)	1128	82.21
Driving time (%) $V \leq 30$ km/h/$30 < V \leq 60$ km/h	42.93	48.76
Driving time (%) $60 < V \leq 100$ km/h/$V > 100$ km/h	8.31	0
Cruising time (s)/(%)	109	7.94
Time spent accelerating (s)/(%)	544	39.65
Acceleration time (%) $0 < a \leq 1.0$ m/s^2/$a > 1.0$ m/s^2	33.16	6.49
Time spent decelerating (s)/(%)	475	34.62
Decel. time (%) $a < -1.0$ m/s^2/$-1.0 \leq a < 0$ m/s^2	8.67	25.95
Idling time (s)/(%)	244	17.78 (19.61)
No. of accelerations/Positive acceleration σ (m/s^2)	57	0.451
Accelerations per km/per min	4.75	2.49
Number of stops/Max. interm. stop duration (s)	15	38
Stops per km/Average stop duration (s)	1.25	16.27
RPA (m/s^2)/PKE (m/s^2)	0.185	4.536

A.9 U.S. FTP-75 Cycle for Passenger Cars and Light-Duty Trucks

Derived from the FTP-72 by adding a third hot-started phase of 505 s, identical to the first (cold-started) one. The third phase starts after the engine has stopped for 10 minutes. Valid from model year 1975 (Fig. A.9 and Table A.9).

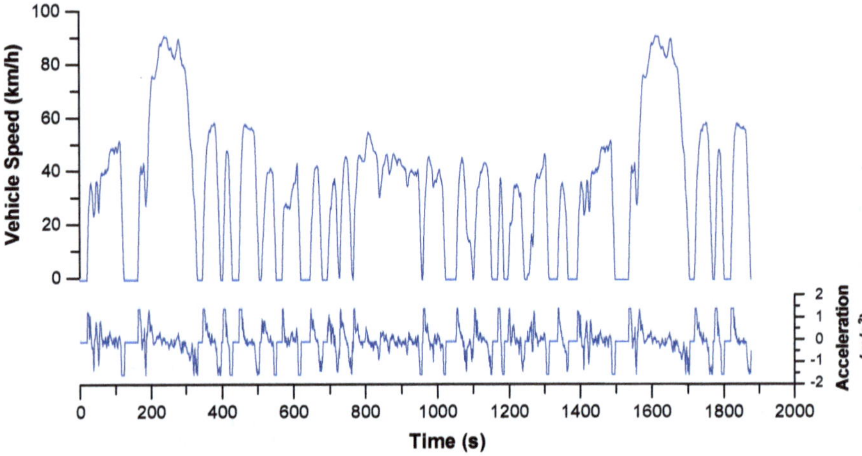

Fig. A.9 Vehicle speed and acceleration versus time of the U.S. FTP-75

Table A.9 Technical specifications of the U.S. FTP-75

Distance (m)/Duration (s or min)	17,769.4	1877 (31.28)
Maximum/Average vehicle speed (km/h)	91.25	34.08
Average driving speed (km/h)/Speed σ (km/h)	41.57	25.67
Average/Maximum acceleration (m/s^2)	0.511	1.475
Average/Minimum deceleration (m/s^2)	−0.576	−1.475
Driving time (s)/(%)	1539	81.99
Driving time (%) V $\leq$ 30 km/h/30 < V $\leq$ 60 km/h	40.65	47.20
Driving time (%) 60 < V $\leq$ 100 km/h/V > 100 km/h	12.15	0
Cruising time (s)/(%)	145	7.73
Time spent accelerating (s)/(%)	739	39.37
Acceleration time (%) 0 < a $\leq$ 1.0 m/s^2/a > 1.0 m/s^2	32.34	7.03
Time spent decelerating (s)/(%)	655	34.90
Decel. time (%) a < −1.0 m/s^2/−1.0 $\leq$ a < 0 m/s^2	8.79	26.11
Idling time (s)/(%)	338	18.01 (19.61)
No. of accelerations/Positive acceleration σ (m/s^2)	78	0.463
Accelerations per km/per min	4.39	2.49
Number of stops/Max. interm. stop duration (s)	19	38
Stops per km/Average stop duration (s)	1.07	17.79
RPA (m/s^2)/PKE (m/s^2)	0.184	4.508

A.10 U.S. Highway Fuel Economy Test—HFET

Rural/motorway driving cycle developed in the early 70s, used for fuel economy purposes (Fig. A.10 and Table A.10).

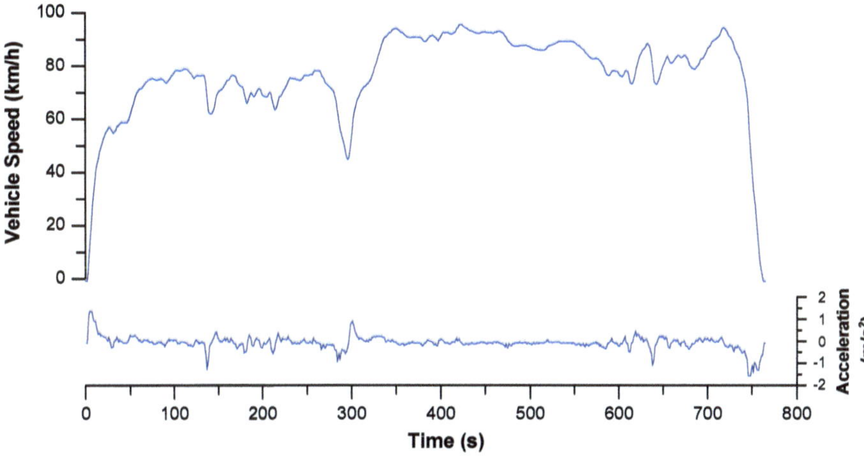

Fig. A.10 Vehicle speed and acceleration versus time of the U.S. HFET

Table A.10 Technical specifications of the U.S. HFET

Distance (m)/Duration (s or min)	16,506.6	765 (12.75)
Maximum/Average vehicle speed (km/h)	96.40	77.68
Average driving speed (km/h)/Speed σ (km/h)	78.09	16.24
Average/Maximum acceleration (m/s^2)	0.194	1.431
Average/Minimum deceleration (m/s^2)	−0.221	−1.475
Driving time (s)/(%)	761	99.47
Driving time (%) V $\leq$ 30 km/h/30 < V $\leq$ 60 km/h	2.61	7.97
Driving time (%) 60 < V $\leq$ 100 km/h/V > 100 km/h	89.41	0
Cruising time (s)/(%)	126	16.47
Time spent accelerating (s)/(%)	338	44.18
Acceleration time (%) 0 < a $\leq$ 1.0 m/s^2/a > 1.0 m/s^2	43.40	0.78
Time spent decelerating (s)/(%)	297	38.82
Decel. time (%) a < −1.0 m/s^2/−1.0 $\leq$ a < 0 m/s^2	1.70	37.12
Idling time (s)/(%)	4	0.52 (0.65)
No. of accelerations/Positive acceleration σ (m/s^2)	26	0.221
Accelerations per km/per min	1.58	2.04
Number of stops/Max. interm. stop duration (s)	0	–
Stops per km/Average stop duration (s)	0	0
RPA (m/s^2)/PKE (m/s^2)	0.071	1.830

A.11 U.S. SFTP US06

Supplemental, to the FTP-75, US06 simulates aggressive, high-speed and high-acceleration driving behavior on a motorway; based on data from driving conditions in Baltimore, Atlanta, Los Angeles and Spokane in the early 90s. Among the cycles with the highest vehicle speed and acceleration. Implemented gradually from 2000 (Fig. A.11 and Table A.11).

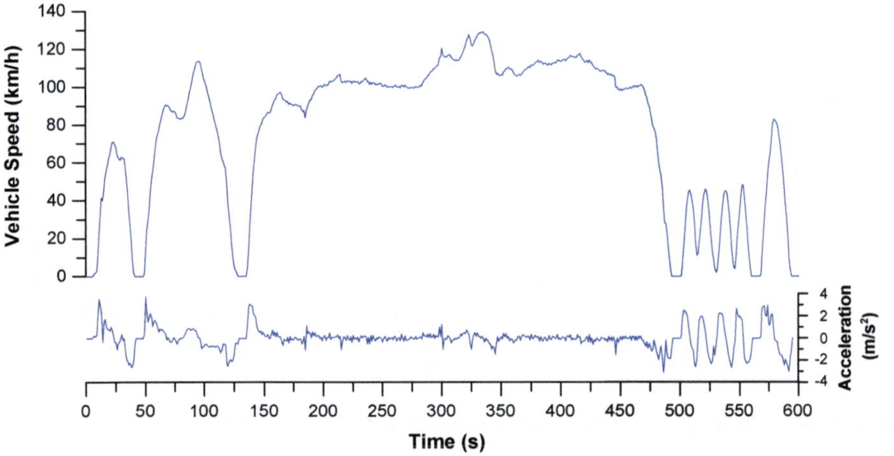

Fig. A.11 Vehicle speed and acceleration versus time of the U.S. SFTP US06

Table A.11 Technical specifications of the U.S. SFTP US06

Distance (m)/Duration (s or min)	12,887.6	600 (10)
Maximum/Average vehicle speed (km/h)	129.23	77.33
Average driving speed (km/h)/Speed σ (km/h)	82.70	39.44
Average/Maximum acceleration (m/s^2)	0.670	3.755
Average/Minimum deceleration (m/s^2)	−0.728	−3.085
Driving time (s)/(%)	561	93.50
Driving time (%) $V \leq 30$ km/h/$30 < V \leq 60$ km/h	18.50	10.67
Driving time (%) $60 < V \leq 100$ km/h/$V > 100$ km/h	26.17	44.67
Cruising time (s)/(%)	33	5.50
Time spent accelerating (s)/(%)	275	45.83
Acceleration time (%) $0 < a \leq 1.0$ m/s^2/$a > 1.0$ m/s^2	35.33	10.50
Time spent decelerating (s)/(%)	253	42.17
Decel. time (%) $a < -1.0$ m/s^2/$-1.0 \leq a < 0$ m/s^2	11.50	30.67
Idling time (s)/(%)	39	6.50 (7.83)
No. of accelerations/Positive acceleration σ (m/s^2)	34	0.815
Accelerations per km/per min	2.64	3.40
Number of stops/Max. interm. stop duration (s)	6	7
Stops per km/Average stop duration (s)	0.47	6.50
RPA (m/s^2)/PKE (m/s^2)	0.222	5.454

A.12 U.S. SFTP SC03

Supplemental, to the FTP-75, SC03 cycle simulates engine load and emissions associated with the use of the air conditioning system; based on data from the early 90s. Implemented from 2000 (Fig. A.12 and Table A.12).

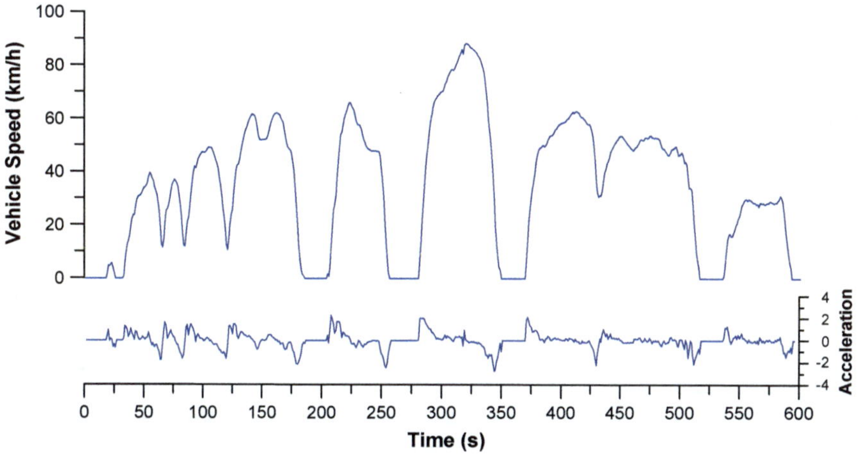

Fig. A.12 Vehicle speed and acceleration versus time of the U.S. SFTP SC03

Table A.12 Technical specifications of the U.S. SFTP SC03

Distance (m)/Duration (s or min)	5760.6	600 (10)
Maximum/Average vehicle speed (km/h)	88.19	34.56
Average driving speed (km/h)/Speed σ (km/h)	42.32	24.66
Average/Maximum acceleration (m/s^2)	0.502	2.280
Average/Minimum deceleration (m/s^2)	−0.604	−2.727
Driving time (s)/(%)	490	81.67
Driving time (%) V $\leq$ 30 km/h/30 < V $\leq$ 60 km/h	43.00	43.00
Driving time (%) 60 < V $\leq$ 100 km/h/V > 100 km/h	14.00	0
Cruising time (s)/(%)	23	3.83
Time spent accelerating (s)/(%)	255	42.50
Acceleration time (%) 0 < a $\leq$ 1.0 m/s^2/a > 1.0 m/s^2	35.33	7.17
Time spent decelerating (s)/(%)	212	35.33
Decel. time (%) a < −1.0 m/s^2/−1.0 $\leq$ a < 0 m/s^2	8.67	26.67
Idling time (s)/(%)	110	18.33 (20.17)
No. of accelerations/Positive acceleration σ (m/s^2)	23	0.512
Accelerations per km/per min	3.99	2.30
Number of stops/Max. interm. stop duration (s)	7	23
Stops per km/Average stop duration (s)	1.22	15.71
RPA (m/s^2)/PKE (m/s^2)	0.218	5.343

A.13 New York City Cycle—NYCC

Urban driving schedule with heavy congestion, very low speeds and many stops; almost one third of the time is spent idling. Forms part of the evaporative emissions testing in the United Stated from 1996 (Fig. A.13 and Table A.13).

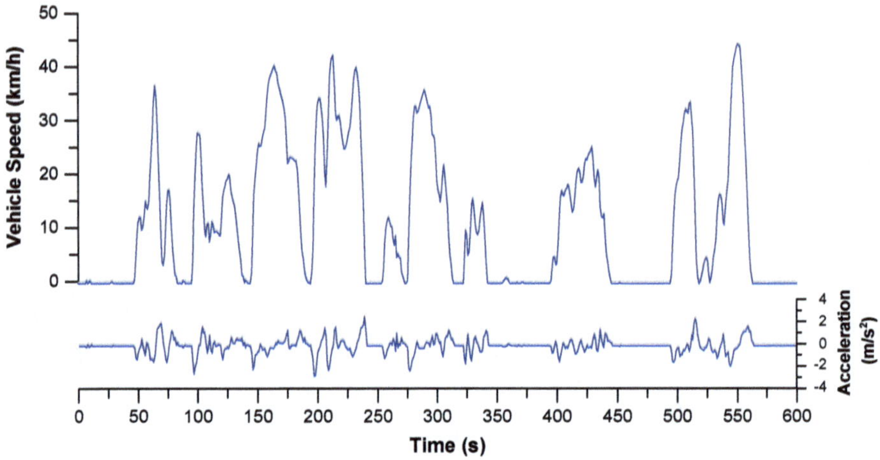

Fig. A.13 Vehicle speed and acceleration versus time of the New York City cycle

Table A.13 Technical specifications of the New York City cycle

Distance (m)/Duration (s or min)	1898.4	600 (10)
Maximum/Average vehicle speed (km/h)	44.58	11.39
Average driving speed (km/h)/Speed σ (km/h)	16.79	12.87
Average/Maximum acceleration (m/s^2)	0.621	2.682
Average/Minimum deceleration (m/s^2)	−0.605	−2.638
Driving time (s)/(%)	407	67.83
Driving time (%) V $\leq$ 30 km/h/30 < V $\leq$ 60 km/h	86.83	13.17
Driving time (%) 60 < V $\leq$ 100 km/h/V > 100 km/h	0	0
Cruising time (s)/(%)	12	2.00
Time spent accelerating (s)/(%)	195	32.50
Acceleration time (%) 0 < a $\leq$ 1.0 m/s^2/a > 1.0 m/s^2	25.50	7.00
Time spent decelerating (s)/(%)	200	33.33
Decel. time (%) a < −1.0 m/s^2/−1.0 $\leq$ a < 0 m/s^2	7.67	25.67
Idling time (s)/(%)	193	32.17 (40.83)
No. of accelerations/Positive acceleration σ (m/s^2)	27	0.568
Accelerations per km/per min	14.22	2.70
Number of stops/Max. interm. stop duration (s)	15	41
Stops per km/Average stop duration (s)	7.90	12.87
RPA (m/s^2)/PKE (m/s^2)	0.345	8.00

A.14 California LA-92

Also known as the Unified Cycle Driving Schedule (UCDS). More aggressive and high speed than the federal FTP-75. Presented here is the complete 1735-s version; it is the supplemental test cycle for Class 3 HDVs (GVWR 10–14,000 lbs) that are chassis tested (Fig. A.14 and Table A.14).

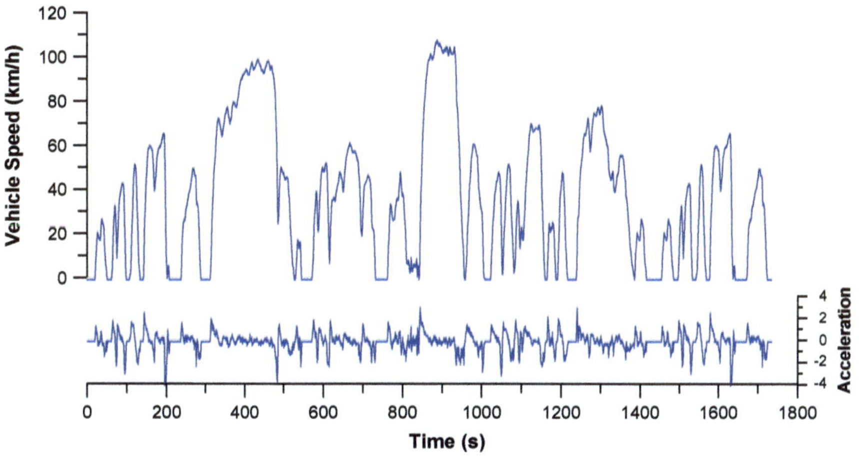

Fig. A.14 Vehicle speed and acceleration versus time of the California LA-92

Table A.14 Technical specifications of the California LA-92

Distance (m)/Duration (s or min)	17,706.0	1735 (28.92)
Maximum/Average vehicle speed (km/h)	108.14	36.74
Average driving speed (km/h)/Speed σ (km/h)	44.51	30.88
Average/Maximum acceleration (m/s^2)	0.685	3.085
Average/Minimum deceleration (m/s^2)	−0.796	−3.934
Driving time (s)/(%)	1432	82.54
Driving time (%) V $\leq$ 30 km/h/30 < V $\leq$ 60 km/h	45.99	32.16
Driving time (%) 60 < V $\leq$ 100 km/h/V > 100 km/h	18.21	3.63
Cruising time (s)/(%)	206	11.87
Time spent accelerating (s)/(%)	659	37.98
Acceleration time (%) 0 < a $\leq$ 1.0 m/s^2/a > 1.0 m/s^2	28.30	9.68
Time spent decelerating (s)/(%)	567	32.68
Decel. time (%) a < −1.0 m/s^2/−1.0 $\leq$ a < 0 m/s^2	10.32	22.36
Idling time (s)/(%)	303	17.46 (19.54)
No. of accelerations/Positive acceleration σ (m/s^2)	83	0.518
Accelerations per km/per min	4.69	2.87
Number of stops/Max. interm. stop duration (s)	16	38
Stops per km/Average stop duration (s)	0.90	18.94
RPA (m/s^2)/PKE (m/s^2)	0.250	6.133

A.15 U.S. Inspection and Maintenance IM240 Cycle

Special-purpose cycle in a number of U.S. states for emissions testing of in-use light-duty vehicles in inspection and maintenance programs (Fig. A.15 and Table A.15).

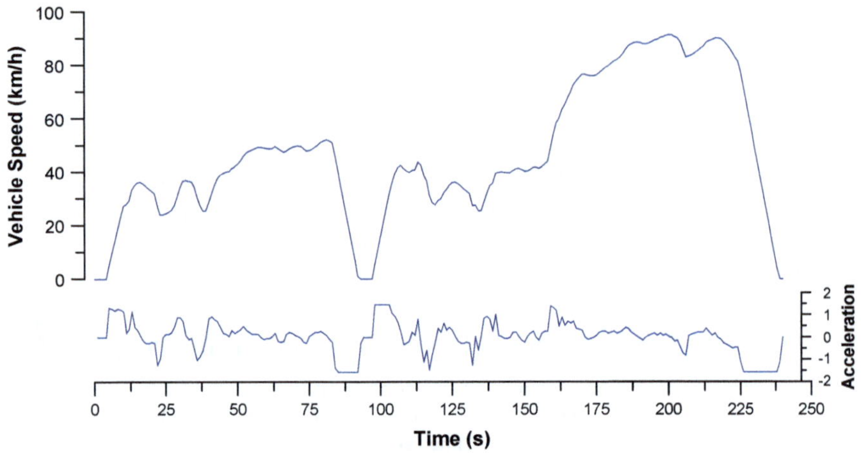

Fig. A.15 Vehicle speed and acceleration versus time of the U.S. IM240

Table A.15 Technical specifications of the U.S. IM240

Distance (m)/Duration (s or min)	3151.9	240 (4)
Maximum/Average vehicle speed (km/h)	91.25	47.28
Average driving speed (km/h)/Speed σ (km/h)	49.12	25.23
Average/Maximum acceleration (m/s^2)	0.439	1.475
Average/Minimum deceleration (m/s^2)	−0.684	−1.565
Driving time (s)/(%)	231	96.25
Driving time (%) V $\leq$ 30 km/h/30 < V $\leq$ 60 km/h	22.92	49.17
Driving time (%) 60 < V $\leq$ 100 km/h/V > 100 km/h	27.92	0
Cruising time (s)/(%)	11	4.58
Time spent accelerating (s)/(%)	134	55.83
Acceleration time (%) 0 < a $\leq$ 1.0 m/s^2/a > 1.0 m/s^2	48.33	7.50
Time spent decelerating (s)/(%)	86	35.83
Decel. time (%) a < −1.0 m/s^2/−1.0 $\leq$ a < 0 m/s^2	12.08	23.75
Idling time (s)/(%)	9	3.75
No. of accelerations/Positive acceleration σ (m/s^2)	15	0.416
Accelerations per km/per min	4.76	3.75 (5.00)
Number of stops/Max. interm. stop duration (s)	2	4
Stops per km/Average stop duration (s)	0.63	4.50
RPA (m/s^2)/PKE (m/s^2)	0.211	5.260

A.16 Japanese 10 Mode—J10

Run six times after a 40-s warm-up period at 40 km/h. Emissions sampling took place at the last 5 runs. Cycle replaced by the 10–15 Mode. Similar in structure (and simplicity) to the ECE-15 urban part of the European NEDC. One fourth of the time is spent cruising. Implemented in 1973 (Fig. A.16 and Table A.16).

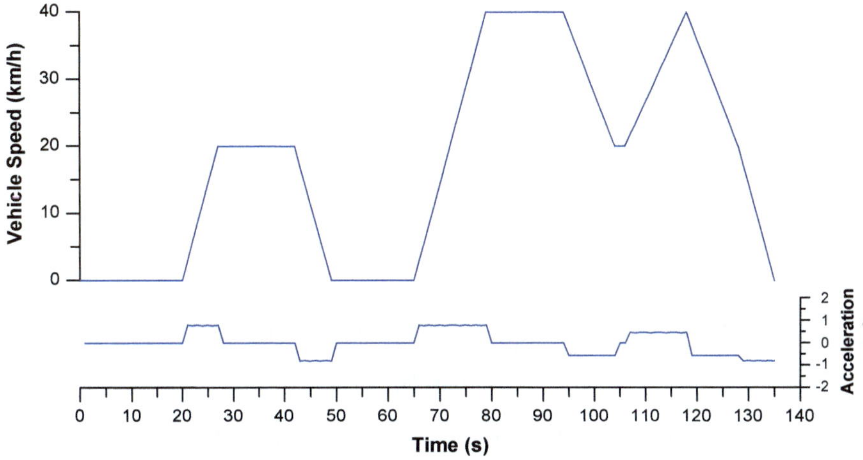

Fig. A.16 Vehicle speed and acceleration versus time of the Japanese J10

Table A.16 Technical specifications of the Japanese J10

Distance (m)/Duration (s or min)	663.9	135 (2.25)
Maximum/Average vehicle speed (km/h)	40.00	17.70
Average driving speed (km/h)/Speed σ (km/h)	24.14	14.58
Average/Maximum acceleration (m/s^2)	0.673	0.806
Average/Minimum deceleration (m/s^2)	−0.654	−0.806
Driving time (s)/(%)	99	73.33
Driving time (%) $V \leq 30$ km/h/$30 < V \leq 60$ km/h	75.56	24.44
Driving time (%) $60 < V \leq 100$ km/h/$V > 100$ km/h	0	0
Cruising time (s)/(%)	32	23.70
Time spent accelerating (s)/(%)	33	24.44
Acceleration time (%) $0 < a \leq 1.0$ m/s^2/$a > 1.0$ m/s^2	24.44	0
Time spent decelerating (s)/(%)	34	25.19
Decel. time (%) $a < -1.0$ m/s^2/$-1.0 \leq a < 0$ m/s^2	0	25.19
Idling time (s)/(%)	36	26.67 (28.15)
No. of accelerations/Positive acceleration σ (m/s^2)	4	0.160
Accelerations per km/per min	6.03	1.78
Number of stops/Max. interm. stop duration (s)	2	17
Stops per km/Average stop duration (s)	3.01	18.00
RPA (m/s^2)/PKE (m/s^2)	0.198	4.820

A.17 Japanese 11 Mode—J11

Modal cycle, supplementary to the J10 and J10-15 in terms of cold-starting emissions. Valid from 1975 to 2008 (replaced by the JC08) (Fig. A.17 and Table A.17).

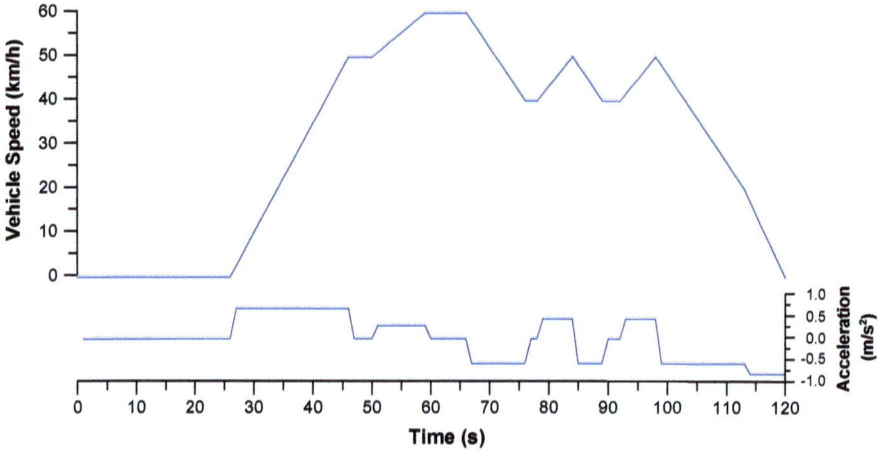

Fig. A.17 Vehicle speed and acceleration versus time of the Japanese J11

Table A.17 Technical specifications of the Japanese J11

Distance (m)/Duration (s or min)	1020.8	120 (2)
Maximum/Average vehicle speed (km/h)	60.00	30.63
Average driving speed (km/h)/Speed σ (km/h)	39.10	21.41
Average/Maximum acceleration (m/s^2)	0.542	0.694
Average/Minimum deceleration (m/s^2)	−0.601	−0.794
Driving time (s)/(%)	94	78.33
Driving time (%) V $\leq$ 30 km/h/30 < V $\leq$ 60 km/h	42.50	57.50
Driving time (%) 60 < V $\leq$ 100 km/h/V > 100 km/h	0	0
Cruising time (s)/(%)	16	13.33
Time spent accelerating (s)/(%)	41	34.17
Acceleration time (%) 0 < a $\leq$ 1.0 m/s^2/a > 1.0 m/s^2	34.17	0
Time spent decelerating (s)/(%)	37	30.83
Decel. time (%) a < −1.0 m/s^2/−1.0 $\leq$ a < 0 m/s^2	0	30.83
Idling time (s)/(%)	26	21.67 (22.50)
No. of accelerations/Positive acceleration σ (m/s^2)	5	0.158
Accelerations per km/per min	4.90	2.50
Number of stops/Max. interm. stop duration (s)	1	–
Stops per km/Average stop duration (s)	0.98	26.00
RPA (m/s^2)/PKE (m/s^2)	0.210	5.289

A.18 Japanese 15 Mode—J15

Modal extra-urban segment added from 1991 to the J10 cycle, forming the J10-15. Discontinued from 2011 (Fig. A.18 and Table A.18).

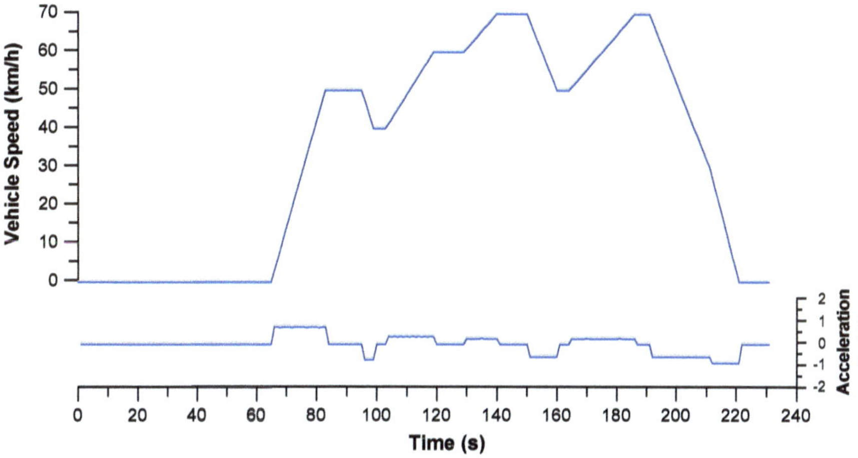

Fig. A.18 Vehicle speed and acceleration versus time of the Japanese J15

Table A.18 Technical specifications of the Japanese J15

Distance (m)/Duration (s or min)	2173.7	231 (3.85)
Maximum/Average vehicle speed (km/h)	70.00	33.88
Average driving speed (km/h)/Speed σ (km/h)	50.16	27.37
Average/Maximum acceleration (m/s^2)	0.415	0.778
Average/Minimum deceleration (m/s^2)	−0.631	−0.833
Driving time (s)/(%)	156	67.53
Driving time (%) V $\leq$ 30 km/h/30 < V $\leq$ 60 km/h	41.56	38.96
Driving time (%) 60 < V $\leq$ 100 km/h/V > 100 km/h	19.48	0
Cruising time (s)/(%)	45	19.48
Time spent accelerating (s)/(%)	67	29.00
Acceleration time (%) 0 < a $\leq$ 1.0 m/s^2/a > 1.0 m/s^2	29.00	0
Time spent decelerating (s)/(%)	44	19.04
Decel. time (%) a < −1.0 m/s^2/−1.0 $\leq$ a < 0 m/s^2	0	19.04
Idling time (s)/(%)	75	32.47 (32.90)
No. of accelerations/Positive acceleration σ (m/s^2)	5	0.220
Accelerations per km/per min	2.30	1.30
Number of stops/Max. interm. stop duration (s)	2	–
Stops per km/Average stop duration (s)	0.92	37.50
RPA (m/s^2)/PKE (m/s^2)	0.149	3.772

A.19 Japanese 10-15 Mode—J10-15

Modal cycle with constant accelerations and high percentage of cruising time (similar to the European NEDC); consists of 3 repetitions of the 10 mode and one of the 15 mode. Replaced by the JC08 cycle from 2011 (Fig. A.19 and Table A.19).

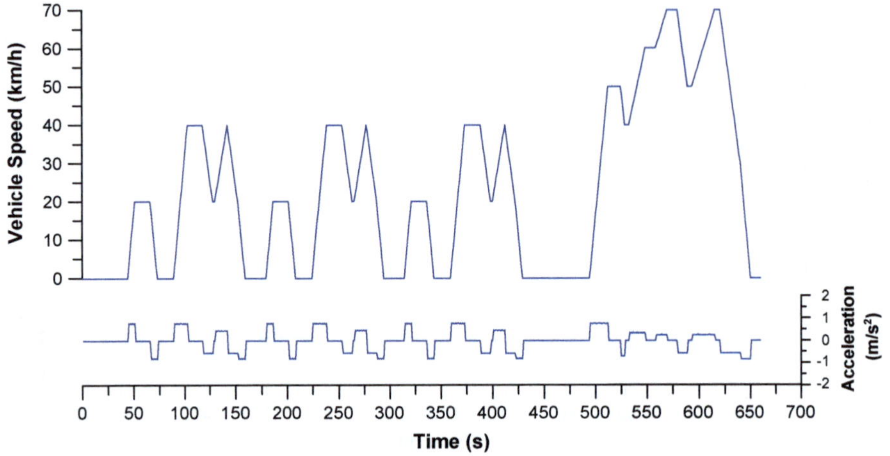

Fig. A.19 Vehicle speed and acceleration versus time of the Japanese J10-15

Table A.19 Technical specifications of the Japanese J10-15

Distance (m)/Duration (s or min)	4165.4	660 (11)
Maximum/Average vehicle speed (km/h)	70.00	22.72
Average driving speed (km/h)/Speed σ (km/h)	33.10	21.69
Average/Maximum acceleration (m/s^2)	0.569	0.806
Average/Minimum deceleration (m/s^2)	−0.647	−0.833
Driving time (s)/(%)	453	68.64
Driving time (%) V $\leq$ 30 km/h/30 < V $\leq$ 60 km/h	64.55	28.64
Driving time (%) 60 < V $\leq$ 100 km/h/V > 100 km/h	6.82	0
Cruising time (s)/(%)	141	21.36
Time spent accelerating (s)/(%)	166	25.15
Acceleration time (%) 0 < a $\leq$ 1.0 m/s^2/a > 1.0 m/s^2	25.15	0
Time spent decelerating (s)/(%)	146	22.12
Decel. time (%) a < −1.0 m/s^2/−1.0 $\leq$ a < 0 m/s^2	0	22.12
Idling time (s)/(%)	207	31.37 (32.42)
No. of accelerations/Positive acceleration σ (m/s^2)	17	0.225
Accelerations per km/per min	4.08	1.55
Number of stops/Max. interm. stop duration (s)	8	65
Stops per km/Average stop duration (s)	1.92	25.88
RPA (m/s^2)/PKE (m/s^2)	0.172	4.273

A.20 Japanese JC08

Highly transient with minimum cruising time (in contrast to its predecessor, the J10-15) but long idling period, indicative of heavy congestion in a Japanese big city. Emission measurements conducted twice, once with a cold start (25% weighting) and once with a hot start (75%). Fully phased-in from October 2011 (Fig. A.20 and Table A.20).

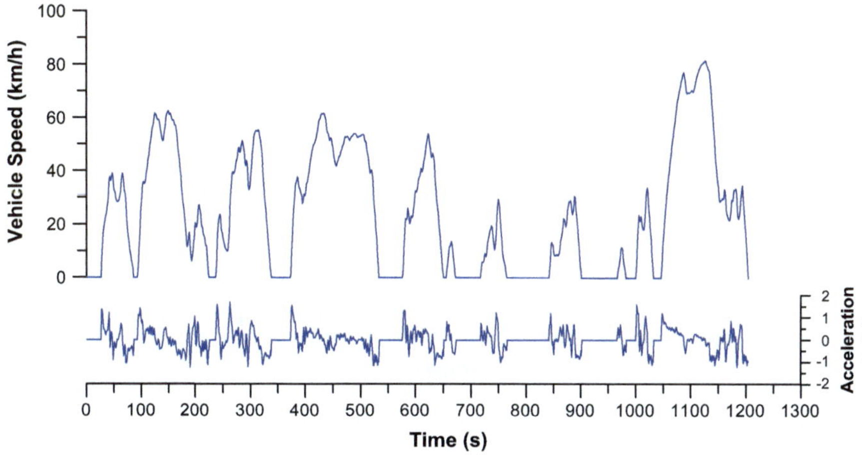

Fig. A.20 Vehicle speed and acceleration versus time of the Japanese JC08

Table A.20 Technical specifications of the Japanese JC08

Distance (m)/Duration (s or min)	8159.4	1204 (20.07)
Maximum/Average vehicle speed (km/h)	81.60	24.40
Average driving speed (km/h)/Speed σ (km/h)	34.24	23.08
Average/Maximum acceleration (m/s^2)	0.426	1.694
Average/Minimum deceleration (m/s^2)	−0.458	−1.222
Driving time (s)/(%)	858	71.26
Driving time (%) V $\leq$ 30 km/h/30 < V $\leq$ 60 km/h	62.13	30.23
Driving time (%) 60 < V $\leq$ 100 km/h/V > 100 km/h	7.64	0
Cruising time (s)/(%)	18	1.50
Time spent accelerating (s)/(%)	435	36.13
Acceleration time (%) 0 < a $\leq$ 1.0 m/s^2/a > 1.0 m/s^2	33.22	2.91
Time spent decelerating (s)/(%)	405	33.64
Decel. time (%) a < −1.0 m/s^2/−1.0 $\leq$ a < 0 m/s^2	1.66	31.97
Idling time (s)/(%)	346	28.74 (29.65)
No. of accelerations/Positive acceleration σ (m/s^2)	46	0.342
Accelerations per km/per min	5.64	2.29
Number of stops/Max. interm. stop duration (s)	11	76
Stops per km/Average stop duration (s)	1.35	31.45
RPA (m/s^2)/PKE (m/s^2)	0.186	4.607

A.21 Australian CUEDC Petrol

Developed in 2005; used for emissions inventory purposes (Fig. A.21 and Table A.21).

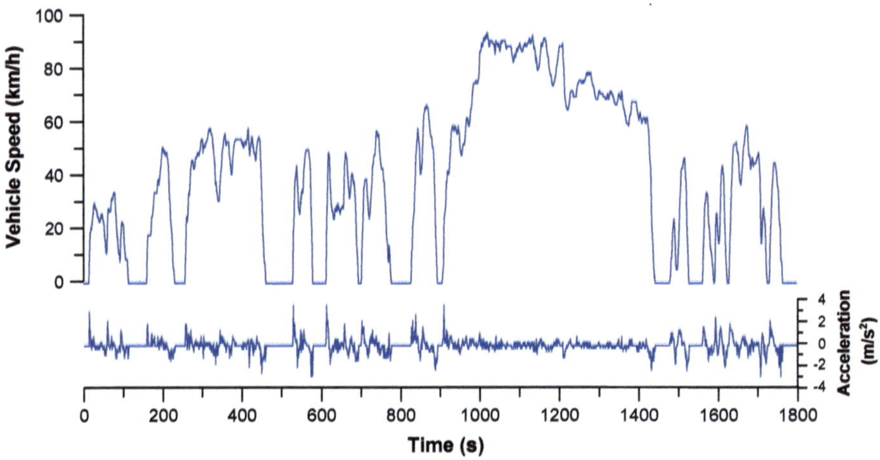

Fig. A.21 Vehicle speed and acceleration versus time of the Australian CUEDC petrol

Table A.21 Technical specifications of the Australian CUEDC petrol

Distance (m)/Duration (s or min)	19,441.7	1797 (29.95)
Maximum/Average vehicle speed (km/h)	94.00	38.95
Average driving speed (km/h)/Speed σ (km/h)	48.94	29.49
Average/Maximum acceleration (m/s^2)	0.651	3.611
Average/Minimum deceleration (m/s^2)	−0.635	−2.778
Driving time (s)/(%)	1430	79.58
Driving time (%) $V \leq 30$ km/h/$30 < V \leq 60$ km/h	41.35	33.33
Driving time (%) $60 < V \leq 100$ km/h/$V > 100$ km/h	25.32	0
Cruising time (s)/(%)	454	25.26
Time spent accelerating (s)/(%)	482	26.82
Acceleration time (%) $0 < a \leq 1.0$ m/s^2/$a > 1.0$ m/s^2	21.76	5.06
Time spent decelerating (s)/(%)	494	27.49
Decel. time (%) $a < -1.0$ m/s^2/$-1.0 \leq a < 0$ m/s^2	5.56	21.92
Idling time (s)/(%)	367	20.42 (21.48)
No. of accelerations/Positive acceleration σ (m/s^2)	50	0.539
Accelerations per km/per min	2.57	1.67
Number of stops/Max. interm. stop duration (s)	13	67
Stops per km/Average stop duration (s)	0.67	28.23
RPA (m/s^2)/PKE (m/s^2)	0.171	4.194

A.22 Australian CUEDC Diesel MC

Developed in the late 90s; used for emissions inventory purposes. Applicable to diesel passenger cars (Fig. A.22 and Table A.22).

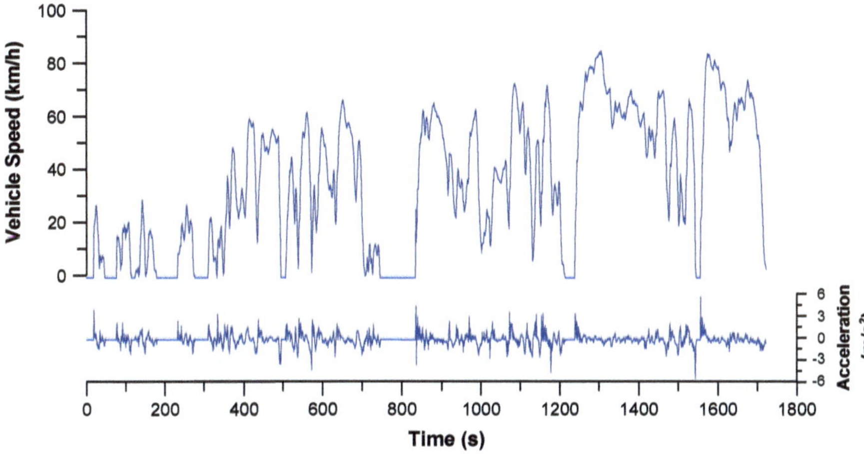

Fig. A.22 Vehicle speed and acceleration versus time of the Australian CUEDC diesel MC

Table A.22 Technical specifications of the Australian CUEDC diesel MC

Distance (m)/Duration (s or min)	16,904.0	1722 (28.70)
Maximum/Average vehicle speed (km/h)	85.26	35.34
Average driving speed (km/h)/Speed σ (km/h)	41.91	25.80
Average/Maximum acceleration (m/s^2)	0.639	5.809
Average/Minimum deceleration (m/s^2)	−0.560	−5.242
Driving time (s)/(%)	1452	84.32
Driving time (%) $V \leq 30$ km/h/$30 < V \leq 60$ km/h	44.02	33.80
Driving time (%) $60 < V \leq 100$ km/h/$V > 100$ km/h	22.18	0
Cruising time (s)/(%)	1	0.06
Time spent accelerating (s)/(%)	678	39.37
Acceleration time (%) $0 < a \leq 1.0$ m/s^2/$a > 1.0$ m/s^2	30.89	8.48
Time spent decelerating (s)/(%)	773	44.89
Decel. time (%) $a < -1.0$ m/s^2/$-1.0 \leq a < 0$ m/s^2	7.49	37.40
Idling time (s)/(%)	270	15.68 (17.31)
No. of accelerations/Positive acceleration σ (m/s^2)	86	0.722
Accelerations per km/per min	5.09	3.00
Number of stops/Max. interm. stop duration (s)	10	88
Stops per km/Average stop duration (s)	0.59	27.00
RPA (m/s^2)/PKE (m/s^2)	0.269	6.487

A.23 Worldwide WLTC Class 1

Representative of low-speed and low-powered vehicles used in India, WLTC Class 1 corresponds to vehicles with power to mass ratio less than 22 W/kg. The certification procedure includes phases low (first 589 s), middle and again low (Fig. A.23 and Table A.23).

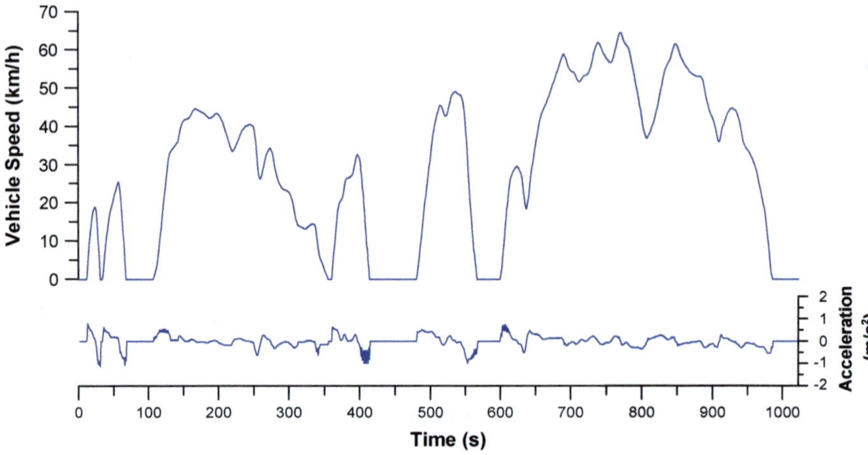

Fig. A.23 Vehicle speed and acceleration versus time of the WLTC Class 1

Table A.23 Technical specifications of the WLTC Class 1

Distance (m)/Duration (s or min)	8097.6	1022 (17.03)
Maximum/Average vehicle speed (km/h)	64.40	28.52
Average driving speed (km/h)/Speed σ (km/h)	35.12	20.30
Average/Maximum acceleration (m/s^2)	0.215	0.806
Average/Minimum deceleration (m/s^2)	−0.219	−1.111
Driving time (s)/(%)	830	81.21
Driving time (%) $V \leq 30$ km/h/$30 < V \leq 60$ km/h	49.12	47.36
Driving time (%) $60 < V \leq 100$ km/h/$V > 100$ km/h	3.52	0
Cruising time (s)/(%)	31	3.03
Time spent accelerating (s)/(%)	404	39.53
Acceleration time (%) $0 < a \leq 1.0$ m/s^2/$a > 1.0$ m/s^2	39.53	0
Time spent decelerating (s)/(%)	395	38.65
Decel. time (%) $a < -1.0$ m/s^2/$-1.0 \leq a < 0$ m/s^2	0.29	38.36
Idling time (s)/(%)	192	18.79 (19.96)
No. of accelerations/Positive acceleration σ (m/s^2)	22	0.163
Accelerations per km/per min	2.72	1.29
Number of stops/Max. interm. stop duration (s)	6	66
Stops per km/Average stop duration (s)	0.74	32.00
RPA (m/s^2)/PKE (m/s^2)	0.083	2.107

A.24 Worldwide WLTC Class 2

Representative of vehicles in India and low-speed/power vehicles in Europe and Japan, WLTC Class 2 corresponds to vehicles with power to mass ratio between 22 and 34 W/kg (Fig. A.24 and Table A.24).

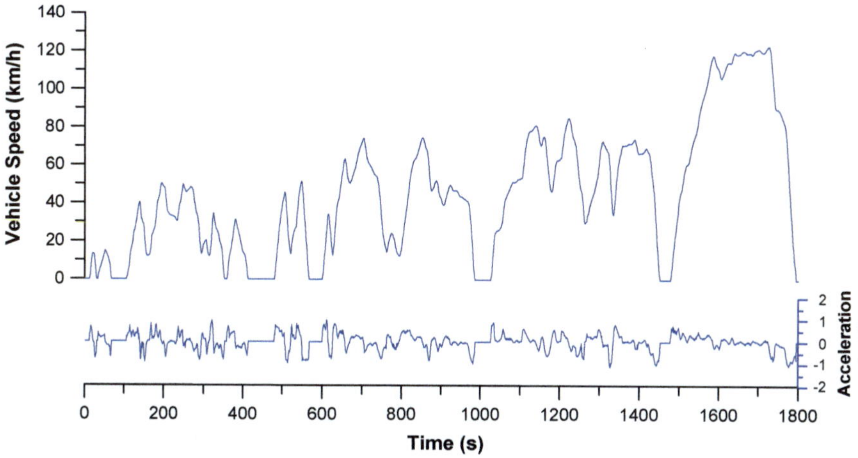

Fig. A.24 Vehicle speed and acceleration versus time of the WLTC Class 2

Table A.24 Technical specifications of the WLTC Class 2

Distance (m)/Duration (s or min)	22,649.2	1800 (30)
Maximum/Average vehicle speed (km/h)	123.10	45.30
Average driving speed (km/h)/Speed σ (km/h)	51.97	34.08
Average/Maximum acceleration (m/s^2)	0.276	0.972
Average/Minimum deceleration (m/s^2)	−0.338	−1.111
Driving time (s)/(%)	1569	87.17
Driving time (%) $V \leq 30$ km/h/$30 < V \leq 60$ km/h	36.94	31.39
Driving time (%) $60 < V \leq 100$ km/h/$V > 100$ km/h	22.06	9.61
Cruising time (s)/(%)	71	3.94
Time spent accelerating (s)/(%)	825	45.83
Acceleration time (%) $0 < a \leq 1.0$ m/s^2/$a > 1.0$ m/s^2	45.83	0
Time spent decelerating (s)/(%)	673	37.39
Decel. time (%) $a < -1.0$ m/s^2/$-1.0 \leq a < 0$ m/s^2	0.44	36.94
Idling time (s)/(%)	231	12.83 (13.33)
No. of accelerations/Positive acceleration σ (m/s^2)	51	0.355
Accelerations per km/per min	2.25	1.70
Number of stops/Max. interm. stop duration (s)	8	67
Stops per km/Average stop duration (s)	0.35	28.88
RPA (m/s^2)/PKE (m/s^2)	0.119	3.036

A.25 Worldwide WLTC Class 3-2

Representative of vehicles in Europe and Japan, WLTC Class 3-2 corresponds to vehicles with power to mass ratio greater than 34 W/kg, and maximum speed higher than 120 km/h (Fig. A.25 and Table A.25).

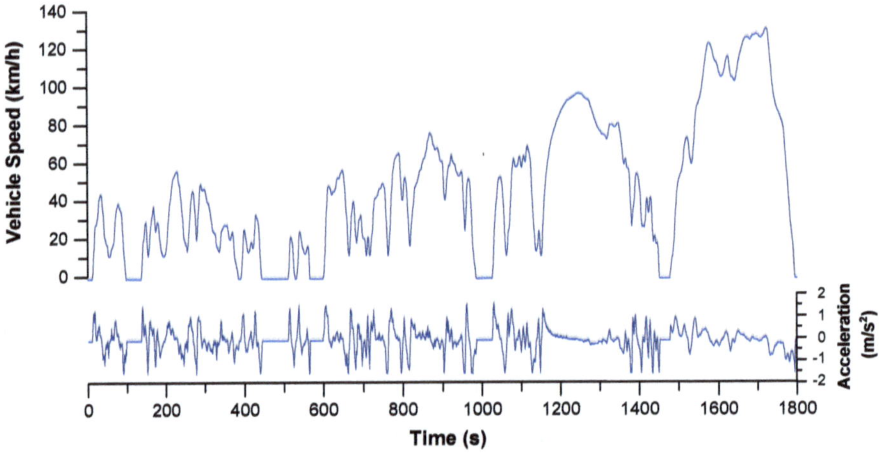

Fig. A.25 Vehicle speed and acceleration versus time of the WLTC Class 3-2

Table A.25 Technical specifications of the WLTC Class 3-2

Distance (m)/Duration (s or min)	23,266.3	1800 (30)
Maximum/Average vehicle speed (km/h)	131.30	46.53
Average driving speed (km/h)/Speed σ (km/h)	53.21	36.10
Average/Maximum acceleration (m/s^2)	0.406	1.667
Average/Minimum deceleration (m/s^2)	−0.445	−1.500
Driving time (s)/(%)	1574	87.44
Driving time (%) V $\leq$ 30 km/h/30 < V $\leq$ 60 km/h	40.22	27.94
Driving time (%) 60 < V $\leq$ 100 km/h/V > 100 km/h	21.72	10.11
Cruising time (s)/(%)	66	3.67
Time spent accelerating (s)/(%)	789	43.83
Acceleration time (%) 0 < a $\leq$ 1.0 m/s^2/a > 1.0 m/s^2	39.83	4.00
Time spent decelerating (s)/(%)	719	39.94
Decel. time (%) a < −1.0 m/s^2/−1.0 $\leq$ a < 0 m/s^2	5.00	34.94
Idling time (s)/(%)	226	12.56 (13.56)
No. of accelerations/Positive acceleration σ (m/s^2)	68	0.367
Accelerations per km/per min	2.92	2.27
Number of stops/Max. interm. stop duration (s)	8	66
Stops per km/Average stop duration (s)	0.344	28.25
RPA (m/s^2)/PKE (m/s^2)	0.159	3.986

Mopeds and Motorcycles

A.26 U.S. Motorcycles UDDS (Engine Capacity < 170 cm^3)

Similar to the passenger cars FTP-72 but with lower speeds during the second micro-trip (Fig. A.26 and Table A.26).

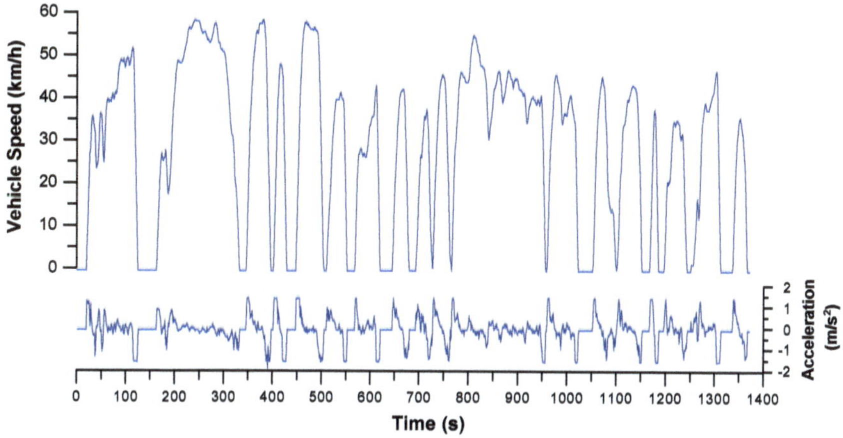

Fig. A.26 Vehicle speed and acceleration versus time of the U.S. UDDS for motorcycles with engine capacity lower than 170 cm^3

Table A.26 Technical specifications of the U.S. UDDS for motorcycles with engine capacity lower than 170 cm^3

Distance (m)/Duration (s or min)	10,864.80	1372 (22.87)
Maximum/Average vehicle speed (km/h)	58.70	28.51
Average driving speed (km/h)/Speed σ (km/h)	34.67	19.22
Average/Maximum acceleration (m/s^2)	0.483	1.500
Average/Minimum deceleration (m/s^2)	−0.551	−1.806
Driving time (s)/(%)	1128	82.22
Driving time (%) V ≤ 30 km/h/30 < V ≤ 60 km/h	44.90	55.10
Driving time (%) 60 < V ≤ 100 km/h/V > 100 km/h	0	0
Cruising time (s)/(%)	109	7.94
Time spent accelerating (s)/(%)	543	39.58
Acceleration time (%) 0 < a ≤ 1.0 m/s^2/a > 1.0 m/s^2	33.89	5.69
Time spent decelerating (s)/(%)	476	34.69
Decel. time (%) a < −1.0 m/s^2/−1.0 ≤ a < 0 m/s^2	8.02	26.68
Idling time (s)/(%)	244	17.78 (19.61)
No. of accelerations/Positive acceleration σ (m/s^2)	56	0.438
Accelerations per km/per min	5.15	2.45
Number of stops/Max. interm. stop duration (s)	15	38
Stops per km/Average stop duration (s)	1.38	16.27
RPA (m/s^2)/PKE (m/s^2)	0.180	4.385

A.27 Indian Motorcycle IDC

Representative of low-powered motorcycles in India, the modal IDC is run six times. Developed in the mid 80s (Fig. A.27 and Table A.27).

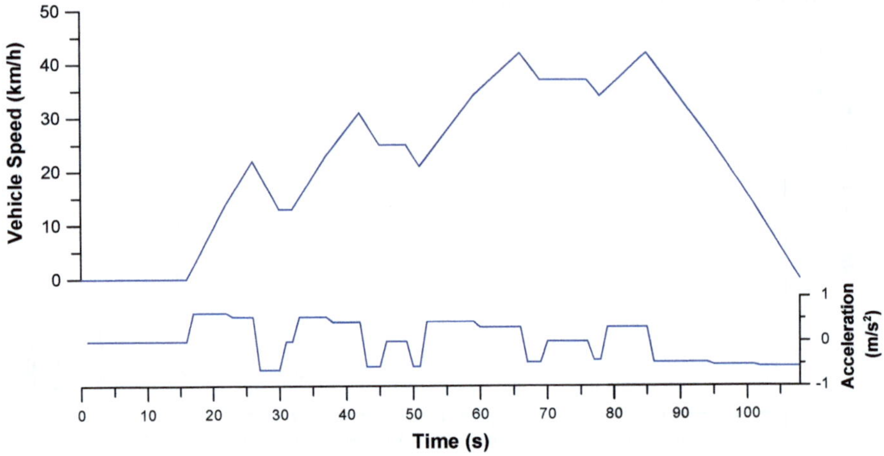

Fig. A.27 Vehicle speed and acceleration versus time of the Indian motorcycle IDC

Table A.27 Technical specifications of the Indian motorcycle IDC

Distance (m)/Duration (s or min)	657.9	108 (1.8)
Maximum/Average vehicle speed (km/h)	42.00	21.93
Average driving speed (km/h)/Speed σ (km/h)	25.75	13.70
Average/Maximum acceleration (m/s^2)	0.456	0.650
Average/Minimum deceleration (m/s^2)	−0.518	−0.625
Driving time (s)/(%)	92	85.19
Driving time (%) V $\leq$ 30 km/h/30 < V $\leq$ 60 km/h	65.74	34.25
Driving time (%) 60 < V $\leq$ 100 km/h/V > 100 km/h	0	0
Cruising time (s)/(%)	13	12.04
Time spent accelerating (s)/(%)	42	38.89
Acceleration time (%) 0 < a $\leq$ 1.0 m/s^2/a > 1.0 m/s^2	38.89	0
Time spent decelerating (s)/(%)	37	34.26
Decel. time (%) a < −1.0 m/s^2/−1.0 $\leq$ a < 0 m/s^2	0	34.26
Idling time (s)/(%)	16	14.81 (15.74)
No. of accelerations/Positive acceleration σ (m/s^2)	4	0.118
Accelerations per km/per min	6.08	2.22
Number of stops/Max. interm. stop duration (s)	1	–
Stops per km/Average stop duration (s)	1.52	16
RPA (m/s^2)/PKE (m/s^2)	0.195	4.874

A.28 Worldwide Motorcycle WMTC Class 1

Representative of urban driving with frequent stops; highly transient. For motorcycles with engine capacity up to 150 cm^3 and maximum speed lower than 100 km/h (Fig. A.28 and Table A.28).

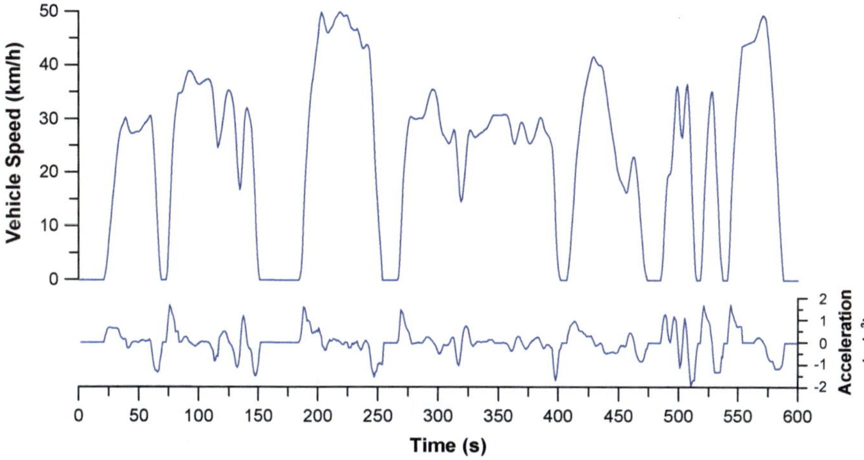

Fig. A.28 Vehicle speed and acceleration versus time of the WMTC Class 1

Table A.28 Technical specifications of the WMTC Class 1

Distance (m)/Duration (s or min)	3837.8	600 (10)
Maximum/Average vehicle speed (km/h)	50.00	23.03
Average driving speed (km/h)/Speed σ (km/h)	27.91	15.42
Average/Maximum acceleration (m/s^2)	0.435	1.722
Average/Minimum deceleration (m/s^2)	−0.523	−1.917
Driving time (s)/(%)	495	82.50
Driving time (%) V $\leq$ 30 km/h/30 < V $\leq$ 60 km/h	63.83	36.17
Driving time (%) 60 < V $\leq$ 100 km/h/V > 100 km/h	0	0
Cruising time (s)/(%)	28	4.67
Time spent accelerating (s)/(%)	255	42.50
Acceleration time (%) 0 < a $\leq$ 1.0 m/s^2/a > 1.0 m/s^2	37.17	5.33
Time spent decelerating (s)/(%)	212	35.33
Decel. time (%) a < −1.0 m/s^2/−1.0 $\leq$ a < 0 m/s^2	6.50	28.83
Idling time (s)/(%)	105	17.50 (19.67)
No. of accelerations/Positive acceleration σ (m/s^2)	30	0.418
Accelerations per km/per min	7.82	3.00
Number of stops/Max. interm. stop duration (s)	8	33
Stops per km/Average stop duration (s)	2.08	13.13
RPA (m/s^2)/PKE (m/s^2)	0.187	4.545

A.29 Worldwide Motorcycle WMTC Class 2-1

For motorcycles with maximum speed lower than 115 km/h (Fig. A.29 and Table A.29).

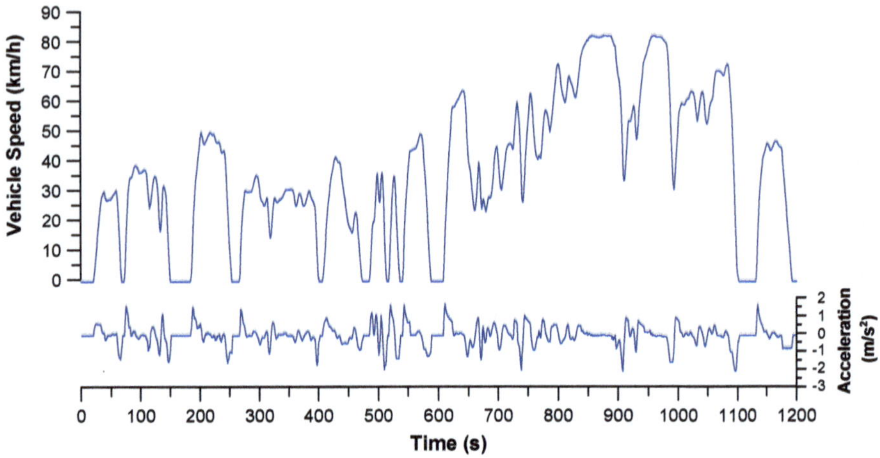

Fig. A.29 Vehicle speed and acceleration versus time of the WMTC Class 2-1

Table A.29 Technical specifications of the WMTC Class 2-1

Distance (m)/Duration (s or min)	12,286.8	1200 (20)
Maximum/Average vehicle speed (km/h)	82.50	36.86
Average driving speed (km/h)/Speed σ (km/h)	42.09	24.13
Average/Maximum acceleration (m/s^2)	0.426	1.750
Average/Minimum deceleration (m/s^2)	−0.513	−2.000
Driving time (s)/(%)	1051	87.58
Driving time (%) $V \leq 30$ km/h/$30 < V \leq 60$ km/h	40.83	39.25
Driving time (%) $60 < V \leq 100$ km/h/$V > 100$ km/h	19.92	0
Cruising time (s)/(%)	76	6.33
Time spent accelerating (s)/(%)	533	44.42
Acceleration time (%) $0 < a \leq 1.0$ m/s^2/$a > 1.0$ m/s^2	40.42	4.00
Time spent decelerating (s)/(%)	442	36.83
Decel. time (%) $a < -1.0$ m/s^2/$-1.0 \leq a < 0$ m/s^2	5.83	31.00
Idling time (s)/(%)	149	12.42 (13.75)
No. of accelerations/Positive acceleration σ (m/s^2)	57	0.383
Accelerations per km/per min	4.64	2.85
Number of stops/Max. interm. stop duration (s)	10	33
Stops per km/Average stop duration (s)	0.81	14.90
RPA (m/s^2)/PKE (m/s^2)	0.181	4.509

A.30 Worldwide Motorcycle WMTC Class 2-2

For motorcycles with maximum speed between 115 and 130 km/h (Fig. A.30 and Table A.30).

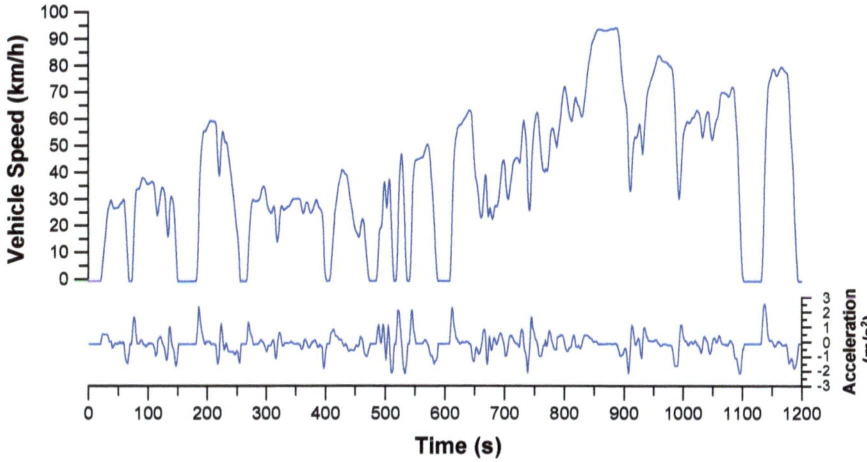

Fig. A.30 Vehicle speed and acceleration versus time of the WMTC Class 2-2

Table A.30 Technical specifications of the WMTC Class 2-2

Distance (m)/Duration (s or min)	13,178.1	1200 (20)
Maximum/Average vehicle speed (km/h)	94.90	39.53
Average driving speed (km/h)/Speed σ (km/h)	44.93	26.10
Average/Maximum acceleration (m/s^2)	0.467	2.694
Average/Minimum deceleration (m/s^2)	−0.543	−2.000
Driving time (s)/(%)	1056	88.00
Driving time (%) V ≤ 30 km/h/30 < V ≤ 60 km/h	38.92	37.58
Driving time (%) 60 < V ≤ 100 km/h/V > 100 km/h	23.50	0
Cruising time (s)/(%)	59	4.92
Time spent accelerating (s)/(%)	536	44.67
Acceleration time (%) 0 < a ≤ 1.0 m/s^2/a > 1.0 m/s^2	39.08	5.58
Time spent decelerating (s)/(%)	461	38.42
Decel. time (%) a < −1.0 m/s^2/−1.0 ≤ a < 0 m/s^2	7.25	31.17
Idling time (s)/(%)	144	12.00 (13.33)
No. of accelerations/Positive acceleration σ (m/s^2)	60	0.519
Accelerations per km/per min	4.55	3.00
Number of stops/Max. interm. stop duration (s)	11	31
Stops per km/Average stop duration (s)	0.83	13.09
RPA (m/s^2)/PKE (m/s^2)	0.203	5.00

A.31 Worldwide Motorcycle WMTC Class 3-1

For motorcycles with maximum speed between 130 and 140 km/h (Fig. A.31 and Table A.31).

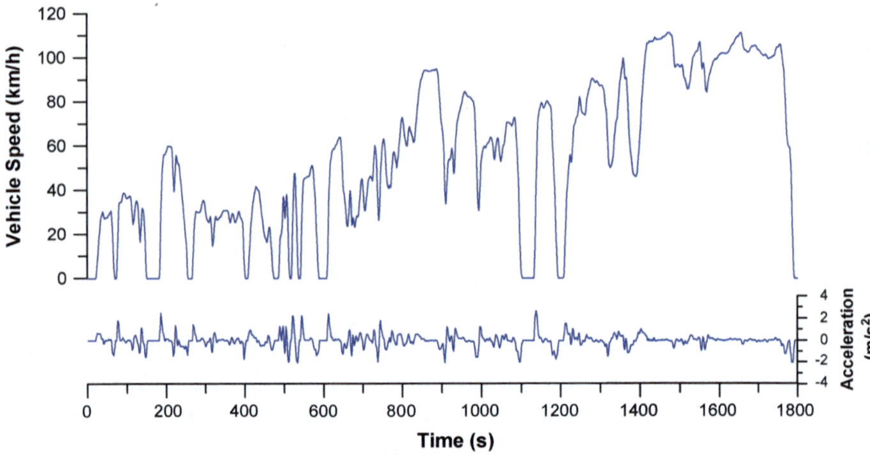

Fig. A.31 Vehicle speed and acceleration versus time of the WMTC Class 3-1

Table A.31 Technical specifications of the WMTC Class 3-1

Distance (m)/Duration (s or min)	27,614.7	1800 (30)
Maximum/Average vehicle speed (km/h)	111.30	55.23
Average driving speed (km/h)/Speed σ (km/h)	60.58	34.09
Average/Maximum acceleration (m/s^2)	0.387	2.694
Average/Minimum deceleration (m/s^2)	−0.484	−2.000
Driving time (s)/(%)	1641	91.16
Driving time (%) $V \leq 30$ km/h/$30 < V \leq 60$ km/h	27.50	28.67
Driving time (%) $60 < V \leq 100$ km/h/$V > 100$ km/h	30.00	13.83
Cruising time (s)/(%)	111	6.17
Time spent accelerating (s)/(%)	850	47.22
Acceleration time (%) $0 < a \leq 1.0$ m/s^2/$a > 1.0$ m/s^2	43.00	4.22
Time spent decelerating (s)/(%)	680	37.78
Decel. time (%) $a < -1.0$ m/s^2/$-1.0 \leq a < 0$ m/s^2	6.06	31.72
Idling time (s)/(%)	159	8.83 (9.83)
No. of accelerations/Positive acceleration σ (m/s^2)	85	0.461
Accelerations per km/per min	3.08	2.83
Number of stops/Max. interm. stop duration (s)	12	31
Stops per km/Average stop duration (s)	0.43	13.25
RPA (m/s^2)/PKE (m/s^2)	0.154	3.853

A.32 Worldwide Motorcycle WMTC Class 3-2

For motorcycles with maximum speed greater than 140 km/h (Fig. A.32 and Table A.32).

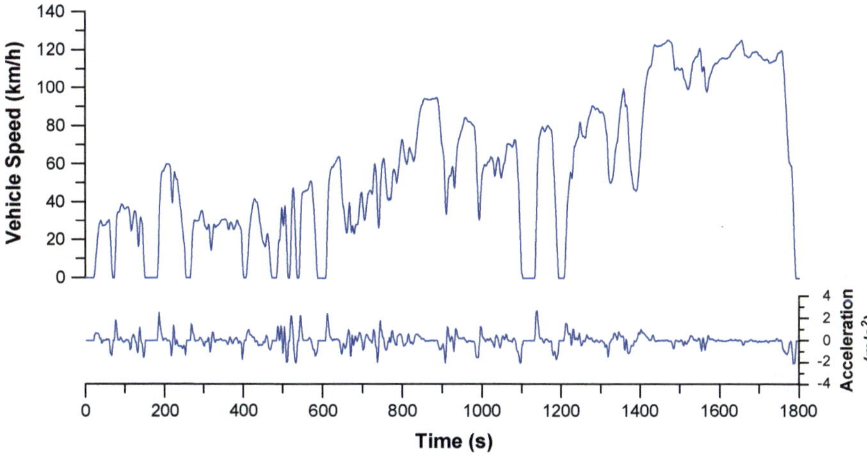

Fig. A.32 Vehicle speed and acceleration versus time of the WMTC Class 3-2

Table A.32 Technical specifications of the WMTC Class 3-2

Distance (m)/Duration (s or min)	28,915.4	1800 (30)
Maximum/Average vehicle speed (km/h)	125.30	57.83
Average driving speed (km/h)/Speed σ (km/h)	63.43	37.87
Average/Maximum acceleration (m/s^2)	0.387	2.694
Average/Minimum deceleration (m/s^2)	−0.491	−2.000
Driving time (s)/(%)	1641	91.17
Driving time (%) V ≤ 30 km/h/30 < V ≤ 60 km/h	27.50	28.67
Driving time (%) 60 < V ≤ 100 km/h/V > 100 km/h	24.78	19.06
Cruising time (s)/(%)	103	5.72
Time spent accelerating (s)/(%)	860	47.78
Acceleration time (%) 0 < a ≤ 1.0 m/s^2/a > 1.0 m/s^2	43.55	4.22
Time spent decelerating (s)/(%)	678	37.67
Decel. time (%) a < −1.0 m/s^2/−1.0 ≤ a < 0 m/s^2	6.11	31.55
Idling time (s)/(%)	159	8.83 (9.83)
No. of accelerations/Positive acceleration σ (m/s^2)	85	0.459
Accelerations per km/per min	2.94	2.83
Number of stops/Max. interm. stop duration (s)	12	31
Stops per km/Average stop duration (s)	0.42	13.25
RPA (m/s^2)/PKE (m/s^2)	0.154	3.854

Heavy-Duty Vehicles

A.33 European FIGE

Developed in the early 90s; used as the basis for the engine-dynamometer ETC valid in Europe through emission standards Euro III, IV and V. Highway oriented with limited urban segment (Fig. A.33 and Table A.33).

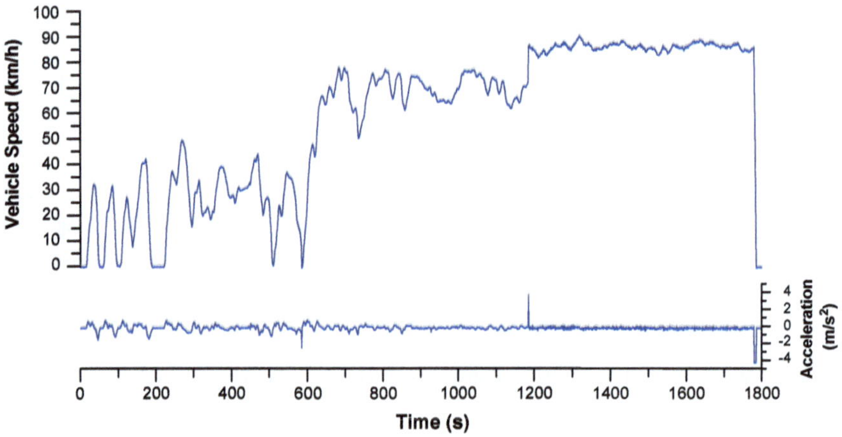

Fig. A.33 Vehicle speed and acceleration versus time of the European FIGE

Table A.33 Technical specifications of the European FIGE

Distance (m)/Duration (s or min)	29,482.4	1800 (30)
Maximum/Average vehicle speed (km/h)	91.1	58.96
Average driving speed (km/h)/Speed σ (km/h)	61.03	28.73
Average/Maximum acceleration (m/s^2)	0.195	3.833
Average/Minimum deceleration (m/s^2)	−0.221	−4.028
Driving time (s)/(%)	1739	96.61
Driving time (%) V $\leq$ 30 km/h/30 < V $\leq$ 60 km/h	22.44	14.61
Driving time (%) 60 < V $\leq$ 100 km/h/V > 100 km/h	62.94	0
Cruising time (s)/(%)	147	8.17
Time spent accelerating (s)/(%)	846	47.00
Acceleration time (%) 0 < a $\leq$ 1.0 m/s^2/a > 1.0 m/s^2	46.94	0.06
Time spent decelerating (s)/(%)	746	41.44
Decel. time (%) a < −1.0 m/s^2/−1.0 $\leq$ a < 0 m/s^2	1.06	40.38
Idling time[a] (s)/(%)	61	3.39 (4.94)
No. of accelerations/Positive acceleration σ (m/s^2)	65	0.217
Accelerations per km/per min	2.20	2.17
Number of stops/Max. interm. stop duration (s)	6	18
Stops per km/Average stop duration (s)	0.20	10.17
RPA (m/s^2)/PKE (m/s^2)	0.070	1.788

[a]Speed less than 0.5 km/h, since there are no points during the cycle with zero speed

A.34 Braunschweig

Bus cycle developed in the mid 70s, based on the U.S. FTP-75 (Fig. A.34 and Table A.34).

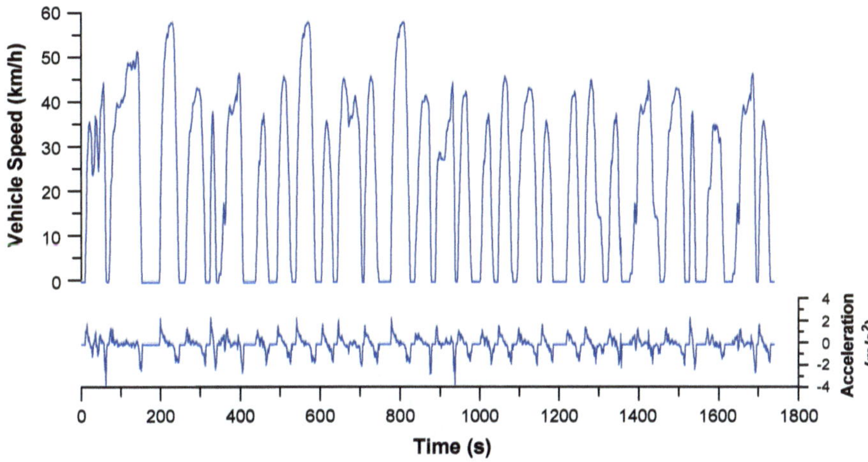

Fig. A.34 Vehicle speed and acceleration versus time of the Braunschweig cycle

Table A.34 Technical specifications of the Braunschweig cycle

Distance (m)/Duration (s or min)	10,873.1	1740 (29)
Maximum/Average vehicle speed (km/h)	58.20	22.50
Average driving speed (km/h)/Speed σ (km/h)	29.48	18.31
Average/Maximum acceleration (m/s^2)	0.543	2.417
Average/Minimum deceleration (m/s^2)	−0.697	−3.583
Driving time (s)/(%)	1328	76.32
Driving time (%) V $\leq$ 30 km/h/30 < V $\leq$ 60 km/h	56.61	43.39
Driving time (%) 60 < V $\leq$ 100 km/h/V > 100 km/h	0	0
Cruising time (s)/(%)	101	5.80
Time spent accelerating (s)/(%)	690	39.66
Acceleration time (%) 0 < a $\leq$ 1.0 m/s^2/a > 1.0 m/s^2	33.62	6.03
Time spent decelerating (s)/(%)	537	30.86
Decel. time (%) a < −1.0 m/s^2/−1.0 $\leq$ a < 0 m/s^2	9.14	21.72
Idling time (s)/(%)	412	23.68 (26.95)
No. of accelerations/Positive acceleration σ (m/s^2)	83	0.462
Accelerations per km/per min	7.63	2.86
Number of stops/Max. interm. stop duration (s)	30	44
Stops per km/Average stop duration (s)	2.76	13.73
RPA (m/s^2)/PKE (m/s^2)	0.235	5.675

A.35 Paris ADEME

Bus cycle developed around 2000, based on bus route No. 21 in Paris (Fig. A.35 and Table A.35).

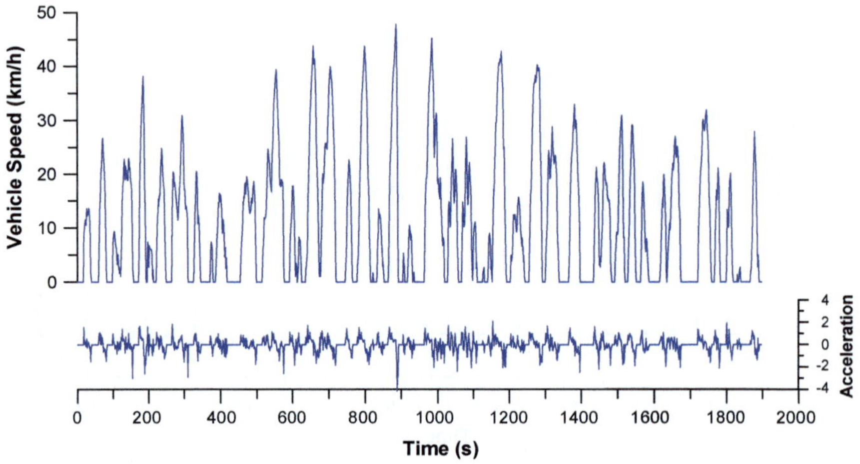

Fig. A.35 Vehicle speed and acceleration versus time of the Paris ADEME cycle

Table A.35 Technical specifications of the Paris ADEME cycle

Distance (m)/Duration (s or min)	5677.9	1898 (31.63)
Maximum/Average vehicle speed (km/h)	47.90	10.77
Average driving speed (km/h)/Speed σ (km/h)	15.50	11.78
Average/Maximum acceleration (m/s^2)	0.514	2.083
Average/Minimum deceleration (m/s^2)	−0.544	−4.083
Driving time (s)/(%)	1319	69.49
Driving time (%) $V \leq 30$ km/h/$30 < V \leq 60$ km/h	91.57	8.43
Driving time (%) $60 < V \leq 100$ km/h/$V > 100$ km/h	0	0
Cruising time (s)/(%)	18	0.95
Time spent accelerating (s)/(%)	669	35.25
Acceleration time (%) $0 < a \leq 1.0$ m/s^2/$a > 1.0$ m/s^2	31.51	3.74
Time spent decelerating (s)/(%)	632	33.30
Decel. time (%) $a < -1.0$ m/s^2/$-1.0 \leq a < 0$ m/s^2	5.27	28.03
Idling time (s)/(%)	579	30.50 (35.72)
No. of accelerations/Positive acceleration σ (m/s^2)	70	0.362
Accelerations per km/per min	12.33	2.21
Number of stops/Max. interm. stop duration (s)	40	44
Stops per km/Average stop duration (s)	7.04	14.48
RPA (m/s^2)/PKE (m/s^2)	0.276	6.548

A.36 U.S. Heavy-Duty Urban Dynamometer Driving Schedule—UDDS

Cycle with long idling period and high speeds. Developed in the late 70s and employed for many years for evaporative emissions testing of HDVs in the U.S. (Fig. A.36 and Table A.36).

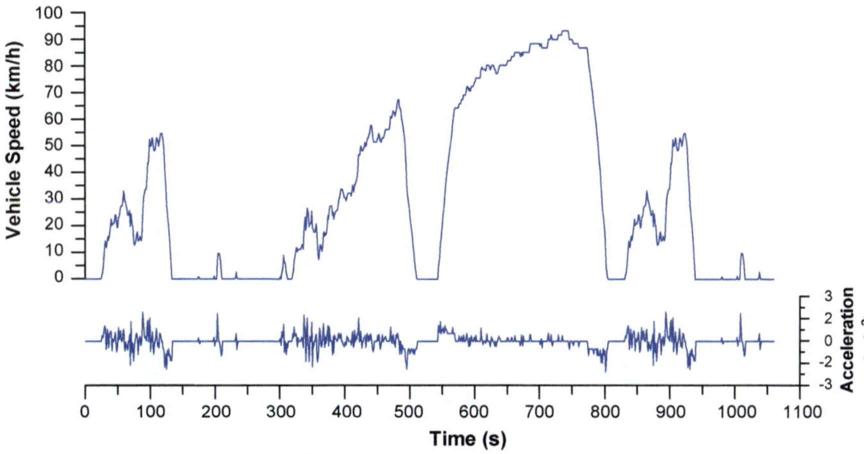

Fig. A.36 Vehicle speed and acceleration versus time of the U.S. HD UDDS

Table A.36 Technical specifications of the U.S. HD UDDS

Distance (m)/Duration (s or min)	8935.1	1060 (17.67)
Maximum/Average vehicle speed (km/h)	93.34	30.35
Average driving speed (km/h)/Speed σ (km/h)	44.55	31.92
Average/Maximum acceleration (m/s^2)	0.483	1.958
Average/Minimum deceleration (m/s^2)	−0.577	−2.070
Driving time (s)/(%)	722	68.11
Driving time (%) $V \leq 30$ km/h/$30 < V \leq 60$ km/h	60.09	17.55
Driving time (%) $60 < V \leq 100$ km/h/$V > 100$ km/h	22.36	0
Cruising time (s)/(%)	195	18.40
Time spent accelerating (s)/(%)	287	27.08
Acceleration time (%) $0 < a \leq 1.0$ m/s^2/$a > 1.0$ m/s^2	24.62	2.45
Time spent decelerating (s)/(%)	240	22.64
Decel. time (%) $a < -1.0$ m/s^2/$-1.0 \leq a < 0$ m/s^2	3.02	19.62
Idling time (s)/(%)	338	31.89 (34.72)
No. of accelerations/Positive acceleration σ (m/s^2)	23	0.382
Accelerations per km/per min	2.57	1.30
Number of stops/Max. interm. stop duration (s)	12	65
Stops per km/Average stop duration (s)	1.34	28.17
RPA (m/s^2)/PKE (m/s^2)	0.143	3.541

A.37 U.S. Central Business District—CBD

Highly repetitive test schedule, part of the Advanced Design Bus (ADB) 3-phase cycle (Fig. A.37 and Table A.37).

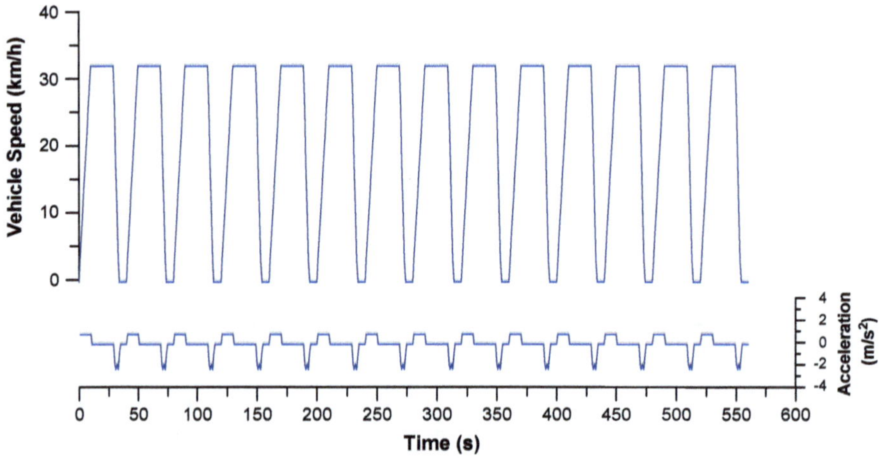

Fig. A.37 Vehicle speed and acceleration versus time of the U.S. CBD

Table A.37 Technical specifications of the U.S. CBD

Distance (m)/Duration (s or min)	3292.0	560 (9.33)
Maximum/Average vehicle speed (km/h)	32.19	21.16
Average driving speed (km/h)/Speed σ (km/h)	24.90	13.11
Average/Maximum acceleration (m/s^2)	0.894	0.894
Average/Minimum deceleration (m/s^2)	−1.788	−2.235
Driving time (s)/(%)	476	85.00
Driving time (%) $V \le 30$ km/h/$30 < V \le 60$ km/h	50.00	50.00
Driving time (%) $60 < V \le 100$ km/h/$V > 100$ km/h	0	0
Cruising time (s)/(%)	266	47.50
Time spent accelerating (s)/(%)	140	25.00
Acceleration time (%) $0 < a \le 1.0$ m/s^2/$a > 1.0$ m/s^2	25.00	0
Time spent decelerating (s)/(%)	70	12.50
Decel. time (%) $a < -1.0$ m/s^2/$-1.0 \le a < 0$ m/s^2	10.00	2.50
Idling time (s)/(%)	84	15.00 (17.50)
No. of accelerations/Positive acceleration σ (m/s^2)	28	0
Accelerations per km/per min	8.51	3.00
Number of stops/Max. interm. stop duration (s)	14	6
Stops per km/Average stop duration (s)	4.25	6.00
RPA (m/s^2)/PKE (m/s^2)	0.187	4.406

A.38 U.S. Arterial

Part of the Advanced Design Bus (ADB) 3-phase cycle (Fig. A.38 and Table A.38).

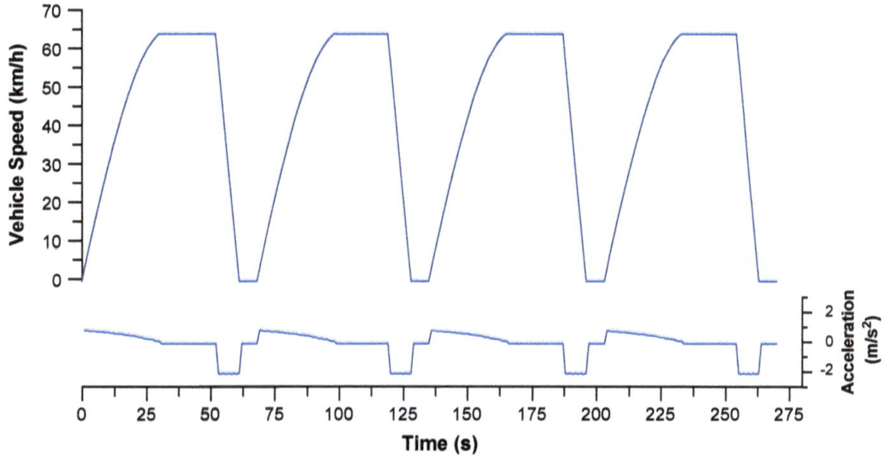

Fig. A.38 Vehicle speed and acceleration versus time of the U.S. Arterial

Table A.38 Technical specifications of the U.S. Arterial

Distance (m)/Duration (s or min)	3155.4	270 (4.50)
Maximum/Average vehicle speed (km/h)	64.37	42.07
Average driving speed (km/h)/Speed σ (km/h)	46.94	23.93
Average/Maximum acceleration (m/s^2)	0.596	0.894
Average/Minimum deceleration (m/s^2)	−1.987	−2.012
Driving time (s)/(%)	242	89.63
Driving time (%) V $\leq$ 30 km/h/30 < V $\leq$ 60 km/h	32.59	26.67
Driving time (%) 60 < V $\leq$ 100 km/h/V > 100 km/h	40.74	0
Cruising time (s)/(%)	86	31.85
Time spent accelerating (s)/(%)	120	44.44
Acceleration time (%) 0 < a $\leq$ 1.0 m/s^2/a > 1.0 m/s^2	44.44	0
Time spent decelerating (s)/(%)	36	13.33
Decel. time (%) a < −1.0 m/s^2/−1.0 $\leq$ a < 0 m/s^2	13.33	0
Idling time (s)/(%)	28	10.37 (11.85)
No. of accelerations/Positive acceleration σ (m/s^2)	12	0.218
Accelerations per km/per min	3.80	2.67
Number of stops/Max. interm. stop duration (s)	4	7
Stops per km/Average stop duration (s)	1.27	7.00
RPA (m/s^2)/PKE (m/s^2)	0.210	5.253

A.39 U.S. Commuter

Part of the Advanced Design Bus (ADB) 3-phase cycle (Fig. A.39 and Table A.39).

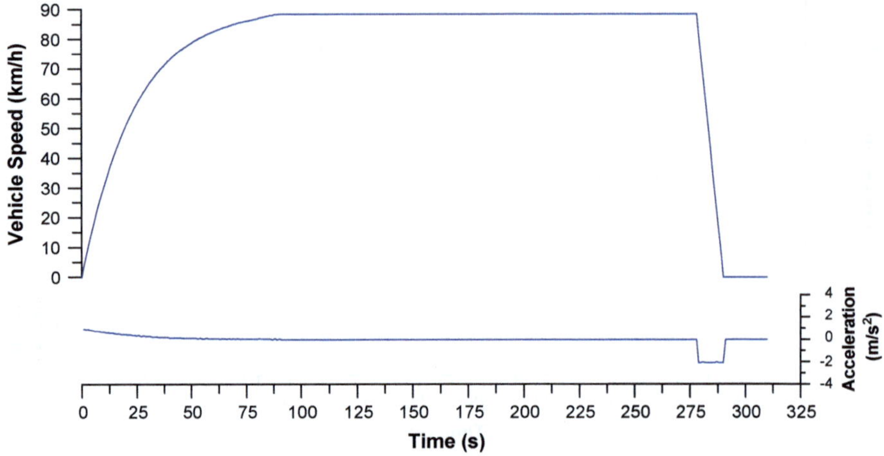

Fig. A.39 Vehicle speed and acceleration versus time of the U.S. Commuter

Table A.39 Technical specifications of the U.S. Commuter

Distance (m)/Duration (s or min)	6434.2	310 (5.17)
Maximum/Average vehicle speed (km/h)	88.51	74.72
Average driving speed (km/h)/Speed σ (km/h)	79.87	26.69
Average/Maximum acceleration (m/s^2)	0.279	0.939
Average/Minimum deceleration (m/s^2)	−2.049	−2.056
Driving time (s)/(%)	290	93.55
Driving time (%) $V \leq 30$ km/h/$30 < V \leq 60$ km/h	10.97	6.77
Driving time (%) $60 < V \leq 100$ km/h/$V > 100$ km/h	82.26	0
Cruising time (s)/(%)	190	61.29
Time spent accelerating (s)/(%)	88	28.39
Acceleration time (%) $0 < a \leq 1.0$ m/s^2/$a > 1.0$ m/s^2	28.39	0
Time spent decelerating (s)/(%)	12	3.87
Decel. time (%) $a < -1.0$ m/s^2/$-1.0 \leq a < 0$ m/s^2	3.87	0
Idling time (s)/(%)	20	6.45 (6.77)
No. of accelerations/Positive acceleration σ (m/s^2)	3	0.263
Accelerations per km/per min	0.47	0.58
Number of stops/Max. interm. stop duration (s)	1	–
Stops per km/Average stop duration (s)	0.16	20.00
RPA (m/s^2)/PKE (m/s^2)	0.048	1.218

A.40 Manhattan Bus

Highly transient cycle, typical of a bus route in heavy congestion. Very low average vehicle and driving speeds, with frequent stops and accelerations, and high RPA (Fig. A.40 and Table A.40).

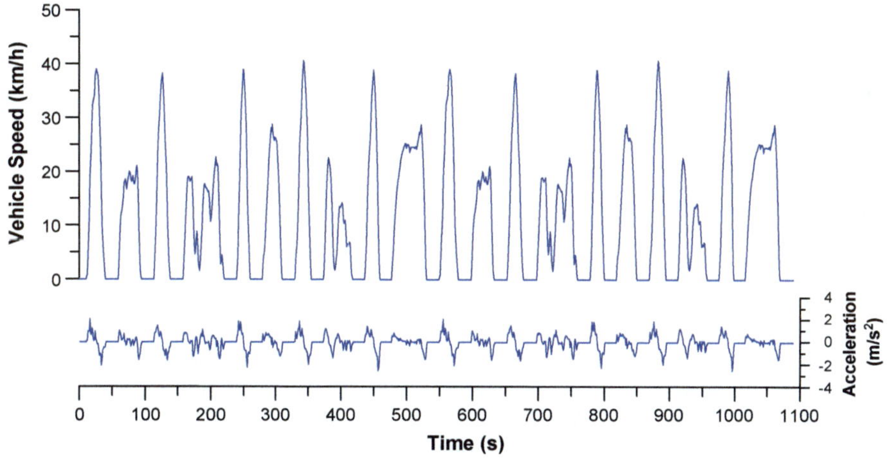

Fig. A.40 Vehicle speed and acceleration versus time of the Manhattan Bus

Table A.40 Technical specifications of the Manhattan Bus

Distance (m)/Duration (s or min)	3324.4	1089 (18.15)
Maximum/Average vehicle speed (km/h)	40.72	10.99
Average driving speed (km/h)/Speed σ (km/h)	16.71	11.81
Average/Maximum acceleration (m/s^2)	0.540	2.056
Average/Minimum deceleration (m/s^2)	−0.668	−2.503
Driving time (s)/(%)	716	65.75
Driving time (%) V $\leq$ 30 km/h/30 < V $\leq$ 60 km/h	91.92	8.08
Driving time (%) 60 < V $\leq$ 100 km/h/V > 100 km/h	0	0
Cruising time (s)/(%)	18	1.65
Time spent accelerating (s)/(%)	386	35.44
Acceleration time (%) 0 < a $\leq$ 1.0 m/s^2/a > 1.0 m/s^2	31.04	4.41
Time spent decelerating (s)/(%)	312	28.65
Decel. time (%) a < −1.0 m/s^2/−1.0 $\leq$ a < 0 m/s^2	7.53	21.12
Idling time (s)/(%)	373	34.25 (38.66)
No. of accelerations/Positive acceleration σ (m/s^2)	42	0.401
Accelerations per km/per min	12.63	2.31
Number of stops/Max. interm. stop duration (s)	21	19
Stops per km/Average stop duration (s)	6.32	17.76
RPA (m/s^2)/PKE (m/s^2)	0.308	7.303

A.41 New York Bus

Highly transient, typical of a bus route in heavy congestion. Very low average vehicle and driving speeds, with frequent accelerations and stops. For two thirds for the time, the bus idles. Notice the very high RPA (Fig. A.41 and Table A.41).

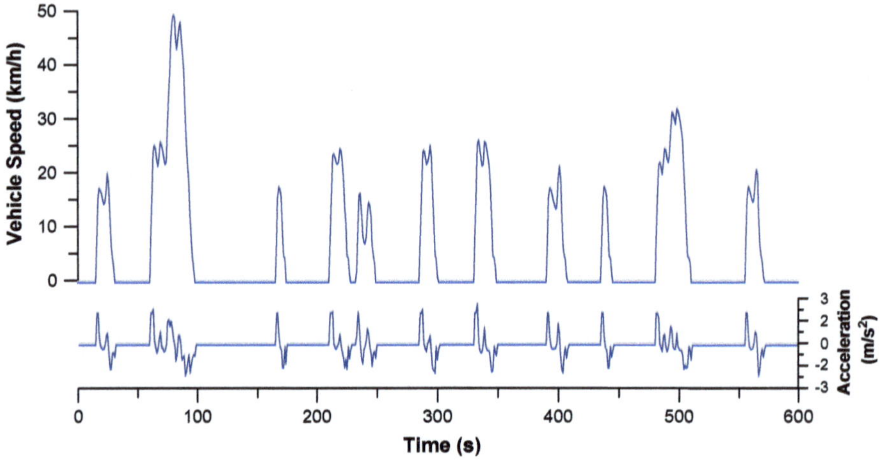

Fig. A.41 Vehicle speed and acceleration versus time of the New York Bus

Table A.41 Technical specifications of the New York Bus

Distance (m)/Duration (s or min)	989.4	600 (10)
Maximum/Average vehicle speed (km/h)	49.57	5.94
Average driving speed (km/h)/Speed σ (km/h)	17.12	10.48
Average/Maximum acceleration (m/s^2)	1.168	2.772
Average/Minimum deceleration (m/s^2)	−0.673	−2.056
Driving time (s)/(%)	208	34.67
Driving time (%) V $\leq$ 30 km/h/30 < V $\leq$ 60 km/h	96.50	3.50
Driving time (%) 60 < V $\leq$ 100 km/h/V > 100 km/h	0	0
Cruising time (s)/(%)	0	0
Time spent accelerating (s)/(%)	76	12.67
Acceleration time (%) 0 < a $\leq$ 1.0 m/s^2/a > 1.0 m/s^2	6.67	6.00
Time spent decelerating (s)/(%)	132	22.00
Decel. time (%) a < −1.0 m/s^2/−1.0 $\leq$ a < 0 m/s^2	5.17	16.83
Idling time (s)/(%)	392	65.33 (67.66)
No. of accelerations/Positive acceleration σ (m/s^2)	29	0.827
Accelerations per km/per min	2.93	0.29
Number of stops/Max. interm. stop duration (s)	12	67
Stops per km/Average stop duration (s)	12.13	32.67
RPA (m/s^2)/PKE (m/s^2)	0.459	9.868

A.42 New York City Composite

Based on the same database as the heavy-duty UDDS. Includes congested urban and freeway parts (Fig. A.42 and Table A.42).

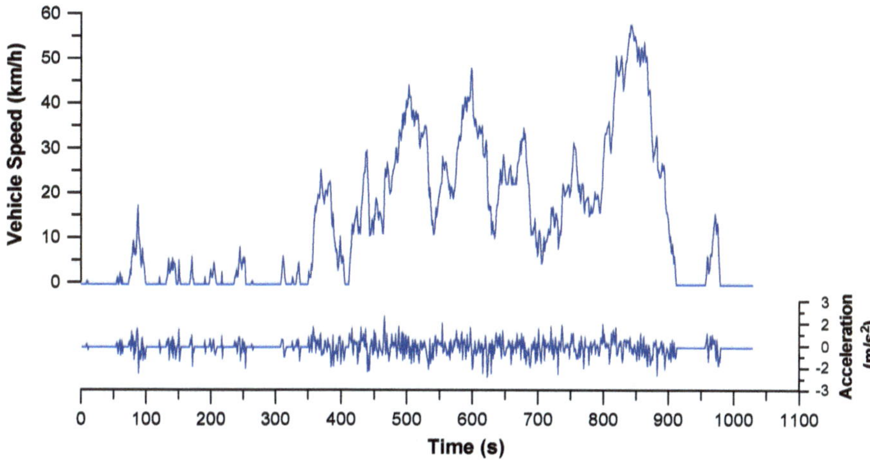

Fig. A.42 Vehicle speed and acceleration versus time of the New York City Composite

Table A.42 Technical specifications of the New York City Composite

Distance (m)/Duration (s or min)	4031.9	1029 (17.15)
Maximum/Average vehicle speed (km/h)	57.94	14.11
Average driving speed (km/h)/Speed σ (km/h)	20.50	15.23
Average/Maximum acceleration (m/s^2)	0.460	2.070
Average/Minimum deceleration (m/s^2)	−0.522	−1.958
Driving time (s)/(%)	708	68.80
Driving time (%) V $\leq$ 30 km/h/30 < V $\leq$ 60 km/h	82.80	17.20
Driving time (%) 60 < V $\leq$ 100 km/h/V > 100 km/h	0	0
Cruising time (s)/(%)	50	4.86
Time spent accelerating (s)/(%)	350	34.01
Acceleration time (%) 0 < a $\leq$ 1.0 m/s^2/a > 1.0 m/s^2	31.78	2.24
Time spent decelerating (s)/(%)	308	29.93
Decel. time (%) a < −1.0 m/s^2/−1.0 $\leq$ a < 0 m/s^2	3.79	26.14
Idling time (s)/(%)	321	31.20 (35.08)
No. of accelerations/Positive acceleration σ (m/s^2)	27	0.332
Accelerations per km/per min	6.70	1.57
Number of stops/Max. interm. stop duration (s)	18	50
Stops per km/Average stop duration (s)	4.46	17.83
RPA (m/s^2)/PKE (m/s^2)	0.244	5.950

A.43 California HHDDT Creep

Part of CARB's heavy heavy-duty diesel truck (HHDDT) cycles. Very low speeds and covered distance (Fig. A.43 and Table A.43).

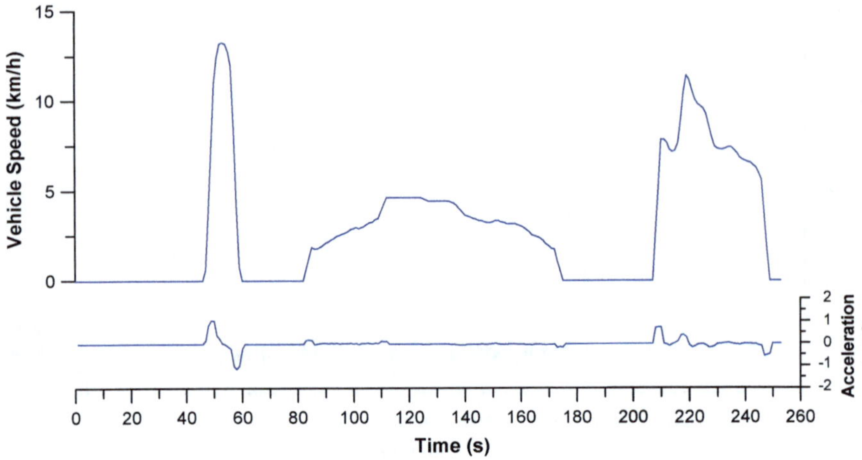

Fig. A.43 Vehicle speed and acceleration versus time of the CARB HHDDT creep

Table A.43 Technical specifications of the CARB HHDDT creep

Distance (m)/Duration (s or min)	199.7	253 (4.22)
Maximum/Average vehicle speed (km/h)	13.26	2.84
Average driving speed (km/h)/Speed σ (km/h)	4.83	3.30
Average/Maximum acceleration (m/s^2)	0.183	1.028
Average/Minimum deceleration (m/s^2)	−0.114	−1.131
Driving time (s)/(%)	149	58.89
Driving time (%) V ≤ 30 km/h/30 < V ≤ 60 km/h	100	0
Driving time (%) 60 < V ≤ 100 km/h/V > 100 km/h	0	0
Cruising time (s)/(%)	29	11.46
Time spent accelerating (s)/(%)	46	18.18
Acceleration time (%) 0 < a ≤ 1.0 m/s^2/a > 1.0 m/s^2	17.39	0.79
Time spent decelerating (s)/(%)	74	29.24
Decel. time (%) a < −1.0 m/s^2/−1.0 ≤ a < 0 m/s^2	0.395	28.85
Idling time (s)/(%)	104	41.10 (43.87)
No. of accelerations/Positive acceleration σ (m/s^2)	4	0.279
Accelerations per km/per min	20.02	0.95
Number of stops/Max. interm. stop duration (s)	4	32
Stops per km/Average stop duration (s)	20.02	26.00
RPA (m/s^2)/PKE (m/s^2)	0.079	1.706

A.44 California HHDDT Transient

Part of CARB's heavy heavy-duty diesel truck cycles; transient segment (Fig. A.44 and Table A.44).

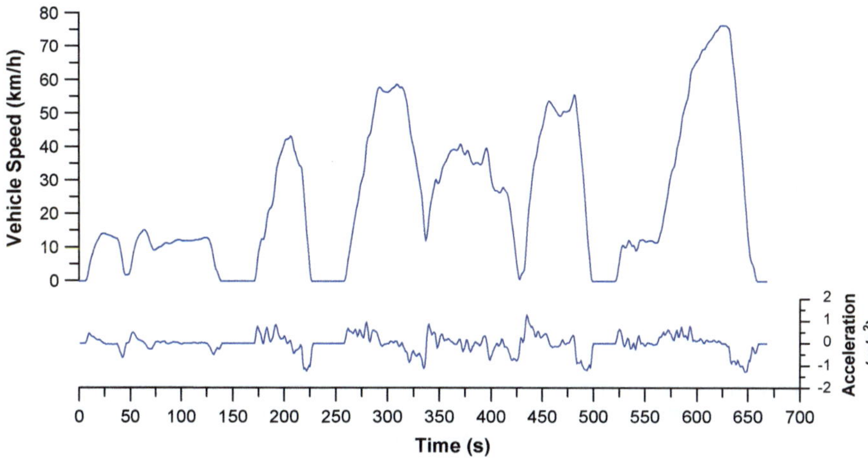

Fig. A.44 Vehicle speed and acceleration versus time of the CARB HHDDT transient

Table A.44 Technical specifications of the CARB HHDDT transient

Distance (m)/Duration (s or min)	4588.2	668 (11.13)
Maximum/Average vehicle speed (km/h)	76.44	24.73
Average driving speed (km/h)/Speed σ (km/h)	29.08	21.53
Average/Maximum acceleration (m/s^2)	0.286	1.296
Average/Minimum deceleration (m/s^2)	−0.379	−1.239
Driving time (s)/(%)	568	85.03
Driving time (%) V $\leq$ 30 km/h/30 < V $\leq$ 60 km/h	61.68	31.89
Driving time (%) 60 < V $\leq$ 100 km/h/V > 100 km/h	6.44	0
Cruising time (s)/(%)	31	4.64
Time spent accelerating (s)/(%)	306	45.81
Acceleration time (%) 0 < a $\leq$ 1.0 m/s^2/a > 1.0 m/s^2	45.36	0.45
Time spent decelerating (s)/(%)	231	34.58
Decel. time (%) a < −1.0 m/s^2/−1.0 $\leq$ a < 0 m/s^2	2.99	31.59
Idling time (s)/(%)	100	14.97 (17.06)
No. of accelerations/Positive acceleration σ (m/s^2)	25	0.241
Accelerations per km/per min	5.45	2.25
Number of stops/Max. interm. stop duration (s)	5	32
Stops per km/Average stop duration (s)	1.09	20.00
RPA (m/s^2)/PKE (m/s^2)	0.154	3.872

A.45 California HHDDT Cruise

Part of CARB's heavy heavy-duty diesel truck cycles; highway segment (Fig. A.45 and Table A.45).

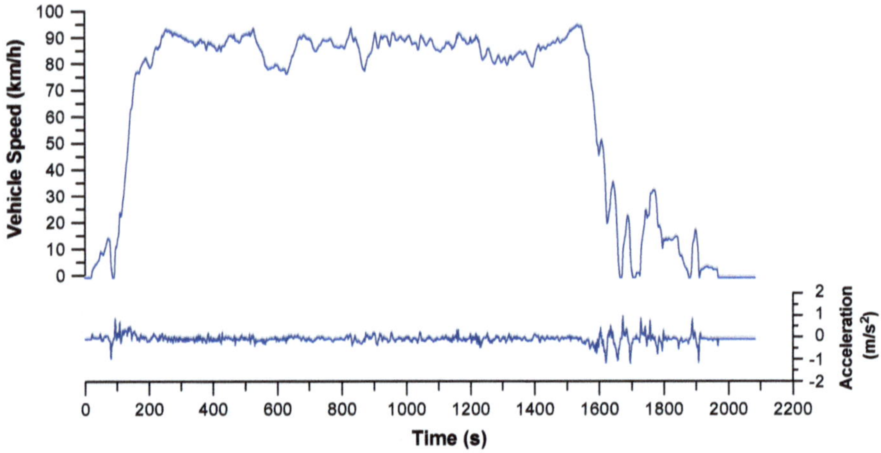

Fig. A.45 Vehicle speed and acceleration versus time of the CARB HHDDT cruise

Table A.45 Technical specifications of the CARB HHDDT cruise

Distance (m)/Duration (s or min)	37,130.9	2083 (34.72)
Maximum/Average vehicle speed (km/h)	95.43	64.17
Average driving speed (km/h)/Speed σ (km/h)	69.55	35.41
Average/Maximum acceleration (m/s^2)	0.119	1.046
Average/Minimum deceleration (m/s^2)	−0.128	−1.095
Driving time (s)/(%)	1922	92.27
Driving time (%) $V \leq 30$ km/h/$30 < V \leq 60$ km/h	26.36	4.27
Driving time (%) $60 < V \leq 100$ km/h/$V > 100$ km/h	69.37	0
Cruising time (s)/(%)	168	8.07
Time spent accelerating (s)/(%)	907	43.54
Acceleration time (%) $0 < a \leq 1.0$ m/s^2/$a > 1.0$ m/s^2	43.49	0.05
Time spent decelerating (s)/(%)	847	40.66
Decel. time (%) $a < -1.0$ m/s^2/$-1.0 \leq a < 0$ m/s^2	0.19	40.47
Idling time (s)/(%)	161	7.73 (8.50)
No. of accelerations/Positive acceleration σ (m/s^2)	73	0.140
Accelerations per km/per min	1.97	2.10
Number of stops/Max. interm. stop duration (s)	6	9
Stops per km/Average stop duration (s)	0.16	26.83
RPA (m/s^2)/PKE (m/s^2)	0.044	1.126

A.46 California Orange County Bus

Highly transient bus cycle with almost half the time spent accelerating, and quite prolonged idling period distributed among equal-duration stops (Fig. A.46 and Table A.46).

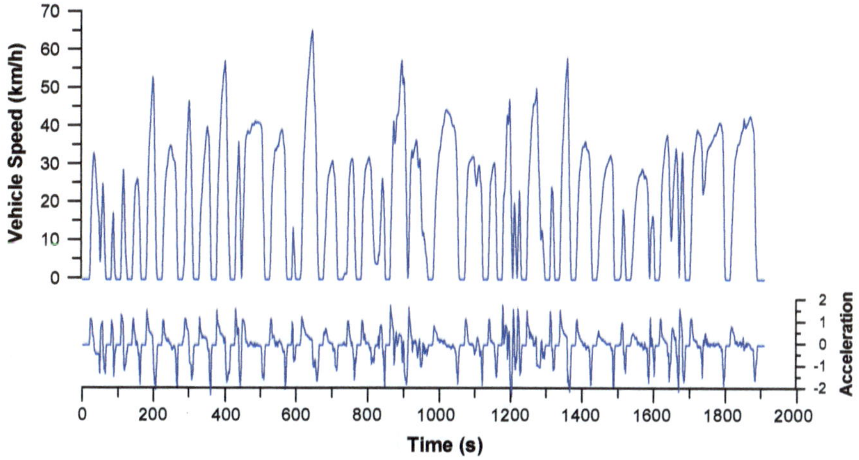

Fig. A.46 Vehicle speed and acceleration versus time of the Orange County Bus

Table A.46 Technical specifications of the Orange County Bus

Distance (m)/Duration (s or min)	10,526.1	1909 (31.82)
Maximum/Average vehicle speed (km/h)	65.39	19.85
Average driving speed (km/h)/Speed σ (km/h)	24.70	16.57
Average/Maximum acceleration (m/s^2)	0.453	1.815
Average/Minimum deceleration (m/s^2)	−0.636	−2.293
Driving time (s)/(%)	1534	80.36
Driving time (%) V $\leq$ 30 km/h/30 < V $\leq$ 60 km/h	66.16	33.37
Driving time (%) 60 < V $\leq$ 100 km/h/V > 100 km/h	0.47	0
Cruising time (s)/(%)	12	0.63
Time spent accelerating (s)/(%)	889	46.57
Acceleration time (%) 0 < a $\leq$ 1.0 m/s^2/a > 1.0 m/s^2	40.13	6.44
Time spent decelerating (s)/(%)	633	33.16
Decel. time (%) a < −1.0 m/s^2/−1.0 $\leq$ a < 0 m/s^2	8.91	24.25
Idling time (s)/(%)	375	19.64 (25.62)
No. of accelerations/Positive acceleration σ (m/s^2)	88	0.400
Accelerations per km/per min	8.36	2.77
Number of stops/Max. interm. stop duration (s)	32	11
Stops per km/Average stop duration (s)	3.04	11.72
RPA (m/s^2)/PKE (m/s^2)	0.233	5.647

A.47 SmartWay Neighborhood Refuse Truck

Part of the U.S. SmartWay program. Cycle with many stops and long idling period (Fig. A.47 and Table A.47).

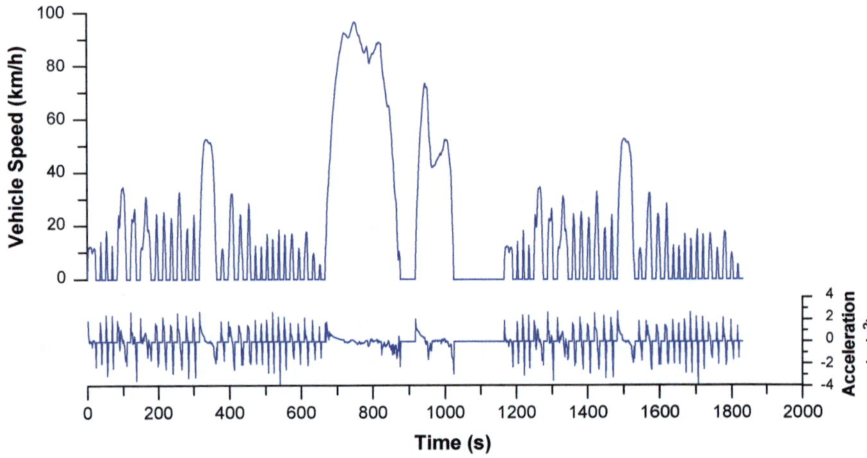

Fig. A.47 Vehicle speed and acceleration versus time of the SmartWay Neighborhood Refuse

Table A.47 Technical specifications of the SmartWay Neighborhood Refuse

Distance (m)/Duration (s or min)	9150.6	1832 (30.53)
Maximum/Average vehicle speed (km/h)	96.56	17.98
Average driving speed (km/h)/Speed σ (km/h)	32.58	26.23
Average/Maximum acceleration (m/s^2)	0.772	2.727
Average/Minimum deceleration (m/s^2)	−0.897	−3.807
Driving time (s)/(%)	1011	55.19
Driving time (%) $V \leq 30$ km/h/$30 < V \leq 60$ km/h	77.95	12.12
Driving time (%) $60 < V \leq 100$ km/h/$V > 100$ km/h	9.93	0
Cruising time (s)/(%)	62	3.38
Time spent accelerating (s)/(%)	510	27.84
Acceleration time (%) $0 < a \leq 1.0$ m/s^2/$a > 1.0$ m/s^2	18.34	9.50
Time spent decelerating (s)/(%)	439	23.96
Decel. time (%) $a < -1.0$ m/s^2/$-1.0 \leq a < 0$ m/s^2	9.01	14.96
Idling time (s)/(%)	821	44.81 (48.09)
No. of accelerations/Positive acceleration σ (m/s^2)	65	0.648
Accelerations per km/per min	7.10	2.13
Number of stops/Max. interm. stop duration (s)	60	142
Stops per km/Average stop duration (s)	6.56	13.68
RPA (m/s^2)/PKE (m/s^2)	0.235	5.357

A.48 Japanese JE05

Highly transient, with only 3 s constant speed driving. One quarter of the time is spent idling. Used for fuel consumption purposes too (Fig. A.48 and Table A.48).

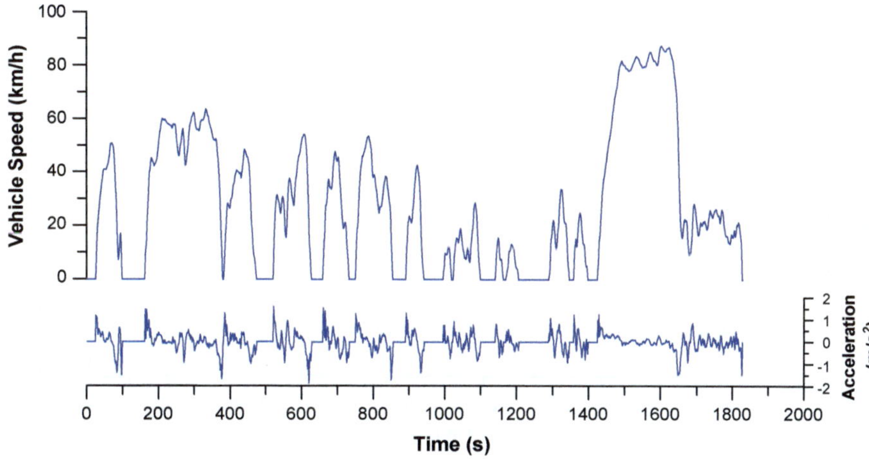

Fig. A.48 Vehicle speed and acceleration versus time of the Japanese JE05

Table A.48 Technical specifications of the Japanese E05

Distance (m)/Duration (s or min)	13,892.1	1829 (30.48)
Maximum/Average vehicle speed (km/h)	87.60	27.34
Average driving speed (km/h)/Speed σ (km/h)	36.16	25.66
Average/Maximum acceleration (m/s^2)	0.314	1.592
Average/Minimum deceleration (m/s^2)	−0.328	−1.831
Driving time (s)/(%)	1383	75.62
Driving time (%) V $\leq$ 30 km/h/30 < V $\leq$ 60 km/h	61.02	27.72
Driving time (%) 60 < V $\leq$ 100 km/h/V > 100 km/h	11.26	0
Cruising time (s)/(%)	3	0.16
Time spent accelerating (s)/(%)	705	38.55
Acceleration time (%) 0 < a $\leq$ 1.0 m/s^2/a > 1.0 m/s^2	37.18	1.37
Time spent decelerating (s)/(%)	675	36.91
Decel. time (%) a < −1.0 m/s^2/−1.0 $\leq$ a < 0 m/s^2	1.91	34.99
Idling time (s)/(%)	446	24.38 (24.15)
No. of accelerations/Positive acceleration σ (m/s^2)	62	0.283
Accelerations per km/per min	4.46	2.03
Number of stops/Max. interm. stop duration (s)	13	83
Stops per km/Average stop duration (s)	0.94	34.3
RPA (m/s^2)/PKE (m/s^2)	0.126	3.141

A.49 Australian CUEDC ME

For emissions inventory purposes. The cycle corresponds to heavy buses (Fig. A.49 and Table A.49).

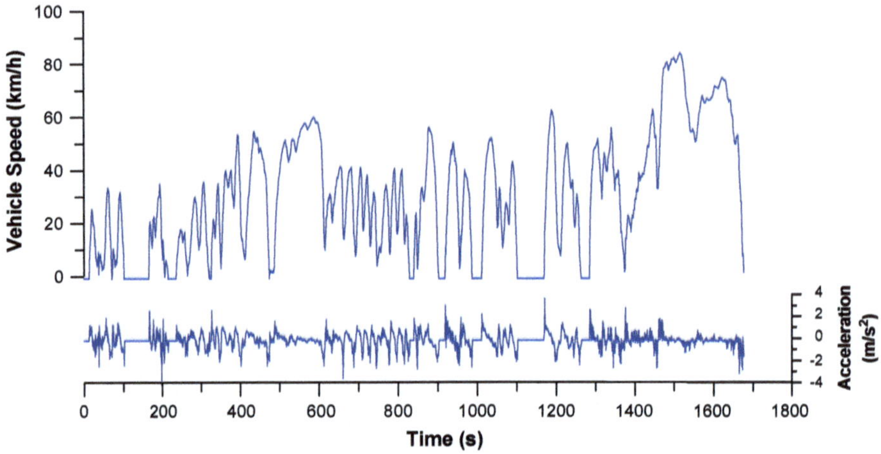

Fig. A.49 Vehicle speed and acceleration versus time of the Australian CUEDC ME

Table A.49 Technical specifications of the Australian CUEDC ME

Distance (m)/Duration (s or min)	14,387.2	1677 (27.95)
Maximum/Average vehicle speed (km/h)	84.95	30.89
Average driving speed (km/h)/Speed σ (km/h)	36.09	22.71
Average/Maximum acceleration (m/s^2)	0.531	3.801
Average/Minimum deceleration (m/s^2)	−0.582	−3.945
Driving time (s)/(%)	1435	85.57
Driving time (%) V $\leq$ 30 km/h/30 < V $\leq$ 60 km/h	49.43	39.77
Driving time (%) 60 < V $\leq$ 100 km/h/V > 100 km/h	10.79	0
Cruising time (s)/(%)	3	0.18
Time spent accelerating (s)/(%)	749	44.66
Acceleration time (%) 0 < a $\leq$ 1.0 m/s^2/a > 1.0 m/s^2	38.52	6.14
Time spent decelerating (s)/(%)	683	40.73
Decel. time (%) a < −1.0 m/s^2/−1.0 $\leq$ a < 0 m/s^2	8.35	32.38
Idling time (s)/(%)	242	14.43 (15.27)
No. of accelerations/Positive acceleration σ (m/s^2)	80	0.512
Accelerations per km/per min	5.56	2.86
Number of stops/Max. interm. stop duration (s)	9	67
Stops per km/Average stop duration (s)	0.63	26.89
RPA (m/s^2)/PKE (m/s^2)	0.230	5.588

A.50 Australian CUEDC NCH

For emissions inventory purposes. The cycle corresponds to heavy goods vehicles with GVW > 25,000 kg (Fig. A.50 and Table A.50).

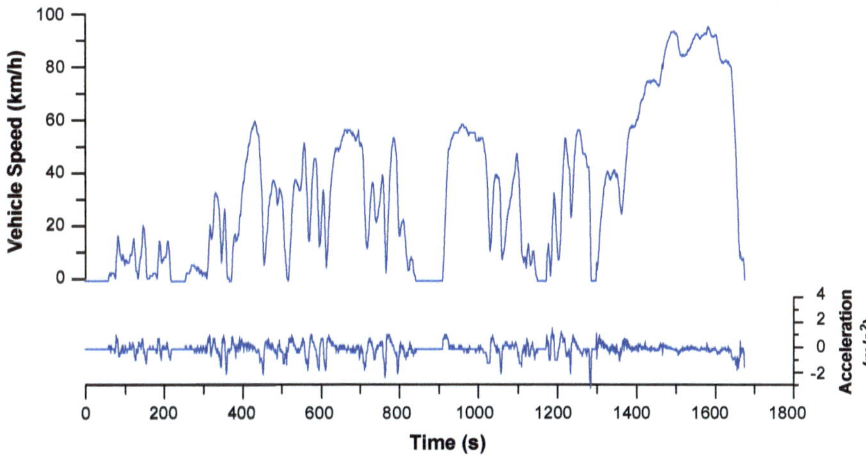

Fig. A.50 Vehicle speed and acceleration versus time of the Australian CUEDC NCH

Table A.50 Technical specifications of the Australian CUEDC NCH

Distance (m)/Duration (s or min)	15,524.3	1675 (27.92)
Maximum/Average vehicle speed (km/h)	96.14	33.37
Average driving speed (km/h)/Speed σ (km/h)	37.86	28.37
Average/Maximum acceleration (m/s^2)	0.414	1.667
Average/Minimum deceleration (m/s^2)	−0.502	−3.056
Driving time (s)/(%)	1476	88.12
Driving time (%) V $\leq$ 30 km/h/30 < V $\leq$ 60 km/h	50.27	34.81
Driving time (%) 60 < V $\leq$ 100 km/h/V > 100 km/h	14.92	0
Cruising time (s)/(%)	312	18.63
Time spent accelerating (s)/(%)	638	38.09
Acceleration time (%) 0 < a $\leq$ 1.0 m/s^2/a > 1.0 m/s^2	36.06	2.03
Time spent decelerating (s)/(%)	526	31.40
Decel. time (%) a < −1.0 m/s^2/−1.0 $\leq$ a < 0 m/s^2	4.96	26.44
Idling time (s)/(%)	199	11.88 (13.91)
No. of accelerations/Positive acceleration σ (m/s^2)	63	0.301
Accelerations per km/per min	4.06	2.26
Number of stops/Max. interm. stop duration (s)	6	67
Stops per km/Average stop duration (s)	0.39	33.17
RPA (m/s^2)/PKE (m/s^2)	0.143	3.561

A.51 Chinese C-WTVC

'Softer' version of the worldwide WTVC, employed for fuel consumption purposes in China (Fig. A.51 and Table A.51).

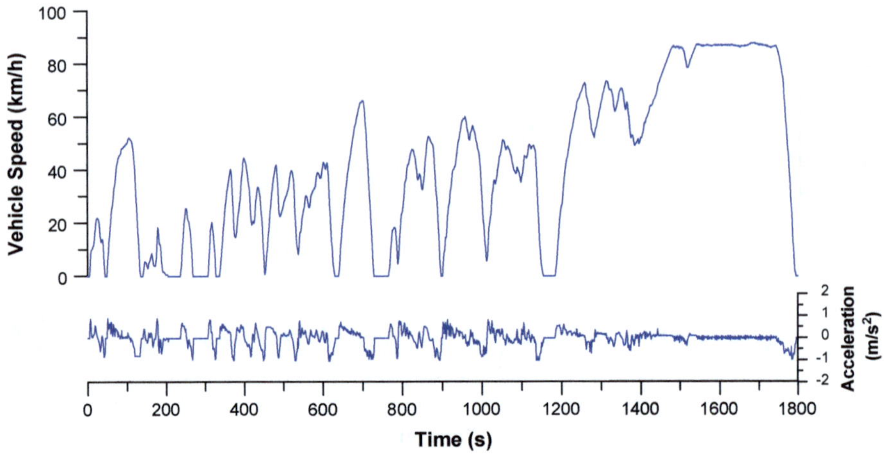

Fig. A.51 Vehicle speed and acceleration versus time of the Chinese C-WTVC

Table A.51 Technical specifications of the Chinese C-WTVC

Distance (m)/Duration (s or min)	20,509.7	1800 (30)
Maximum/Average vehicle speed (km/h)	87.80	41.02
Average driving speed (km/h)/Speed σ (km/h)	45.46	28.61
Average/Maximum acceleration (m/s^2)	0.238	0.917
Average/Minimum deceleration (m/s^2)	−0.321	−1.033
Driving time (s)/(%)	1624	90.22
Driving time (%) $V \leq 30$ km/h/$30 < V \leq 60$ km/h	38.22	34.44
Driving time (%) $60 < V \leq 100$ km/h/$V > 100$ km/h	27.33	0
Cruising time (s)/(%)	21	1.17
Time spent accelerating (s)/(%)	920	51.11
Acceleration time (%) $0 < a \leq 1.0$ m/s^2/$a > 1.0$ m/s^2	51.11	0
Time spent decelerating (s)/(%)	693	37.94
Decel. time (%) $a < -1.0$ m/s^2/$-1.0 \leq a < 0$ m/s^2	0.67	37.28
Idling time (s)/(%)	176	9.78 (11.44)
No. of accelerations/Positive acceleration σ (m/s^2)	55	0.192
Accelerations per km/per min	2.68	1.83
Number of stops/Max. interm. stop duration (s)	11	38
Stops per km/Average stop duration (s)	0.54	16.00
RPA (m/s^2)/PKE (m/s^2)	0.098	2.482

A.52 Worldwide WTVC

Highly transient cycle with many accelerations and high speeds. Used as the basis for the legislated in many countries engine-dynamometer WHTC. Also referred to as WHVC (Fig. A.52 and Table A.52).

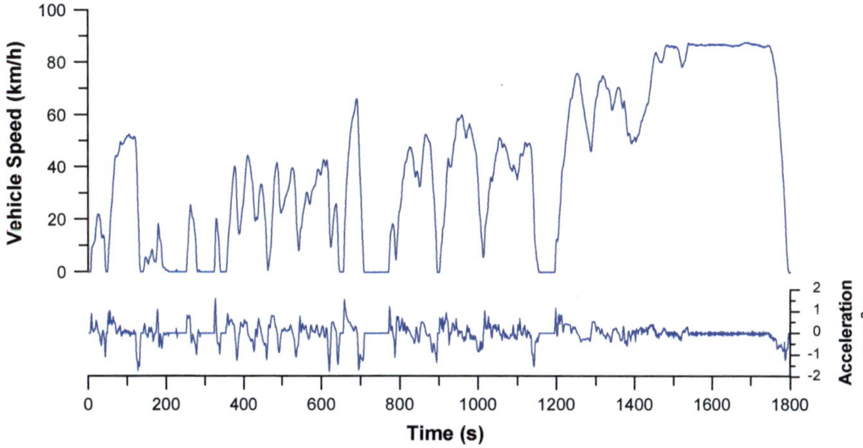

Fig. A.52 Vehicle speed and acceleration versus time of the worldwide WTVC

Table A.52 Technical specifications of the worldwide WTVC

Distance (m)/Duration (s or min)	20,072.1	1800 (30)
Maximum/Average vehicle speed (km/h)	87.80	40.14
Average driving speed (km/h)/Speed σ (km/h)	46.11	29.50
Average/Maximum acceleration (m/s^2)	0.279	1.594
Average/Minimum deceleration (m/s^2)	−0.320	−1.731
Driving time (s)/(%)	1567	87.06
Driving time (%) V $\leq$ 30 km/h/30 < V $\leq$ 60 km/h	39.89	33.28
Driving time (%) 60 < V $\leq$ 100 km/h/V > 100 km/h	26.83	0
Cruising time (s)/(%)	22	1.22
Time spent accelerating (s)/(%)	825	45.83
Acceleration time (%) 0 < a $\leq$ 1.0 m/s^2/a > 1.0 m/s^2	45.11	0.72
Time spent decelerating (s)/(%)	720	40.00
Decel. time (%) a < −1.0 m/s^2/−1.0 $\leq$ a < 0 m/s^2	2.72	37.28
Idling time (s)/(%)	233	12.94 (14.78)
No. of accelerations/Positive acceleration σ (m/s^2)	62	0.254
Accelerations per km/per min	3.09	2.07
Number of stops/Max. interm. stop duration (s)	12	62
Stops per km/Average stop duration (s)	0.60	19.42
RPA (m/s^2)/PKE (m/s^2)	0.107	2.704

Speed/Acceleration Distributions of All Cycles

See Figs. A.53, A.54, A.55, A.56, A.57, A.58 and A.59.

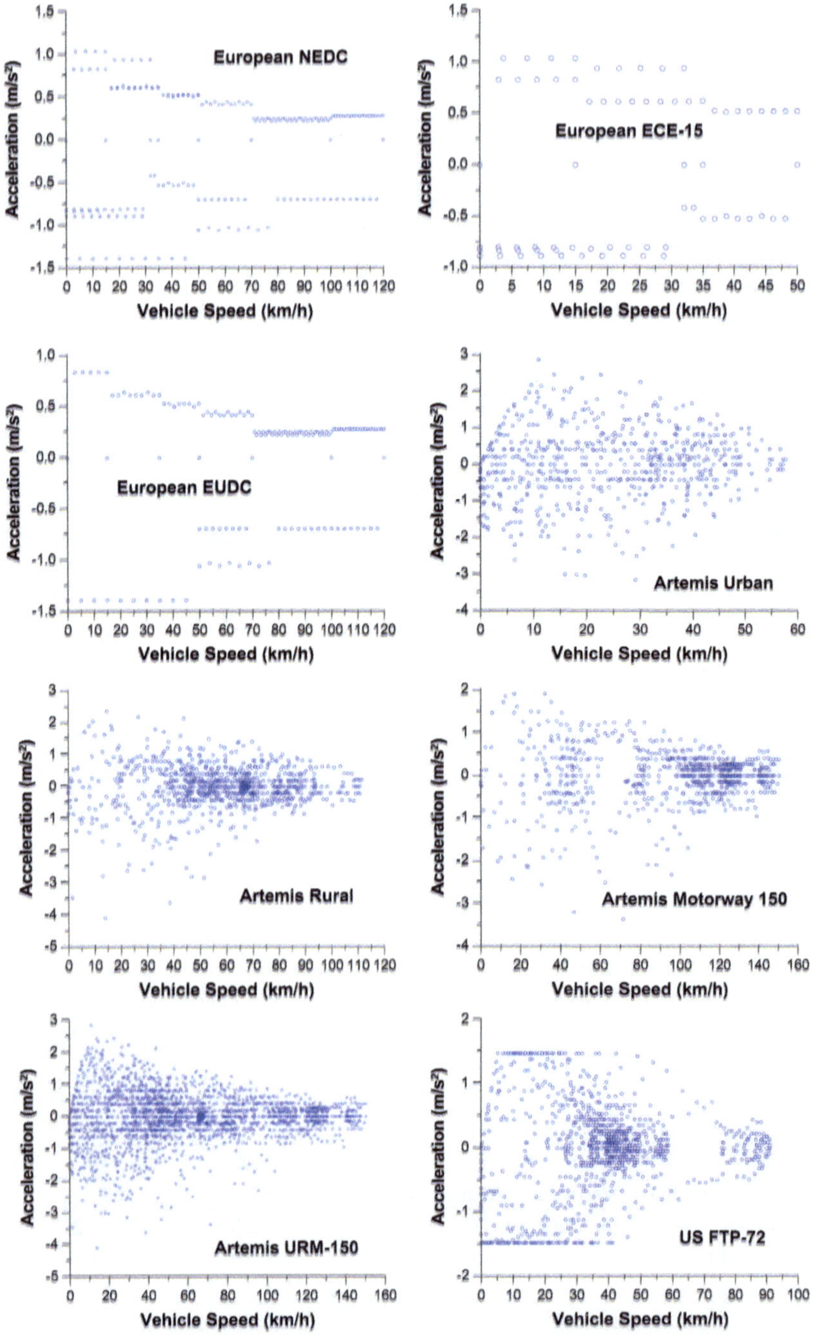

Fig. A.53 Speed/acceleration distribution of chassis-dynamometer cycles

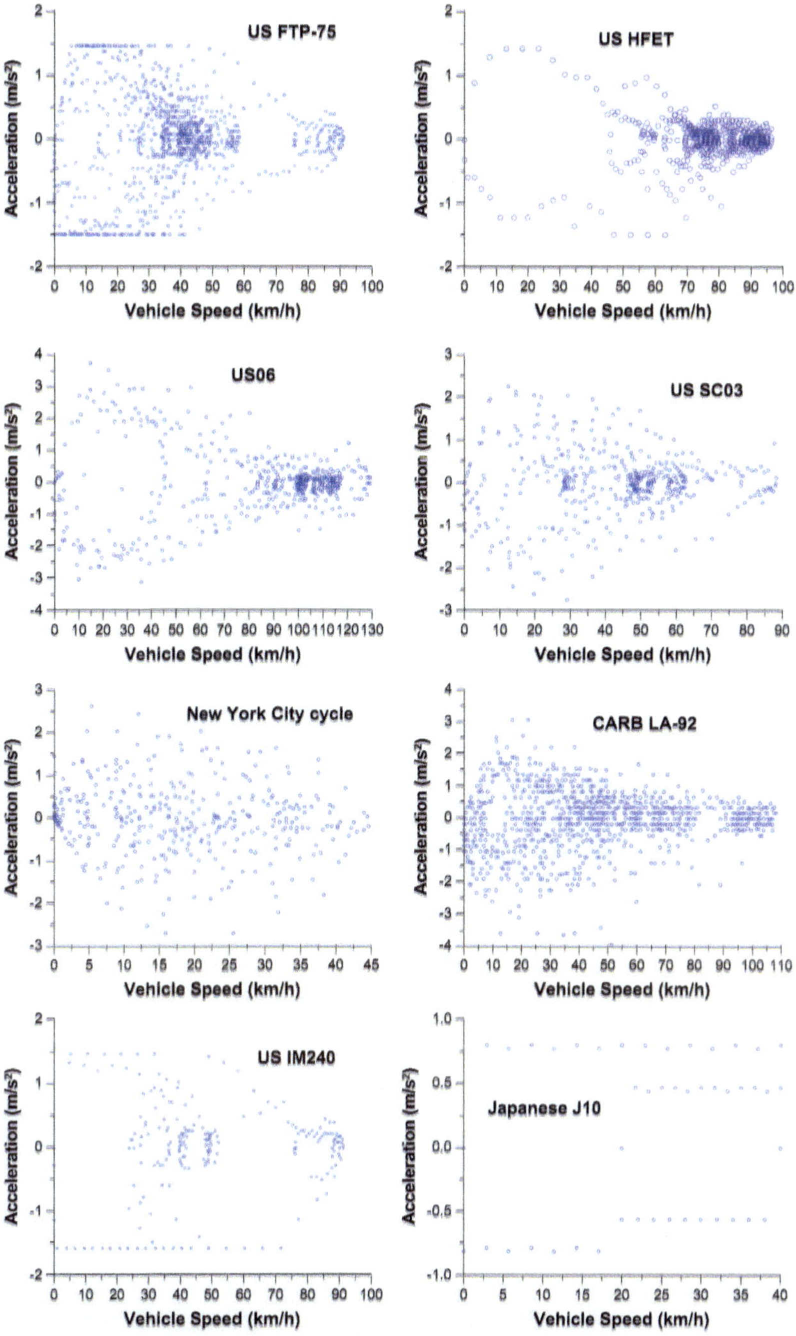

Fig. A.54 Speed/acceleration distribution of chassis-dynamometer cycles

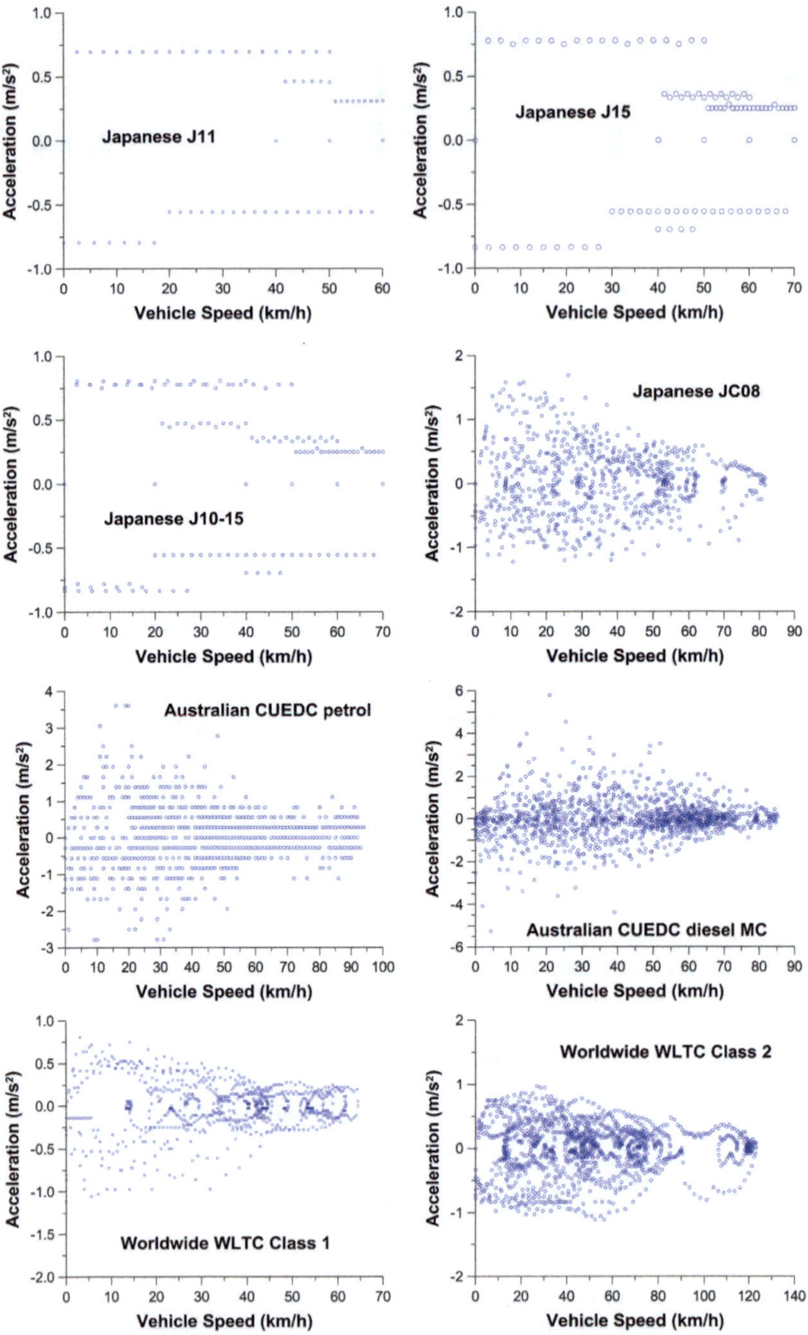

Fig. A.55 Speed/acceleration distribution of chassis-dynamometer cycles

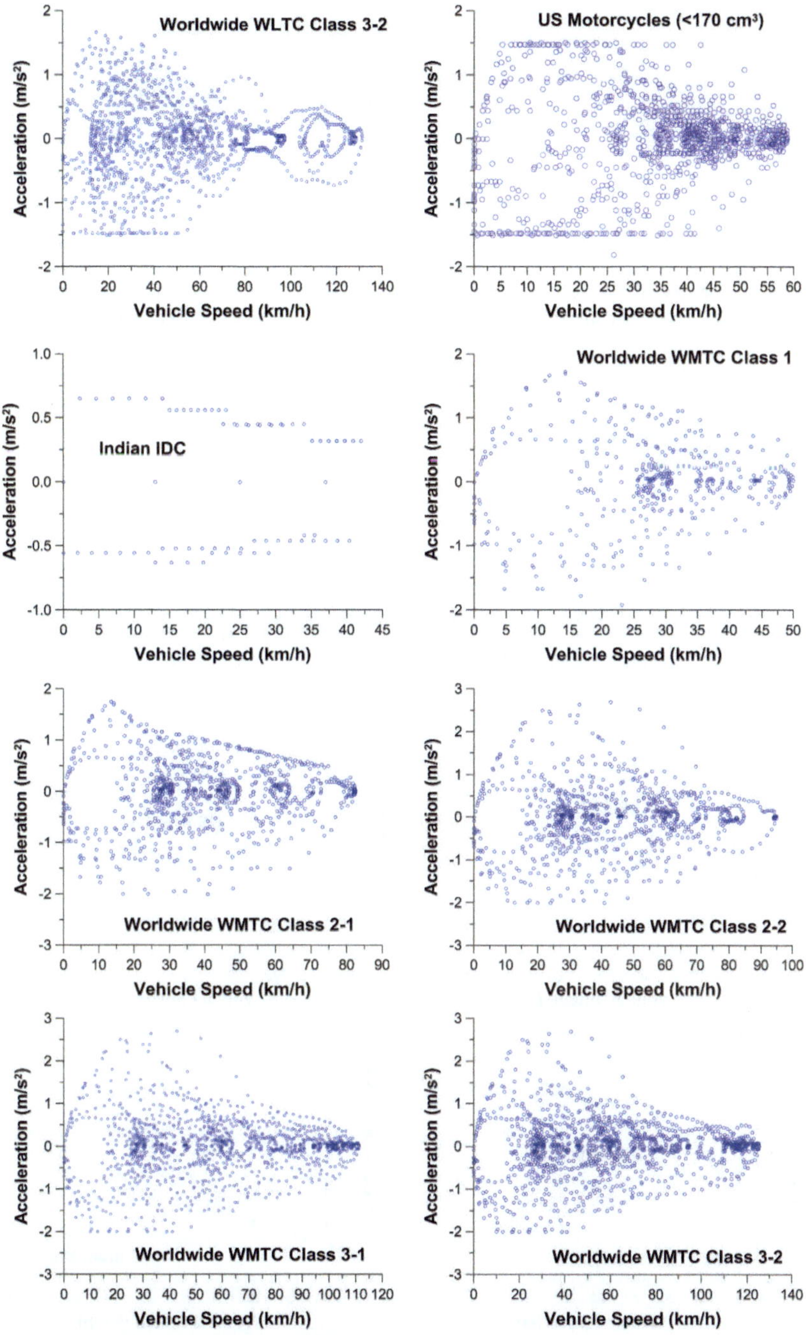

Fig. A.56 Speed/acceleration distribution of chassis-dynamometer cycles

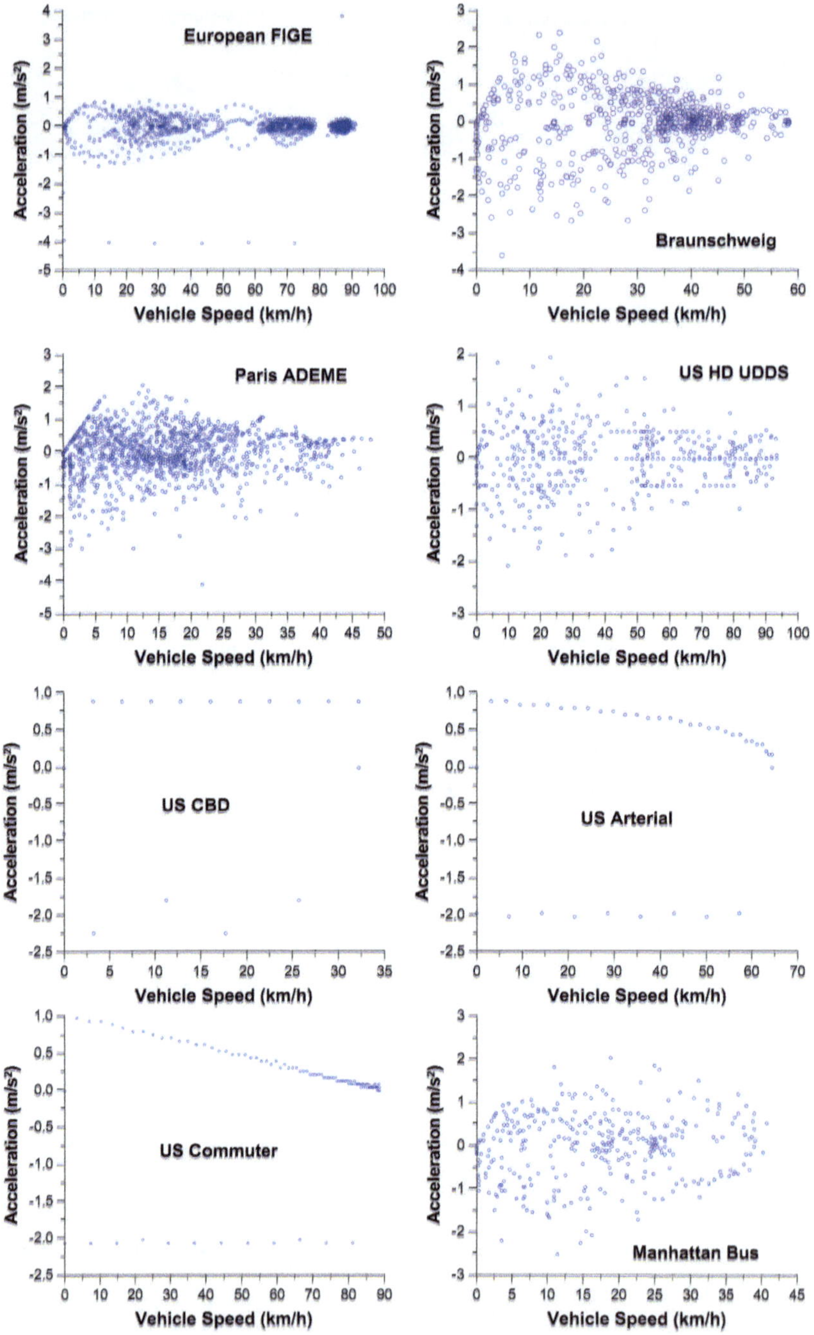

Fig. A.57 Speed/acceleration distribution of chassis-dynamometer cycles

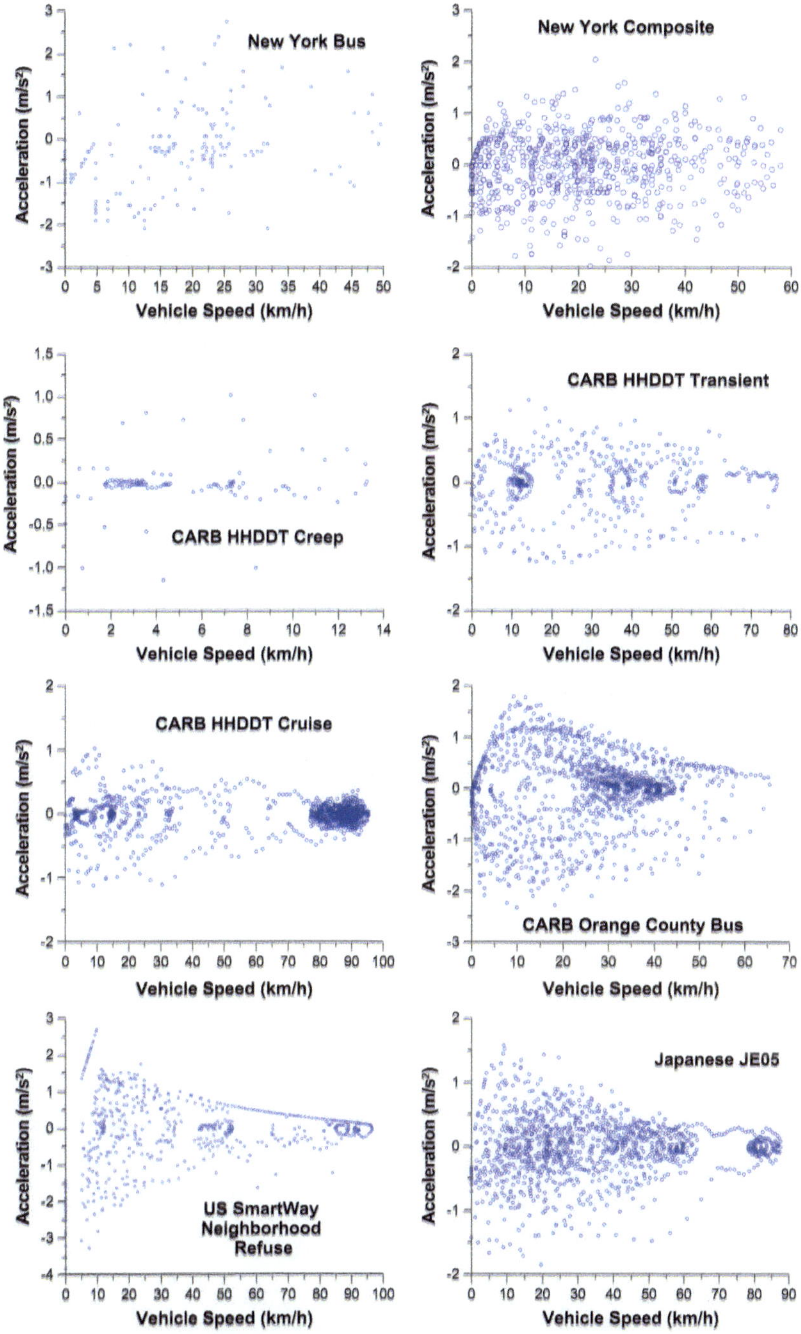

Fig. A.58 Speed/acceleration distribution of chassis-dynamometer cycles

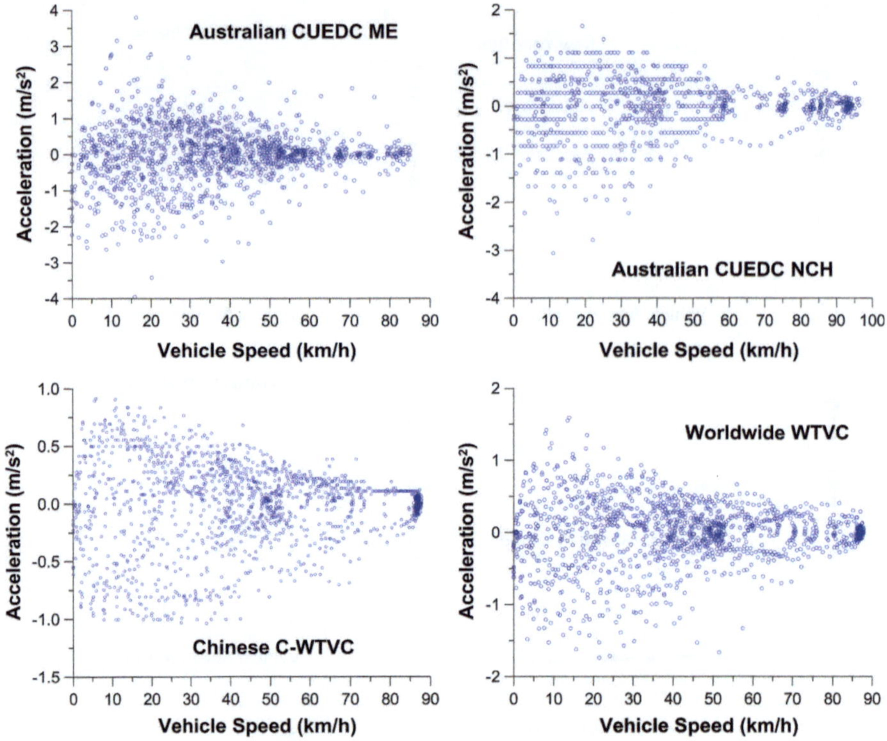

Fig. A.59 Speed/acceleration distribution of chassis-dynamometer cycles

MIX
Papier aus verantwortungsvollen Quellen
Paper from responsible sources
FSC® C105338

If you have any concerns about our products,
you can contact us on
ProductSafety@springernature.com

In case Publisher is established outside the EU,
the EU authorized representative is:
Springer Nature Customer Service Center GmbH
Europaplatz 3, 69115 Heidelberg, Germany

Printed by Libri Plureos GmbH
in Hamburg, Germany